Fluid Film Lubrication

Fluid film bearings are among the best devices for overcoming friction and eliminating wear. They are machine elements and, together with shafts, gears, and cams, constitute the building blocks engineers use in the design and construction of mechanical devices.

This book offers a systematic treatment of the fundamentals of fluid film lubrication and fluid film bearings. The introduction places fluid film bearings within the broader context of tribology, a subject that encompasses friction, lubrication, and wear. The early chapters provide a thorough discussion of classical lubrication theory. The remainder of the book is devoted to critical aspects of fluid film lubrication and bearing design. These later chapters consider the more advanced topics of inertia, thermal and turbulence effects, lubrication of counterformal contacts, and non-Newtonian lubricants. Also included are areas in which future developments are likely and/or desirable, such as lubrication with emulsions. The large number of references throughout the book will guide the reader in further study.

Graduate and senior undergraduate students, researchers, and practicing engineers will appreciate this clear, thorough discussion of fluid film lubrication and fluid film bearings.

Andras Z. Szeri is the Robert Lyle Spencer Professor of Mechanical Engineering and Chair of the Department of Mechanical Engineering at the University of Delaware.

*To my wife Mary for her continued encouragement,
and to my children, Maria, Cora, and Andrew, for
they have always made me proud.*

Fluid Film Lubrication

Theory and Design

ANDRAS Z. SZERI

University of Delaware

PUBLISHED BY THE PRESS SYNDICATE OF THE UNIVERSITY OF CAMBRIDGE
The Pitt Building, Trumpington Street, Cambridge, United Kingdom

CAMBRIDGE UNIVERSITY PRESS
The Edinburgh Building, Cambridge CB2 2RU, UK
40 West 20th Street, New York NY 10011–4211, USA
477 Williamstown Road, Port Melbourne, VIC 3207, Australia
Ruiz de Alarcón 13, 28014 Madrid, Spain
Dock House, The Waterfront, Cape Town 8001, South Africa

http://www.cambridge.org

First published 1998
First paperback edition 2005

Typeset in Times Roman 10/12pt, in LATEX 2ε [TB]

A catalogue record for this book is available from the British Library

Library of Congress Cataloguing in Publication Data

Szeri, A. Z.
    Fluid film lubrication: theory and design / Andras Z. Szeri.
        p.   cm.
    Includes bibliographical references (p.).
    ISBN 0 521 48100 7 (hardback)
    1. Fluid-film bearings.   I. Title.
    TJ1073.5.S97
    621.8'22—dc21   1998   98-11559
                                                        CIP

ISBN 0 521 48100 7 hardback
ISBN 0 521 61945 9 paperback

Table of Contents

Preface

Fluid film bearings are machine elements which should be studied within the broader context of tribology, "the science and technology of interactive surfaces in relative motion and of the practices related thereto."* The three subfields of tribology – friction, lubrication, and wear – are strongly interrelated. Fluid film bearings provide but one aspect of lubrication. If a bearing is not well designed, or is operated under other than the design conditions, other modes of lubrication, such as boundary lubrication, might result, and frictional heating and wear would also have to be considered.

Chapter 1 defines fluid film bearings within the context of the general field of tribology, and is intended as an introduction; numerous references are included, however, should a more detailed background be required. Chapters 2, 3, and 4 outline classical lubrication theory, which is based on isothermal, laminar operation between rigid bearing surfaces. These chapters can be used as text for an advanced undergraduate or first-year graduate course. They should, however, be augmented with selections from Chapter 8, to introduce the students to the all important rolling bearings, and from Chapter 9, to make the student realize that no bearing operation is truly isothermal. Otherwise, the book will be useful to the industrial practitioner and the researcher alike. Sections in small print may be omitted on first reading – they are intended for further amplification of topics. In writing this book, my intent was to put essential information into a rational framework for easier understanding. So the objective was to teach, rather than to compile all available information into a handbook. I have also included thought-provoking topics; for example lubrication with emulsions, the treatment of which has not yet reached maturity. I expect significant advances in this area as it impacts on the environment.

The various chapters were read by Dr. M. L. Adams, Case Western Reserve University, Dr. M. Fillon, University of Poitiers, France, Dr. S. Jahanmir, National Institute for Standards and Technology, Dr. F. E. Kennedy, Dartmouth College, Mr. O. Pinkus, Sigma Inc., Dr. K. R. Rajagopal, Texas A & M University, Dr. A. J. Szeri, University of California at Berkeley, and Dr. J. A Tichy, Rensselaer Polytechnic Institute. However, in spite of the considerable assistance I received from various colleagues, the mistakes are mine alone.

The typing was expertly done by my daughter Maria Szeri-Leon and son-in-law Jorge Leon. I am grateful to them for their diligence and perseverance; not even their wedding interrupted the smooth flow of the project. I would also like to thank Ms. Florence Padgett, Editor, Cambridge University Press, for suggesting the project and for having confidence in me. My thanks are also due to Ms. Ellen Tirpak, Senior Project Manager, TechBooks, for providing expert editing of the manuscript.

* British Lubrication Engineering Working Group, 1966.

CHAPTER 1

Introduction

The term *tribology*, meaning the *science* and *technology of friction, lubrication,* and *wear*, is of recent origin (Lubrication Engineering Working Group, 1966), but its practical aspects reach back to prehistoric times. The importance of tribology has greatly increased during its long history, and modern civilization is surprisingly dependent on sound tribological practices.

The field of tribology affects the performance and life of all mechanical systems and provides for reliability, accuracy, and precision of many. Tribology is frequently the pacing item in the design of new mechanical systems. Energy loss through friction in tribo-elements is a major factor in limits on energy efficiency. Strategic materials are used in many tribo-elements to obtain the required performance.

Experts estimate that in 1978 over 4.22×10^6 Tjoule (or four quadrillion Btu) of energy were lost in the United States due to simple friction and wear – enough energy to supply New York City for an entire year (Dake, Russell, and Debrodt, 1986). This translates to a \$20 billion loss, based on oil prices of about \$30 per barrel. Most frictional loss occurs in the chemical and the primary metal industries. The metalworking industry's share of tribological losses amount to 2.95×10^4 Tjoule in friction and 8.13×10^3 Tjoule in wear; it has been estimated that more than a quarter of this loss could be prevented by using surface modification technologies to reduce friction and wear in metal working machines. The unsurpassed leader in loss due to wear is mining, followed by agriculture.

1.1   Historical Background

There is little evidence of tribological practices in the early Stone Age. Nevertheless, we may speculate that the first fires made by humans were created by using the heat of friction. In later times hand- or mouth-held bearings were developed for the spindles of drills, which were used to bore holes and start fires. These bearings were often made of wood, antlers, or bone; their recorded use covers some four millennia. Among the earliest-made bearings were door sockets, first constructed of wood or stone and later lined with copper, and potter's wheels, such as the one unearthed in Jericho, dated 2000 BC. The wheel contained traces of bitumen, which might have been used as a lubricant.

Lubricants were probably used on the bearings of chariots, which first appeared ca. 3500 BC (McNeill, 1963). One of the earliest recorded uses of a lubricant, probably water, was for transportation of the statue of Ti ca. 2400 BC. Considerable development in tribology occurred in Greece and Rome beginning in the fourth century BC, during and after the time of Aristotle. Evidence of advanced lubrication practices during Roman times is provided by two pleasure boats that sank in Lake Nemi, Italy, ca. AD 50; they contain what might be considered prototypes of three kinds of modern rolling-element bearings. The Middle Ages saw a further improvement in the application of tribological principles, as evidenced by the development of machinery such as the water mill. An excellent account of the history of tribology up to the time of Columbus is given by Dowson (1973). See also Dowson's *History of Tribology* (Dowson, 1979).

The basic laws of friction were first deduced correctly by da Vinci (1519), who was interested in the music made by the friction of the heavenly spheres. They were rediscovered in 1699 by Amontons, whose observations were verified by Coulomb in 1785. Coulomb was able to distinguish between static friction and kinetic friction but thought incorrectly that friction was due only to the interlocking of surface asperities. It is now known that friction is caused by a variety of surface interactions. These surface interactions are so complex, however, that the friction coefficient in dry sliding still cannot be predicted.

The scientific study of lubrication began with Rayleigh, who, together with Stokes, discussed the feasibility of a theoretical treatment of film lubrication. Reynolds (1886) went even further; he detailed the theory of lubrication and discussed the importance of boundary conditions. Notable subsequent work was done by Sommerfeld and Michell, among others. However, for many years the difficulty of obtaining two-dimensional solutions to Reynolds' pressure equations impeded the application of lubrication theory to bearing design. This impediment was finally removed with the arrival of the digital computer (Raimondi and Boyd, 1958).

In contrast to friction, the scientific study of wear is more recent. As sliding wear, a term often used to define progressive removal of material due to relative motion at the surface, is caused by the same type of interaction as friction, the quantitative prediction of wear rate is fraught with the same difficulties. The situation is even more gloomy, as under normal conditions the value of the coefficient of friction between different metal pairs changes by one order of magnitude at most, while corresponding wear rates can change by several orders. Although there have been attempts to predict wear rate, Archard's formula (Archard, 1953) being perhaps the most noteworthy in this direction, for the foreseeable future at least, the designer will have to rely on experimentation and handbook data (see Peterson and Winer, 1980).

1.2   Tribological Surfaces

Even early attempts to develop a theory of friction recognized the fact that all practically prepared surfaces are rough on the microscopic scale. The aspect ratio and the absolute height of the hills, or *asperities*, and valleys one observes under the microscope vary greatly, depending on material properties and on the method of surface preparation. Roughness height may range from 0.05 μm or less on polished surfaces to 10 μm on medium-machined surfaces, to even greater values on castings. Figure 1.1 shows a size comparison of the various surface phenomena of interest in tribology.

When two solid surfaces are brought into close proximity, actual contact will be made only by the asperities of the two surfaces, specifically along areas over which the atoms of one asperity surface are within the repulsive fields of the other.[1] The *real area of contact A_r*, which is the totality of the individual asperity contact areas, is only a fraction of the *apparent area of contact*, perhaps as small as 1/100,000 at light loads. The areas of individual asperity contacts are typically 1 to 5 μm across and 10 to 50 μm apart.

---

[1] The equilibrium spacing of atoms is on the order of 0.2–0.5 nm (2–5 Angstrom); at distances less than the equilibrium spacing, the repulsive forces dominate, while at greater distances the forces of attraction are influential. The equilibrium spacing changes with temperature; macroscopically we recognize this change as thermal expansion.

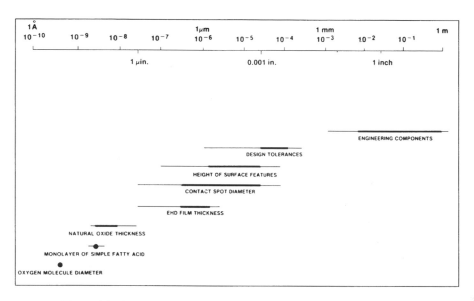

Figure 1.1. Comparative size of surface-related phenomena. (Reprinted with permission from Williamson, J. B. P. The shape of surfaces. In Booser, E. R. *CRC Handbook of Lubrication.* Copyright CRC Press, Boca Raton, Florida, © 1984.)

The topography of engineering surfaces indicates features of four different length scales: (1) *error of form* is a gross deviation from shape of the machine element, (2) *waviness* is of a smaller scale and may result from heat treatment or from vibration of the workpiece or the tool during machining, (3) *roughness* represents closely spaced irregularities and includes features that are intrinsic to the process that created the surface, and (4) surface features on the *atomic scale* are important for the recording industry and in precision machining.

One of the methods used for describing surface roughness consists of drawing a fine stylus across it. The stylus is usually a conical diamond with a radius of curvature at its tip of the order of 2 μm. The movement of the stylus is amplified, and both vertical and horizontal movements are recorded electronically for subsequent statistical analysis. The instrument designed to accomplish this is the *profilometer*. Clearly, such an instrument is limited in resolution by the diameter and the radius of curvature of the tip of the stylus. A profilometer trace[2] of an engineering surface is shown in Figure 1.2.

Two modern instruments, the scanning *electron microscope* and the *transmission electron microscope* (Sherrington and Smith, 1988), have resolution higher than profilometers and are employed extensively in surface studies. *Optical interferometers*, which can record surface profiles without distortion or damage, have recently come into use thanks to advances in microprocessors. Vertical resolution of the order of 1 nm has been achieved by optical interferometers, although the maximum measurable height is somewhat limited by the depth of focus of these instruments (Bhushan, Wyant, and Meiling, 1988). The *atomic*

---

[2] That the vertical amplification is typically 10–1000 times greater than the horizontal one has led to the popular misconception that engineering surfaces support steep gradients. Machined surfaces have aspect ratios normally found in the topography of the Earth, the slopes rarely exceeding 5–10°; Figure 1.2 is a distortion of this.

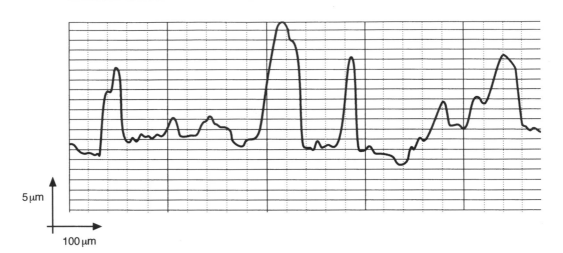

5 µm

100 µm

Figure 1.2. Profilometer trace of a rolled metal specimen. The vertical magnification is 20 times the horizontal magnification.

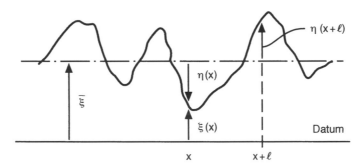

Figure 1.3. Schematics of a surface showing mean surface height, $\bar{\xi}$, and surface deviation from mean height, $\eta(x)$.

force microscope measures the forces between a probe tip and the surface and has been used for topographical measurement of surfaces on the nanometers scale. Its modification, known as the *friction force microscope* (Ruan and Bhushan, 1994) is used for friction studies on the atomic scale. Details of these recent additions to the arsenal of the surface scientist can be found in the excellent review article by Bhushan, Israelachvili, and Landman (1995).

To discuss surface roughness quantitatively, let $\xi(x)$ represent the height of the surface above an arbitrary datum at the position x, and let $\bar{\xi}$ be its mean value as depicted in Figure 1.3. Furthermore, denote by $|\eta(x)|$ the vertical distance between the actual surface at x and the mean. Surface roughness is often characterized in terms of the *arithmetic average*, R_a, of the absolute value of surface deviations from the mean

$$R_a = \frac{1}{L} \int_{-L/2}^{L/2} |\eta(x)| \, dx, \tag{1.1}$$

or in terms of its *standard deviation* [i.e., root mean square (rms)], R_q, defined by

$$R_q^2 = \frac{1}{L} \int_{-L/2}^{L/2} \eta^2(x)\, dx. \tag{1.2}$$

where L is the sample length.

The rms value, Eq. (1.2), is some 10–20% greater than the R_a value for many common surfaces; for surfaces with Gaussian distribution $R_q = 1.25 R_a$. Typical values of R_a for metals prepared by various machining methods are: turned, 1–6 μm; course ground, 0.5–3 μm; fine ground, 0.1–0.5 μm; polished, 0.06–0.1 μm; and super finished, 0.01–0.06 μm.

Another quantity used in characterizing surfaces is the *autocorrelation function*, $R(\ell)$, it has the definition (see Figure 1.3)

$$R(\ell) = \frac{1}{L} \int_{-L/2}^{L/2} \eta(x)\eta(x + \ell)\, dx. \tag{1.3}$$

$R(\ell)$ attains its maximum value at $\ell = 0$, equal to R_q^2, then vanishes rapidly as ℓ is increased. Its normalized value, $r(\ell) = R(\ell)/R_q^2$, is called the *autocorrelation coefficient*. Peklenik (1968) analyzed surfaces that were produced by different machining techniques and proposed a surface classification based on the shape of the correlation function and the magnitude of the correlation length $\lambda_{0.5}$, defined by $R(\lambda_{0.5}) = 0.5$.

The Fourier cosine transform, $P(\omega)$, of the autocorrelation function

$$P(\omega) = \frac{2}{\pi} \int_0^{\infty} R(\ell) \cos(\omega \ell)\, d\ell, \tag{1.4}$$

is a quantity particularly suitable to the study of machined surfaces (see Figure 1.5), since it clearly depicts and separates strong surface periodicities that may result from the machining process (i.e., waviness).

There are other numerical characteristics of surfaces in use; to define these we make recourse to probability theory. To this end consider the random *variable* ξ, representing the height of the surface at some position x relative to an arbitrary datum, and examine the event $\xi < y$, signifying that the random variable ξ has a value less than the number y. The probability of this event occurring, designated by $P(\xi < y)$, is a function of y. Define the *integral distribution function* by $F(y) = P(\xi < y)$, then $F(-\infty) = 0$, $F(+\infty) = 1$ and $0 \leq F(y) \leq 1$. The random variable ξ is considered known if its integral distribution, $F(y)$, is given.

For any two numbers y_2 and y_1, where $y_2 > y_1$, the probability of the event $\xi < y_2$ is given by the sum of the probabilities that $\xi < y_1$ and $y_1 \leq \xi < y_2$, or

$$\begin{aligned} P(\xi < y_2) &= P(\xi < y_1 \quad \text{or} \quad y_1 \leq \xi < y_2) \\ &= P(\xi < y_1) + P(y_1 \leq \xi < y_2). \end{aligned} \tag{1.5}$$

From Eq. (1.5) we find that

$$\begin{aligned} P(y_1 \leq \xi < y_2) &= P(\xi < y_2) - P(\xi < y_1) \\ &= F(y_2) - F(y_1), \end{aligned} \tag{1.6}$$

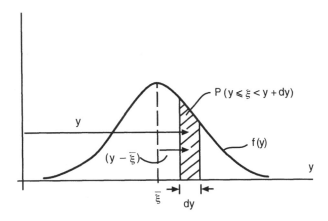

Figure 1.4. Illustration of the probabilistic terminology used.

In the case of a continuous random quantity the distribution function is differentiable. Define the *probability density function* or *probability distribution* by

$$f(y) = \lim_{\Delta y \to 0} \frac{F(y + \Delta y) - F(y)}{\Delta y}. \tag{1.7}$$

From here we can show that the probability that the random variable ξ has a value between y and $y + dy$ is

$$P(y \leq \xi < y + dy) = F(y + dy) - F(y)$$
$$= f(y) \, dy,$$

and that the probability that ξ is located between the numbers a and b is

$$P(a \leq \xi < b) = \int_a^b f(y) \, dy.$$

Instead of the probability density function $f(y)$ itself, its various moments are often employed. The first *initial moment*, given by

$$\bar{\xi} = \int_{-\infty}^{\infty} y f(y) \, dy, \tag{1.8}$$

is the *mean value* of the random variable ξ (Figure 1.4). It is equivalent to R_a of Eq. (1.1).

The fluctuation about the mean can now be defined by $\eta = \xi - \bar{\xi}$; this is the (random) quantity appearing in Eqs. (1.1) and (1.2).

The first *central moment*, i.e., the moment about the mean, of the probability density function is zero. Its second central moment

$$\sigma^2 = \int_{-\infty}^{\infty} (y - \bar{\xi})^2 f(y) \, dy \tag{1.9}$$

is nonnegative, and it is called the *variance* of the random variable ξ. The square root of the variance is termed the *standard deviation* and is equivalent to the rms. of the deviation from the mean, $\sigma = R_q$.

Many variables that express the results of physical, biological, or medical experiments are, at least to first approximation, distributed according to

$$f(y) = \frac{1}{\sigma\sqrt{2\pi}} \exp[-(y - \bar{\xi})^2/2\sigma^2], \tag{1.10}$$

the so-called *normal* or *Gaussian* distribution. For this reason, the normal distribution has played an important role in the development of statistical theory, and one frequently encounters Eq. (1.10) in applications. We note from Eq. (1.10) that if the random variable ξ is normally distributed, it is characterized completely by its mean value $\bar{\xi}$ and its standard deviation σ. The simplicity in representation this affords is the reason why there is often great compulsion to declare a distribution Gaussian even though it may deviate from Eq. (1.10).

Other statistical quantities in use for surface characterization are the third and fourth (nondimensional) central moments, the *skewness, Sk*, and the *kurtosis* or "hump," K, respectively

$$Sk = \frac{1}{\sigma^3} \int_{-\infty}^{\infty} (y - \bar{\xi})^3 f(y)\, dy, \qquad K = \frac{1}{\sigma^4} \int_{-\infty}^{\infty} (y - \bar{\xi})^4 f(y)\, dy. \tag{1.11}$$

Both Sk and K are dimensionless numbers; $Sk = 0$ indicates perfect symmetry, while K is small for a flat, broad distribution. For normal distribution $Sk = 0$ and $K = 3$.

There are many ways to statistically characterize surface roughness. Which of the characterizations is best is application dictated.

It has been shown recently (Sayles and Thomas, 1978) that the value of the various averages defined here changes with the sampling length L, i.e., surface roughness is a nonstationary random function of position. It is then more amenable to treatment by fractal methods (Majumdar and Bhushan, 1990; Wang and Komvopoulos, 1994).

Figure 1.5 shows statistical characteristics of some machined surfaces:

Manufacturing processes	R_a (μm)	σ (μm)	Sk	K	Peak to valley height (μm)	Figure
Shaping, fine	8.0	11.0	0	2.8	47.0	1.5 (a)
Milling	2.3	2.7	+0.22	2.4	13.0	1.5 (b)
Surface grinding	1.0	1.3	+0.17	3.1	15.0	1.5 (c)
Superfinish	0.18	0.25	+0.32	5.9	1.6	1.5 (d)

The asperity-height distribution of many engineering surfaces is approximately Gaussian. Several surface-finishing processes, such as bead-blasting, which are the cumulative result of a large number of random happenings, will encourage a Gaussian distribution.[3] Other processes, including wear, will destroy it. Figure 1.6 follows such a process. A mild steel pad lubricated with SAE-20 oil was worn against a finely ground hard steel flat (N.B., when plotted on probability paper, the Gaussian distribution appears as a straight line).

---

[3] Let the n random variables $\xi_1, \ldots, \xi_n$ be independent. Then the *central limit theorem* asserts, under very general conditions, that in the limit as $n \to \infty$ the standardized sum $(\xi - \bar{\xi})/\sigma$ approaches Gaussian distribution (Cramer, 1955). Here $\bar{\xi} = \bar{\xi}_1 + \cdots + \bar{\xi}_n, \sigma^2 = \sigma_1^2 + \cdots + \sigma_n^2$ and $\xi = \xi_1 + \cdots + \xi_n$.

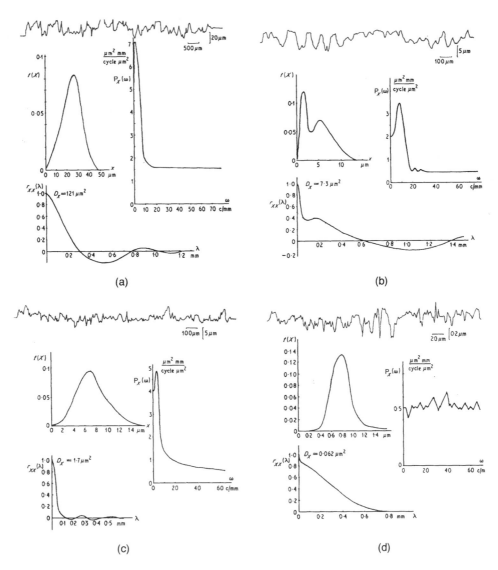

Figure 1.5. Examples of engineering surfaces: their distributions, autocorrelation functions, power spectra. (Reprinted by permission of the Council of the Institution of Mechanical Engineers from Peklenik, J. New developments in surface characterization and measurements by means of random process analysis, *Proc. Inst. Mech. Engrs.* **182**, Pt. 3K, 108–126, 1968.)

1.3   Friction

If two solid bodies, in direct or indirect surface contact, are made to slide relative to one another there is always a resistance to the motion called friction. Friction is beneficial in many instances, and we may even try to increase it. However, in other cases friction is energy consuming, and we endeavor to decrease it, although it may never be eliminated entirely.

Friction is present in all machinery, and it converts part of the useful kinetic energy to heat, thus decreasing the overall efficiency of the machine. About 30% of the power

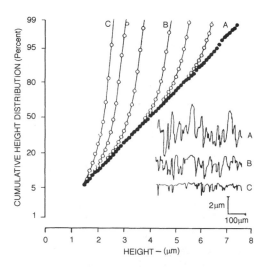

Figure 1.6. The effect of wear. The initial height distribution (A) and six non-Gaussian distributions (open circles) of a bead blasted surface represent, from right to left, progressive states. Height distributions of this form are typical of those created by stratified secondary preparation processes. (Reprinted with permission from Williamson, J. B. P. The shape of surfaces. In Booser, E. R. *CRC Handbook of Lubrication.* Copyright CRC Press, Boca Raton, Florida, © 1984.)

in an automobile (Hershey, 1966) and about 1.5% in a modern turbojet engine is wasted through friction. The two journal bearings of a large generator dissipate perhaps 0.75 MW or more. In 1951, G. Vogelpohl estimated that one-third to one-half of the world's energy production is consumed by friction (Fuller, 1956). Not all friction is undesirable, however, and in numerous instances we promote it, e.g., in brakes.

Laws of Friction

The two basic laws of friction:

1. Friction force is proportional to the normal force between surfaces,
2. Friction force is independent of the (apparent) area of contact,

were first deduced by da Vinci (1519) and discussed by Amontons (1699). Coulomb (1785) verified these laws experimentally.[4] Coulomb's observation that "kinetic friction is nearly independent of the sliding speed" is at times referred to as the third law of friction. The laws of friction have remained intact for more than 400 years, and even modern experimental research supports them in numerous cases.

This is not true, however, for the origin of friction as discussed by Coulomb. At first Coulomb inclined toward the view that friction is produced by molecular adhesion between the interacting surfaces, which is somewhat in line with present-day theories. Later Coulomb rejected this in favor of the view that friction is produced by interlocking surface asperities. According to this theory, the frictional force is the force required to lift the load

---

[4] To derive Amontons' laws, we need the assumption that the real area of contact is proportional to the normal load $A_r = qW$, where q is a constant. If now we denote the friction force per unit area by τ, we have for the friction force $F = \tau A_r$, and Amontons' laws follow at once. In the adhesion theory of friction of Bowden and Tabor (1986), the constant q is made equal to the yield pressure p_0.

over the asperities. Considering that sliding down the asperities releases as much energy
as was spent on climbing up, Coulomb's friction is nondissipative, as was first pointed out
by Leslie in 1804.

Most current theories recognize that frictional force in metals arises from three sources:
(1) the force necessary to shear *adhesive junctions*, formed at the real area of contact
between the asperities; (2) the deformation force, due to the *ploughing* of the asperities of
the harder metal through the asperities of the softer one; and (3) *asperity deformation*, which
is responsible for the static coefficient of friction—Suh (1986) lists the force required for
this as the third source of frictional force. Though these three forces, and the three effects
causing them, are not independent, it is customary to treat friction as a result of adhesion
interactions, plowing interactions, and asperity deformations. In elastomers, elastic and
viscoelastic effects dominate, while in ceramics the type of bonding (ionic in MgO and
Al_2O_3 and covalent in TiC, diamond, and SiC) limits plastic flow and the high plastic
strains associated with junction growth, at room temperature.

The idea of formation of adhesive junctions (cold welding) over the area of real contact
seems frivolous at first, until one considers ultraclean metallic surfaces. When such surfaces
are brought together in high vacuum ($P < 10^{-8}$ Pa), the atoms of the real area of contact
approach one another across the interface. When they are within 2 nm (20 Angstrom), long
distance, weak van der Waals forces are first experienced. As the interfacial distance is
decreased to 0.2–0.1 nm, a full metallic bond will form and the pieces weld together. The
experiments of Buckley (1977) have been concerned with the force required to overcome
this so-called *cold welding*. The adhesive forces are sometimes greater than the forces
necessary to press the metals together. However, the metallic bond is completely broken if
extended to 0.5 nm, thus a surface film of this thickness signifies that only weak van der
Waals forces are acting. As a result, one should expect considerable reduction in adhesive
strength. These ideas recently have been confirmed by molecular dynamics simulations
(Landman, Luedtke, and Ringer, 1992).

Two elementary methods of measuring static friction, both considered by Leonardo da
Vinci, are illustrated in Figure 1.7. Though these methods are quick and convenient, they
have had limited success due to the response of the systems being too slow for variations
in the coefficient of friction to be detected. Once the body has started moving it will
accelerate under constant force, for in general $f_{static} > f_{kinetic}$. Even such a variation in
friction can hardly be detected by these simple methods. More sophisticated devices for
measuring friction are described by Bowden and Tabor (1986), who identify cleanliness

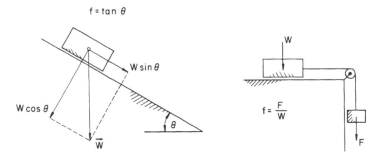

Figure 1.7. Elementary methods of measuring friction.

Table 1.1. *Coefficients of static and dynamic friction*[a,b]

	Static		Dynamic	
Materials	Dry	Greasy[c]	Dry	Greasy[c]
Hard steel on hard steel	0.78(1)	0.11(1,a)	0.42(2)	0.03(5,h)
		0.23(1,b)		0.08(5,c)
		0.15(1,c)		0.08(5,i)
		0.11(1,d)		0.06(5,j)
		0.01(17,p)		0.08(5,d)
		0.01(18,h)		0.11(5,k)
				0.10(5,l)
				0.11(5,m)
				0.12(5,a)
Mild steel on mild steel	0.74(19)		0.57(3)	0.09(3,a)
				0.19(3,u)
Hard steel on graphite	0.21(1)	0.09(1,a)		
Hard steel on Babbitt (ASTM 1)	0.70(11)	0.23(1,b)	0.33(6)	0.16(1,b)
		0.15(1,c)		0.06(1,c)
		0.08(1,d)		0.11(1,d)
		0.09(1,e)		
Hard steel on Babbitt (ASTM 8)	0.42(11)	0.17(1,b)	0.35(11)	0.14(1,b)
		0.11(1,c)		0.07(1,c)
		0.09(1,d)		0.07(1,d)
		0.08(1,e)		0.08(11,h)
Hard steel on Babbitt (ASTM 10)		0.25(1,b)		0.13(1,b)
		0.12(1,c)		0.06(1,c)
		0.10(1,d)		0.06(1,d)
		0.11(1,e)		
Mild steel on cadmium silver				0.10(2,f)
Mild steel on phosphor bronze			0.34(3)	0.17(2,f)
Mild steel on copper lead				0.15(2,f)
Mild steel on cast iron		1.83(15,c)	0.23(6)	0.13(2,f)
Mild steel on lead	0.95(11)	0.5(1,f)	0.95(11)	0.30(11,f)
Nickel on mild steel			0.64(3)	0.18(3,x)
Aluminum on mild steel	0.61(8)		0.47(3)	
Magnesium on mild steel			0.42(3)	
Magnesium on magnesium	0.6(22)	0.08(22,y)		
Teflon on Teflon	0.04(22)			0.04(22,f)
Teflon on steel	0.04(22)			0.04(22,f)
Tungsten carbide on tungsten carbide	0.2(22)	0.12(22,a)		
Tungsten carbide on steel	0.5(22)	0.08(22,a)		
Tungsten carbide on copper	0.35(23)			
Tungsten carbide on iron	0.8(23)			
Bonded carbide on copper	0.35(23)			
Bonded carbide on iron	0.8(23)			
Cadmium on mild steel			0.46(3)	
Copper on mild steel			0.36(3)	0.18(17,a)
Nickel on nickel	1.10(16)		0.53(3)	0.12(3,w)

Table 1.1. (*cont.*)

Materials	Static		Dynamic	
	Dry	Greasy[c]	Dry	Greasy[c]
Brass on mild steel	0.51(8)		0.44(6)	
Brass on cast iron			0.30(6)	
Zinc on cast iron	0.85(8)		0.21(7)	
Magnesium on cast iron			0.25(7)	
Copper on cast iron	1.05(16)		0.29(7)	
Tin on cast iron			0.32(7)	
Lead on cast iron			0.43(7)	
Aluminium on aluminium	1.05(16)		1.4(3)	
Glass on glass	0.94(8)	0.01(10,p)	0.40(3)	0.09(3,a)
		0.01(10,q)		0.12(3,v)
Carbon on glass			0.18(3)	
Granite on mild steel			0.39(3)	
Glass on nickel	0.78(8)		0.56(3)	
Copper on glass	0.68(8)		0.53(3)	
Cast iron on cast iron	1.10(16)		0.15(9)	0.07(9,d)
				0.06(9,n)
Bronze on cast iron			0.22(9)	0.08(9,n)
Oak on oak (parallel to grain)	0.62(9)		0.48(9)	0.16(9,r)
Oak on oak (perpendicular)	0.54(9)		0.32(9)	0.07(9,s)
Leather on oak (parallel)	0.61(9)		0.52(9)	
Cast iron on oak			0.49(9)	0.08(9,n)
Leather on cast iron			0.56(9)	0.36(9,t)
				0.13(9,n)
Laminated plastic on steel			0.35(12)	0.05(12,t)
Fluted rubber bearing on steel				0.05(13,t)

[a]From Baumeister, T. *Handbook of Mechanical Engineers*, 7th ed. Copyright McGraw Hill Book Co., © 1967. With permission.

[b]Key to lubricants used: a, oleic acid; b, Atlantic spindle oil (light mineral); c, castor oil; d, lard oil; e, Atlantic spindle oil plus 2% oleic acid; f, medium mineral oil; g, medium mineral oil plus 1/2% oleic acid; h, stearic acid; i, grease (zinc oxide base); j, graphite; k, turbine oil plus 1% graphite; 1, turbine oil plus 1% stearic acid; m, turbine oil (medium mineral); n, olive oil; p, palmitic acid; q, ricinoleic acid; r, dry soap; s, lard; t, water; u, rape oil; v, 3-in-1 oil; w, octyl alcohol; x, triolein; y, 1% lauric acid in paraffin oil.

[c]Note that "Greasy" is not sufficient to describe surface conditions. The friction coefficient depends on other factors such as speed, load, environment, etc. Thus Table 1.1 is, necessarily, an oversimplification.

of the surface as the single most important factor in achieving repeatable friction results. Surface contaminants, even when present in a layer only one molecule thick, are capable of drastically modifying the friction coefficient because of the reduction in adhesive interactions. Table 1.1 lists f_{static} and $f_{dynamic}$ for various surface pairs under both dry and greasy (lubricated) conditions.

Asperity Contact

For the sake of this illustration, assume that the two surfaces in contact have hemispherical-shaped asperities of radii r_1 and r_2, respectively. As the normal load is slowly increased, contact is first made by the most prominent asperities. According to Hertz (Johnson, 1992), the deformation of these asperities is initially elastic and the region of contact is a circle with radius

$$a = \left(\frac{3WR}{2E'} \right)^{1/3}.$$ (1.12)

Here W is the load, E' is the effective contact modulus[5], R is the effective radius

$$\frac{1}{E'} = \frac{1}{2} \left(\frac{1 - v_1^2}{E_1} + \frac{1 - v_2^2}{E_2} \right), \qquad \frac{1}{R} = \frac{1}{r_1} + \frac{1}{r_2},$$ (1.13)

and E_1, v_1 and E_2, v_2 are Young's modulus and Poisson's ratio for the two solids, respectively.

At this elastic stage the real area of contact is proportional to $W^{2/3}$ and, using Eq. (1.12), the mean pressure over the contact circle is given by

$$\bar{p} = \frac{2}{3\pi} \left(\frac{3WE'^2}{2R^2} \right)^{1/3}.$$ (1.14)

As the load is increased the mean pressure $\bar{p}$ will also increase, until the elastic limit (of the softer of the two materials) is reached. This will first occur at the point Z, located at $z/a = 0.48$ (when $v = 0.3$) below the center of the contact circle, as indicated in Figure 1.8; here the shear stress first achieves its yield value k (Johnson, 1992). The yield value[6] of the shear stress is equal to $0.5\sigma_{yp}$, where σ_{yp} is the *yield stress in uniaxial tension*. At the instant of reaching the yield value of shear, the mean contact pressure attains the value $\bar{p} \approx 1.1\sigma_{yp}$.

Though the elastic limit is attained at $Z = 0.48$ as the mean contact pressure reaches $\bar{p} = 1.1\sigma_{yp}$, plastic flow is not yet possible due to the constraining influence of the surrounding material in which deformation is still mainly elastic. Consequently, if the load is removed at this stage, only a slight amount of residual deformation is noticeable.

As the normal load is further increased, the zone of plastic deformation propagates outward from the point where it first occurred, until it eventually reaches the surface; the value of the mean pressure is now $\bar{p} = p_0 \approx 3\sigma_{yp}$. The mean pressure at this point is essentially the *indentation hardness* value, H, of the material. This is why, for ductile metals, $H \approx 3\sigma_{yp}$. For hard tool steel $E = 200$ GPa, $\sigma_{yp} \approx 1.96$ GPa; an asperity of radius $r = 1$ μm will deform plastically when the load is less than 10^{-5} N.

Addition of a tangential force to the normal load introduces several effects. The location of the maximum shear stress moves closer to the surface. Friction also increases the

---

[5] Note that Johnson (1992) uses $E^* = E'/2$ instead of our E'. The latter, our notation, is generally employed in discussions on elastohydrodynamic lubrication (EHL).

[6] According to the Tresca yield criterion, a ductile material will yield under a slowly applied complex state of stress when the maximum shear stress equals that which exists at yielding in a static tensile test of the metal, i.e., at $\tau_{max} = \sigma_{yp}/2$. Plastic flow in metals occurs along crystal planes so that a critical shear stress criterion is preferable to other yield criteria.

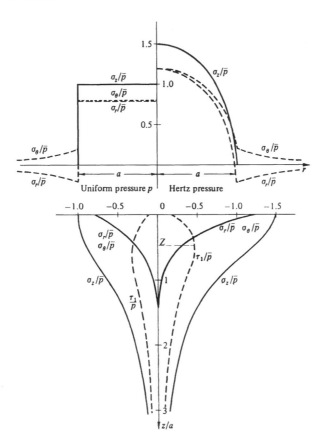

Figure 1.8. Stress distribution at the surface and along the axis of symmetry caused by (left) uniform pressure and (right) Hertz pressure acting on a circular area of radius a. (Reprinted with permission from Johnson, K. L. *Contact Mechanics*. Copyright Cambridge University Press, © 1992.)

maximum value reached by von Mises yield parameter

$$J_2 = \frac{1}{6}\{(\sigma_x - \sigma_y)^2 + (\sigma_y - \sigma_z)^2 + (\sigma_z - \sigma_x)^2\} + \tau_{xy}^2 + \tau_{yz}^2 + \tau_{zx}^2$$

so that yielding will occur at lower loads. This is shown in Figure 1.9, where the von Mises yield parameter is plotted against f for $\nu = 0.3$.

For single asperity contact, Eq. (1.12), the contact area is proportional to the 2/3 power of the load and not to its first power, as apparently is required by Amontons' laws (see footnote 4). The asperities of real surfaces, however, are not of uniform height, as indicated by the profilometer trace in Figure 1.2, and do not all engage immediately as load is first applied. Upon increasing the load, the number of asperities that take active part in carrying the load will also increase. If, as in the probabilistic contact model of Greenwood and Williamson (Greenwood, 1992), the number of asperity contacts is allowed to increase with increasing load in such a manner that the average size of each asperity contact can remain constant, the real area of contact becomes proportional to the load itself rather than to its 2/3 power. Thus, even though the deformation is elastic, the Greenwood and Williamson model supports Amontons' law.

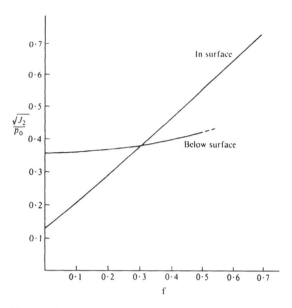

Figure 1.9. Variation of the maximum von Mises yield parameter, in and below the surface. With friction in excess of 0.5, there is no longer a clear maximum. (Reprinted by permission of the Council of the Institution of Mechanical Engineers from Hamilton, G. M. Explicit equations for the stress beneath a sliding spherical contact. *Proc. I. Mech. E.*, **197**, 53–59, 1983.)

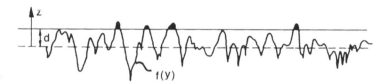

Figure 1.10. Loading a rough surface, of asperity height distribution $f(y)$, against a smooth plane a distance d apart.

To illustrate this, we follow Greenwood (1992) and consider a rough surface with asperity height distribution $f(y)$, located at distance d from a perfectly smooth surface, as in Figure 1.10. If there are a total of N asperities (per unit area), the number of asperity contacts with the plane is

$$n = N \times P(d \le \xi) = N \times [1 - P(\xi < d)]$$

$$= N \times \left[\int_{-\infty}^{\infty} f(\hat{y})\, d\hat{y} - \int_{-\infty}^{d} f(\hat{y})\, d\hat{y} \right] \tag{1.15a}$$

$$= N \times \int_{d}^{\infty} f(\hat{y})\, d\hat{y}.$$

For a single asperity, we can obtain both the area of contact $\pi a^2 = \pi r \delta$, and the load $w = (2/3)E'\sqrt{r\delta^3}$, in terms of the asperity compression $\delta = (z - d)$ and the radius of curvature r. Then the total area of contact, A_r, and the total load, W, are given, respectively, by (Johnson, 1992)

$$A_r = N\pi r \int_y^\infty (\hat{y} - d) f(\hat{y}) \, d\hat{y} \tag{1.15b}$$

and

$$W = \frac{2}{3} N E' r^{1/2} \int_d^\infty (\hat{y} - d)^{3/2} f(\hat{y}) \, d\hat{y}. \tag{1.15c}$$

These equations can be evaluated once the probability distribution is known. Greenwood and Williamson (Greenwood, 1992) chose an exponential distribution, $f(y) = \exp(-\lambda/y)$, and found

$$n = \frac{N}{\lambda} e^{-\lambda d}, \tag{1.16a}$$

$$A_r = \frac{N\pi r}{\lambda^2} e^{-\lambda d}, \tag{1.16b}$$

and

$$W = \frac{1}{2} N E' \left(\frac{\pi r}{\lambda^5} \right)^{1/2} e^{-\lambda d}, \tag{1.16c}$$

so that

$$A_r = \sqrt{\frac{2\lambda \pi r}{E'}} \, W \tag{1.16d}$$

and the area of contact is directly proportional to load, as required by Amontons' laws.

Though the above derivation was for fully elastic deformation, we can glimpse at the onset of plastic flow (Greenwood, 1992). The fraction of asperity contacts at which plastic deformation occurs is proportional to δ_Y/σ, where σ is the standard deviation of asperity height, Eq. (1.9), and δ_Y is the asperity compression at first yield, i.e., when $\bar{p} \approx H/3$. Greenwood and Williamson (1966) defined the so-called *plasticity index* ψ to be inversely proportional to $\sqrt{\sigma/\delta_Y}$, so that

$$\psi = \frac{E'}{2H} \sqrt{\frac{\sigma}{r}}. \tag{1.17}$$

$\Psi \approx 1$ corresponds to 1% of the total contact area at yielded contacts. Asperity contact will become plastic when the plasticity index exceeds unity (note that Ψ is independent of the average pressure $\bar{p}$). For metal surfaces produced by normal engineering methods, $0.1 < \Psi < 100$.

Adhesion Theory of Friction

In metals it has been found that the pressure, p_0, that the asperities can support when subjected to localized plastic deformation is approximately constant. In the plastic range, then, if we double the load, the area of contact must also double in order to maintain a constant yield pressure. Let $A_1, A_2, A_3, \ldots$ represent a series of areas of contact, supporting loads $W_1, W_2, W_3, \ldots$. If W is the total load and A_{ro} is the total area of real contact under normal load, we have

$$
\begin{aligned}
&= W_1 + W_2 + W_3 + \cdots \\
&= p_0 A_1 + p_0 A_2 + p_0 A_3 + \cdots \\
&= p_0 (A_1 + A_2 + A_3 + \cdots) \\
&= p_0 A_{ro}.
\end{aligned}
\tag{1.18}
$$

From this it follows that the area of real contact A_{ro} is dependent neither on the size nor on the shape of the area of apparent contact. It is determined only by the yield pressure p_0 and the load (Bowden and Tabor, 1956).

> Note that the real area of contact is proportional to the load when the deformation is plastic, as is required by Amontons' law. However, fully plastic deformation of the asperities seems feasible when the surfaces are used only a limited number of times. It is not realistic to expect the same surface, say the surface of a cylinder, to deform plastically during every one of the millions of times the other surface, the piston in this case, makes a pass. One is inclined to think that after a short run-in period almost all the plastic (irreversible) deformations that were to take place have done so, and that after run-in the load carrying deformation will be mainly elastic. On the second and successive passes the material is subjected to the combined action of contact stresses and residual stresses from previous passes; the effect of the latter is such as to make yielding less likely.[7] The trouble with this line of thought is that, at least according to the Hertz analysis for an isolated, single asperity, Eq. (1.12), the area of contact in elastic deformation is proportional to the 2/3 power of the load, which would negate Amontons' law. If, as in the probabilistic contact model of Greenwood and Williamson (Greenwood, 1992), the number of asperity contacts is allowed to increase with increasing load in such a manner that the average size of each asperity contact can remain constant, the real area of contact becomes proportional to the load itself rather than to its 2/3 power, Eq. (1.16d). Thus, even though the deformation is elastic, the Greenwood and Williamson model supports Amontons' law.

As the surfaces make contact only at the tips of their asperities, the pressures are extremely high. Over the regions of intimate contact strong adhesion takes place, and the specimens become, in effect, a continuous body (cold welding). As the surfaces are made to slide over one another, the just welded junctions are sheared. Let s represent the shear strength of the material and A_{ro} the area of real contact. We may then write $A_{ro} \times s$ for the shear force. For the *coefficient of friction* we have

$$
\begin{aligned}
f &= \frac{F}{W} = \frac{A_{ro}s}{A_{ro}p_0} = \frac{s}{p_0} \\
&= \frac{\text{junction shear strength}}{\text{yield pressure}}.
\end{aligned}
\tag{1.19}
$$

This conclusion will be considerably altered in practice. For most materials, s is of order $0.2 \times p_0$, so that according to this model $f = 0.2$. This is far too small. For identical metals in normal atmosphere $f = O(1)$ and for clean metals in vacuum f can reach 10 or larger. This compelled Bowden and Tabor to modify Eq. (1.19) and the argument leading up to it.

The asperities are already loaded to their elastic limit by the normal load as the interfacial tangential force is applied. To support the combined normal and tangential forces, the plasticity condition at the junction would now be exceeded, which cannot be, unless there

---

[7] This is known as the *shakedown principle* (Johnson, 1992).

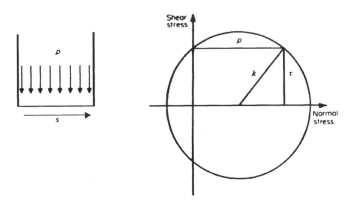

Figure 1.11. Simple loading of an asperity and the corresponding Mohr's circle. (Reprinted with permission from Arnell, R. D., Davies, P. B., Halling, J. and Whomes, T. L. *Tribology Principles and Design Applications.* Copyright Springer Verlag, © 1991.)

were an appropriate increase of the real area of contact, from A_{ro} to some A_r. This mechanism, referred to as junction growth, has a profound effect on the value of the coefficient of friction.

Junction Growth

In the previous section the real area of contact A_{ro} was determined solely by the normal load and the yield pressure, $A_{ro} = W/p_0$. When a tangential load is also applied, it is more appropriate to calculate the real area of contact on the basis of the combined tangential and normal loading.

When subjecting a slab of solid material to the combination of (a) simple shear τ and (b) normal loading p, the Tresca yield criterion, viz., $k \equiv \tau_{\max} = \sigma_{yp}/2$, takes the form (Figure 1.11)

$$4\tau^2 + p^2 = \sigma_{yp}^2. \tag{1.20}$$

Substituting $p = W/A_r$ and $\tau = F/A_r$ into Eq. (1.20), we find that

$$\frac{A_r}{A_{ro}} = \sqrt{1 + \left(\frac{2F}{W}\right)^2}. \tag{1.21}$$

Here, A_r is the area of real contact under the combined normal and tangential loads and $A_{ro} = W/\sigma_{yp}$. Equation (1.21) verifies our earlier assertion on junction growth. The coefficient of friction is now given by

$$f = \frac{1/2}{\sqrt{\left(\frac{k}{\tau}\right)^2 - 1}}. \tag{1.22}$$

There is nothing in this model, Eq. (1.21), to limit junction growth, which continues indefinitely if (1) the surfaces are perfectly clean and (2) the metals are very ductile (Tabor, 1981).

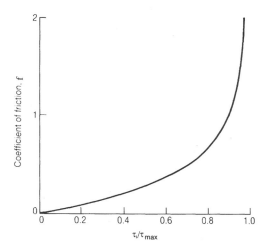

Figure 1.12. Junction growth: variation of the coefficient of friction with the interfacial yield stress τ_i.

When present, contaminants will limit junction growth. To illustrate this mechanism, let τ_i represent the interfacial yield stress, which is less than or equal to the shear yield stress of the asperity material in bulk, k. We now write $F = A_r\tau_i$ for the tangential force and obtain from the Tresca yield criterion, Eq. (1.20), an improved formula for the coefficient of friction

$$f = \frac{1/2}{\sqrt{\left(\dfrac{k}{\tau_i}\right)^2 - 1}}. \tag{1.23}$$

The variation of f with τ_i is illustrated in Figure 1.12.

In analogy with combined loading and shear, p and τ acting on an actual adhesive junction are assumed by Childs (1992) to obey

$$p^2 + \alpha\tau^2 = \beta k^2, \tag{1.24}$$

where α and β are constants. In place of Eqs. (1.22) and (1.23), Childs obtains

$$\frac{A_r}{A_{ro}} = \sqrt{1 + \alpha\left(\frac{F}{W}\right)^2}, \qquad f = \frac{\tau}{k}\left[\beta - \alpha\left(\frac{\tau}{k}\right)^2\right]^{-1/2}. \tag{1.25}$$

Equation (1.25) suggests that junction growth is not the only possible mechanism for reducing friction. If τ/k is of order unity and $\beta > \alpha + 1$, then $f < 1$ results from the second part of Eq. (1.25). It is also indicated that a small amount of weakening at the interface, caused by a thin contaminant film of shear strength τ, can produce a significant reduction in the coefficient of friction. This is the principle underlying boundary lubrication, and lubrication by soft metal films.

Ploughing

For a hard conical asperity riding on a softer metal, we can illustrate the magnitude of the ploughing term by considering a conical asperity of semi-angle θ, shown in Figure 1.13.

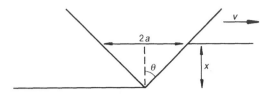

Figure 1.13. A conical asperity of semi-angle θ indents a softer metal: a model for ploughing.

The pressure needed to make the softer material flow ahead of the advancing hard asperity can be taken to be the hardness, H, of the softer material. The normal and tangential forces supported by the asperity may then be calculated, respectively, as

$$
W = \left(\frac{\pi r^2}{2}\right) H = \frac{1}{2} H \pi h^2 \tan^2 \theta,
$$
$$
F_p = r h H = H h^2 \tan \theta.
$$
(1.26)

Under these idealized conditions, the coefficient of friction due to ploughing is

$$
f_p = \frac{F_p}{W} = \frac{2 \cot \theta}{\pi}.
$$
(1.27)

Under normal circumstances the slope of the asperities rarely exceeds 5–10°, fixing the range of θ at 85–80°; this yields the ploughing component of the coefficient of friction lying between 0.07 and 0.14. As the ploughing component is small, it may be considered additive to the adhesion component. A more thorough plasticity theory is called for to nonlinearly combine the two contributions to friction (Suh, 1986).

Friction of Metals

The coefficient of friction is given as the sum of three terms, f_a, f_p, and f_d the adhesion component, the ploughing component, and the deformation component, respectively. According to Suh (1986), the adhesion component for metals varies from about 0 to 0.4, depending on the presence of a contaminant layer covering the asperities. The ploughing component is smaller than the adhesion component, usually not exceeding 0.1, except for identical metals sliding against one another, having wear particles trapped between the surfaces. The friction component due to asperity deformation, f_d, can be as large as 0.4 to 0.75 in special cases, and is believed to be responsible for the static coefficient of friction (Suh, 1986).

Suh (1986) defines six stages of surface interaction that characterize time-dependent friction behavior of metals (Figure 1.14). The relative importance of f_a, f_p, and f_d changes from stage to stage. Stage I is characterized by ploughing of the surface by asperities; the coefficient of friction is largely independent of material combinations. In Stage II the value of friction is beginning to rise. The slope is steeper if wear particles generated by asperity deformation and fracture become trapped. The swift increase in the number of wear particles gives the friction curve a steep slope during Stage III; another contributor here is adhesion due to the rapid increase of clean surfaces. Friction remains constant during Stage IV as adhesion and asperity deformation are constant now, as is the number of trapped wear particles. Stage V occurs when a very hard stationary slider is slid against a soft specimen. The asperities of the slider are gradually removed, creating a mirror finish of the hard surface. A decrease in friction will result in this case, due to the decrease in asperity deformation

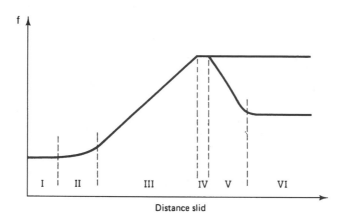

Figure 1.14. Six stages in the frictional force versus distance slid relationship. (Reprinted with permission from Suh, N. P. *Tribophysics.* Copyright Prentice-Hall, © 1986.)

and plowing. In Stage VI the hard surface is mirror smooth and the value of the friction levels off. Stages V and VI do not occur if it is the hard slider that moves and the soft specimen that is stationary (Suh, 1986).

Friction of Polymers

The interfacial bonding of polymers is of van der Waals type, the same as in the bulk of the material. The tendency for shear to occur is in the bulk material (Briscoe and Tabor, 1978). Two notable exceptions are Teflon (PTFE) and Ultra High Molecular Weight Polyethylene (UHMWPE). When a polymer is sliding on metal or another polymer, the deformation is mostly elastic, with virtually no plastic flow. For a given surface roughness the value of the plasticity index for polymers is an order less than for metals.

Most polymers exhibit lower coefficient of friction at high normal load (Archard, 1953). At low sliding velocity the coefficient of friction is generally low, it increases with sliding speed to a maximum, then decreases again (Ettles, 1981).

Another important factor to take into account is that many polymers are viscoelastic. Viscoelastic materials are both solid-like and fluid-like in their response to stress; the work of deformation is neither completely stored, as in elastic solids, nor completely dissipated, as in viscous fluids. These materials show a marked increase of flow stress with strain rate. The coefficient of friction of polymers, sliding against one another or against a metal, varies so widely that tabulation of friction coefficients would be meaningless. For viscoelastic materials Amontons' laws are not applicable.

Voyutski (1963) found that in most cases interfacial bonding is strongly strengthened by diffusion of polymer chains across the interface. Polymers generally soften early or have low melting points and poor heat conductivity. At high relative speed, the surface layers often melt. However, friction still has both an adhesive component and a deformation component.

Friction of Ceramics

The bonding within ceramics is largely covalent and very strong, but the force of adhesion across the interface between contacting ceramics is of the van der Waals type and is partly ionic. Friction is generally lower for ceramics than for metals owing to the weaker interfacial bonds. These interfacial bonds will, at least in the absence of high temperatures, be weaker than the bulk (Hutchings, 1992). Therefore, when surfaces are pulled or slid apart, the break tends to occur along the original interface. If the ceramic is brittle, surface cracks may develop at the rear of the moving junction, increasing the rate of energy dissipation. The reader may wish to consult Jahanmir (1994) for further details on the tribological properties of ceramics.

Thermal Effects of Friction

Almost all of frictional energy is dissipated in the form of heat. Continuous rubbing of surfaces can build up not only significant temperatures but also large temperature gradients in the contacting bodies. The nonuniform thermal expansion that accompanies this can lead to loss of dimensional tolerance in the case of machine components.

It is relatively easy to set up energy conservation equations once the geometry and thermal characteristics of the bodies are known and to calculate the average temperature field for given energy input. But heat does not enter the system in any easily definable manner. Furthermore, it is generated instantaneously at the random asperity contacts of the two surfaces. The instantaneous and random asperity temperatures, the so-called *flash temperatures* that result from asperity interaction, are significantly higher than the *nominal temperature* given by our steady-state energy conservation equations. A solution for flash temperatures can be found in a book by Carslaw and Jaeger (1959), but this solution is far too detailed to be reproduced here. For some recent work and bibliography, the reader is referred to Tian and Kennedy (1993) and Tian and Kennedy (1994). The first of these papers deals with the nominal surface temperature rise, while the second discusses local flash temperature rise; the total frictional temperature rise is considered to be given by the sum of these.

1.4   Wear

Wear is the progressive loss of substance of a body, due to relative motion at its surface. Many different types of wear have been identified. Godfrey (1980), e.g., recognized a dozen different types of wear, though at times it is not easy to differentiate between them. In this introduction to wear we follow Rabinowicz (1965) and list four main types: sliding wear, abrasive wear, corrosion, and surface fatigue. Only sliding wear and abrasive wear will be discussed in some detail.

Research into wear follows two lines, wear modeling and the study of damaged surfaces. Wear modeling, with the objective of engineering prediction of wear rates, is the older of the two aspects of wear research. The study of damaged surfaces owes its existence to significant recent advances in experimental methods and microscopy equipment.

Sliding Wear

In contrast to that of friction, the scientific study of sliding wear is recent. As friction and sliding wear are caused by the same type of surface interaction, the quantitative prediction of wear rates is fraught with the same difficulties as that of friction. However, the situation is even more bleak, as under normal conditions the value of the coefficient of friction between different metal pairs changes by one order of magnitude at most, while corresponding wear rates can change by several orders. There have been numerous attempts made in the past to predict wear rate, Archard's formula (Archard, 1953) being one of the most noteworthy in this direction.

To arrive at Archard's wear rate equation, consider the asperities of opposing surfaces as they make contact with one another while supporting the yield pressure p_0 of the softer material. Asperity contacts are assumed to occur uniformly over circular contact areas of radius a. The elemental load carried by each asperity is then $\delta W = \pi a^2 p_0$, and the total load supported by n asperities is the sum

$$W = \pi p_0 \sum_n a^2. \tag{1.28}$$

Now picture the asperities as having hemispherical tips of volume $(2/3)\pi a^3$. On separating from its main body an asperity tip will contribute $\delta V = (2/3)\pi a^3$ to the wear volume.[8] However, not every asperity will break off on contact. If the probability of a particular asperity breaking off on first contact is κ, then the wear volume per sheared distance per asperity is

$$\delta Q = \kappa \left(\frac{\delta V}{2a} \right)$$
$$= \frac{\kappa \pi}{3} a^2. \tag{1.29}$$

Summing Eq. (1.29) for n asperities, substituting for $\sum_n a^2$ from Eq. (1.28), and recognizing that the yield pressure equals the hardness value H, we have the wear volume per distance slid, i.e., the *wear rate*, as

$$Q = K \frac{W}{H}, \tag{1.30a}$$

where the wear coefficient $K = \kappa/3$ is a pure number. It is found that $K < 1$ always, and that the value of K can vary by several orders of magnitude when conditions change,[9] even for the same material pair. Equation (1.30) is *Archard's wear rate formula*.

When written in the form[10]

$$V \propto \frac{W}{H} L, \tag{1.30b}$$

where $V = QL$ is the volume of material worn, Archard's wear rate formula has the following interpretation, called *Archard's laws* for sliding wear:[11]

(1)   The wear volume is proportional to the distance slid,
(2)   The wear volume is proportional to the total load,
(3)   The wear volume is inversely proportional to the hardness of the softer material.

We find confirmation of the first of these relationships in the original paper of Archard and Hirst (1956). The relevant figure is reproduced in Figure 1.15.

[8] That the asperities are spherical and that the asperity is completely removed is a gross simplification of actual conditions.

[9] For lubricated sliding wear, Archard's wear rate formula is supplied with another coefficient $\alpha < 1$, which characterizes the ratio of the area of metal to metal contact to the apparent area.

[10] This was developed for metals assuming plastic deformation. It may or may not apply to other materials.

[11] It should be obvious that, on using Archard's equation in the form of Eq. (1.30), we can substitute wear rate in place of wear volume.

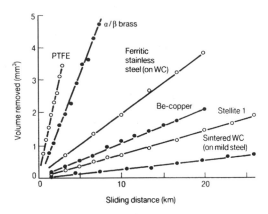

Figure 1.15. Variation of wear rate, for various materials. (Reprinted with permission from Archard, J. F. and Hirst, W. The wear of metals under unlubricated conditions. *Proc. Roy. Soc.*, **A 236**, 397–410, 1956.)

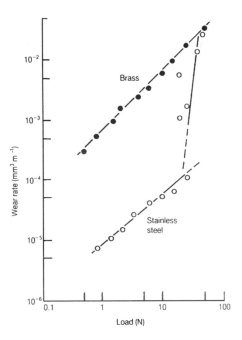

Figure 1.16. There is transition from mild to severe wear in stainless steel, while brass obeys Archard's law in the whole load range. (Reprinted with permission from Archard, J. F. and Hirst, W. The wear of metals under unlubricated conditions. *Proc. Roy. Soc.*, **A 236**, 397–410, 1956.)

The second relationship, viz., that the wear volume (or, the wear rate), is proportional to the applied load, holds for metals in certain load ranges. Often at low load the wear rate is small (*mild wear*), and increases linearly with load, up to a *critical load*. On reaching the critical load, the mechanism that produces mild wear becomes unstable and transition to *severe wear* takes place, the wear coefficient often changing by several orders of magnitude. This situation is well illustrated in Figure 1.16, which shows transition from mild to severe wear in stainless steel. Brass, on the other hand, undergoes no transition and obeys Archard's first law throughout the load range of Figure 1.16.

It is now generally recognized that the wear of a soft material against a hard one can be characterized as *mild* or *severe*. In severe wear there is metallic contact of newly exposed surfaces and severe surface damage. Severe wear takes place under conditions where a protective oxide film is unable to form during the time available between interasperity contacts. Mild wear, on the other hand, is manifested when, at light load and speed, there is sufficient time between asperity interactions for oxide formation even at the prevailing "low" reaction rate. Mild wear will also be encountered at high temperatures, caused either by high rate of frictional heating (Figure 1.17), or by external heating, when the reaction rate is sufficiently high to promote formation of a protective oxide film in the time available.

There are, hence, two transitions. The T_1 transition is from mild to severe wear, and occurs as the load (or the speed) is increased. The rate of exposure of virgin metal surface during severe wear is opposed by the increasing rate of contamination of the surfaces by reaction with the ambient atmosphere. As the temperature increases above some critical T_2, 250–350°C in Lancaster's experiments for a carbon steel pin sliding against tool steel, a second transition takes place, this time from severe to mild wear. Here T_2 is a nominal surface temperature, the flash temperature, that characterizes individual asperity interactions

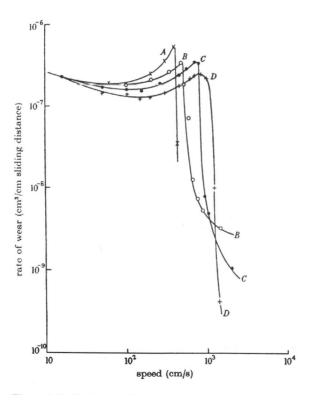

Figure 1.17. Variation of the rate of wear for pins with frictional heating. Apparent contact area increasing from A to D. (Reprinted with permission from Lancaster, J. K. The formation of surface films at the transition between mild and severe metallic wear. *Proc. Roy. Soc.*, **A 273**, 466–483, 1963.)

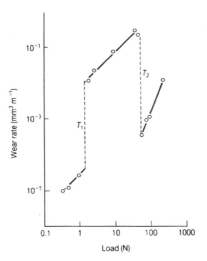

Figure 1.18. Variation of wear rate with load, for carbon steel pin riding on tool steel. (Reprinted with permission from Welsh, N. C. The dry wear of steels. *Phil. Trans. Roy. Soc.*, **A 257**, 31–70, 1965.)

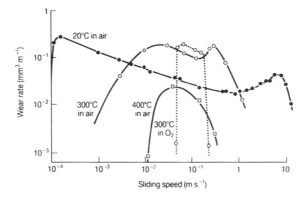

Figure 1.19. The variation of wear rate with sliding speed at different ambient temperatures for brass sliding against steel, — in air; ......, in oxygen. (Reprinted with permission from Lancaster, J. K. The formation of surface films at the transition between mild and severe metallic wear. *Proc. Roy. Soc.*, **A 273**, 466–483, 1963.)

and has no effect on the transition. Figure 1.18 shows the variation in the wear rate with varying load for a carbon steel pin sliding against tool steel.

Qualitatively, transition remains the same irrespective of whether heating occurs by external means or as a consequence of friction (Lancaster, 1963). The variation of wear rate with sliding speed at different temperatures is illustrated in Figure 1.19. At low speed, mild wear results from the 'large' times available for oxidation; at high speeds, mild wear is the consequence of increased rates of oxidation. The magnitudes of critical loads and speeds will also depend on temperature. Changing the surrounding atmosphere to pure oxygen will also vary the critical conditions (speed and load) for transition.

Classification of sliding wear into mild wear and severe wear infers linking the wear mechanism to elastic contacts and plastic contacts, respectively (Kragelskii and Marchenko, 1982). Resulting from mild wear, the surface roughness is of the order of $R_a \approx 0.5\,\mu$m and the wear debris particle dimension is in the range $d_p = 0.01 - 1\,\mu$m. Mild wear dominates both at low speed and high temperature, where oxidation can keep balance with the rate of exposure of fresh surfaces, allowing the formation of an oxide layer. This last condition facilitates a lower coefficient of friction. On the other hand, severe wear will create a rough surface, with the arithmetic average of roughness R_a reaching $25\,\mu$m. The wear debris is now made up of particles of dimension $d_p \approx 200\,\mu$m, and surface oxidation can no longer keep up with the exposure of fresh surfaces; therefore, the coefficient of friction is usually large. Further increasing the load increases frictional work, leading to increased temperature and, thus, to increased rate of oxidation. Formation of oxide film that can be attained between successive asperity interactions grows exponentially with load as the load is increased, other things being equal, while the time between collisions decreases linearly (Arnell et al. 1991) and a second transition from severe wear to mild wear may occur.

In an effort to develop a simple design guide, Lim and Ashby (1978) graphed the various wear regimes for given pairs of materials, thereby producing *wear maps*. The vertical axis of the graphs is normalized pressure and the horizontal axis is normalized sliding velocity. A wear map for soft carbon steel sliding on the same, in air at room temperature, is shown in Figure 1.20.

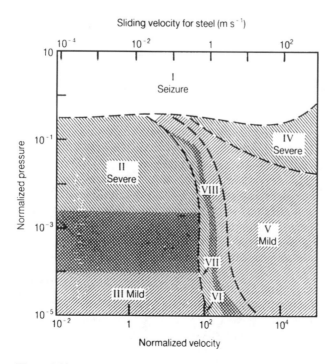

Figure 1.20. Wear map for soft carbon steels at room temperature in air. (Printed with permission from Lim, S. C. and Ashby, M. F. Wear mechanism maps. *Acta Metall.*, **35**, 1–24. 1987.)

Table 1.2. *Wear coefficient $K \times 10^6$ for sliding wear, Eq. (1.30)*

Lubrication	Metal-on-metal			Metal-on-nonmetal	
	Identical	Soluble	Intermediate	Insoluble	Nonmetal-on-metal
None	1500	500	100	15	3
Poor	300	100	20	3	1.5
Good	30	10	2	0.3	1
Excellent	1	0.3	0.1	0.03	0.05

(Reprinted with permission from Rabinowicz, E. Wear coefficients, in Booser, E. R., *CRC Handbook of Lubrication*. Copyright CRC Press, Boca Raton, Florida, © 1984.)

Region I in Figure 1.20 is characterized by high contact pressure and gross seizure of the surfaces. Region II represents severe wear at high loads and low speed, while Region III has mild wear at low load and speed. Thermal effects, which were hitherto unimportant, become significant in Regions IV and V, both of which have increased reaction rate with the ambient atmosphere. Regions VI, VII, and VIII represent narrow transition regimes. For more detailed discussion of the significance and characteristics of the various regions of the wear map of Figure 1.20, the reader is referred to Lim and Ashby (1987).

In the model described above, the idea that wear particle size is related to contact patch size has been advanced. Rabinowicz (1965) suggested that energy considerations govern wear particle size. He noted that the strain energy associated with a plastic contact of elastic bodies is $H^2/2E$ per unit volume of the material. The surface energy of the wear particle is $2\pi a^2 \gamma$, where γ is the surface energy per unit of surface area. Rabinowicz then required that the stored elastic energy in the particle volume exceed the surface energy

$$\frac{2}{3}\pi a^3 \frac{H^2}{2E} > 2\pi a^2 \gamma$$

or

$$a > \frac{6E\gamma}{H^2}.$$

Furthermore if $E/H = k$, a constant, then

$$a > \frac{k\gamma}{H}.$$

Rabinowicz provided evidence to support his arguments, one implication of which is that there exists a minimum size of particles that can be generated through deformation.

Instead of attempting to find and tabulate the wear coefficient for pairs of metals (and conditions), we follow Rabinowicz and divide sliding systems into a limited number of categories, then give appropriate wear coefficient data for each category. The categories are: identical, soluble, intermediate, insoluble, nonmetal-on-nonmetal, and metal-on-nonmetal. For metals, the two principal factors that determine the wear coefficient are (1) the degree of lubrication and (2) the metallurgical compatibility, as indicated by the mutual solubility. Wear coefficients for adhesive wear for the five categories are given in Table 1.2, while Figure 1.21, after Rabinowicz, illustrates compatibility for metal pairs. The corresponding compatibility relationship is listed in Table 1.3.

Table 1.3. *Compatibility relationship for metals*

Symbol	Metallurgical solubility	Metallurgical compatibility	Sliding compatibility	Anticipated wear
○	100%	Identical	Very poor	Very high
⊙	>1%	Soluble	Poor	High
○	0.1–1%	Intermed. soluble	Intermediate	Intermediate
◗	<0.1%	Interm. insoluble	Intermed. or good	Intermed. or low
●	Two liquid phases	Insoluble	Very good	Very low

(Reprinted with permission from Rabinowicz, E. Wear coefficients, in Booser, E. R., *CRC Handbook of Lubrication.* Copyright CRC Press, Boca Raton, Florida, © 1984.)

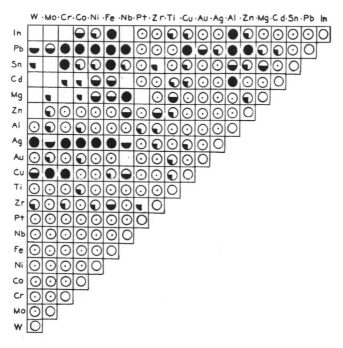

Figure 1.21. Compatibility diagram for metal pairs. The significance of the symbols is shown in Table 1.3. (Reprinted with permission from Rabinowicz, E. Wear coefficients. In Booser, E. R., *CRC Handbook of Lubrication.* Copyright CRC Press, Boca Raton, Florida, © 1984.)

Abrasive Wear

It is usual to distinguish between two-body abrasion, in which the asperities of the harder surface abrade the softer surface, and three-body abrasion, in which hard particles trapped between two surfaces abrade one or possibly both surfaces.

The mechanics of two-body abrasion closely resembles that of ploughing, as discussed earlier, except that in ploughing material is pushed aside while in abrasive wear some cutting is also involved. Referring to Figure 1.13, we find the normal load supported as in Eq. (1.26). The volume of material displaced by the cone while creating a groove of length

ℓ is $\ell h^2 \tan\theta$. If a fraction, ε, of the displaced material becomes wear debris, then the wear volume produced in unit ploughing distance is

$$q = \varepsilon h^2 \tan\theta.$$

Substituting now for h^2 from Eq. (1.26), we obtain

$$q = \frac{2\varepsilon W}{\pi H \tan\theta}.$$

Summing for all asperities engaged in abrading the surface, we find that the total wear volume removed per unit sliding distance is

$$Q = \frac{\hat{K} W}{H}. \qquad (1.31)$$

In Eq. (1.31) W is the applied normal load and $\hat{K}$ is a coefficient (Hutchings, 1992). We note that formally Eq. (1.31) is identical to Archard's rate equation for adhesive wear, Eq. (1.30).

We conclude this section with the observation that there has been considerable interest in recent years in methods of predicting wear rates. Investigations have concentrated on two problem areas, *running in*, which is a time-dependent phenomenon, and *steady state*, which is homogeneous in time.

1.5   Effect of Lubrication

Lubrication is used to reduce/prevent wear and lower friction. The behavior of sliding surfaces is strongly modified with the introduction of a lubricant between them. If we plot, for example, for a journal bearing, the coefficient of friction against $\mu N/P$, where μ is the lubricant viscosity, N is the shaft speed, and $P = W/LD$ is the specific load, we find that at large values of $\mu N/P$, the friction coefficient, is low and is proportional to $\mu N/P$. This is the regime of *thick-film* lubrication. Upon decreasing $\mu N/P$, the friction passes through a minimum value, as indicated schematically in Figure 1.22, and we enter into the regime of *mixed lubrication*. For even smaller values of $\mu N/P$, the coefficient of friction increases rapidly, marking the complete breakdown of the lubricant film in this so-called *boundary lubrication* regime. Table 1.4 compares average values of the coefficient of friction in the various lubrication regimes.

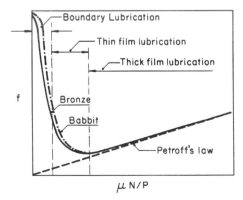

Figure 1.22. Lubrication regimes.

Table 1.4. *Average values of the coefficient of friction*

Lubrication type	Friction coefficient	Degree of wear
Unlubricated	0.5–2.0	Heavy
Boundary and thin film	0.05–0.15	Slight
Thick film	0.001	None

Thick-film Lubrication

When the minimum film thickness exceeds, say, 2.5 μm, Petroff's law[12] is approximately obeyed by a lightly loaded journal bearing. The coefficient of friction is small and depends on no other material property of the lubricant than its bulk viscosity. This type of lubrication is called thick-film lubrication. In many respects this is the simplest and most desirable kind of lubrication to have.

Mixed Lubrication

The low $\mu N/P$ branch of the curve in Figure 1.22 represents varying degrees of thin-film lubrication, a name given by Hersey (1966) to the lubrication regime in which the coefficient of friction depends on surface roughness and on a lubricant property that Hersey terms "oiliness." The transition from thick-film lubrication to mixed lubrication takes place around the minimum point of the f-$\mu N/P$ curve. As the value of $\mu N/P$ is made smaller, the film becomes thinner and some of the opposing asperities touch. The friction coefficient now depends on the surface roughness, the material properties of the solids, and the material properties of the lubricant. For increasing smoothness the minimum point of the f-$\mu N/P$ curve shifts to the left.

Boundary Lubrication

In boundary lubrication, the film is so thin that its properties are no longer the same as those of the bulk. If the speed is reduced or the load is increased, the lubricant film becomes thinner than the height of some of the asperities. If these asperities are covered by a suitable molecular layer of lubricant, they will not weld together. This will be the case if the lubricant contains small amounts of surface-active materials. Typical active materials are long-chain fatty acids, alcohols, and esters. For this type of lubrication the friction coefficient is typically 0.1. Mixed lubrication can simply be viewed as a mixture of hydrodynamic lubrication and boundary lubrication. Mixed lubrication is a term often applied to conditions to the left of the minimum point of the f-$\mu N/P$ curve.

This concept of boundary lubrication must be revised when applied to mineral oils. It has been found that mineral oils under contact pressures of the order of 0.5–3 GPa increase their viscosity 100- or even 1000-fold. The oil, which is trapped between the elastically

[12]Petroff postulates a uniform shear stress $\tau_w = \mu U/P$ acting on the journal. The coefficient of friction $f = F/W$ is then given by $f = 2\pi^2 \mu RN/PC$. In conventional oil-lubricated bearings $C/R \sim 1/500$, and we write $f = \text{const} \times \mu N/P$. In older literature, the symbol Z (for the German *Zähigheit*) is used for viscosity and Petroff's law is written as $f \propto ZN/P$.

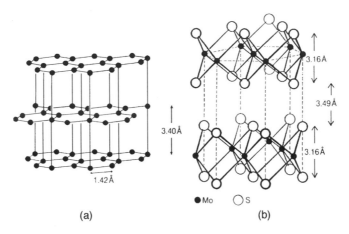

Figure 1.23. The structures of (a) graphite and (b) molybdenum disulfide. (Reprinted with permission from Hutchings, I. M. *Tribology: Friction and Wear of Engineering Materials.* Copyright CRC Press, Boca Raton, Florida, © 1992.)

deformed solid surfaces, behaves as a virtual solid, preventing contact of the asperities. For this type of lubrication to occur the surfaces must be smooth and well aligned.

Solid Lubrication

Some metals show strong mechanical anisotropy, being strong in compression but weak in shear along certain directions. These metals may acquire low-friction sliding surfaces under loading. For special materials such as molybdenum disulfide (MoS_2) and graphite, the crystal structure is in the form of tightly bonded layers lying on one another, as shown in Figure 1.23.

In both graphite and molybdenum disulfide the bond between the interlayer atoms is due to covalent forces and is very strong. The lamella-to-lamella forces, on the other hand, are van der Waals type weak forces.[13] When rubbed against a metal these solids will transfer material onto a metallic substrate. The transfer film will be oriented so that the weak bond is on its outer face. Thus, in sliding contact only this weak bond need be broken where asperities on the harder body contact the film-covered substrate. For MoS_2 the transfer process is more or less independent of ambient conditions surrounding the bodies. For graphite, however, moisture (or other contaminants such as simple hydrocarbons) must be available to weaken the interlayer bond strength of the solid. This effect is of importance in high-altitude rockets. In the reduced moisture levels at high altitude, brush wear in the electric generators becomes extreme and large amounts of carbon dust are formed. Very small amounts of organic compounds or moisture will inhibit this wear (Bisson and Anderson, 1964). Not all lamellar solids exhibit small friction; some are more isotropic.

---

[13]Note that the distance between the lamellae is 0.34 nm for carbon and 0.349 nm in the molybdenum disulfide, putting them safely out of range of short distance, i.e., strong forces. The interlayer atomic distance in the graphite, on the other hand, is only 0.142 nm.

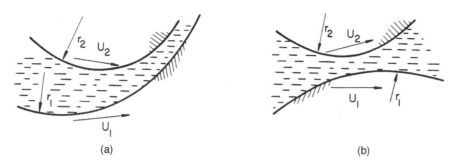

Figure 1.24. (a) Conformal and (b) counterformal geometry.

1.6   Fluid Film Bearings

Bearings are machine elements whose function is to promote smooth relative motion at low friction between solid surfaces. The surfaces might be in direct contact, but if they are not, the lubricant film separating them can be liquid, gaseous, or solid.

Considering the global geometry of the surfaces, a bearing may be *conformal* or *counterformal*. The conformal condition is depicted in part (a) of Figure 1.24. In conformal bearings, a prototype of which is the journal bearing, the apparent area of contact is large. The maximum film pressure is of the same order of magnitude as the *specific bearing load*, defined by $P = W/A$, where W is the external load and A is the projected (normal to the load) bearing area.[14] The film pressures are relatively small in conformal bearings and the lubricant film is thick. In consequence, the bulk deformation of the bearing surfaces is relatively unimportant or at least less important than the deformation of the asperities, say, during the running-in period.

The counterformal condition, illustrated in part (b) of Figure 1.24, is typically found in gear lubrication and in rolling-contact bearings. There the contact stresses are extremely high and the film is thin. The lubricant oil exhibits properties in this high-pressure contact zone that might be significantly different from its properties in bulk; it behaves as a virtual solid. The elastic deformation of the solid surfaces forms an essential component of the analysis of these bearings. One of the first investigator to study the elastic deformation of contacts was Hertz, and the counterformal condition is often referred to as the *Hertzian* condition.

When there is a continuous fluid film separating the solid surfaces we speak of fluid film bearings. There are two principal ways of creating and maintaining a load-carrying film between solid surfaces in relative motion. We call a bearing *self-acting*, and say that it operates in the *hydrodynamic mode* of lubrication, when the film is generated and maintained by the viscous drag of the surfaces themselves, as they are sliding relative to one another. The bearing is *externally pressurized*, and it operates in the *hydrostatic mode*, when the film is created and maintained by an external pump that forces the lubricant between the solid surfaces. The term fluid is used here to designate either a liquid or a gaseous substance, but there are some fundamental differences between liquid-lubricated bearings and gas-lubricated bearings because of the compressibility of gases. For this reason this book will discuss both liquid-lubricated bearings and gas-lubricated

---

[14]For journal bearings, irrespective of the arc length, the specific bearing load is defined as P/LD, where L is the length and D is the diameter of the bearing.

Table 1.5. *Comparison of liquid film bearings*

Lubrication mode	Film thickness (μm)	Friction coefficient
Hydrostatic	50–5	10^{-6}–10^{-3}
Hydrodynamic	10–1	10^{-3}–10^{-2}
Elastohydrodynamics	1–0.1	10^{-3}–10^{-2}

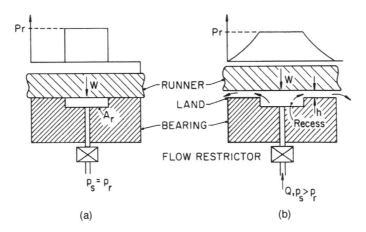

Figure 1.25. Hydrostatic bearing schematics (a) before and (b) after lift-off.

bearings, although in separate chapters. However, for the purpose of this introduction it is sufficient to illustrate the method of operation of liquid film bearings only. Liquid-lubricated bearings are compared on the basis of film thickness and coefficient of friction in Table 1.5.

Hydrostatic Bearings

Hydrostatic lubrication is, in principle, the simplest mode of liquid film lubrication. The load-carrying film is both created and maintained by external means, the essential property of this lubrication mode being that the load-carrying surface is floated, irrespective of whether there is relative motion.

Part (a) of Figure 1.25 illustrates the geometry of a hydrostatic bearing at the commencement of its operation. The lubricant is supplied to the recess at the supply pressure p_S either directly by a pump or via a manifold and flow restrictor. The load-carrying runner will be supported by and will rest on the land as long as the supply pressure is below the value W/A_R, where A_R is the recess area. The runner will lift-off as soon as the supply pressure reaches W/A_R. At this stage lubricant flow out of the recess and over the land will commence, and the solid surfaces will be separated by a continuous lubricant film, as shown in part (b) of Figure 1.25. The recess pressure is now given by $p_R = p_S - \Delta p$, where Δp represents line losses, including those encountered in the flow control devices. Varying the supply pressure once lift-off has

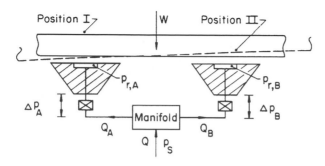

Figure 1.26. Operation with flow restrictors.

occurred will simply vary the film thickness h but will leave the recess pressure unchanged.

Generally speaking, hydrostatic systems use several evenly spaced pads, as shown in Figure 1.26, so that asymmetric load distributions may be managed. Under normal conditions, depicted by position I in Figure 1.26, the external load W is distributed equally between the two hydrostatic pads A and B with uniform and equal film thickness. Tilting the runner to position II will have the effect of decreasing the film thickness h_A over pad A and increasing the film thickness h_B over B.

Because of the increased resistance to flow over pad A, the rate of lubricant flow Q_A will decrease. This will cause the recess pressure $p_{R,A}$ to increase, as there is now less pressure drop across the flow restrictor in line A. On the other hand, the increase in film thickness over pad B will lead to a decrease in the recess pressure $p_{R,B}$.

Operation with flow restrictors may be summarized as follows:

$$p_S = p_{R,A} + \Delta p_A = p_{R,B} + \Delta p_B = \text{const.}$$

$$\text{Pad } A \begin{cases} Q_A^{II} < Q_A^{I} \\ \Delta p_A^{II} < \Delta p_A^{I} \\ p_{R,A}^{II} > p_{R,A}^{I} \end{cases} \qquad \text{Pad } B \begin{cases} Q_B^{II} > Q_B^{I} \\ \Delta p_B^{II} > \Delta p_B^{I} \\ p_{R,B}^{II} < p_{R,B}^{I} \end{cases}$$

The net effect of the action of the two pads will be a restoring moment on the runner.

An externally pressurized bearing equipped with *flow restrictors* is called a *compensated* bearing. Flow restrictor design influences bearing stiffness, required supply pressure, required pumping power, and lubricant flow.

Among the advantages of hydrostatic bearings are:

(1) Low friction (vanishing with relative speed)
(2) Unaffected by discontinuous motion of the runner
(3) Exact positioning of the runner, as film thickness can be controlled accurately
(4) Capable of supporting heavy loads
(5) Only moderate temperature rise across pad because of large film thickness
(6) Continuous outflow of lubricant prevents ingress of dirt

Among the disadvantages of externally pressurized systems we list:

(1)   Requirement for auxiliary external equipment
(2)   Risk of lubrication failure because of failure of auxiliary equipment
(3)   High power consumption by pumps
(4)   Necessity of constant supervision
(5)   High initial cost

Despite their many disadvantages, externally pressurized bearings are widely used. Machine tools using this type of bearing can grind parts round within 0.05 μm, with a 0.025 μm surface finish. The lubricant can be either gaseous or liquid. Gaseous lubricant is often used in lightly loaded applications (high-speed drills, gyroscopes, torque meters), whereas liquid systems are normally used for heavier loads. Examples of the latter are the bearings of the 200-inch optical telescope at Mount Palomar, which weighs 500 tons, and the 140-foot radio telescope at Green Bank, with a support weight of 2000 tons in addition to wind loads. Oil lifts, working in the hydrostatic mode, are often built into the hydrodynamic bearings of large rotating apparatus to aid starting and stopping.

Hydrodynamic Bearings

To understand the operation of bearings in the hydrodynamic mode, consider fluid flow between two solid surfaces. Both surfaces extend to infinity in the direction perpendicular to the plane of Figure 1.27 and are slightly inclined toward one another. A typical value for the slope of the solid surfaces in industrial bearings is 0.001 radian. The top surface in Figure 1.27 is sliding in its own plane, with a velocity, U, relative to the lower surface, dragging the lubricant with it into the convergent (in the direction of relative motion) gap. No external pressure gradient is imposed on the flow.

If the fluid pressure were uniform everywhere in the film, with the flow induced solely by the motion of the slider, the flow rate from left to right would be given by $Q = Uh/2$. But the film thickness h varies along the gap, whereas the principle of mass conservation requires Q to be constant.

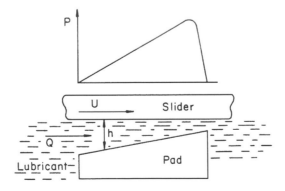

Figure 1.27. Schematics of a plane slider.

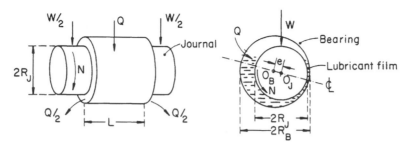

Figure 1.28. Schematics and nomenclature of a journal bearing.

The apparent inconsistency is removed if we allow for the existence of a pressure flow of varying strength and direction, such that it aids shear flow where $hU/2$ is small (i.e., where the gap is narrow) and hinders it where it is large. In this way we arrive at a uniform flow rate so that mass conservation is satisfied. The type of pressure curve that would induce the desired pressure flow is shown in Figure 1.27, it is readily measured in hydrodynamic bearings.

The plane slider of Figure 1.27 is the prototype of the hydrodynamic *thrust bearing*. One of the more frequent functions of hydrodynamic bearings is to support rotating shafts; thrust bearings are employed in these cases if the load vector is parallel to the axis of rotation.

If the motion that the bearing must accommodate is rotational and the load vector is perpendicular to the axis of rotation, the hydrodynamic bearings employed are *journal bearings*.

When the lubricant is incompressible, the film temperature is nearly uniform, and the load vector is fixed in space, as shown in Figure 1.28, the operating conditions of a journal bearing of aspect ratio L/D can be uniquely characterized with reference to a single dimensionless parameter. This parameter, the *Sommerfeld number*, is defined as

$$S = \frac{\mu N}{P}\left(\frac{R}{C}\right)^2,$$

where C represents the radial clearance, i.e., the difference between the bearing and journal radii $C = R_B - R_J$, $P = W/LD$ is the specific bearing load, N is the rotational speed, and μ is the dynamic viscosity of the lubricant. Following accepted practice, we put $R_B = R$ in the definition of the Sommerfeld number.

The journal will be concentric with the bearing under the condition $P \to 0$ or $N \to \infty$, i.e., when $S \to \infty$. On decreasing the speed or increasing the load – in general, on decreasing the value of the parameter S – the journal will occupy an eccentric position of eccentricity e, its center sinking below the center of the bearing. As $S \to 0$ the solid surfaces approach one another and the lubricant film eventually fails, resulting in high temperatures and considerable wear due to rubbing. The movement of the journal center is generally not along a straight line. But its locus, well represented by a semicircle of diameter C, can be predicted in many cases to the required accuracy.

The oil required for hydrodynamic lubrication can be fed from an oil reservoir under gravity, it may be supplied from a sump by rings, disks, or wicks. The bearing might even be made of a porous metal impregnated with oil, which "bleeds" oil to the bearing surface as the journal rotates. Most porous metal bearings, however, operate in the mixed or even in the boundary lubrication regime.

Hydrodynamic bearings vary enormously both in their size and in the load they support. At the low end of the specific-load scale we find bearings used by the jeweler, and at the high end we find the journal bearings of a large turbine generator set, which might be 0.8 m in diameter and carry a specific load of 3 MPa, or the journal bearings of a rolling mill, for which a specific load of 30 MPa is not uncommon. Gas bearings, on the other hand, operate at low specific load (0.03 MPa) but often at high speeds. The high-speed, air turbine-driven dental drill is capable of 500,000 rev/min.

Elastohydrodynamic Lubrication

The term elastohydrodynamic lubrication (EHL) is reserved for hydrodynamic lubrication applied to lubricant films between elastically deforming solids. The principles of EHL are readily applicable to such diverse objects as gears, rolling-element bearings, and human and animal joints. In general, bearings that are lubricated in the EHL mode are of low geometric conformity, and, in the absence of a lubricant film and of elastic deformation, the opposing surfaces would contact in a point (ball bearings) or along a line (gears or roller bearings).

If the solid surfaces that are lubricated in the EHL mode have large elastic modulus, the contact pressures will be large, perhaps of the order of 1 GPa. The film thickness will be correspondingly small, of the order of 1 μm. Under such conditions the material properties of the lubricant will be distinctly different from its properties in bulk. This change in lubricant properties, when coupled with the effects of elastic deformation of the solid surfaces, yields film thicknesses one or two orders of magnitude larger than those estimated from constant viscosity theory applied to nondeforming surfaces.

EHL theory may be viewed as a combination of hydrodynamic lubrication, allowance for the pressure dependence of viscosity, and elastic deformation of the bounding surfaces. The subject of elastohydrodynamic lubrication is outlined in Chapter 8.

1.7   Bearing Selection

While promoting smooth relative motion of the contact surfaces, bearings will necessarily constrain such motions to occur (1) about a point, (2) about a line, (3) along a line, or (4) in a plane.

Our main concern here is with motions that are constrained to proceed about a line and in a plane. When continuous, both of these types of motion include rotation, the former situation calling for journal bearings and the latter for thrust bearings.

There are four methods of supporting the contact load in the cases above: (1) letting the surfaces rub against one another, (2) separating the surfaces by a fluid film, (3) rolling one surface over another, and (4) separating the surfaces by electromagnetic forces.

A fifth method, that of introducing a flexible member between the surfaces, is not suitable for continuous motion and is mentioned here only for the sake of completeness. Of the four methods of dealing with contact forces, our interest is with dry rubbing bearings, fluid film bearings, and rolling-element bearings. These may be journal bearings or thrust bearings, according to the constraining influence they are required to supply.

Rubbing Bearings

The friction and wear of rubbing surfaces has been discussed in earlier sections of this chapter. It is sufficient to only consider frictional heating here. The rate of heat generated is given by FU, where F is the frictional force and U is the relative velocity of the surfaces. If A is the projected bearing area, then the risk of overheating is some function of

$$\frac{FU}{A} = fPU, \tag{1.32}$$

where $P = W/A$ is the specific bearing load. Overheating may result in seizure of the surfaces and should be avoided. The performance of dry rubbing bearings is also limited by an upper value of P, above which fatigue or extrusion of the bearing material may occur, and by a wear rate [which again is proportional to the product PU, Eq. (1.31)], as excessive wear leads to slackness of the machine elements. To avoid excessive wear, rubbing bearings are, in general, lubricated by a thin film of solid lubricant.

Dry rubbing and boundary-lubricated bearings are best suited to low speeds and intermittent duty. They provide high stiffness once loaded heavily, but they are the least-stiff bearings under light load. Safe maximum PU values have been determined experimentally, but these must be used with caution.

Rolling-Element Bearings

Bearings based on rolling-sliding are called rolling-element bearings. In a typical application this type of bearing usually provides for more precise shaft positioning than hydrodynamic bearings (but where extreme precision in positioning is required, externally pressurized bearings are called for). These bearings perform well when subjected to repeated starts-stops under full load. They are best suited to low-speed high-load situations. Rolling-element bearings have little inherent damping capacity, leading to excessive plastic deformation of raceways and premature fatigue under repeated shock loads.

With well-designed rolling-element bearings, wear is microscopic through most of the life of the bearing. The ultimate limit of bearing performance is often surface fatigue of the elements or the raceway. Rolling contact fatigue is characterized by the fairly rapid formation of large wear fragments, signaling the end of bearing life. Up until such time, there is no detectable wear in well-lubricated rolling-contact bearings. Therefore, it is inappropriate to discuss wear rate in these bearings, and the term *bearing life* is preferable. Bearing life is a function of the load and the total number of revolutions. Empirically, it has been found that the bearing life, L, is inversely proportional to the third power of the load W.

Fluid Film Bearings

Wear is not a consideration in fluid film bearings, since in a well-designed fluid film bearing the surfaces are always separated by a continuous fluid film. One of the exceptions to this rule is the heavily loaded self-acting (hydrodynamic) bearing during starting-stopping.

If the bearing is externally pressurized, the load capacity is virtually independent of surface speeds, although at high speeds a decrease of load capacity might be expected because of thermal and/or inertial effects.

If the bearing is of the self-acting type, the load capacity increases with surface speed initially. However, when the rate of heat generation becomes high enough to influence lubricant viscosity, the load capacity drops off. If the bearing is self-contained, an additional constraint is presented by the minimum film thickness limit at low speeds and by the heat removal (maximum Babbitt temperature) limit at high speeds. Hydrodynamic bearings are best suited by their load-speed character to applications where the load increases with speed,

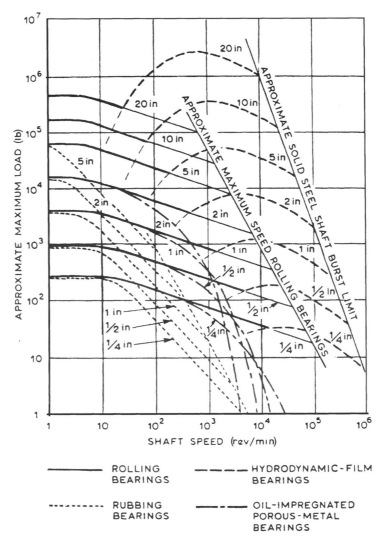

Figure 1.29. Load-speed characteristics of journal bearings. (From the Tribology Subseries Item No. 65,007, *General Guide to the Choice of Journal Bearing Type*, by permission of Engineering Sciences Data Unit Ltd., London.)

whereas hydrostatic bearings are most useful in situations where the load is independent of speed.

The performance of steadily loaded journal bearings and of thrust bearings is shown in Figures 1.29 and 1.30, respectively. These figures are reproduced from a paper by Neale (1967) and show only general trends. The reader interested in bearing selection is referred to the *Tribology Handbook* by Neale (1973) and to an exhaustive article in *Machine Design* (1978). Considerations other than load and speed may often have overriding importance in bearing selection. Tables 1.6 and 1.7 may be used to advantage in bearing selection.

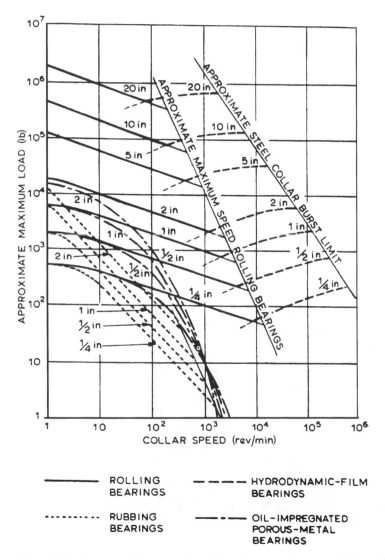

Figure 1.30. Load-speed characteristics of thrust bearings. (From the Tribology Subseries Item No. 65,007, *General Guide to the Choice of Thrust Bearing Type*, by permission of Engineering Sciences Data Unit Ltd., London.)

Table 1.6. *Advantages and limitations of journal bearings*

Condition	General comments	Journal bearing type						
		Rubbing bearings	Oil-impregnated porous metal bearings	Rolling-element bearings	Hydrodynamic fluid film bearings	Hydrostatic fluid film bearings	Self-acting gas bearings	Externally pressurized gas bearings
High temperature	Attention to differential expansions and their effect on fits and clearances is necessary	Normally satisfactory depending on material	Attention to oxidation resistance of lubricant is necessary	Up to 100°C no limitations; from 100 to 250°C stabilized bearings and special lubrication procedures are probably required	Attention to oxidation resistance of lubricant is necessary		Excellent	Excellent
Low temperature	Attention to differential expansions and starting torques is necessary		Lubricant may impose limitations; consideration of starting torque is necessary	Below −30°C special lubricants are required; consideration of starting torque is necessary	Lubricant may impose limitations; consideration of starting torque is necessary	Lubricant may impose limitations	Excellent; thorough drying of gas is necessary	
External vibration	Attention to the possibility of fretting damage is necessary (except for hydrostatic bearings)	Normally satisfactory when peak of impact load exceeds load-carrying capacity	Normally satisfactory except when peak of impact load exceeds load-carrying capacity	May impose limitation; consult manufacturer	Satisfactory	Excellent	Normally satisfactory	Excellent
Space requirements		Small radial extent	Small radial extent	Bearings of many different proportions; small axial extent	Small radial extent but total space requirement depends on the lubrication feed system		Small radial extent	Small radial extent, but total space requirement depends on the gas feed system

Dirt or dust	Normally satisfactory; sealing is advantageous	Sealing is important		Satisfactory; filtration of lubricant is important	Sealing important	Satisfactory
Vacuum	Excellent	Lubricant may impose limitations			Not normally applicable	Not applicable when vacuum has to be maintained
Wetness and humidity	Normally satisfactory depending on material	Normally satisfactory; sealing advantageous	Normally satisfactory, but specia. attention to sealing may be necessary	Satisfactory		Satisfactory
Radiation	Satisfactory	Lubr cant may impose limitations				Excellent
Low starting torque	Not normally recommended	Satisfactory	Good	Satisfactory	Excellent	Satisfactory / Excellent
Low running torque						
Accuracy of radial location	Poor	Good		Excellent	Good	Excellent
Life	Finite but predictable			Theoretically infinite but affected by infinite filtration and number of stops and starts	Theoretically infinite	Theoretically infinite but affected by number of stops and starts / Theoretically infinite

Table 1.6. (cont.)

Condition	General comments	Journal bearing type						
		Rubbing bearings	Oil-impregnated porous metal bearings	Rolling-element bearings	Hydrodynamic fluid film bearings	Hydrostatic fluid film bearings	Self-acting gas bearings	Externally pressurized gas bearings
Combination of axial and load-carrying capacity		A thrust face must be provided to carry the axial loads		Most types capable of dual duty	A thrust face must be provided to carry the axial loads			
Silent running		Good for steady loading	Excellent	Usually satisfactory; consult manufacturer	Excellent	Excellent except for possible pump noise	Excellent	Excellent except for possible compressor noise
Simplicity of lubrication		Excellent		Excellent with self-contained grease or oil lubrication	Self-contained assemblies can be used with certain limits of load, speed, and diameter; beyond this, oil circulation is necessary	Auxiliary high pressure is necessary	Excellent	Pressurized supply of dry clean gas is necessary
Availability of standard parts		Good to excellent depending on type	Excellent		Good		Not available	

Prevention of contamination product and surroundings	Improved performance can be obtained by allowing a process liquid to lubricate and cool the bearing, but wear debris may impose limitations	Normally satisfactory, but attention to sealing is necessary, except where a process liquid can be used as a lubricant				Excellent
Frequent stop-starts	Excellent	Good	Excellent	Good	Excellent	Poor
Frequent change of rotating direction		Generally good		Generally good		
Running costs	Very low		Depends on Complexity of lubrication system		Cost of lubricant supply has to be considered	Nil
						Cost of gas supply has to be considered

(From ESDU, 1965, by permission of Engineering Sciences Data Unit Ltd., London.)

Table 1.7. *Advantages and limitations of thrust bearings*

Condition	General comments	Thrust bearing type						
		Rubbing bearings	Oil-impregnated porous metal bearings	Rolling-element bearings	Hydrodynamic fluid film bearings	Hydrostatic fluid film bearings	Self-acting gas bearings	Externally pressurized gas bearings
High temperature	Attention to differential expansions and their effect upon axial clearance is necessary	Normally satisfactory depending on material	Attention to oxidation resistance of lubricant is necessary	Up to 100°C no limitations; from 100 to 250°C stabilized bearings and special lubrication procedures are probably required	Attention to oxidation resistance of lubrication is necessary	Attention to oxidation resistance of lubrication is necessary	Excellent	Excellent
Low temperature	Attention to differential expansions and starting torques is necessary		Lubricant may impose limitations; consideration of starting torque is necessary	Below −30°C special lubricants are required; consideration of starting torque is necessary	Lubricant may impose limitations; consideration of starting torque is necessary	Lubricant may impose limitations	Excellent; thorough drying of gas is necessary	Excellent; thorough drying of gas is necessary
External vibration	Attention to the possibility of fretting damage is necessary (except for hydrostatic bearings)	Normally satisfactory except when peak of impact load exceeds load-carrying capacity	Normally satisfactory except when peak of impact load exceeds load-carrying capacity	May impose limitations; consult manufacturer	Satisfactory	Excellent	Normally satisfactory	Excellent
Space requirements		Small radial extent	Small radial extent	Bearings of many different proportions are available	Small radial extent but total space requirement depends on the lubrication feed system	Small radial extent but total space requirement depends on the lubrication feed system	Small radial extent	Small radial total space requirement depends on gas feed system
Dirt or dust		Normally satisfactory; sealing advantageous	Normally satisfactory; sealing advantageous	Sealing is important	Satisfactory; filtration of lubricant is important	Satisfactory; filtration of lubricant is important	Sealing important	Satisfactory

Vacuum	Excellent	Lubricant may impose limitations		Not normally applicable	Not applicable when vacuum has to be maintained
Wetness and humidity	Attention to possibility of metallic corrosion is necessary	Normally satisfactory depending on material	Normally satisfactory; sealing advantageous	Normally satisfactory, but special attention to sealing is perhaps necessary	Satisfactory
Radiation	Satisfactory	Satisfactory	Lubricant may impose limitations		Excellent
Low starting torque	Not normally recommended	Satisfactory	Good	Satisfactory	Satisfactory
Low running torque	Satisfactory	Satisfactory	Excellent	Excellent	Excellent
Accuracy of radial location	Good	Good	Excellent	Good	Excellent
Life	Finite but can be estimated	Theoretically infinite but affected by filtration and number of stops and starts	Theoretically infinite	Theoretically infinite but affected by number of stops and starts	Theoretically infinite
Combination of axial and load-carrying capacity	A journal bearing surface must be provided to carry the radial loads		Some types capable of dual duty	A journal bearing surface must be provided to carry the radial loads	

Table 1.7. (cont.)

Condition	General comment	Thrust bearing type						
		Rubbing bearings	Oil-impregnated porous metal bearings	Rolling-element bearings	Hydrodynamic fluid film bearings	Hydrostatic fluid film bearings	Self-acting gas bearings	Externally pressurized gas bearings
Silent running		Good for steady loading	Excellent	Usually satisfactory; consult manufacturer	Excellent	Excellent, except for possible pump noise	Excellent	Excellent, except for possible compressor noise
Simplicity of lubrication		Excellent		Excellent with self-contained grease lubrication; with large sizes or high speeds, oil lubrication might be necessary	Self-contained assemblies can be used with certain limits of load, speed, and diameter; beyond this, oil circulation is necessary	Auxiliary high pressure is necessary	Excellent	Pressurized supply of dry, clean gas is necessary
Availability of standard parts		Good to excellent depending on type	Excellent		Good		Poor	

Prevention of contamination of product and surroundings	Performance can be improved by allowing a process liquid to lubricate and cool the bearing, but wear debris may impose limitations		Normally satisfactory, but attention to sealing is necessary, except where a process liquid can be used as a lubricant			Excellent
Tolerance to manufacturing and assembly inaccuracies	Good	Satisfactory	Poor	Satisfactory	Poor	Satisfactory
Type of motion — Frequent start-stops	Excellent		Good	Excellent		Excellent
Type of motion — Unidirectional	Suitable					
Type of motion — Bidirectional	Suitable		Some types are suitable / Unsuitable	Suitable	Some types are suitable / Unsuitable	Suitable
Type of motion — Oscillatory						
Running costs	Very low		Depends on complexity of lubrication system	Cost of lubricant supply has to be considered	Nil	Cost of gas supply has to be considered

(From ESDU, 1967, by permission of Engineering Sciences Data Unit Ltd., London.)

1.8   Nomenclature

A	projected bearing area
A_r	area of contact
C	journal bearing radial clearance
D	diameter
E	elastic modulus
F	friction force
L	bearing length
N	shaft revolution
O_B, O_J	bearing, journal center
P	specific bearing load
Q	lubricant flow rate
R	radius
R_B, R_J	bearing, journal radius
S	Sommerfeld number
U	velocity
V	wear volume
W	external load
Ψ	plasticity number
e	journal eccentricity
k	shear strength
f	coefficient of friction (F/W)
p	pressure
p_0	yield pressure
p_R, p_S	supply, recess pressure
Δp	pressure drop
μ	viscosity
a	radius
h	asperity height, film thickness
ℓ	asperity spacing
$\bar{p}$	average pressure,
κ_N	probability
σ	standard deviation
σ_Y	limiting tensile stress
τ	tangential stress
τ_c	shear strength
τ_j	junction failure stress
τ_m	maximum shear stress

1.9   References

Amontons, G. 1699. De la Resistance Causee dans les Machines. *Roy. Soc.* (Paris), 206.

Archard, J. F. 1953. Contact and rubbing of flat surfaces. *J. Appl. Phys.*, **24**, 981–988.

Archard, J. F. and Hirst, W. 1956. The wear of metals under unlubricated conditions. *Proc. Roy. Soc.*, **A 236**, 397–410.

Archard, J. F. 1957. *Proc. Roy. Soc.*, **A 243**, 190.

Arnell, R. D., Davies, P. B., Halling, J. and Whomes, T. L. 1991. *Tribology Principles and Design Applications*. Springer Verlag, New York.

Bhushan, B., Israelachvili, J. N. and Landman, U. 1995. Nanotribology: friction, wear and lubrication at the atomic scale. *Nature*, **374**, 607–616.

Bhushan, B., Wyant, J. C. and Meiling, J. 1988. New three-dimensional non-contact digital optical profiler. *Wear*, **122**, 301–312.

Bisson, E. E. and Anderson, W. J. 1964. *Advanced Bearing Technology*. NASA Spec. Publ. SP-38.

Booser, E. R. 1984. *Handbook of Lubrication. Volume II*. CRC Press, Boca Raton, Florida.

Bowden, F. P. and Tabor, D. 1986. *The Friction and Lubrication of Solids*, Vol. 1 (paperback ed.). Oxford Univ. Press, Oxford.

Bowden, F. P. and Tabor, D. 1964. *The Friction and Lubrication of Solids*, Vol. 2. Oxford Univ. Press, Oxford.

Bowden, F. P. and Rowe, G. W. 1956. The adhesion of clean metals. *Proc. Roy. Soc.*, **A 233**, 429–442.

Briscoe, B. J. and Tabor, D. 1978. The friction and wear of polymers: The role of mechanical properties. Polymer J., **10**, 74–96.

Buckley, D. H. 1977. The metal-to-metal interface and its effect on adhesion and friction. *J. Coll. Interface Sci.*, **58**, 36–53.

Childs, T. C. C. 1992. Deformation and flow of metals in sliding friction. In *Fundamentals of Friction in Macroscopic and Microscopic Processes*. NATO ASI Series, Kluwer, Dordrecht.

Coulomb, C. A. 1785. Theorie des Machines Simples. *Mem. Math. Phys. Acad. Sci.*, **10**, 161.

Cramer, H. 1955. *The Elements of Probability Theory*. Wiley, New York.

da Vinci, L. 1519. *The Notebooks of Leonardo da Vinci*, E. Macurdy (ed.). Reynal & Hitchcock, New York, 1938.

Dake, L. S., Russel, J. A. and Debrodt, D. C. 1986. A review of DOE ECUT Tribology Surveys. *ASME Journal of Tribology*, **108**, 497–501.

Dowson, D. 1973. Tribology before Columbus. *Mech. Eng.*, **95**, 12–20.

Dowson, D. 1979. *History of Tribology*. Lomgman, London and New York.

Ettles, C. M. 1987. Polymer and elastomer friction in the thermal control regime. *ASLE Trans.*, 30, 149–159.

General guide to the choice of journal bearing type. 1965. *Tribology Subseries Item No. 65007*. Engineering Sciences Data Unit Ltd., 251-259, Regent Street, London.

General guide to the choice of thrust bearing type. 1967. *Tribology Subseries Item No. 65033*. Engineering Sciences Data Unit Ltd., 251-259, Regent Street, London.

Fuller, D. D. 1956. *Theory and Practice of Lubrication for Engineers*. Wiley, New York.

Godfrey, D. 1980. *Wear Control Handbook*. ASME New York.

Greenwood, J. A. and Williamson, J. B. 1966. Contact of nominally flat surfaces. *Proc. Roy. Soc.*, **A 295**, 300–319.

Greenwood, J. A. 1992. Contact of rough surfaces. In *Fundamentals of Friction: Macroscopic and Microscopic Processes*. (I. L. Singer and H. M. Pollock, eds.), pp. 37–56. Kluwer, Dordrecht.

Hamilton, G. M. 1983. Explicit equations for the stress beneath a sliding spherical contact. *Proc. I. Mech. E.*, **197**, 53–59.

Hutchings, I. M. 1992. *Tribology: Friction and Wear of Engineering Materials*. CRC Press, Boca Raton.

Hersey, M. D. 1966. *Theory and Research in Lubrication*. Wiley, New York.

Jahanmir, S. 1994. *Friction and Wear of Ceramics*. Marcel Dekker, New York.

Johnson, K. L. 1992. *Contact Mechanics* (paperback ed.). Cambridge University Press.

Kragelskii, I. V. and Marchenko, E. A. 1982. Wear of machine components. *ASME Journal of Tribology*, **104**, 1–7.

Lancaster, J. K. 1963. The formation of surface films at the transition between mild and severe metallic wear. *Proc. Roy. Soc.*, **A 273**, 466–483.

Landman, U., Luedtke, W. D. and Ringer, E. M. 1992. Molecular dynamics simulations of adhesive contact formation and friction. In *Fundamentals of Friction in Macroscopic and Microscopic Processes*. NATO ASI Series, Kluwer, Dordrecht.

Lim, S. C. and Ashby, M. F. 1987. Wear mechanism maps. *Acta Metall.*, **35**, 1–24.

Lubrication Engineering Working Group, Dept. of Ed. and Sci. 1966. *Lubrication (Tribology)*. H.M.S.O., London.

McNeill, W. 1963. *The Rise of the West, A History of the Human Community*. University of Chicago Press.

Majumdar, A. and Bhushan, B. 1990. Role of fractal geometry in roughness characterization and contact mechanics of surfaces. *ASME Journal of Tribology*, **112**, 205–216.

Merchant, M. E. 1940. The mechanism of static friction. *J. Appl. Phys.*, **11**, 230.

Neale, M. M. 1967. Selection of bearings and lubricants. *Proc. Inst. Mech. Eng.* **182**, pt. 3A, 547–556.

Neale, M. J. 1978. *Tribology Handbook*. Butterworth, London.

Mechanical Drives. 1978. *Mach. Design*, **50**, no. 15, 137–236.

Peterson, M. B. and Winer, W. O. 1980. *Wear Control Handbook*. ASME Press.

Peklenik, J. 1968. New developments in surface characterization and measurements by means of random process analysis. *Proc. Inst. Mech. Engrs.*, **182**, Pt. 3K, 108–126.

Rabinowicz, E. 1965. *Friction and Wear of Materials*. Wiley, New York.

Raimondi, A. A. and Boyd, J. 1958. A solution of the finite journal bearing and its application to analysis and design. *ASLE Trans.*, **1**, 159–209.

Reynolds, O. 1886. On the theory of lubrication and its application to Mr. Tower's experiments. *Philos. Trans. Roy. Soc.*, **177**, 157–234.

Ruan, J.-A. and Bhushan, B. 1994. Atomic scale friction measurements using friction force microscopy. *ASME Journal of Lubrication Technology*, **116**, 378–388.

Sayle, R. S. and Thomas, T. R. 1979. Measurements of the statistical microgeometry of engineering surfaces. *ASME Journal of Lubrication Technology*, **101**, 409–418.

Sherrington, I. and Smith, E. H. 1988. Modern measurement techniques in surface metrology, *Wear*. **125**, 271–288.

Suh, N. P. 1986. *Tribophysics*. Prentice-Hall, Englewood Cliffs.

Tabor, D. 1939. Junction growth in metallic friction, the role of combined stresses and surface contamination. *Proc. Roy. Soc.*, **A 251**, 378.

Tabor, D. 1981. Friction – The present state of our understanding. *ASME Journal of Lubrication Technology*, **103**, 169–179.

Tian, X. and Kennedy, F. E. 1993. Contact surface temperature models of finite bodies in dry and boundary lubricated sliding systems. *ASME Journal of Tribology*, **115**, 411–418.

Tian, X. and Kennedy, F. E. 1994. Maximum and average flash temperatures in sliding contacts. *ASME Journal of Tribology*, **116**, 187–174.

Voyutski, S. S. 1963. *Autoadhesion and Adhesion of Polymers*. Wiley, New York.

Wang, S. and Komvopoulos, K. 1994. A fractal theory of the interfacial temperature distribution in the slow sliding regime: Parts I & II. *ASME Journal of Tribology*, **116**, 812–832.

Welsh, N. C. 1965. The dry wear of steels. *Phil. Trans. Roy. Soc.*, **A 257**, 31–70.

CHAPTER 2

Basic Equations

2.1   Fluid Mechanics

The equations employed to describe the flow of lubricants in bearings result from simplifications of the governing equations of fluid mechanics. It is appropriate, therefore, to devote a chapter to summarizing pertinent results from fluid mechanics. This discussion will not be limited only to concepts necessary to understand the classical theory of lubrication. A more than elementary discussion of fluid behavior is called for here, as various nonlinear effects will be studied in later chapters.

Our discussion begins with the mathematical description of motion, followed by the definition of stress. Cauchy's equations of motion will be obtained by substituting the rate of change of linear momentum of a fluid particle and the forces acting on it, into Newton's second law of motion. This will yield three equations, one in each of the three coordinate directions. But these three equations will contain twelve unknowns: three velocity components, (u, v, w) and nine stress components $(T_{xx}, T_{xy}, \ldots, T_{zz})$. To render the problem well posed, i.e., to have the number of equations agree with the number of unknowns so that solutions might be obtained, we will need to find additional equations. A fourth equation is easy to come by, by way of the principle of conservation of mass. The situation further improves on recognizing that only six of the nine stress components are independent, due to symmetry of the stress matrix. However, on specifying incompressibility of the fluid (incompressible lubricants are the only type treated in this chapter) a tenth unknown, the fluid pressure, makes its debut. In summary, we end up with four equations in ten unknowns.

Up to this point this discussion applies equally well not only to all incompressible fluids but to all incompressible continuous media, irrespective of other material properties. However, if we wish to obtain the six additional equations that are still needed to close the problem, we may no longer retain this generality. Accounting for the material behavior of the particular fluid under consideration is, in fact, the ruse that will yield the required number of additional equations. Material behavior of a class of fluids is postulated in a *constitutive theory*, the mathematical statement of which is the set of *constitutive equations* for the class.[1]

The simplest constitutive theory postulates that the normal stress components are equal, while the shear stress components vanish identically; this defines what is called an *inviscid* or *ideal fluid*. When the corresponding constitutive equations are substituted into Cauchy's equations of motion, *Euler's equations* result. These equations have relative simplicity. However, inviscid fluids do not abound in engineering practice, and Euler's equations do not hold in the vicinity of solid surfaces, where viscous effects are all important. In engineering practice Euler's equations are useful mostly in external-flow problems.

A somewhat more complicated constitutive theory states that the stress is linearly dependent on the rate of deformation (a more complete definition will be given below). This is the constitutive theory of a *Newtonian fluid*. When the corresponding constitutive

---

[1] Incompressibility of the material is a constitutive assumption; we have already made use of it.

53

equations are substituted into Cauchy's equations of motion, the *Navier-Stokes equations* result. Newtonian theory is extremely useful as its predictions agree well with experimental data on a large class of common fluids, such as water or air.

Not all fluids are *linearly viscous* (a term used to describe Newtonian behavior); in particular, polymers and other man made fluids are not. Some naturally occurring fluids, such as crude oil, also belong to the class of *non-Newtonian fluids*. In some of these the current state of stress at a point is a nonlinear function of the current state of deformation. In others, the current state of stress at a point is determined by the *history* of the deformation within a neighborhood of that point. There are materials which are both solid-like and fluid-like in their response, in the sense that the work of deformation is neither completely stored, as in solids, nor is it completely dissipated, as in fluids. These substances, displaying both elasticity and viscosity, are called *viscoelastic* materials.

Though we are aware of the wide variety of fluid behaviors and of the fact that many non-Newtonian fluids are actual or potential lubricants, the present chapter is concerned only with Newtonian behavior. Non-Newtonian lubricants will be discussed in later chapters.

Kinematics

Picture a body of fluid as it is moving through physical space. Think of this fluid as consisting entirely of infinitesimal *particles* or *material points* (the terms particle and material point will be used interchangeably). Each such material point occupies one, and only one, *spatial point* at any given time. To specify the location of spatial points, we employ an orthogonal Cartesian coordinate system (x_1, x_2, x_3).

For the purpose of future identification, we assign permanent names to each of the infinity of fluid particles. This is best achieved by momentarily freezing the flow, initializing our clock, and assigning, as permanent particle names, the coordinates of the spatial point that the material point happens to occupy in this frozen configuration (Figure 2.1). We call this frozen configuration the *reference configuration* of the fluid. The coordinates of a particle (material point) in the reference configuration are called its *reference coordinates*, or *Lagrangian coordinates* or *material coordinates*, and are designated by the capital letters (X, Y, Z), or by (X_1, X_2, X_3) when employing index notation. They are assigned to, and

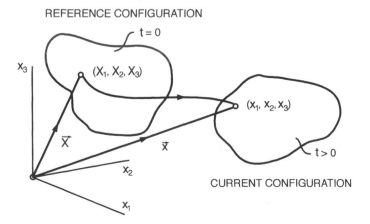

Figure 2.1. Motion of a fluid body.

will be retained by, fluid particles as their permanent name. Particle position in the reference configuration is thus specified by position vector X that has components X, Y, and Z; this is symbolized by writing $X = (X, Y, Z)$ or $X = (X_1, X_2, X_3)$.

Upon unfreezing the fluid body, motion resumes. The particle that occupied the spatial position $X = (X_1, X_2, X_3)$ at $t = 0$, now, at time $t > 0$, occupies another position $x = (x_1, x_2, x_3)$ (Figure 2.1). The continuous sequence of configurations that results from increasing the parameter t is called a motion.

Motion is, therefore, defined by the mapping

$$x_i = \chi_i(X, t), \qquad t \geq 0, \quad i = 1, 2, 3, \tag{2.1}$$

where it is understood that Eq. (2.1) is the index-notation representation of the vector equation, $x = \chi(X, t)$, $t \geq 0$.

> When it is not essential to discriminate among the various components of a vector, say, the position vector $x = (x_1, x_2, x_3)$, i.e., when all coordinate directions have equal importance in a mathematical statement, we shall write x_i to represent the components of the vector. Should it become necessary to state the number of the components explicitly, we shall write x_i, $i = 1, 2, 3$. Here i, the index, *ranges* over 1, 2, 3 and the symbolism is called *index notation*. We shall continue using x_1, or x_2, or x_3, however, when wishing to draw attention to a particular component.

Equations (2.1) signify that the particle that occupied point X in the reference configuration is located in point x at time t. The coordinates x_i are called *Eulerian* or *spatial coordinates*. If X is fixed while t varies, Eqs. (2.1) describe the *path* of the particle X. If, on the other hand, t is kept constant, Eqs. (2.1) represent the mapping of the reference configuration onto the configuration at time t.

We shall insist on Eqs. (2.1) being invertible, so as to assure that no two particles occupy the same spatial point simultaneously, or that no particle can be found at two different locations at the same time. In other words, we study only motions that can be characterized by *deformation functions* χ_i that possess single-valued inverse (this by no means constrains us in our investigations). We shall assume, further, that a sufficient number of derivatives of the χ_i and its inverse the χ_i^{-1} exist and are continuous. For such motions we may invert Eqs. (2.1) and write

$$X_i = \chi_i^{-1}(x, t). \tag{2.2}$$

Equation (2.2), or its vector representation $X = \chi^{-1}(x, t)$, informs us that the spatial point x at time t is occupied by the particle that was located in position X in the reference configuration.

There are two viewpoints in fluid mechanics: (1) the *Lagrangian view* fixes attention on a given material point, i.e., holds X constant and describes the changes this material point experiences while moving through space (varying x), and (2) the *Eulerian view* fixes attention on a spatial point, i.e., holds x constant and investigates how conditions change at that particular spatial point as various particles (varying X) stream through it. Accordingly, variables associated with material points are given the qualifier *material* or *Lagrangian*, while variables associated with spatial points are named *spatial* or *Eulerian*. The two types of variables are, of course, related to one another in unique fashion through the motion, Eqs. (2.1), of the fluid: e.g., the velocity at a fixed point x in space will vary with time, provided that the velocity of the string of particles $\ldots, X_k, X_{k+1}, \ldots$ streaming through spatial location x varies from particle to particle.

Velocity is defined as the time rate of change of particle position. In Lagrangian representation we write

$$V_i(X, t) = \frac{\partial x_i(X, t)}{\partial t}. \tag{2.3}$$

Here the V_i, $i = 1, 2, 3$, are the components of the velocity of a given particle (note that particle identity X is held constant during differentiation) as it travels along its path. By means of the motion of the fluid, Eq. (2.2), we can transform the Lagrangian variables V_i into the Eulerian variables v_i, where the v_i represent the velocity at the spatial point x:

$$V(X, t) = V[\chi^{-1}(x, t), t] = v(x, t). \tag{2.4}$$

Acceleration is the time rate of change of velocity. It is the vectorial sum of its two components: (1) the *local acceleration* and (2) the *convective acceleration*. Local acceleration can be observed at a fixed point x of space, when the velocity V of the different fluid particles that pass through that point varies from particle to particle. It is, therefore, the time derivative of the velocity $v(x, t)$ with x held constant. Convective acceleration, on the other hand, is the rate of change of particle velocity as experienced by a particle moving along its own path through the prevailing velocity field. It is, therefore, the product of the velocity of the particle and the spatial rate of change of velocity.

The fact that there are two types of accelerations is acknowledged mathematically by the statement that the velocity is both an explicit and implicit function of time, so that

$$a_i = \frac{dv_i}{dt} = \left(\frac{\partial v_i}{\partial t} + \frac{\partial v_i}{\partial x_1} \frac{dx_1}{dt} + \frac{\partial v_i}{\partial x_2} \frac{dx_2}{dt} + \frac{\partial v_i}{\partial x_3} \frac{dx_3}{dt} \right). \tag{2.5}$$

Here $a = (a_1, a_2, a_3)$ is the acceleration vector. The first term on the right-hand side of Eq. (2.5) is the local acceleration, and the sum of the last three terms represents convective acceleration.

On recognizing that

$$\frac{dx_1}{dt} = v_1, \qquad \frac{dx_2}{dt} = v_2, \qquad \frac{dx_3}{dt} = v_3, \tag{2.6}$$

Eq. (2.5) can be written as

$$\frac{dv_i}{dt} = \left(\frac{\partial v_i}{\partial t} + v_1 \frac{\partial v_i}{\partial x_1} + v_2 \frac{\partial v_i}{\partial x_2} + v_3 \frac{\partial v_i}{\partial x_3} \right)$$

$$= \frac{\partial v_i}{\partial t} + \sum_{j=1}^{3} v_j \frac{\partial v_i}{\partial x_j}.$$

To abbreviate the notation, we adopt the summation convention: whenever an index appears exactly twice within the same term, summation is implied with respect to that index over its range. The abbreviated form of Eq. (2.5) is then

$$\frac{dv_i}{dt} = \frac{\partial v_i}{\partial t} + v_j \frac{\partial v_i}{\partial x_j}. \tag{2.7}$$

The operator (d/dt) is called the material or total derivative. In vectorial form

$$a = \frac{dv}{dt} = \frac{\partial v}{\partial t} + (v \cdot \nabla)v. \tag{2.8}$$

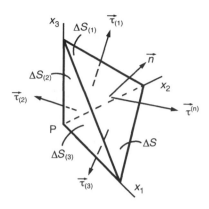

Figure 2.2. Stress equilibrium.

Stress

Consider the fluid body in its present configuration and consider an interior part, b, of this body that has volume, ΔV, enclosed by surface, ΔS. The various forces that may act on the interior part b are of two types: (1) *body forces*, which are long-range forces that act through the mass centers of the fluid, and (2) *surface forces*, which are short-range forces that act across surfaces or interfaces. Body forces and surface forces are defined by means of integrals, as follows.

First we assume the existence of a vector field, the *body force per unit mass* $f(x, t)$. The total body force on b is then

$$F_B = \int_{\Delta V} f\rho \, dv. \tag{2.9}$$

Next we assume the existence of another vector field, the surface force per unit area, called *surface traction* (stress vector), τ. The total surface force acting on b is given by

$$F_S = \oint_{\Delta S} \tau \, ds. \tag{2.10}$$

The surface traction, τ, is, in general, not perpendicular to the elemental surface, ΔS, on which it acts (Figure 2.2). Furthermore, both its direction and its magnitude depend on the orientation of ΔS. To show this, we write Newton's second law for a fluid particle. The fluid particle chosen for this purpose is in the form of a small tetrahedron,[2] cut out by three coordinate planes through a point P, and by the inclined surface element ΔS; the latter has orientation n, as indicated in Figure 2.2. Let $\tau^{(n)}$ be the *surface traction* acting on ΔS and let $\tau_{(i)}$ represent the surface traction acting on $\Delta S_{(i)}$, the projection of ΔS onto the $x_{(i)} = $ const. coordinate plane.

Newton's second law, written for the fluid particle in Figure 2.2, takes the following form:

$$\rho \frac{dv}{dt} \Delta V = \tau^{(n)} \Delta S - \tau_{(i)} \Delta S_{(i)} - \rho f \Delta V, \tag{2.11}$$

---

[2] Though simple geometry often makes for simple analysis, the shape of the volume is not central to the argument.

The components of the unit vector $\boldsymbol{n} = (n_i)$ are the direction cosines of its orientation relative to the coordinate axes. Hence, we have $\Delta S_{(i)} = n_i \Delta S$, which allows us to rewrite Eq. (2.11) as

$$\rho(\boldsymbol{a} - \boldsymbol{f})\frac{\Delta V}{\Delta S} = \left[\boldsymbol{\tau}^{(n)} - n_i \boldsymbol{\tau}_{(i)}\right]. \tag{2.12}$$

On shrinking the tetrahedron to the point P we observe that

$$\lim_{\Delta S \to 0} \frac{\Delta V}{\Delta S} = 0,$$

thus the left hand side of Eq. (2.12) vanishes in the limit and yields

$$\boldsymbol{\tau}^{(n)} = n_1 \boldsymbol{\tau}_{(1)} + n_2 \boldsymbol{\tau}_{(2)} + n_3 \boldsymbol{\tau}_{(3)}$$

or, in index notation,

$$\boldsymbol{\tau}^{(n)} = n_i \boldsymbol{\tau}_{(i)}. \tag{2.13}$$

Equation (2.13) shows that the surface traction at a point acting on an arbitrary surface is given by a linear combination of the surface tractions of the coordinate planes through that point. The coefficients in this linear combination are the direction cosines of the surface.

The components of the surface tractions, $\boldsymbol{\tau}_{(i)}$, acting on the coordinate planes will be designated by the capital letter T with appropriate indices. On designating these components we adopt the scheme, illustrated in Figure 2.3, by which the first index specifies the orientation of the plane on which the surface traction acts while the second index specifies the direction of the component itself.

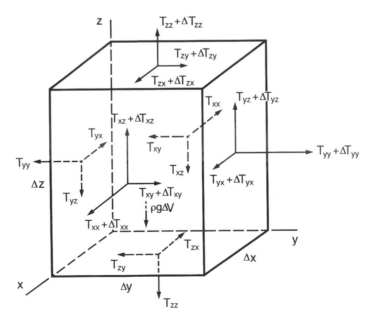

Figure 2.3. Convention for designating stress components.

Thus T_{12} is a stress component that acts on the $x_1 = $ const. coordinate plane and is parallel to the x_2 coordinate axis. Using this scheme we may write

$$\boldsymbol{\tau}_{(x)} = T_{xx}\boldsymbol{e}_x + T_{xy}\boldsymbol{e}_y + T_{xz}\boldsymbol{e}_z,$$
$$\boldsymbol{\tau}_{(y)} = T_{yx}\boldsymbol{e}_x + T_{yy}\boldsymbol{e}_y + T_{yz}\boldsymbol{e}_z, \tag{2.14a}$$
$$\boldsymbol{\tau}_{(z)} = T_{zx}\boldsymbol{e}_x + T_{zy}\boldsymbol{e}_y + T_{zz}\boldsymbol{e}_z,$$

where $(\boldsymbol{e}_x, \boldsymbol{e}_y, \boldsymbol{e}_z)$ are the unit base vectors of the Cartesian coordinate system.

On employing index notation, we have

$$\begin{bmatrix} \boldsymbol{\tau}_{(1)} \\ \boldsymbol{\tau}_{(2)} \\ \boldsymbol{\tau}_{(3)} \end{bmatrix} = \begin{bmatrix} T_{11} & T_{12} & T_{13} \\ T_{21} & T_{22} & T_{23} \\ T_{31} & T_{32} & T_{33} \end{bmatrix} \begin{bmatrix} \boldsymbol{e}_1 \\ \boldsymbol{e}_2 \\ \boldsymbol{e}_3 \end{bmatrix} \tag{2.14b}$$

or

$$\boldsymbol{\tau}_{(i)} = T_{ij}\boldsymbol{e}_j.$$

Substituting for $\boldsymbol{\tau}_{(i)}$ from Eq. (2.14) into Eq. (2.13) and writing $\tau_i^{(n)}$ for the components of $\boldsymbol{\tau}^{(n)}$, we obtain

$$\left[\tau_1^{(n)}, \tau_2^{(n)}, \tau_3^{(n)}\right] = (n_1, n_2, n_3) \begin{bmatrix} T_{11} & T_{12} & T_{13} \\ T_{21} & T_{22} & T_{23} \\ T_{31} & T_{32} & T_{33} \end{bmatrix}, \tag{2.15a}$$

or, on using the summation convention,

$$\tau_i^{(n)} = n_j T_{ji}. \tag{2.15b}$$

We may now drop the superscript n signifying the orientation of the plane and write $\boldsymbol{\tau} = (\tau_1, \tau_2, \tau_3)$ for the surface traction $\boldsymbol{\tau}^{(n)}$. Equation (2.15b) then takes its generic form

$$\tau_i = n_j T_{ji}. \tag{2.15c}$$

The matrix $\boldsymbol{T} = (T_{ij})$ is called the *stress matrix* (or *stress tensor*). To find the components of surface traction acting on an arbitrary surface through a point, we must know the nine components of the stress matrix at that point, as well as the orientation of the surface.

The actual situation is somewhat less demanding than this. We can show that for conventional materials the stress matrix is *symmetric* (exceptions to symmetry of the stress matrix will be noted in later chapters when they actually occur).

To show symmetry of stress, consider the stresses that contribute to a torque, M_z, about the z axis (Figure 2.3):

$$M_z = (T_{xy} - T_{yx})\,\Delta x\,\Delta y\,\Delta z.$$

The *principle of conservation of angular momentum*[3] (Newton's second law, applied to a rotating system) requires that the *torque is balanced by the product of the moment of inertia and the angular acceleration.* In the present context this means that

$$(T_{xy} - T_{yx})\,\Delta x\,\Delta y\,\Delta z = \frac{1}{12}\rho\,\Delta x\,\Delta y\,\Delta z(\Delta x^2 + \Delta y^2)\frac{d\omega_z}{dt}$$

---

[3] In continuum mechanics, symmetry of the stress tensor is assumed (Boltzman postulate), then conservation of angular momentum follows (Serrin, 1959).

or, after division through by the volume element $\Delta V = \Delta x \, \Delta y \, \Delta z$,

$$(T_{xy} - T_{yx}) = \frac{1}{12}\rho(\Delta x^2 + \Delta y^2)\frac{d\omega_z}{dt}.$$

On shrinking the parallelepiped to the point P, the right-hand side vanishes; we have just demonstrated that $T_{xy} = T_{yx}$.

But if $T_{ij} = T_{ji}$, it is sufficient to know six independent components of T. However, then only six independent stress components will appear in Cauchy's equations and it will be necessary for us to come up with only six constitutive equations.

Constitutive Equations

Most problems in engineering fluid mechanics fall into one of two categories: (1) a moving boundary or a pressure gradient induces flow, or (2) a moving fluid exerts force on its boundary. In the first case, the fluid yields to stress by flowing. In the second case, stress is the consequence of flow. We know from experience, however, that not all fluids are alike: some, such as water or air, flow readily, while others, such as honey and pitch, show reluctance to flow. We are thus led to the necessity of knowing, within the context of responding to stress, the behavior of particular fluids.

Perhaps one of the simplest experiments we can perform in studying the force-motion behavior of fluids involves two parallel plates, both submerged in the fluid. The plates are separated by a short distance, Δy, as indicated in Figure 2.4. The idea here is to mark with dye the fluid particles along a generic line at right angles to the plates and then to observe the motion of this line of color (a *material line* as X is held constant during the experiment) as the upper plate is moved with small velocity, Δu, relative to the lower plate.

A short time, Δt, into the experiment the dye line will still be straight but will be inclined to its original direction at the small angle $\Delta \gamma$. For small values of Δu and Δt, the *angular deformation*, $\Delta \gamma$, is given by

$$\Delta \gamma \approx \tan \Delta \gamma = \frac{\Delta u \, \Delta t}{\Delta y}. \tag{2.16}$$

It was first recorded by Newton that in common fluids the surface traction, τ, defined by $\tau = \Delta F / \Delta A$, where ΔF is the force required to move the top plate of area ΔA, is proportional to the time rate of change of the angular deformation:

$$\tau \propto \frac{\Delta \gamma}{\Delta t} = \frac{\Delta u}{\Delta y}.$$

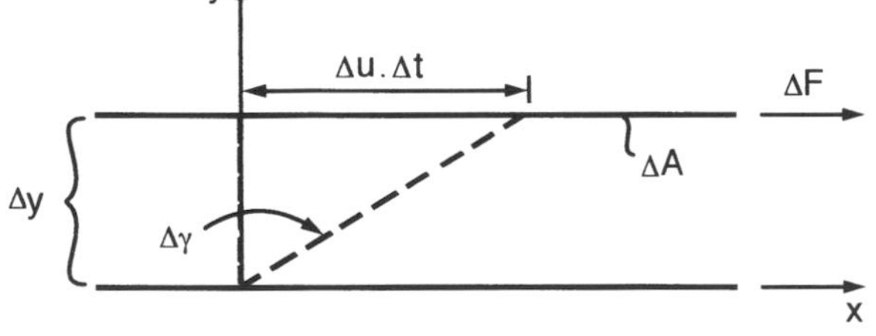

Figure 2.4. Definition of viscosity.

At the limit $\Delta y \to 0$, as the plates are moved closer to one another, we can write the differential form of the above relationship as

$$\tau = \mu \frac{du}{dy}. \tag{2.17}$$

The factor of proportionality, μ, is called the *molecular viscosity*, or simply the *viscosity* of the fluid, and Eq. (2.17) is referred to as *Newton's law of viscosity*. When μ is constant or is dependent on temperature at most, as for water, air, and common petroleum-oil lubricants at low pressures, the fluid is said to be a *Newtonian fluid*. *Non-Newtonian fluids*, and there are a host of them, are said to be *shear thinning* if μ decreases when the fluid is sheared at increasing rate and *shear thickening* in the opposite case, i.e., if the viscosity increases on increasing the rate of shear.

Equation (2.17) is restrictive in that it applies only to a single dimension. We will generalize it to three dimensional flows by replacing:

(1) y, by the triplet (x, y, z);
(2) du/dy, by the nine spatial derivatives of velocity $\partial v_i/\partial x_j$, $i, j = 1, 2, 3$; and
(3) τ, by the nine components of stress T_{ij}, $i, j = 1, 2, 3$.

The nine spatial derivatives of velocity can conveniently be organized into a matrix, the so-called *velocity gradient matrix L*, as follows

$$L - \begin{bmatrix} \dfrac{\partial u}{\partial x} & \dfrac{\partial v}{\partial x} & \dfrac{\partial w}{\partial x} \\[2mm] \dfrac{\partial u}{\partial y} & \dfrac{\partial v}{\partial y} & \dfrac{\partial w}{\partial y} \\[2mm] \dfrac{\partial u}{\partial z} & \dfrac{\partial v}{\partial z} & \dfrac{\partial w}{\partial z} \end{bmatrix}, \qquad L_{ij} = \frac{\partial u_j}{\partial x_i}. \tag{2.18}$$

Based on the above, our extension of Newton's law of viscosity to three dimensions might take the form

$$T = f(L), \tag{2.19}$$

where $f(L)$ is a yet unspecified matrix-valued function of the velocity gradient matrix, except that that this equation implies more than we should want it to. The velocity gradient matrix, L, carries too much information to our liking about the flow: (1) information specifying *rigid body rotation* and (2) information describing rate of deformation, which consists of *stretching* (rate of change of linear dimension) and *shearing* (rate of change of angular deformation) of the particle. Thus Eq. (2.19), as it is now written, states that stress depends on both rotation and deformation. The first part of this statement is unacceptable on physical grounds, as will be shown presently; Eq. (2.19) will thus have to be modified so that the right hand side no longer contains information concerning rigid body rotation.

To illustrate the principle that stress cannot depend on rigid body motion, consider two observers, O and O^*, one stationary and the other rotating, while both observing the same fluid particle at the same instant of time. If stress depended on rigid body rotation then O^*, whose *frame of reference* is fixed to the rotating particle and who, therefore, finds the particle to be at rest, would conclude that the particle was also free of stress. O, whose frame of reference is held stationary, would, on the other hand, see the particle rotate and record a nonzero state of stress. As stress cannot be both zero and nonzero at the same particle and at the same time, the only way out of this dilemma is to postulate that *stress cannot depend on rotation*. Rotation is not *frame indifferent*, as its value depends on the motion of the observer. Deformation, on the other hand, is independent of the motion of the observer, as will be shown below. Physical intuition tells us that stress can depend only on quantities that are frame indifferent, or *frame invariant*, so that stress itself is independent of the motion of the observer. Frame indifferent quantities are also called *objective* quantities.

In an effort to isolate that part of the velocity gradient matrix that describes rigid body rotation, we write it as the sum of its *symmetric* and *skew-symmetric* parts (a process known as *Cartesian decomposition*):

$$L = \frac{1}{2}(L + L^T) + \frac{1}{2}(L - L^T) \tag{2.20a}$$
$$= D + \Omega.$$

Here

$$D = \frac{1}{2}(L + L^T) = D^T \tag{2.20b}$$

and

$$\Omega = \frac{1}{2}(L - L^T) = -\Omega^T. \tag{2.20c}$$

We will show presently that the symmetric matrix, D, called the *stretching matrix (tensor)*, carries information about deformation, i.e., changes in volume and shape, while the skew-symmetric matrix, Ω, called the *spin matrix (tensor)*, describes rigid body rotation.

A quantity is said to be frame indifferent (objective) if it is invariant under all changes of frame. A frame $[x, t]$ consists of an orthogonal Cartesian coordinate system and a clock. Consider two frames $[x, t]$ and $[x^*, t^*]$, related to one another through

$$x^* = c(t) + Q(t) \cdot x \tag{2.21a}$$
$$t^* = t - a$$

This is the most general change of frame that preserves distance, time interval, and the sense of time. Here $Q(t)$ is a time dependent orthogonal matrix, $Q \cdot Q^T = 1$, which specifies the orientation of (O, x) relative to (O^*, x^*), $c(t)$ is the distance between the two frames, and a is a constant. Under change of frame, Eq. (2.21a), objective vectors, ν, and second rank tensors, τ, transform, respectively, as[4]

$$\nu^* = Q \cdot \nu, \tag{2.21b}$$
$$\tau^* = Q \cdot \tau \cdot Q^T.$$

[4] To conform with the usual notation when writing transformation laws, in this section only, all vectors are column vectors.

To demonstrate Eq. (2.21b) for vectors, let $\nu = x_1 - x_2$ represent the distance between points x_1 and x_2. In the starred frame the corresponding formula is $\nu^* = x_1^* - x_2^*$. Substituting from Eq. (2.21a) into this last equation, we are lead to

$$\nu^* = Q \cdot (x_1 - x_2) = Q \cdot \nu.$$

To demonstrate Eq. (2.21b) for the tensor τ, construct a vector, w, according to the recipe $w = \tau \cdot \nu$. In the starred frame we have $w^* = \tau^* \cdot \nu^*$, so that

$$\begin{aligned} w^* &= Q \cdot w = Q \cdot \tau \cdot \nu, \\ w^* &= \tau^* \cdot \nu = \tau^* \cdot Q \cdot \nu, \end{aligned} \qquad (2.21c)$$

where we used the first of Eq. (2.21b). Equating the right hand sides and observing that $Q^{-1} = Q^T$ leads to the second of Eq. (2.21b).

Using Eq. (2.21a), it can be shown that $D^* = Q \cdot D \cdot Q^T$ for the stretching tensor and $\Omega^* = Q \cdot \Omega \cdot Q^T + \dot{Q} \cdot Q$ for the spin tensor. Comparison with Eq. (2.21b) will then show that D is objective but Ω is not (e.g., by selecting $Q = 1$ and $\dot{Q} = -\Omega$ we obtain $\Omega^* = 0$ even though $\Omega \neq 0$). Here $1 = (\delta_{ij})$ is the *identity matrix* (the *Kronecker delta* $\delta_{ij} = 1$ if $i = j$; otherwise its value is zero).

To investigate the physical meaning of the spin matrix Ω in detail, we shall consider rotation of a fluid particle, referring to Figure 2.5.

Figure 2.5 depicts a fluid particle located on the positively directed (x, y) face of a rectangular parallelepiped (Figure 2.3) and observed from the $+z$ (outward normal) direction, at two time instances; at time t the particle is rectangular in shape, but deforms by time $t + \Delta t$. During the time interval Δt, the particle undergoes average rotation, $(\alpha - \beta)/2$, in the positive (counterclockwise) sense. The time rate of change (angular velocity) of this rotation is

$$\omega_z = \frac{1}{2}\left(\frac{d\alpha}{dt} - \frac{d\beta}{dt}\right), \qquad (2.22)$$

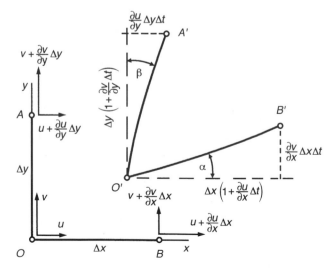

Figure 2.5. Motion of a fluid particle.

where

$$\frac{d\alpha}{dt} = \lim_{\Delta t \to 0} \frac{1}{\Delta t} \left(\tan^{-1} \frac{\frac{\partial v}{\partial x} \Delta x \, \Delta t}{\Delta x + \frac{\partial u}{\partial x} \Delta x \, \Delta t} \right) = \frac{\partial v}{\partial x}, \tag{2.23a}$$

$$\frac{d\beta}{dt} = \lim_{\Delta t \to 0} \frac{1}{\Delta t} \left(\tan^{-1} \frac{\frac{\partial u}{\partial y} \Delta y \, \Delta t}{\Delta y + \frac{\partial v}{\partial y} \Delta y \, \Delta t} \right) = \frac{\partial u}{\partial y}. \tag{2.23b}$$

Substitution into Eq. (2.22) yields

$$\omega_z = \frac{1}{2} \left(\frac{\partial v}{\partial x} - \frac{\partial u}{\partial y} \right). \tag{2.24a}$$

On the negatively directed (x, y) face of the parallelepiped the average rate of rotation appears to be $-\omega_z$ as viewed from the $-z$ (outward normal) direction. (That the same rigid body rotation can be evaluated as either ω_z or $-\omega_z$, depending on the position of the observer, confirms our earlier assertion that rotation is not objective.)

Similarly, motion in the positively oriented coordinate faces (y, z) and (x, z), as observed from the $+x$ and $+y$ coordinate directions, respectively, gives

$$\omega_x = \frac{1}{2} \left(\frac{\partial w}{\partial y} - \frac{\partial v}{\partial z} \right), \tag{2.24b}$$

$$\omega_y = \frac{1}{2} \left(\frac{\partial u}{\partial z} - \frac{\partial w}{\partial x} \right). \tag{2.24c}$$

The vector $\boldsymbol{\omega} = 2(\omega_x, \omega_y, \omega_z)$ is called the *vorticity vector*. It can be shown by simple substitution that $\boldsymbol{\omega} = \text{curl } \boldsymbol{v}$.

The spin matrix $\boldsymbol{\Omega}$ is constructed from the rotations $\omega_x, \omega_y, \omega_z$ as follows:

$$\Omega = \begin{bmatrix} 0 & \omega_z & -\omega_y \\ -\omega_z & 0 & \omega_x \\ \omega_y & -\omega_x & 0 \end{bmatrix} = \frac{1}{2} \begin{bmatrix} 0 & \left(\frac{\partial v}{\partial x} - \frac{\partial u}{\partial y} \right) & \left(\frac{\partial w}{\partial x} - \frac{\partial u}{\partial z} \right) \\ \left(\frac{\partial u}{\partial y} - \frac{\partial v}{\partial x} \right) & 0 & \left(\frac{\partial w}{\partial y} - \frac{\partial v}{\partial z} \right) \\ \left(\frac{\partial u}{\partial z} - \frac{\partial w}{\partial x} \right) & \left(\frac{\partial v}{\partial z} - \frac{\partial w}{\partial y} \right) & 0 \end{bmatrix}. \tag{2.25}$$

In compact notation we write

$$\Omega_{ij} = \frac{1}{2} \left(\frac{\partial v_j}{\partial x_i} - \frac{\partial v_i}{\partial x_j} \right). \tag{2.26}$$

To illustrate that the elements of the stretching matrix $\boldsymbol{D}$ contain information on the rate of deformation, consider again the fluid particle depicted in Figure 2.5. At time t the two material lines $\overline{OA}$ and $\overline{OB}$, which form the sides of the particle, are perpendicular to one another. At the later time $t + \Delta t$ they are shown inclined at angle $(\pi/2 - \gamma_z)$, where $\gamma_z = (\alpha + \beta)$ is the angular deformation. The time rate of γ_z is

$$\frac{d\gamma_z}{dt} = \left(\frac{d\alpha}{dt} + \frac{d\beta}{dt} \right).$$

Substitution into this from Eqs. (2.23a) and (2.23b) yields the rate of the angular deformation, or shearing in the (x, y) plane, as

$$\frac{d\gamma_z}{dt} = \left(\frac{\partial v}{\partial x} + \frac{\partial u}{\partial y} \right). \tag{2.27a}$$

In similar fashion, in the (x, z) and the (y, z) planes, respectively, we have

$$\frac{d\gamma_y}{dt} = \left(\frac{\partial w}{\partial x} + \frac{\partial u}{\partial z} \right), \qquad \frac{d\gamma_x}{dt} = \left(\frac{\partial v}{\partial z} + \frac{\partial w}{\partial y} \right). \tag{2.27b}$$

Note that on the opposing, negatively directed coordinate surfaces we calculate identical shearings (demonstrating that deformation is objective). We equate the shearings to twice the off-diagonal elements of the stretching matrix D:

$$\frac{d\gamma_z}{dt} = 2D_{12}, \qquad \frac{d\gamma_y}{dt} = 2D_{13}, \qquad \frac{d\gamma_x}{dt} = 2D_{23}.$$

Also, because of frame indifference of shearings referred to above, we are allowed to put

$$D_{21} = D_{12}, \qquad D_{31} = D_{13}, \qquad D_{32} = D_{23}.$$

To calculate the diagonal elements of the matrix D, define the *stretching* of an infinitesimal fluid line, originally of length ℓ stretched to length $\ell + \Delta\ell$ in time Δt, as

$$d = \frac{1}{\ell} \lim_{\Delta t \to 0} \frac{\Delta\ell}{\Delta t}.$$

If the line element is in the x-coordinate direction, along the horizontal side of the rectangle in Figure 2.5, we have $\ell_x = \Delta x$ and $\Delta\ell_x = (\partial u/\partial x)\Delta x\, \Delta t$, so that

$$d_x = \frac{1}{\ell_x} \lim_{\Delta t \to 0} \frac{\Delta\ell_x}{\Delta t} = \frac{\partial u}{\partial x}$$

and, similarly, in the other coordinate directions

$$d_y = \frac{\partial v}{\partial y}, \qquad d_z = \frac{\partial w}{\partial z}.$$

Putting $D_{11} = d_x$, $D_{22} = d_y$, and $D_{33} = d_z$, the stretching matrix has the form

$$D = \frac{1}{2} \begin{bmatrix} 2\dfrac{\partial u}{\partial x} & \left(\dfrac{\partial u}{\partial y} + \dfrac{\partial v}{\partial x} \right) & \left(\dfrac{\partial u}{\partial z} + \dfrac{\partial w}{\partial x} \right) \\[2ex] \left(\dfrac{\partial u}{\partial y} + \dfrac{\partial v}{\partial x} \right) & 2\dfrac{\partial v}{\partial y} & \left(\dfrac{\partial w}{\partial y} + \dfrac{\partial v}{\partial z} \right) \\[2ex] \left(\dfrac{\partial u}{\partial z} + \dfrac{\partial w}{\partial x} \right) & \left(\dfrac{\partial w}{\partial y} + \dfrac{\partial v}{\partial z} \right) & 2\dfrac{\partial w}{\partial z} \end{bmatrix}. \tag{2.28}$$

In index notation we write

$$D_{ij} = \frac{1}{2}\left(\frac{\partial v_j}{\partial x_i} + \frac{\partial v_i}{\partial x_j}\right). \tag{2.29}$$

Substituting for L, Ω, and D from Eqs. (2.18), (2.26), and (2.29), respectively, into Eq. (2.20a), we find that the latter is identically satisfied.

To emphasize that stress depends on the objective part of motion, we rewrite Eq. (2.19) so that it contains only that part of the velocity gradient matrix L, that is free of rigid body rotation:

$$T = f(D). \tag{2.30}$$

When f is linear, we expect Eq. (2.30) to specify the constitutive equation of a Newtonian fluid. Nonlinear f will characterize a *Stokesian fluid*.

We shall now discuss what forms of f are physically acceptable (of course, it is one thing to postulate a physically correct f and quite another to find a fluid that obliges to behave accordingly). To start off the discussion, we first insist that (i) *stress is objective*, and (ii) *the constitutive equation is frame invariant*. (We refer to this as the *principle of material frame-indifference, or PMI.*) In the starred reference frame Eq. (2.30) must then retain its form

$$T^* = f(D^*). \tag{2.31a}$$

Since both T and D are objective, we have $T^* = Q \cdot T \cdot Q^T$ and $D^* = Q \cdot D \cdot Q^T$ from Eq. (2.21c). Substitution into Eq. (2.31) yields

$$Q \cdot f \cdot Q^T = f(Q \cdot D \cdot Q^T). \tag{2.32}$$

A function f satisfying condition (2.32) for all symmetric matrices D and all orthogonal matrices Q (i.e., for all rotations of the frame) is said to be an *isotropic function*; the functional dependence it specifies is *invariant under rotation*. All fluids of the type Eq. (2.30) are isotropic, i.e., show no directional dependence.

If two matrices are related through an isotropic function, as T and D are, they can be shown to possess the same principal directions (eigenvectors), i.e., any orthogonal transformation that reduces D to diagonal form will likewise diagonalize T. To show this let $\bar{D}$ represent the stretching tensor when written relative to its principal coordinate system and let $\bar{T}$ be the corresponding transformation of T. From Eq. (2.30) we have

$$\bar{T} = f(\bar{D}). \tag{2.31b}$$

The orthogonal transformation

$$Q = \begin{bmatrix} 1 & 0 & 0 \\ 0 & -1 & 0 \\ 0 & 0 & -1 \end{bmatrix}$$

leaves $\bar{D}$ unaltered, therefore for this value of Q, Eqs. (2.30) and (2.32) yield

$$Q \cdot \bar{T} \cdot Q^T = f(Q \cdot \bar{D} \cdot Q^T) = f(\bar{D}) = \bar{T}. \tag{2.33}$$

It follows from the first and the last of Eq. (2.33) that

$$\begin{bmatrix} \bar{T}_{11} & -\bar{T}_{12} & -\bar{T}_{13} \\ -\bar{T}_{21} & \bar{T}_{22} & \bar{T}_{23} \\ -\bar{T}_{31} & \bar{T}_{32} & \bar{T}_{33} \end{bmatrix} = \begin{bmatrix} \bar{T}_{11} & \bar{T}_{12} & \bar{T}_{13} \\ \bar{T}_{21} & \bar{T}_{22} & \bar{T}_{23} \\ \bar{T}_{31} & \bar{T}_{32} & \bar{T}_{33} \end{bmatrix},$$

thus $\bar{T}_{12} = \bar{T}_{21} = \bar{T}_{13} = \bar{T}_{31} = 0$. Furthermore, on using

$$Q = \begin{bmatrix} -1 & 0 & 0 \\ 0 & -1 & 0 \\ 0 & 0 & 1 \end{bmatrix}$$

it can be shown that $\bar{T}_{23} = \bar{T}_{32} = 0$. Equation (2.31b) then takes the form

$$\begin{bmatrix} t_1 & 0 & 0 \\ 0 & t_2 & 0 \\ 0 & 0 & t_3 \end{bmatrix} = f \begin{bmatrix} d_1 & 0 & 0 \\ 0 & d_2 & 0 \\ 0 & 0 & d_3 \end{bmatrix}, \quad \text{or} \quad t_i = f_i(d_1, d_2, d_3), \tag{2.34a}$$

where (t_1, t_2, t_3) and (d_1, d_2, d_3) are the eigenvalues of the matrices T and D, respectively. It is permissible to rewrite Eq. (2.34a) as

$$t_{(i)} = \alpha + \beta d_{(i)} + \gamma d_{(i)}^2, \tag{2.34b}$$

where α, β, and γ are functions of (d_1, d_2, d_3). Transferring Eq. (2.34b) to a generic coordinate system, we arrive at

$$T = \alpha \mathbf{1} + \beta D + \gamma D^2, \tag{2.34c}$$

the *constitutive equation for a Stokesian fluid*. For incompressible fluid we put $\alpha = -p$, where p is the fluid pressure.

We are now ready to define a *Stokesian fluid* as follows:

(1) T is a continuous function of the stretching matrix D, and is independent of all other kinematic quantities.
(2) T does not depend explicitly on the position x (spatial homogeneity).
(3) There is no preferred direction in space (isotropy).
(4) When $D = 0$, T reduces to $-p\mathbf{1}$.

Constitutive equation (2.34c) satisfies all postulates for a Stokesian fluid.

Specifying further, that the dependence of T on D is linear and the fluid is incompressible, Eq. (2.34c) simplifies to

$$T = -p\mathbf{1} + 2\mu D. \tag{2.35}$$

This equation defines an *incompressible Newtonian fluid*. In full coordinate notation it has the form

$$\begin{bmatrix} T_{xx} & T_{xy} & T_{xz} \\ T_{yx} & T_{yy} & T_{yz} \\ T_{zx} & T_{zy} & T_{zz} \end{bmatrix} = \begin{bmatrix} -p & 0 & 0 \\ 0 & -p & 0 \\ 0 & 0 & -p \end{bmatrix}$$

$$+ 2\mu \begin{bmatrix} \dfrac{\partial u}{\partial x} & \dfrac{1}{2}\left(\dfrac{\partial u}{\partial y} + \dfrac{\partial v}{\partial x}\right) & \dfrac{1}{2}\left(\dfrac{\partial u}{\partial z} + \dfrac{\partial w}{\partial x}\right) \\[3mm] \dfrac{1}{2}\left(\dfrac{\partial u}{\partial y} + \dfrac{\partial v}{\partial x}\right) & \dfrac{\partial v}{\partial y} & \dfrac{1}{2}\left(\dfrac{\partial v}{\partial z} + \dfrac{\partial w}{\partial y}\right) \\[3mm] \dfrac{1}{2}\left(\dfrac{\partial u}{\partial z} + \dfrac{\partial w}{\partial x}\right) & \dfrac{1}{2}\left(\dfrac{\partial v}{\partial z} + \dfrac{\partial w}{\partial y}\right) & \dfrac{\partial w}{\partial z} \end{bmatrix}.$$

$$\tag{2.36}$$

The behavior of most common fluids under moderate conditions is described well by the Newtonian constitutive equations (2.36). There is, however, an abundance of fluids whose behavior differs from Newtonian in essential ways. Molten polymers and oil-in-water emulsions, for example, are not represented by Eq. (2.36).

Cauchy's Equations of Motion

We will now apply Newton's second law of motion to a fluid particle. As we are working with orthogonal Cartesian coordinates and since in that coordinate system the rectangular parallelepiped is the most easily described volume element, the fluid particle chosen for our analysis has this shape. The sides of the parallelepiped are the coordinate planes, and its volume is $\Delta V = \Delta x \, \Delta y \, \Delta z$, where Δx, Δy, and Δz are its linear dimensions in the x, y, and z directions, respectively (Figure 2.3). The shape of the particle is, however, in no way central to the argument we are about to develop.

In most of engineering fluid mechanics, we are concerned either with stress caused by flow or flow generated by stress. In either case, there is a strong cause-effect relationship between stress and deformation. From this it follows that as velocity varies, stress will also vary both with position and with time. In the sequel we also agree, for bookkeeping purposes during formulation, that stress components and all other flow variables increase both with time and with the relevant coordinate. To indicate this clearly, we assign, e.g., the value T_{ij} to the stress component acting on the $x_i = $ const. coordinate plane (in the x_j direction) and the value $T_{ij} + \Delta T_{ij}$ to the component acting on a parallel plane, a distance Δx_i apart. The stress components are labeled and drawn according to this scheme in Figure 2.3. We also note that a stress component is drawn in the positive (negative) direction when acting on a coordinate surface of positive (negative) orientation.

Let us now apply Newton's second law, i.e., $\boldsymbol{F} = m\boldsymbol{a}$, to the infinitesimal fluid particle in Figure 2.3. Force balance in the x, y, and z directions, respectively, yields the equations

$$\rho \frac{du}{dt} \Delta V = \Delta T_{xx} \, \Delta y \, \Delta z + \Delta T_{yx} \, \Delta x \, \Delta z + \Delta T_{zx} \, \Delta x \, \Delta y + \rho f_x \, \Delta V,$$

$$\rho \frac{dv}{dt} \Delta V = \Delta T_{xy} \, \Delta y \, \Delta z + \Delta T_{yy} \, \Delta x \, \Delta z + \Delta T_{zy} \, \Delta x \, \Delta y + \rho f_y \, \Delta V, \qquad (2.37)$$

$$\rho \frac{dw}{dt} \Delta V = \Delta T_{xz} \, \Delta y \, \Delta z + \Delta T_{yz} \, \Delta x \, \Delta z + \Delta T_{zz} \, \Delta x \, \Delta y + \rho f_z \, \Delta V.$$

Here f_x, f_y, f_z are components of the (prescribed) body force.

The small quantities ΔT_{xx}, ΔT_{yx}, ΔT_{zx} appearing in the first of Eq. (2.37) represent the change suffered by the stress components T_{xx}, T_{yx}, T_{zx} in the $x > 0$ direction over a distance Δx, Δy, Δz, respectively. Thus we may write

$$\Delta T_{xx} = \frac{\partial T_{xx}}{\partial x} \Delta x, \qquad \Delta T_{yx} = \frac{\partial T_{yx}}{\partial y} \Delta y, \qquad \Delta T_{zx} = \frac{\partial T_{zx}}{\partial z} \Delta z, \qquad (2.38)$$

and similarly for the other incremental terms in the second and third of Eq. (2.37).

Substituting for the stress increments from Eq. (2.38) into Eq. (2.37) we obtain, after division by ΔV, *Cauchy's equations of motion*

$$\rho \frac{du}{dt} = \frac{\partial T_{xx}}{\partial x} + \frac{\partial T_{yx}}{\partial y} + \frac{\partial T_{zx}}{\partial z} + \rho f_x,$$

$$\rho \frac{dv}{dt} = \frac{\partial T_{xy}}{\partial x} + \frac{\partial T_{yy}}{\partial y} + \frac{\partial T_{zy}}{\partial z} + \rho f_y, \qquad (2.39)$$

$$\rho \frac{dw}{dt} = \frac{\partial T_{xz}}{\partial x} + \frac{\partial T_{yz}}{\partial y} + \frac{\partial T_{zz}}{\partial z} + \rho f_z.$$

Here it is understood that the stress matrix is symmetric, $T_{ij} = T_{ji}$. Consequently, Eq. (2.39) contains only six independent stress components.

Navier-Stokes Equations

The *Navier-Stokes equations* are the general equations of motion for a Newtonian fluid. For an incompressible fluid,[5] $\rho = $ const., they are obtained by substitution of Eq. (2.35) into Cauchy's equations of motion (2.39). In case the viscosity, μ, is also constant, the Navier-Stokes equations have the form

$$\rho \left(\frac{\partial u}{\partial t} + u \frac{\partial u}{\partial x} + v \frac{\partial u}{\partial y} + w \frac{\partial u}{\partial z} \right) = -\frac{\partial p}{\partial x} + \mu \left(\frac{\partial^2 u}{\partial x^2} + \frac{\partial^2 u}{\partial y^2} + \frac{\partial^2 u}{\partial z^2} \right) + \rho f_x,$$

$$\rho \left(\frac{\partial v}{\partial t} + u \frac{\partial v}{\partial x} + v \frac{\partial v}{\partial y} + w \frac{\partial v}{\partial z} \right) = -\frac{\partial p}{\partial y} + \mu \left(\frac{\partial^2 v}{\partial x^2} + \frac{\partial^2 v}{\partial y^2} + \frac{\partial^2 v}{\partial z^2} \right) + \rho f_y, \qquad (2.40)$$

$$\rho \left(\frac{\partial w}{\partial t} + u \frac{\partial w}{\partial x} + v \frac{\partial w}{\partial y} + w \frac{\partial w}{\partial z} \right) = -\frac{\partial p}{\partial z} + \mu \left(\frac{\partial^2 w}{\partial x^2} + \frac{\partial^2 w}{\partial y^2} + \frac{\partial^2 w}{\partial z^2} \right) + \rho f_z.$$

By multiplying the first, second, and third of Eq. (2.40) by the unit base vectors and summing the result, we obtain Eq. (2.40) in vector notation:

$$\rho \left[\frac{\partial \boldsymbol{v}}{\partial t} + (\boldsymbol{v} \cdot \nabla) \boldsymbol{v} \right] = -\nabla p + \mu \nabla^2 \boldsymbol{v} + \rho \boldsymbol{f}. \qquad (2.41a)$$

In index notation the Navier-Stokes equations for incompressible, constant viscosity fluid have the form

$$\rho \left(\frac{\partial v_i}{\partial t} + v_j \frac{\partial v_i}{\partial x_j} \right) = -\frac{\partial p}{\partial x_i} + \mu \frac{\partial^2 v_i}{\partial x_j \partial x_j} + \rho f_i. \qquad (2.41b)$$

We see that the Navier-Stokes equations of motion (2.40) contain four unknowns, the three velocity components u, v, w, and the pressure p, but that there are only three equations. In order to render the problem mathematically well posed, i.e., to have the number of unknowns equal the number of equations, we need one more equation involving the velocity components or their derivatives. This equation is provided by the principle of conservation of mass. The resulting equation is known as the equation of continuity.

---

[5] To keep the presentation simple, discussion of compressible fluids is delayed until Chapter 11.

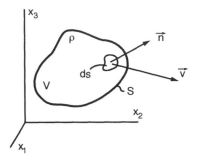

Figure 2.6. Mass-flow rate across boundary.

Equation of Continuity

The *equation of continuity* is the mathematical statement of the *principle of mass conservation: mass can neither be created nor can it be destroyed.* The mathematical statement of this principle is the assertion that the net through-flow of mass across any closed surface equals the depletion of mass inside that surface.

Let S be a closed surface enclosing the simply connected arbitrary volume V (Figure 2.6). The mass-flow across an element of the surface area, da, is given by $\rho v_n \, da$, where $v_n = \boldsymbol{v} \cdot \boldsymbol{n}$ is the component of the velocity normal to the area element and $\boldsymbol{n}$ is the outward unit-normal. The net through-flow across the closed surface S can then be calculated from

$$\oint_S (\rho \boldsymbol{v}) \cdot \boldsymbol{n} \, ds.$$

As we are permitted neither to create nor to destroy mass, there can be an increase of mass within a container completely filled with fluid at all times only if the density of the fluid increases. Thus the rate at which mass increases inside S can be calculated from

$$\int_V \frac{\partial \rho}{\partial t} dV.$$

The principle of mass conservation can now be stated as

$$\oint_S (\rho \boldsymbol{v}) \cdot \boldsymbol{n} \, dS + \int_V \frac{\partial \rho}{\partial t} dV = 0. \tag{2.42}$$

We can bring both integrands in Eq. (2.42) under a common integral sign by means of the *divergence theorem*

$$\oint_S (\rho \boldsymbol{v}) \cdot \boldsymbol{n} \, dS = \int_V \mathrm{div}(\rho \boldsymbol{v}) \, dV,$$

which allows Eq. (2.42) to be cast in the form

$$\int_V \left[\frac{\partial \rho}{\partial t} + \mathrm{div}(\rho \boldsymbol{v}) \right] dV = 0. \tag{2.43}$$

In this expression, the volume is arbitrary; it then follows that for Eq. (2.43) to hold, the

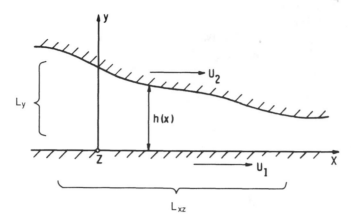

Figure 2.7. Bearing surfaces, coordinate axes, and length scales.

integrand itself must vanish. This leads to the *equation of continuity*,

$$\frac{\partial \rho}{\partial t} + \mathrm{div}(\rho v) = 0, \tag{2.44a}$$

for *compressible fluids.*

For *incompressible fluids* the density is constant. Division of Eq. (2.44a) by $\rho \neq 0$ yields

$$\mathrm{div}\, v = 0, \quad \text{or} \quad \frac{\partial v_i}{\partial x_i} = 0. \tag{2.44b}$$

2.2 Lubrication Approximation

To find solutions of the full Navier-Stokes and continuity equations is far from elementary, and in applications we look for ways to simplify these equations. Such simplification is made particularly easy in lubrication due to the geometry of lubricant films: under normal conditions the in-plane dimension of the film is significantly greater than its thickness. If L_{xz} designates the *length scale* of the lubricant film in the (x, z) plane of Figure 2.7 and L_y its length scale across its thickness in the y direction, then for typical lubricant films we have

$$(L_y / L_{xz}) = O(10^{-3}).$$

Having two greatly differing length scales is what makes the analysis of fluid film bearings relatively simple. We will make use of this property to estimate the order of magnitude of the various terms of the governing equations. The equations will then be simplified by deleting those terms that were judged too small to have significant effect.

To perform the analysis indicated in the previous paragraph, we must first normalize the variables participating in the governing equations.

The definition

$$\bar{x} = x/L_{xz}, \qquad \bar{y} = y/L_y, \qquad \bar{z} = z/L_{xz},$$

assures us that the range of the *nondimensional coordinates* $\bar{x}, \bar{y}, \bar{z}$ is from 0 to 1. We normalize the velocity similarly: if the maximum value of the velocity in the (x, z) plane of the film is U_*, the nondimensional velocity components, $\bar{u} = u/U_*$ and $\bar{w} = w/U_*$, will again vary within the range 0 to 1. Obviously, the *velocity scale* in the direction across the film will not be equal to U_*. In fact, intuition tells us that it will be considerably smaller than U_*, as we expect the flow to be approximately two-dimensional, parallel to the (x, z) plane.

Let us denote the velocity scale in the direction across the film by V_*. We can get an estimate for the magnitude of V_* relative to U_* from the equation of continuity. Substituting

$$x = L_{xz}\bar{x}, \qquad y = L_y\bar{y}, \qquad z = L_{xz}\bar{z},$$
$$u = U_*\bar{u}, \qquad v = V_*\bar{v}, \qquad w = U_*\bar{w}, \tag{2.45}$$

into Eq. (2.44b), we obtain

$$\frac{\partial \bar{u}}{\partial \bar{x}} + \frac{V_* L_{xz}}{U_* L_y}\frac{\partial \bar{v}}{\partial \bar{y}} + \frac{\partial \bar{w}}{\partial \bar{z}} = 0.$$

The three terms of this equation will be of the same order of magnitude if

$$\frac{V_* L_{xz}}{U_* L_y} = O(1) \quad \text{and we choose} \quad V_* = \left(\frac{L_y}{L_{xz}}\right)U_*. \tag{2.46}$$

Thus if $(L_y/L_{xz}) = O(10^{-3})$, the magnitude of the velocity scale in the y direction will be 10^{-3} times the magnitude of the velocity scale along x or z.

We now make use of the nondimensional quantities just defined and substitute Eqs. (2.45) and (2.46) into the Navier-Stokes equations. After some tedious algebra we find the nondimensional form of these equations; to achieve this we must also normalize the pressure, p, and the time, t. Define the nondimensional quantities $\bar{p}$ and $\bar{t}$ by[6]

$$\bar{p} = \text{Re}\left(\frac{L_y}{L_{xz}}\right)\frac{p}{\rho U_*^2}, \qquad \bar{t} = \Omega t. \tag{2.47}$$

Here

$$\text{Re} = U_* L_y/\nu$$

is the Reynolds number.

The nondimensionalized (normalized) form of the Navier-Stokes equations are

$$\overset{*}{\Omega}\frac{\partial \bar{u}}{\partial \bar{t}} + \overset{*}{\text{Re}}\left(\bar{u}\frac{\partial \bar{u}}{\partial \bar{x}} + \bar{v}\frac{\partial \bar{u}}{\partial \bar{y}} + \bar{w}\frac{\partial \bar{u}}{\partial \bar{z}}\right) = -\frac{\partial \bar{p}}{\partial \bar{x}} + \frac{\partial^2 \bar{u}}{\partial \bar{y}^2} + \left(\frac{L_y}{L_{xz}}\right)^2\left(\frac{\partial^2 \bar{u}}{\partial \bar{x}^2} + \frac{\partial^2 \bar{u}}{\partial \bar{z}^2}\right), \tag{2.48a}$$

$$\left(\frac{L_y}{L_{xz}}\right)^2\left[\overset{*}{\Omega}\frac{\partial \bar{v}}{\partial \bar{t}} + \overset{*}{\text{Re}}\left(\bar{u}\frac{\partial \bar{v}}{\partial \bar{x}} + \bar{v}\frac{\partial \bar{v}}{\partial \bar{y}} + \bar{w}\frac{\partial \bar{v}}{\partial \bar{z}}\right) - \frac{\partial^2 \bar{v}}{\partial \bar{y}^2}\right.$$
$$\left.-\left(\frac{L_y}{L_{xz}}\right)^2\left(\frac{\partial^2 \bar{v}}{\partial \bar{x}^2} + \frac{\partial^2 \bar{v}}{\partial \bar{z}^2}\right)\right] = -\frac{\partial \bar{p}}{\partial \bar{y}}, \tag{2.48b}$$

$$\overset{*}{\Omega}\frac{\partial \bar{w}}{\partial \bar{t}} + \overset{*}{\text{Re}}\left(\bar{u}\frac{\partial \bar{w}}{\partial \bar{x}} + \bar{v}\frac{\partial \bar{w}}{\partial \bar{y}} + \bar{w}\frac{\partial \bar{w}}{\partial \bar{z}}\right) = -\frac{\partial \bar{p}}{\partial \bar{z}} + \frac{\partial^2 \bar{w}}{\partial \bar{y}^2} + \left(\frac{L_y}{L_{xz}}\right)^2\left(\frac{\partial^2 \bar{w}}{\partial \bar{x}^2} + \frac{\partial^2 \bar{w}}{\partial \bar{z}^2}\right). \tag{2.48c}$$

---

[6] Had we put $\bar{p} = p/\rho U^2$, the pressure would disappear from the nondimensional Navier-Stokes equations on taking the limit $(L_y/L_{xz})\text{Re} \to 0$, leaving four scalar equations to be satisfied by three unknown velocity components.

Here we neglected the body force. Also, we employed the *reduced frequency* and the *reduced Reynolds number*, which are defined, respectively, by

$$\overset{*}{\Omega} = \frac{L_y^2 \Omega}{\nu} \quad \text{and} \quad \overset{*}{\text{Re}} = \text{Re}\left(\frac{L_y}{L_{xz}}\right). \tag{2.49}$$

The nondimensional continuity equation is

$$\frac{\partial \bar{u}}{\partial \bar{x}} + \frac{\partial \bar{v}}{\partial \bar{y}} + \frac{\partial \bar{w}}{\partial \bar{z}} = 0, \tag{2.50}$$

thus the continuity equation is invariant under the transformation specified by Eqs. (2.45) and (2.46), as expected.

Because for lubricant films $(L_y/L_{xz}) = O(10^{-3})$, we can dispense with Eq. (2.48b) completely, as

$$\frac{\partial \bar{p}}{\partial \bar{y}} = O(10^{-6}), \quad \text{while} \quad \frac{\partial \bar{p}}{\partial \bar{x}} \approx \frac{\partial \bar{p}}{\partial \bar{z}} = O(1). \tag{2.51}$$

Thus, to order $(L_y/L_{xz})^2$, the pressure is constant across the lubricant film.

Inertia due to local acceleration will survive only if $\overset{*}{\Omega} \approx O(1)$. Assuming that $\Omega \approx (U_*/L_{xz})$, we obtain

$$\overset{*}{\Omega} \approx \overset{*}{\text{Re}},$$

in which case Eqs. (2.48a) and (2.48b) show that inertia terms will survive, in general, if

$$\overset{*}{\text{Re}} \geq O(1). \tag{2.52}$$

Translated into practical language, condition (2.52) suggests that inertia effects will be significant in superlaminar flow, $\text{Re} \geq O(10^3)$, and/or when there is a rapid change in the cross section (large local value of L_y/L_{xz}).

On taking the limit $(L_y/L_{xz})^2 \to 0$ in Eqs. (2.48a) and (2.48c) we obtain the reduced equations

$$\text{Re}^*\left(\frac{\partial \bar{u}}{\partial \bar{t}} + \bar{u}\frac{\partial \bar{u}}{\partial \bar{x}} + \bar{v}\frac{\partial \bar{u}}{\partial \bar{y}} + \bar{w}\frac{\partial \bar{u}}{\partial \bar{z}}\right) = -\frac{\partial \bar{p}}{\partial \bar{x}} + \frac{\partial^2 \bar{u}}{\partial \bar{y}^2}, \tag{2.53}$$

$$\text{Re}^*\left(\frac{\partial \bar{w}}{\partial \bar{t}} + \bar{u}\frac{\partial \bar{w}}{\partial \bar{x}} + \bar{v}\frac{\partial \bar{w}}{\partial \bar{y}} + \bar{w}\frac{\partial \bar{w}}{\partial \bar{z}}\right) = -\frac{\partial \bar{p}}{\partial \bar{z}} + \frac{\partial^2 \bar{w}}{\partial \bar{y}^2}, \tag{2.54}$$

$$\frac{\partial \bar{u}}{\partial \bar{x}} + \frac{\partial \bar{v}}{\partial \bar{y}} + \frac{\partial \bar{w}}{\partial \bar{z}} = 0. \tag{2.55}$$

Equations (2.53) to (2.55) will be made use of later, when discussing inertia effects (see Chapter 5). Classical theory further simplifies these equations by requiring that $\text{Re}^* \to 0$.

The Reynolds Equation

In this section we develop the (classical) *Reynolds theory of lubrication.* The principal simplifying assumptions of the theory derive from the observation that the lubricant flow, at least in a first approximation, is isoviscous and laminar and that it takes place in an "almost parallel" thin film of negligible curvature. The *Reynolds equation* in lubricant pressure is the mathematical statement of the classical theory of lubrication as it was formulated by Osborne Reynolds just over a century ago.

The Reynolds theory of lubrication can be obtained on the assumptions of

(1) constant viscosity, Newtonian lubricant;
(2) thin film geometry, $(L_y/L_{xz}) \to 0$;
(3) negligible inertia, $\overset{*}{Re}$; and
(4) negligible body force.

In thrust bearings the lubricant film is bounded by plane surfaces. In journal bearings the bounding surfaces are no longer plane, nevertheless, in relation to its thickness the film curves only very gently. In fact, the ratio of film thickness to radius of curvature in most practical bearings is at most of order (L_y/L_{xz}). Therefore, assumption (2) is also a statement on film curvature. As consequence of assumption (2), it is permissible to describe fluid film lubrication relative to an orthogonal Cartesian coordinate system. The y axis of this Cartesian system is in the direction of the minimum film dimension, its (x, z) plane coincides with the "plane" of the lubricant film.

In journal bearings we select the bearing surface to be the $y = 0$ "plane." Though, in reality, all vectors normal to this surface intersect in the center of the bearing, in the approximate world of lubrication theory we consider these normal vectors to be parallel to one another; i.e., we focus on such short distances along these vectors that the fact that they intersect at, what seems to us a very great distance, remains unnoticed. This is the essence of neglecting film curvature.

When assumptions (1)–(4) are applied to the equations of motion, Eqs. (2.53) and (2.54), and to the equation of continuity, Eq. (2.55), we obtain, now in terms of the *primitive variables*,

$$\frac{\partial p}{\partial x} = \mu \frac{\partial^2 u}{\partial y^2}, \tag{2.56}$$

$$\frac{\partial p}{\partial z} = \mu \frac{\partial^2 w}{\partial y^2}, \tag{2.57}$$

$$\frac{\partial u}{\partial x} + \frac{\partial v}{\partial y} + \frac{\partial w}{\partial z} = 0. \tag{2.58}$$

The equations of motion, Eqs. (2.56) and (2.57), may now be integrated twice with respect to y, since by assumption (2) and Eq. (2.51) neither $\partial p/\partial x$ nor $\partial p/\partial z$ varies across the film

$$u = \frac{1}{2\mu} \frac{\partial p}{\partial x} y^2 + Ay + B, \tag{2.59a}$$

$$w = \frac{1}{2\mu} \frac{\partial p}{\partial z} y^2 + Cy + D. \tag{2.59b}$$

A, B, C, and D are either constants or, at most, functions of x and z. Their value must be chosen so that u and w satisfy prescribed boundary conditions in y.

The boundary conditions for u and w are

$$u = U_1, w = 0 \quad \text{at} \quad y = 0,$$
$$u = U_2, w = 0 \quad \text{at} \quad y = h, \tag{2.60}$$

where U_1 and U_2 represent the of the velocity of the bearing surfaces (Figure 2.7).

Substitution of the velocity from Eqs. (2.59) into the boundary conditions (2.60) evaluates the integration constants in Eqs. (2.59), and yields the velocity distribution

$$u = \frac{1}{2\mu}\frac{\partial p}{\partial x}(y^2 - yh) + \left(1 - \frac{y}{h}\right)U_1 + \frac{y}{h}U_2,$$
$$w = \frac{1}{2\mu}\frac{\partial p}{\partial z}(y^2 - yh). \tag{2.61}$$

The pressure gradient in Eq. (2.61) is, as yet, unknown. But since the pressure p in Eqs. (2.56) and (2.57) is an induced pressure, the sole function of which is to guarantee compliance with the principle of conservation of mass, it can be evaluated from the condition that both u and w satisfy the equation of continuity. This seems like a reasonable scheme, but it has one serious flaw. If u and w are substituted into Eq. (2.58) the resulting single equation will contain two unknowns,[7] v and p, and, unless v is given, we have insufficient information to determine p. This difficulty will be alleviated by integrating, in effect averaging, the equation of continuity across the film; the integrated equation of continuity will contain the velocity component v only in the values it assumes at the boundaries $y = 0$ and $y = h(x, t)$. And as the approach velocity of the surfaces is presumed known during this analysis, integration across the film eliminates one of the two unknowns.

Integrating the equation of continuity (2.58) across the film results in

$$[v]_{y=0}^{h(x,t)} = -\int_0^{h(x,t)} \frac{\partial u}{\partial x}dy - \int_0^{h(x,t)} \frac{\partial w}{\partial z}dy. \tag{2.62}$$

Interchanging integration and differentiation in Eq. (2.62) and substituting for u and w from Eq. (2.61) we obtain[8]

$$[v]_0^{h(x,t)} = -\frac{\partial}{\partial x}\left[\frac{1}{2\mu}\frac{\partial p}{\partial x}\int_0^{h(x,t)}(y^2 - yh)\,dy\right]$$
$$-\frac{\partial}{\partial z}\left[\frac{1}{2\mu}\frac{\partial p}{\partial z}\int_0^{h(x,t)}(y^2 - yh)\,dy\right]$$
$$-\frac{\partial}{\partial x}\int_0^{h(x,t)}\left[\left(1 - \frac{y}{h}\right)U_1 + \frac{y}{h}U_2\right]dy + U_2\frac{\partial h}{\partial x}. \tag{2.63a}$$

Evaluating the integrals and taking into account that

$$[v]_{y=0}^{h(x,t)} = -(V_1 - V_2) = \frac{dh}{dt}, \tag{2.63b}$$

---

[7] The problem originates with the approximation itself; the set of reduced equations, Eqs. (2.56), (2.57), and (2.58), contain four unknowns $\bar{u}$, $\bar{v}$, $\bar{w}$, and $\bar{p}$, but only three equations.

[8] Here we employ Leibnitz's rule for differentiating under the integral sign:

$$\frac{d}{dx}\int_A^B f(x, t)\,dt = \int_A^B \frac{\partial f(x, t)}{\partial x}\,dt + f(x, B)\frac{dB}{dx} - f(x, A)\frac{dA}{dx}.$$

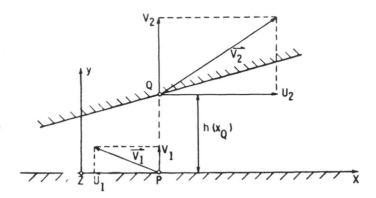

Figure 2.8. Velocities for rigid body translation of bearing surfaces.

where $V_1 - V_2$ is the velocity of approach of the surfaces, we obtain the *Reynolds equation* for lubricant pressure

$$\frac{\partial}{\partial x}\left(\frac{h^3}{\mu}\frac{\partial p}{\partial x}\right) + \frac{\partial}{\partial z}\left(\frac{h^3}{\mu}\frac{\partial p}{\partial z}\right) = 6(U_1 - U_2)\frac{\partial h}{\partial x} + 6h\frac{\partial(U_1 + U_2)}{\partial x} + 12(V_2 - V_1).$$

$$(2.64)$$

It is emphasized that $V_1 = (U_1, V_1)$ and $V_2 = (U_2, V_2)$ are the velocities of "corresponding" points, each fixed to one of the bearing surfaces.[9] The velocities V_1 and V_2 result from rigid body motion, which may include both rotation and translation, of the bearing surfaces. It will be to our advantage in later work to separate rigid body translation from rigid body rotation. Thus we decompose the velocity of surface 2, surface 1 being the reference surface, according to the scheme

$$U_2 = U_{2,r} + U_{2,t},$$
$$V_2 = V_{2,r} + V_{2,t},$$

$$(2.65)$$

so that $(U_{2,r}, V_{2,r})$ and $(U_{2,t}, V_{2,t})$ are caused by rotation and translation, respectively.

For the plane slider in Figure 2.8, and for *thrust bearings* in general, the rotational components of the velocity are identically zero. The translational components are usually prescribed relative to the (x, y, z) coordinate system of the runner (now surface 1); thus

$$U_2 = U_{2,t}, \qquad U_{2,r} \equiv 0,$$
$$V_2 = V_{2,t}, \qquad V_{2,r} \equiv 0.$$

$$(2.66)$$

Interpreting the film thickness as the normal distance between a point fixed on the bearing (point Q in Figure 2.8) and the runner surface (x, z plane), we find that the film thickness changes if and only if the bearing is given a translational velocity $V_{2,t} - V_1 = V_2 - V_1$ in

---

[9] We call two points, one fixed to the bearing surface and the other to the runner surface, corresponding points at the instant when they are located on the same normal to the reference surface. For journal bearings the pad surface is the reference surface. For the plane slider, on the other hand, it is expedient to designate the runner surface as the reference surface.

the y direction, relative to the runner. Such change of film thickness is uniform in x and so is its time rate. Thus

$$U_1 - U_2 = U_1 - U_{2,t}$$
$$V_2 - V_1 = V_{2,t} - V_1 = \frac{\partial h}{\partial t}.$$

We find, furthermore, that both U_1 and U_2 are constant when the surfaces are rigid and thus for thrust bearings the Reynolds equation (2.64) reduces to

$$\frac{\partial}{\partial x}\left(\frac{h^3}{\mu}\frac{\partial p}{\partial x}\right) + \frac{\partial}{\partial z}\left(\frac{h^3}{\mu}\frac{\partial p}{\partial z}\right) = 6(U_1 - U_2)\frac{\partial h}{\partial x} + 12(V_2 - V_1). \tag{2.67}$$

Pressure generation in thrust bearings thus depends on the translational velocity of the bearing surfaces relative to one another but not on the absolute value of the velocity. We are therefore permitted to recast Eq. (2.67) into a form that contains only relative velocities. We will do this in order to bring out the essentials of the analysis. The result is the *Reynolds equation for thrust bearings*; that is, for bearings where only translation of the surfaces is involved:

$$\frac{\partial}{\partial x}\left(\frac{h^3}{\mu}\frac{\partial p}{\partial x}\right) + \frac{\partial}{\partial z}\left(\frac{h^3}{\mu}\frac{\partial p}{\partial z}\right) = 6U_0\frac{\partial h}{\partial x} + 12V_0. \tag{2.68}$$

Here

$$U_0 = U_1 - U_2, \qquad V_0 = V_2 - V_1$$

are the relative velocities in the directions parallel and perpendicular, respectively, to the reference (runner) surface.

To generate positive (load-carrying) pressures in the film, we require that the right-hand side of Eq. (2.68) be negative; that is,

$$U_0\frac{\partial h}{\partial x} < 0, \qquad \frac{\partial h}{\partial t} < 0. \tag{2.69}$$

The first of these conditions specifies a film that is convergent in space (in the direction of relative motion), and the second specifies a film that is convergent in time.

In journal bearings one encounters both rotation and translation of the bearing surfaces. The velocity V_2 of an arbitrary point Q, fixed to the runner surface as in Figure 2.9, is given by the sum of its velocity in rigid body translation $V_{2,t}$ and the velocity $V_{2,r}$ that is caused by rigid body rotation of the journal. From Figure 2.9 we have

$$U_{2,r} = |V_{2,r}|\cos\alpha$$
$$= |V_{2,r}|\left[1 - \frac{1}{2}\left(\frac{\partial h}{\partial x}\right)^2 + \cdots\right] \tag{2.70a}$$
$$\approx |V_{2,r}|$$

as $(\partial h/\partial x) \ll 1$ by assumption (2) of the classical theory.

We may also write

$$V_{2,r} = |V_{2,r}|\sin\alpha$$
$$= U_{2,r}\tan\alpha \tag{2.70b}$$
$$\approx |V_{2,r}|\frac{\partial h}{\partial x}.$$

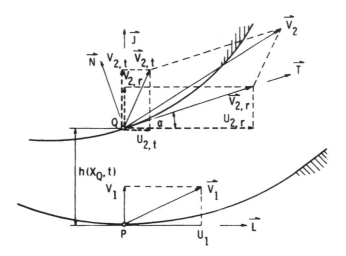

Figure 2.9. Velocities for translation and rotation of bearing surfaces.

On the other hand, $U_{2,r}$ is easily replaced by U_2. In journal bearings

$$\frac{U_{2,t}}{U_{2,r}} = O(10^{-3}),$$

and the following approximation is acceptable in most cases:

$$U_2 = U_{2,r} + U_{2,t}$$

$$= U_{2,r}\left(1 + \frac{U_{2,t}}{U_{2,r}}\right) \tag{2.70c}$$

$$\approx U_{2,r}.$$

Equation (2.70b) then takes the form

$$V_{2,r} \approx U_2\frac{\partial h}{\partial x},$$

$$V_2 \approx V_{2,t} + U_2\frac{\partial h}{\partial x}. \tag{2.71}$$

The first and the last terms on the right-hand side of Eq. (2.64) are combined as follows:

$$6(U_1 - U_2)\frac{\partial h}{\partial x} + 12\frac{dh}{dt} = 6(U_1 - U_2)\frac{\partial h}{\partial x} + 12\left(V_{2,t} + U_2\frac{\partial h}{\partial x} - V_1\right)$$

$$= 6(U_1 + U_2)\frac{\partial h}{\partial x} + 12(V_{2,t} - V_1) \tag{2.72}$$

$$= 6(U_1 + U_2)\frac{\partial h}{\partial x} + 12\frac{\partial h}{\partial t}.$$

As an alternative to Eq. (2.72) we might have started with the interpretation that the film thickness, $h = h(x, t)$, is the normal distance of a fixed point, Q, of the journal to the bearing surface. The film thickness, according to this interpretation, is capable of changing for one or both of two reasons: (1) rigid body translation of the journal center relative to the bearing along a bearing radius (local change) and (2) rotation of the journal (convective change). The latter motion forces Q, and h, to move within a clearance gap of varying width, with velocity $U_{2,r}$. Therefore, recognizing that h is dependent on t both explicitly and implicitly through the motion, we obtain (see Eq. 2.5)

$$\frac{dh}{dt} = \frac{\partial h}{\partial t} + \frac{dx}{dt}\frac{\partial h}{\partial x} = \frac{\partial h}{\partial t} + U_{2,r}\frac{\partial h}{\partial x}$$

and recover Eq. (2.72) approximately. (For thrust bearings the second term on the right-hand side vanishes.)

Taking now Eq. (2.72) into account, Eq. (2.64) reduces to

$$\frac{\partial}{\partial x}\left(\frac{h^3}{\mu}\frac{\partial p}{\partial x}\right) + \frac{\partial}{\partial z}\left(\frac{h^3}{\mu}\frac{\partial p}{\partial z}\right) = 6\frac{\partial}{\partial x}[h(U_1 + U_2)] + 12(V_{2,t} - V_1). \qquad (2.73)$$

Pressure generation in journal bearings has been shown thus to depend on (1) the sum of the tangential velocities $U_0 = U_1 + U_2$ and (2) the difference of the normal velocities $V_0 = V_{2,t} - V_1$. Equation (2.73) may then be recast into a form that contains U_0 and V_0 rather than the absolute value of the individual velocity components.

$$\frac{\partial}{\partial x}\left(\frac{h^3}{\mu}\frac{\partial p}{\partial x}\right) + \frac{\partial}{\partial z}\left(\frac{h^3}{\mu}\frac{\partial p}{\partial z}\right) = 6U_0\frac{\partial h}{\partial x} + 6h\frac{\partial U_0}{\partial x} + 12V_0. \qquad (2.74)$$

Equations (2.68) and (2.74) are formally identical save for the term $6h(\partial U_0/\partial x)$, which must be retained in Eq. (2.74) to account for journal motion under dynamic loading. However, it should be borne in mind that the interpretation put on the velocity U_0 is distinctly different in the two cases. If the bearing surface is stationary, and this will be assumed unless explicitly stated otherwise, then $U_0 = U_2$ for journal bearings and $U_0 = U_1$ for thrust bearings.

2.3   Nomenclature

D	stretching tensor
F_B, F_S	body, surface force
L	velocity gradient tensor
$L_y, L_{x,z}$	characteristic lengths
M	torque
Re, $\overset{*}{\text{Re}}$	Reynolds numbers
$S, \Delta S$	surface
T	stress tensor
V, v	velocity
$V, \Delta V$	volume
X_i, x_i	Cartesian coordinates
e	Cartesian base vectors
f	body force, constitutive function
h	film thickness
n	unit normal vector

p	pressure
t	time
Ω	spin tensor
$\Omega, \overset{*}{\Omega}$	characteristics frequencies
χ	deformation function
ω	vorticity vector
τ	surface traction
μ	viscosity
ρ	density
$(\cdot)_r, (\cdot)_t$	radial, tangential
$\bar{(\cdot)}$	normalized

2.4   References

Aris, R. 1962. *Vectors, Tensors and the Basic Equations of Fluid Mechanics*. Prentice-Hall, Englewood Cliffs.

Cameron, A. 1966. *The Principles of Lubrication*. Wiley, New York.

Dai, R. X., Dong, Q. and Szeri, A. Z. 1992. Approximations in lubrication theory. *ASME Journal of Tribology*, **114**, 9–14.

Leigh, D. C. 1968. *Nonlinear Continuum Mechanics*. McGraw-Hill, New York.

Pinkus, O. and Sternlicht, B. 1961. *Theory of Hydrodynamic Lubrication*. McGraw-Hill, New York.

Sabersky, R. H., Acosta, A. J. and Hauptman, E. G. 1989. *Fluid Flow: A First Course in Fluid Mechanics*. Macmillan, New York.

Serrin, J. 1959. Mathematical Principles in Classical Fluid Mechanics. In *Encyclopedia of Physics*, (S. Flugge and C. Truesdell, eds.). Vol. VIII. Springer-Verlag, Berlin.

Sherman, F. S. 1990. *Viscous Flow*. McGraw-Hill, New York.

Szeri, A. Z. 1980. *Tribology: Friction, Lubrication, and Wear*. Hemisphere, Washington.

Truesdell, C. 1977. *A First Course in Rational Continuum Mechanics*. Vol. 1. Academic Press, New York.

White, F. M. 1991. *Fluid Mechanics*. McGraw-Hill, New York.

CHAPTER 3

Thick-Film Lubrication

Fluid film lubrication naturally divides into two categories. *Thin-film* lubrication is usually met with in counter-formal contacts, principally in rolling bearings and in gears. The thickness of the film in these contacts is of order of 1μm or less, and the conditions are such that the pressure dependence of viscosity and the elastic deformation of the bounding surfaces must both be taken into account.

Thick-film lubrication is encountered in externally pressurized bearings, also called *hydrostatic bearings*, and in self-acting bearings, called *hydrodynamic bearings*. Of the latter, there are two kinds: journal bearings and thrust bearings. The film thickness in these conformal-contact bearings is at least an order of magnitude larger than in counter-formal bearings. In consequence, the prevailing pressures are orders of magnitude smaller, so that neither the pressure dependence of viscosity nor the elastic deformation of the surfaces plays an important roles. If, in addition, the lubricant is linearly viscous and the reduced Reynolds number is small, the classical Reynolds theory, as derived in the previous chapter, will apply.

This chapter discusses isothermal processes only. It should be realized, however, that bearings never operate under truly isothermal conditions, and under near isothermal conditions only in exceptional cases. Viscous dissipation and consequent heating of the lubricant are always present, and the change in viscosity must be accounted for when analyzing thick-film lubrication problems. In restricted cases, where design and operating conditions are such as to suggest "uniform" temperature rise of the lubricant, the "effective viscosity" approach of Chapter 9 might be employed. In other, again very limited, cases, where heat conduction into the bearing surfaces can be neglected, the "adiabatic theory" might be useful. But in the great majority of practical cases, particularly under turbulent flow conditions, full thermohydrodynamic theory, including thermal/elastic deformations of the bearing surfaces, must be employed.

3.1 Externally Pressurized Bearings

Hydrostatic bearings of non-uniform film thickness are discussed by Heller and Shapiro (1968) and by Szeri and Phillips (1974). The effect of fluid inertia is considered by Szeri and Adams (1978), and the coupled effects of nonuniform viscosity and fluid inertia can be found in Gourley (1977).

Here we derive the theory of externally pressurized bearings under the assumptions of (1) steady loading, (2) constant sliding velocity, and (3) uniform film thickness.

The applicable form of the Reynolds equation is obtained from Eq. (2.74) by substitution of $\mu = $ const., $h = $ const. and $U_0 = 0$:

$$\frac{\partial^2 p}{\partial x^2} + \frac{\partial^2 p}{\partial z^2} = 0, \quad \text{or} \quad \hat{\nabla}^2 p = 0, \tag{3.1}$$

where $\hat{\nabla}^2$ is the two-dimensional Laplace operator.

81

The boundary conditions on pressure are

$$p = p_r \quad \text{on} \quad \Gamma_i$$
$$p = p_a \quad \text{on} \quad \Gamma_o. \tag{3.2}$$

Here Γ_i stands for the recess boundary, Γ_o represents the pad external boundary, p_r is the recess pressure, and p_a is the ambient pressure. (Without loss of generality we put $p_a = 0$ for the incompressible lubricant.)

Equation (3.1), together with the boundary conditions in Eq. (3.2), represents a Dirichlet problem of applied mathematics. The solution of this problem is straightforward when obtained numerically, but it is somewhat difficult to obtain analytically (Szeri, 1975). There are two pad geometries, however, for which solutions of Eq. (3.1) exist in closed form, the circular step and the annular geometries.

Pad Characteristics

Irrespective of the geometry or size of a hydrostatic pad, its performance characteristics can be written in the form

$$W = a_f A p_r, \tag{3.3a}$$

$$Q = q_f \frac{h^3}{\mu} p_r, \tag{3.3b}$$

$$H_p = q_f \frac{h^3}{\mu} p_r p_s, \tag{3.3c}$$

$$H_f = h_f \frac{\mu U_M^2 A}{h}, \tag{3.3d}$$

where W is the external load, Q is the flow rate of the lubricant, H_p is the required pumping power, H_f is the frictional loss, A is the pad area (including the area of the recess), and U_M is the maximum sliding velocity of the runner relative to the pad.

The quantities a_f, q_f, and h_f are commonly referred to as the *area factor*, the *flow factor*, and the *friction factor*, respectively. These factors are dimensionless – that is, they are independent of the size of the bearing pad but dependent on its geometry. They may be evaluated for the particular geometry by the designer or, for the more common geometries such as rectangular and sector, can be extracted from the literature (Rippel, 1963; Szeri, 1975).

Exactly how a_f, q_f, and h_f are obtained will be illustrated for the circular step geometry of radii R_1 and $R_2 > R_1$. Equation (3.1) is first written in polar coordinates for simplicity, and then made nondimensional through the substitutions:

$$p = p_r \bar{p}, \qquad r = R_2 \bar{r}, \tag{3.4}$$

where R_2 is the pad outer radius.

Substitution results in the following differential equation and boundary conditions:

$$\frac{d}{d\bar{r}} \left(\bar{r} \frac{d\bar{p}}{d\bar{r}} \right) = 0,$$
$$\bar{p} = 1 \quad \text{at} \quad \bar{r} = \frac{R_1}{R_2}, \tag{3.5}$$
$$\bar{p} = 0 \quad \text{at} \quad \bar{r} = 1,$$

in place of Eqs. (3.1) and (3.2).

The solution of system Eq. (3.5) is

$$\bar{p} = \frac{\ln \bar{r}}{\ln(R_1/R_2)}. \tag{3.6}$$

We are now in the position to evaluate the bearing load capacity W as follows:

$$\begin{aligned} W &= \pi R_1^2 p_r + 2\pi \int_{R_1}^{R_2} rp \, dr \\ &= \pi R_1^2 p_r \left[1 + \frac{2(R_2/R_1)^2}{\ln(R_1/R_2)} \int_{R_1/R_2}^{1} \bar{r} \ln \bar{r} \, d\bar{r} \right] \\ &= \frac{1 - (R_1/R_2)^2}{2 \ln(R_2/R_1)} A p_r. \end{aligned} \tag{3.7}$$

Consistent with the thin-film assumption of lubrication theory, the radial component of the velocity is given by

$$u_r = \frac{1}{2\mu} \frac{dp}{dr} y(y - h), \tag{3.8}$$

so that the flow rate out of the bearing can be calculated from the formula

$$Q = \int_0^h 2\pi r u_r \, dy = \frac{\pi}{6 \ln(R_2/R_1)} \frac{h^3 p_r}{\mu}. \tag{3.9}$$

The pumping power H_p required to pressurize the lubricant to the supply pressure p_s is the product of supply pressure and flow rate:

$$H_p = p_s Q,$$

or, when substituting from Eq. (3.9),

$$H_p = \frac{\pi}{6 \ln(R_2/R_1)} \frac{h^3 p_s p_r}{\mu}. \tag{3.10}$$

In hydrostatic bearings the depth of the recess is much greater than the thickness of the film, and thus viscous dissipation due to shearing motion of the bearing surfaces occurs mainly over the land. There, because of symmetry, the tangential velocity distribution across the gap is approximately linear, and the uniform shear stress is $\tau = \mu r \omega / h$.[1] The power loss from shearing motion can be calculated from

$$H_f = \int_A r \omega \tau \, dA = \frac{1 - (R_1/R_2)^4}{2} \frac{\mu U_M^2 A}{h}. \tag{3.11}$$

Here U_M is the maximum tangential velocity of the runner over the land of the stationary bearing pad.

---

[1] In the classical theory of hydrostatic bearings, the relative velocity, U_0, has no effect on the flow in the direction orthogonal to U_0. Coupling between orthogonal directions is achieved by the nonlinear terms of the equation of motion, and these are neglected in the present analysis.

Equations (3.7)–(3.11) show that, for circular step bearings, the dimensionless performance factors are

$$a_f = \frac{1 - (R_1/R_2)^2}{2\ln(R_2/R_1)}, \tag{3.12a}$$

$$q_f = \frac{\pi}{6\ln(R_2/R_1)}, \tag{3.12b}$$

$$h_f = \frac{1 - (R_1/R_2)^4}{2}. \tag{3.12c}$$

The total power loss in the bearing (excluding line losses and power loss encountered in flow restrictors) is given by

$$H_T = H_p + H_f = q_f \frac{h^3}{\mu} p_r p_s + h_f \frac{\mu U_M^2 A}{h}. \tag{3.13}$$

The dependence of H_T on the film thickness and the viscosity is shown in Figure 3.1 for a certain hydrostatic bearing.

Optimization

The curves of Figure 3.1 suggest the existence of optimum values h_{opt}, μ_{opt} of the film thickness and the viscosity, respectively. If they exist, these optimum values are given by the conditions

$$\frac{\partial H_T}{\partial h} = 0, \tag{3.14a}$$

$$\frac{\partial H_T}{\partial \mu} = 0. \tag{3.14b}$$

Substitution of H_T from Eq. (3.13) into Eqs. (3.14a) and (3.14b) yields the optimum film thickness at constant viscosity and the optimum viscosity at constant film thickness, respectively.[2]

$$h_{\text{opt}} = \left(\frac{h_f \mu^2 U_M^2 A}{3q_f p_r p_s}\right)^{1/4}, \tag{3.15a}$$

$$\mu_{\text{opt}} = \left(\frac{q_f h^4 p_r p_s}{h_f U_M^2 A}\right)^{1/2}. \tag{3.15b}$$

When substituting h_{opt} into Eqs. (3.3c) and (3.3d), we find that

$$\frac{H_f}{H_p} = 3 \tag{3.16a}$$

and Eq. (3.13) yields

$$H_{T,h_{\text{opt}}} = \frac{4}{3^{3/4}}\left(q_f h_f^3 p_r p_s \mu^2 U_M^6 A^3\right)^{1/4}. \tag{3.16b}$$

---

[2] Note that the set of simultaneous Eqs. (3.14) has only the trivial solution $\mu = 0 = h$, and thus we are compelled to optimize the functions $H_T^{(h)} \equiv H_T(h, \mu)|_{\mu=\text{const.}}$ and $H_T^{(\mu)} \equiv H_T(h, \mu)|_{h=\text{const.}}$ separately.

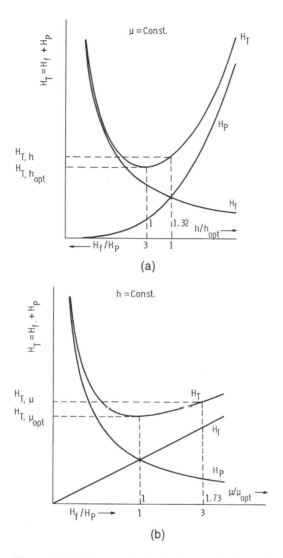

Figure 3.1. Total power loss for hydrostatic bearing as function of (a) the dimensionless film thickness and (b) the dimensionless viscosity.

If, on the other hand, one employs the optimum value of the viscosity in Eqs. (3.3c) and (3.3d), one is led to the conditions

$$\frac{H_f}{H_p} = 1, \tag{3.17a}$$

$$H_{T,\mu_{opt}} = 2\sqrt{q_f h_f A p_r p_s} \, hU_M. \tag{3.17b}$$

Condition (3.16a) states that for a given value of the viscosity, the total power loss has a minimum when the film thickness is chosen such that the frictional power loss equals three times the pumping power loss. Condition (3.17a) asserts that the total power loss is a minimum if, for a given constant film thickness, the viscosity is selected so as to yield equal values of the frictional power loss and the pumping power loss.

Obviously conditions (3.16a) and (3.17a) cannot be satisfied simultaneously (see footnote 2). We then wish to know the maximum variation of H_T when the ratio H_f/H_p is in the interval $1 \leq H_f/H_p \leq 3$. Calculating the total power loss at constant viscosity, and under the requirement $H_p = H_f$, we obtain

$$H_{T,h} = 2\left(q_f h_f^3 p_r p_s \mu^2 U_M^6 A^3\right)^{1/4}. \tag{3.18}$$

We may also calculate the total power loss at constant film thickness for the condition $H_f = 3H_p$, obtaining

$$H_{T,\mu} = \frac{4}{\sqrt{3}} \sqrt{q_f h_f A p_r p_s}\, h U_M. \tag{3.19}$$

Comparison of Eqs. (3.16b) and (3.18) and of Eqs. (3.17b) and (3.19) shows that

$$\left.\begin{array}{c} 1 \leq \dfrac{H_{T,h}}{H_{T,h_{\mathrm{opt}}}} \leq 1.1398 \\[2em] 1.1547 \geq \dfrac{H_{T,\mu}}{H_{T,\mu_{\mathrm{opt}}}} \geq 1 \end{array}\right\} \quad \text{for } 3 \geq \dfrac{H_f}{H_p} \geq 1. \tag{3.20}$$

We have just demonstrated that a plane hydrostatic bearing, irrespective of its geometry, will operate at less than 16% above minimum total power as long as the ratio H_f/H_p is held between the values 1 and 3 (Figure 3.1). It would therefore seem to matter little what value of H_f/H_p we design for within this range. This conclusion is misleading, however, and we do well to design for $H_f/H_p = 1$, for reasons indicated below.

Assuming that all the generated heat is spent on increasing the temperature of the lubricant, the lubricant temperature rise ΔT is given by

$$c\rho Q \Delta T = H_T,$$

so that

$$\Delta T = \frac{1}{\rho c}\left(\frac{h_f}{q_f}\frac{\mu^2 U_M^2 A}{h^4 p_r} + p_s\right). \tag{3.21}$$

Let us assume for simplicity that $p_r = kp_s$, $k = \text{const.}$ then ΔT may easily be optimized with respect to p_s. From the condition

$$\frac{\partial \Delta T}{\partial p_s} = 0$$

we derive the optimum value (in terms of temperature rise) of the supply pressure

$$p_{s,\mathrm{opt}} = \sqrt{\frac{h_f}{q_f}\frac{A}{k}}\, \mu \frac{U_M}{h^2}. \tag{3.22}$$

With $p_s = p_{s,\mathrm{opt}}$ Eqs. (3.3c) and (3.3d) yield Eq. (3.17a) once more, and Eq. (3.21) reduces to

$$\Delta T_{p_s,\mathrm{opt}} = \frac{2}{\rho c} p_s. \tag{3.23}$$

As one might expect, from the fact that Eq. (3.17a) was recovered on optimizing ΔT with respect to p_s, substitution of μ_{opt} from Eq. (3.15b) into Eq. (3.21) yields the result already given by Eq. (3.23).

Operation with Flow Restrictors

To successfully support asymmetric loads, multipad bearings with built-in pressure regulators must be used (see Section 1.6). Regulation of pad pressure is accomplished by use of a flow restrictor between the pressure source and the pad. The most common forms of the control devices (see Chapter 1) used in externally pressurized bearings are (1) viscous restrictors (capillary), (2) turbulent restrictors (orifice), and (3) constant flow devices (valve, pump).

The oil film stiffness of the bearing depends on the control mechanism. When calculating bearing stiffness, the supply system and the type of bearing have to be considered as forming a system. The analysis of such a lubrication system will be shown here for a plane bearing with capillary restrictor (Opitz, 1968).

The resistance to flow over the land is $R_B = \mu/q_f h^3$ from Eq. (3.3c), whereas the resistance of the capillary $R_C = 128\mu\ell^2/\pi d^4$ is given by the Hagen-Poiseuille law. Here ℓ is the length of the capillary and d is its diameter.

From Eq. (3.3a) the arbitrary load W at h is

$$W = a_f A p_r$$
$$= a_f A p_s \frac{R_B}{R_C + R_B},$$
(3.24)

and for the reference load W_0 at h_0 we have

$$W_0 = a_f A p_s \frac{R_{B_o}}{R_C + R_{B_o}}.$$
(3.25)

The ratio of loads W/W_0 can now be calculated:

$$\frac{W}{W_0} = \frac{1+\xi}{1+\xi X^3},$$
(3.26)

where

$$\xi = \frac{R_C}{R_{B_o}} \quad \text{and} \quad X = \frac{h}{h_0}.$$

The numerical value of ξ is thus equal to the ratio of capillary resistance to the resistance over the land for the static load W_0. The ratio of supply pressure to recess pressure is given, under the same condition, by $p_s/p_r = \xi + 1$.

We find the dimensionless bearing stiffness by differentiating Eq. (3.26) with respect to h:

$$\lambda \equiv -\frac{\partial(W/W_0)}{\partial(h/h_0)} = \frac{3\xi(1+\xi)X^2}{(1+\xi X^3)^2}.$$
(3.27)

The ratio X is plotted against W/W_0 in Figure 3.2. Figure 3.3 shows the variation of $\lambda/(1+\xi)$ with X. Inherent control by shallow parallel or tapered recesses is a more recent development (Rowe and O'Donoghue, 1971).

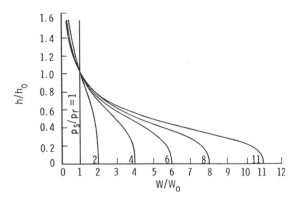

Figure 3.2. Operation with capillary restrictors: film thickness versus load for constant ratios of the supply pressure to the recess pressure.

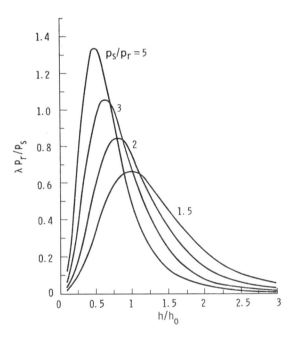

Figure 3.3. Operation with capillary restrictor: dimensionless bearing stiffness versus film thickness.

The feature of shallow recess control is that inlet pressure remains constant while pad co-efficients vary. No external control devices are required, and therefore inherently controlled bearings are very compact and simple.

3.2   Hydrodynamic Bearings

Support of rotating shafts is one of the most common applications of hydrodynamic bearings. The load in such applications is either perpendicular to the axis of rotation or coincident with it. In the former case we speak of a radial load, and the shaft is supported by a journal bearing. Thrust bearings are employed when the load is axial. Although the previously derived Reynolds equation of lubrication is applicable to bearings of either type, these two basic types of hydrodynamic bearings will be discussed under separate headings in this section.

Journal Bearings

In their simplest form, a journal and its bearing consist of two eccentric, rigid, cylinders. The outer cylinder (bearing) is usually held stationary while the inner cylinder (journal) is made to rotate at an angular velocity ω. In addition to this rigid body rotation the journal may also acquire a velocity of translation. The components of the translational velocity are $\dot{e}$ and $e(\dot{\psi} + \dot{\phi})$ measured along the line of centers $\overline{O_B O_J}$ and perpendicular to it, respectively, as depicted in Figure 3.4.

Because the journal is eccentric with the bearing, the clearance gap between the cylindrical surfaces is not uniform along the circumference of the bearing. When the width of this gap is measured along a bearing radius, it is referred to as the film thickness and is customarily given the symbol h.

We will now establish the dependence of h on the angular coordinate θ. Let P be an arbitrary point of the bearing surface. The angular position of P relative to the line of centers $\overline{O_B O_J}$ is characterized by θ, as illustrated in Figure 3.4.

From the triangle $O_J Q O_B$ we may write

$$\frac{\overline{QO_B}}{\sin \gamma} = \frac{R_J}{\sin \theta} = \frac{e}{\sin \delta},$$

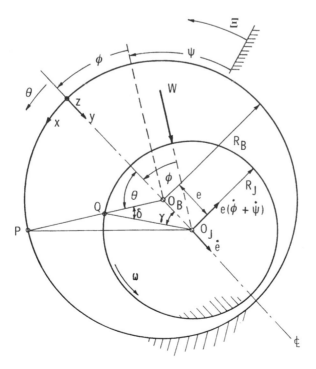

Figure 3.4. Journal bearing geometry and nomenclature.

and as

$$\gamma = \theta - \delta$$

$$= \theta - \arcsin\left(\frac{e}{R_J}\sin\theta\right),$$

the distance $\overline{QO_B}$ is given by

$$\overline{QO_B} = \frac{R_J}{\sin\theta}\sin\left[\theta - \arcsin\left(\frac{e}{R_J}\sin\theta\right)\right]$$

$$= \sqrt{R_J^2 - e^2\sin^2\theta} - e\cos\theta. \tag{3.28}$$

The film thickness, h, at the arbitrary position P of angular coordinate θ is given by

$$h = R_B - \overline{QO_B}$$

$$= C + R_J + e\cos\theta - R_J\sqrt{1 - \left(\frac{e}{R_J}\right)^2\sin^2\theta}.$$

Here we put $R_B = R_J + C$, where C is the radial clearance.

The expression under the square root sign may be expanded in a binomial series as $e/R_J \ll 1$:

$$\sqrt{1 - \left(\frac{e}{R_J}\right)^2\sin^2\theta} = 1 - \frac{1}{2}\left(\frac{e}{R_J}\sin\theta\right)^2 - \frac{1}{8}\left(\frac{e}{R_J}\sin\theta\right)^4 - \cdots.$$

Therefore, to order (e/R_J) the approximate expression for film thickness is

$$h = C + e\cos\theta$$

$$= C(1 + \varepsilon\cos\theta). \tag{3.29}$$

Here $\varepsilon = e/C, 0 \le \varepsilon \le 1$ is the bearing eccentricity ratio. Typically, in liquid-lubricated journal bearings $C/R_J \approx 0.002$, and we find Eq. (3.29) to be of sufficient accuracy in most practical cases (Dai, Dong, and Szeri, 1992).

Having established the dependence of the film thickness on the annular coordinate, we are in a position to evaluate the right-hand side of Eq. (2.72). The angular coordinate θ is related to a fixed direction, Ξ in Figure 3.5, through the attitude angle ϕ and the instantaneous load direction ψ. This permits us to rewrite Eq. (3.29) as

$$h = C + e\cos[\Xi - (\phi + \psi)]. \tag{3.30}$$

Having this new expression for film thickness greatly facilitates calculation of the squeeze film term in Eq. (2.74)

$$V_0 = \frac{\partial h}{\partial t}$$

$$= \frac{de}{dt}\cos[\Xi - (\psi + \phi)] + e\frac{d(\psi + \phi)}{dt}\sin[\Xi - (\psi + \phi)] \tag{3.31}$$

$$= \dot{e}\cos\theta + e(\dot{\phi} + \omega_W)\sin\theta.$$

Here $\omega_W = d\psi/dt$ is the frequency of rotation of the applied load, a dot above a quantity signifies its time rate of change, and we assume that the bearing is stationary.

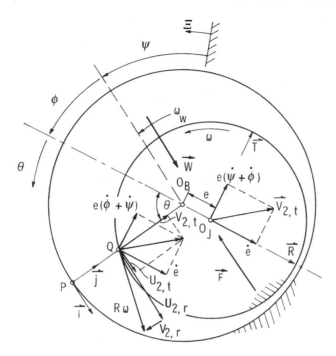

Figure 3.5. Journal velocities and journal bearing nomenclature.

Let $(\boldsymbol{R}, \boldsymbol{T})$ represent a Cartesian coordinate system centered at O_J with unit vectors $\boldsymbol{R}$ and $\boldsymbol{T}$ parallel and perpendicular, respectively, to the line of centers, as illustrated in Figure 3.5. The matrix of transformation from the $(\boldsymbol{R}, \boldsymbol{T})$ coordinate system to the (x, y) coordinate system, which is centered at P and has unit vectors $\boldsymbol{i}$ and $\boldsymbol{j}$, is given by

$$(M) = \begin{pmatrix} \sin\theta & -\cos\theta \\ \cos\theta & \sin\theta \end{pmatrix}.$$

Thus we have

$$\begin{pmatrix} U_{2,t} \\ V_{2,t} \end{pmatrix} = (M) \begin{pmatrix} \dot{e} \\ e(\dot{\phi} + \dot{\psi}) \end{pmatrix},$$

and the first two terms on the right-hand side of Eq. (2.74) simplify as follows (Szeri, 1980):

$$6U_0 \frac{\partial h}{\partial x} + 6h \frac{\partial U_0}{\partial x} = 6 \frac{\partial}{\partial x}\left[hU_{2,r}\left(1 + \frac{U_{2,t}}{U_{2,r}}\right)\right]$$

$$= 6\frac{\partial}{\partial x}\left\{ hR\omega\left[1 + \frac{C}{R}\left(\frac{\dot{\varepsilon}}{\omega}\sin\theta - \varepsilon\frac{\dot{\phi}+\dot{\psi}}{\omega}\cos\theta\right)\right]\right\} \quad (3.32)$$

$$\approx 6R\omega \frac{\partial h}{\partial x}.$$

Approximation (3.32) is good to order C/R, provided that $\dot{\varepsilon}$, $\dot{\phi}$, and $\dot{\psi}$ are all of order ω or smaller.

The sum of Eqs. (3.31) and (3.32) yields the right-hand side of Eq. (2.74) for a stationary bearing

$$\frac{\partial}{\partial x}\left(\frac{h^3}{\mu}\frac{\partial p}{\partial x}\right) + \frac{\partial}{\partial z}\left(\frac{h^3}{\mu}\frac{\partial p}{\partial z}\right) = 6R\omega\frac{\partial h}{\partial x} + 12[\dot{e}\cos\theta + e(\dot{\phi} + \omega_W)\sin\theta]. \tag{3.33}$$

In general both $\dot{e}$ and $\dot{\phi}$ are different from zero, for even when subject to an external load that is constant[3] in both magnitude and direction, the journal center orbits around its *static equilibrium position*. For a well-designed bearing the amplitudes of such orbits are exceedingly small, so that motion of the journal center will not need to be considered when calculating steady-state performance. However, when investigating stability of the orbiting motion of the journal, or when calculating journal response to dynamic loading, $\dot{\varepsilon}$, $\dot{\phi}$, and ω_W must be taken into account. We will consider only steady-state performance of journal bearings in this chapter, so unless otherwise stated $\dot{\varepsilon} = \dot{\phi} = \omega_W = 0$.

It will be to our advantage to bring Eq. (3.33) into a nondimensional form. This may be achieved by the following transformation:

$$x = R\theta, \qquad z = \frac{L}{2}\bar{z}$$
$$h = CH = C(1 + \varepsilon\cos\theta), \qquad p = \mu N\left(\frac{R}{C}\right)^2 \bar{p}. \tag{3.34}$$

Here θ, $\bar{z}$, H, and $\bar{p}$ are the dimensionless circumferential coordinate, the dimensionless axial coordinate, the dimensionless film thickness, and the dimensionless pressure, respectively. We also make the assumption $\mu = $ const., a condition that is rarely, if at all, attained in practice.

The nondimensional pressure equation is

$$\frac{\partial}{\partial\theta}\left(H^3\frac{\partial\bar{p}}{\partial\theta}\right) + \left(\frac{D}{L}\right)^2\frac{\partial}{\partial\bar{z}}\left(H^3\frac{\partial\bar{p}}{\partial\bar{z}}\right) = 12\pi\frac{\partial H}{\partial\theta}. \tag{3.35}$$

Equation (3.35) is usually solved on the computer, using finite difference (Raimondi and Boyd, 1958) or finite-element (Reddi, 1970) methods. Approximate analytical solutions of the full equation are also possible, although these tend to be somewhat complicated (Szeri and Powers, 1967; Safar and Szeri, 1972).

There are two approximations to Eq. (3.35) that have closed-form analytical solutions. Before investigating these, we interpret Eq. (3.35) as a condition for flow continuity. The terms

$$\frac{\partial}{\partial\theta}\left(H^3\frac{\partial\bar{p}}{\partial\theta}\right), \qquad \left(\frac{D}{L}\right)^2\frac{\partial}{\partial\bar{z}}\left(H^3\frac{\partial\bar{p}}{\partial\bar{z}}\right), \qquad 12\pi\frac{\partial H}{\partial\theta}$$

represent the dimensionless rates of change at a point, each in its own direction, of the circumferential pressure flow, the axial pressure flow, and the shear flow, respectively. In a finite bearing these three quantities are all of the same order of magnitude.

If the bearing is "infinitely" long, there is no pressure relief in the axial direction and we have $\partial\bar{p}/\partial\bar{z} = 0$. Axial flow is therefore absent, and changes in shear flow must be balanced

---

[3] Even minute changes that might occur in a nominally constant load, e.g., when the mass center of the shaft does not coincide with its geometric center due to shaft deflection or to manufacturing inaccuracies, will perturb the equilibrium position. So will fluctuations in temperature or shaft speed.

by changes in circumferential pressure flow alone. This condition will also apply in first approximation to finite bearings, leading to the so-called *long-bearing* theory (Reynolds, 1886), if the aspect ratio $L/D > 2$. When this condition is satisfied[4] we are permitted to approximate Eq. (3.35) by

$$\frac{d}{d\theta}\left(H^3 \frac{d\bar{p}}{d\theta}\right) = 12\pi \frac{dH}{d\theta}. \tag{3.36}$$

If, on the other hand, a finite bearing is made progressively shorter while operating at the same speed and the same eccentricity ratio, it will generate lower and lower pressures because of the progressively greater pressure relief in the axial direction. This leads to decreased circumferential pressure flow, while the axial pressure flow will have increased. In such cases we may write Eq. (3.35) in the approximate form

$$\frac{\partial}{\partial \bar{z}}\left(H^3 \frac{\partial \bar{p}}{\partial \bar{z}}\right) = 12\pi \left(\frac{L}{D}\right)^2 \frac{\partial H}{\partial \theta}. \tag{3.37}$$

The ratio L/D at which axial pressure flow first dominates circumferential pressure flow, depends on the eccentricity ratio, $L/D = 0.25$, being a safe figure under normal operating conditions. The theory that is based on Eq. (3.37) is termed the *short-bearing theory* (DuBois and Ocvirk, 1955).

Before the availability of digital computers, the short-bearing and long-bearing approximations represented the sole practical methods for obtaining solutions to bearing problems. They yield good results when applied judiciously and to the appropriate bearing geometry, but great care should be exercised in interpreting results obtained by these approximations – the long-bearing solution should be particularly suspect. These approximations still remain useful in theoretical work or when a large number of solutions of the same bearing configuration are required. Such a situation may arise, for instance, when computing nonlinear journal orbits. During such computations the oil film forces must be evaluated at each time step, leading to a considerable volume of computations.

Short-Bearing Theory

The applicable form of the Reynolds equation is given by Eq. (3.37) and the pressure boundary conditions are

$$\bar{p} = \bar{p}_a \quad \text{at} \quad \bar{z} = \pm 1. \tag{3.38}$$

Notice that it is not possible to prescribe boundary conditions at constant θ. Thus we have no way of specifying arbitrary bearing arc when using the short-bearing approximation. This represents the principal limitation of short-bearing theory.

The solution of Eq. (3.37) that satisfies the conditions specified in Eq. (3.38) is

$$\bar{p} = 6\pi \left(\frac{L}{D}\right)^2 \frac{1}{H^3} \frac{\partial H}{\partial \theta}(\bar{z}^2 - 1) + \bar{p}_a, \tag{3.39}$$

where $\bar{p}_a$ is the dimensionless ambient pressure.

It is seen from Eq. (3.39) that the gauge pressure $\bar{p} - \bar{p}_a$ is a 2π periodic function of θ. It is antisymmetric with respect to the position of minimum film thickness $\theta = \pi$.

---

[4] If the condition on the aspect ratio is not satisfied, the long-bearing theory can lead to serious errors and must be applied judiciously.

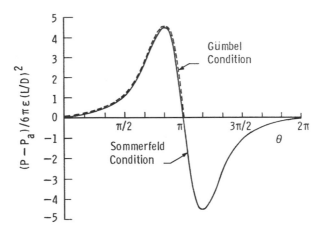

Figure 3.6. Circumferential pressure distribution according to short bearing theory, Eq. (3.39).

According to theory, therefore, below-ambient pressures of the same order of magnitude as above-ambient pressures are generated, as shown in Figure 3.6.

The center of the rotating journal will remain in a fixed position (its static equilibrium position) as long as the external load W is exactly balanced by the resultant pressure force F. Relative to the (R, T) coordinate system of Figure 3.5 we have

$$W = W \cos \phi R - W \sin \phi T, \qquad W = (W \cdot W)^{1/2}$$
$$F = F_R R + F_T T, \qquad F = (F \cdot F)^{1/2} = (F_R^2 + F_T^2)^{1/2}$$

(3.40a)

where R and T are unit vectors directed along the line of centers and perpendicular to it, respectively.

For static equilibrium of the journal

$$W + F = 0$$

or

$$W \cos \phi + F_R = 0,$$
$$-W \sin \phi + F_T = 0.$$

(3.40b)

The components F_R and F_T of the pressure force are given by

$$F_R = \int_{-L/2}^{L/2} \int_0^{R\theta_2} p \cos \theta \, dx \, dz,$$

(3.40c)

$$F_T = \int_{-L/2}^{L/2} \int_0^{R\theta_2} p \sin \theta \, dx \, dz.$$

(3.40d)

Here θ_2 represents the angular position of the trailing edge of the lubricant film, which might or might not be located at $\theta = 360°$.

For the time being we assume that the lubricant cannot withstand tension of any magnitude and that the film ruptures at $\theta_2 = \pi$, where, according to theory, film pressure exactly equals ambient (now zero) pressure. The integrations in Eq. (3.40) are then to be performed over the active bearing arc $0 < \theta < \pi$. If f_R and f_T are the nondimensional radial and tangential force components, respectively, then

$$f_R \equiv \frac{F_R/LD}{\mu N (R/C)^2}$$

$$= -2\pi \left(\frac{L}{D}\right)^2 \int_0^\pi \cos\theta \, \frac{\partial H/\partial \theta}{H^3} \, d\theta \tag{3.41a}$$

and

$$f_T \equiv \frac{F_T/LD}{\mu N (R/C)^2}$$

$$= -2\pi \left(\frac{L}{D}\right)^2 \int_0^\pi \sin\theta \, \frac{\partial H/\partial \theta}{H^3} \, d\theta. \tag{3.41b}$$

To evaluate the integrals in Eq. (3.41), we make the following substitution due to Sommerfeld (1904):

$$H = 1 + \varepsilon \cos\theta = \frac{1 - \varepsilon^2}{1 - \varepsilon \cos\psi}, \tag{3.42a}$$

so that

$$\sin\theta = \frac{\sqrt{1 - \varepsilon^2} \, \sin\psi}{1 - \varepsilon \cos\psi}, \qquad \cos\theta = \frac{\cos\psi - \varepsilon}{1 - \varepsilon \cos\psi}, \tag{3.42b}$$

and

$$d\theta = \frac{\sqrt{1 - \varepsilon^2}}{1 - \varepsilon \cos\psi} d\psi. \tag{3.42c}$$

The integrals in Eq. (3.41) can now be evaluated:

$$\int_0^\pi \cos\theta \, \frac{\partial H/\partial \theta}{H^3} \, d\theta = -\frac{\varepsilon}{(1 - \varepsilon^2)^2} \int_0^\pi \sin\psi (\cos\psi - \varepsilon) \, d\psi = \frac{2\varepsilon^2}{(1 - \varepsilon^2)^2} \tag{3.43}$$

and

$$\int_0^\pi \sin\theta \, \frac{\partial H/\partial \theta}{H^3} \, d\theta = \frac{-\varepsilon}{(1 - \varepsilon^2)^{3/2}} \int_0^\pi \sin^2\psi \, d\psi = -\frac{\pi\varepsilon}{2(1 - \varepsilon^2)^{3/2}}. \tag{3.44}$$

Integrals of the type

$$A_k^{i,j} = \int_{\beta_1}^{\beta_2} \frac{\sin^i \beta \cos^j \beta}{(1 + \varepsilon \cos\beta)^k},$$

useful in journal bearing analysis, have been listed by Gross (1962), Cameron (1966), and others.

Substitution of Eqs. (3.43) and (3.44) into Eq. (3.41) results in the dimensionless force components:

$$f_R = -\left(\frac{L}{D}\right)^2 \frac{4\pi\varepsilon^2}{(1 - \varepsilon^2)^2},$$

$$f_T = \left(\frac{L}{D}\right)^2 \frac{\pi^2\varepsilon}{(1 - \varepsilon^2)^{3/2}}. \tag{3.45}$$

The bearing Sommerfeld number defined by

$$S \equiv \frac{\mu N}{P} \left(\frac{R}{C}\right)^2 = \left(f_R^2 + f_T^2\right)^{-1/2} \tag{3.46}$$

is a dimensionless number, which is used to characterize bearing performance.[5]

For the short bearing with boundary condition $\bar{p} = 0$ at $\theta_2 = \pi$ we have

$$S\left(\frac{L}{D}\right)^2 = \frac{(1 - \varepsilon^2)^2}{\pi \varepsilon \sqrt{\pi^2(1 - \varepsilon^2) + 16\varepsilon^2}} . \tag{3.47}$$

The attitude angle, which is measured from the load line to the line of centers in the direction of journal rotation, is given by

$$\begin{aligned} \phi &= \arctan \left|\frac{f_T}{f_R}\right| \\ &= \arctan \left(\frac{\pi}{4} \frac{\sqrt{1 - \varepsilon^2}}{\varepsilon}\right). \end{aligned} \tag{3.48}$$

The journal locus as given by Eq. (3.48) is almost semicircular in shape and is often referred to as the *equilibrium semicircle*[6] (Figure 3.7).

In short-bearing theory the circumferential pressure gradient is neglected. Implicit in this is the assumption that the distribution of the circumferential velocity across the film is linear, so that the shear stress is $\bar{\tau}_{xy} = \mu U_2 / h$.

The friction force is given by

$$\begin{aligned} F_\mu &= L \int_0^{2\pi R} \tau_{xy} \, dx \\ &= \frac{2\pi^2 \mu L D N}{\sqrt{1 - \varepsilon^2}} \frac{R}{C} \end{aligned}$$

and the dimensionless friction variable by

$$\begin{aligned} c_\mu &\equiv \frac{R}{C} \frac{F_\mu}{W} \\ &= \frac{2\pi^2 S}{\sqrt{1 - \varepsilon^2}} . \end{aligned} \tag{3.49}$$

The total side flow (both sides included) is calculated from the formula

$$Q_s = 2 \int_0^{\pi R} \int_0^{h(x)} w \, dy \, dx \tag{3.50}$$

---

[5] In general, characterization of isothermal bearings requires two independent parameters, the most practical choice being the Sommerfeld number, S, and the ratio, α/β. The latter specifies the position of the load line relative to the leading edge of the bearing arc (Figure 3.9). With the present boundary conditions, $\beta = \pi$ and $\alpha = \pi - \phi$ so that $\alpha/\beta = 1 - \phi/\pi$, and thus α/β is no longer independent of S. This leaves the Sommerfeld number as the sole parameter.

[6] For a semicircle of radius 0.5 centered at $(\varepsilon = 0.5, \phi = 0)$, we have $\tan \phi = \sqrt{1 - \varepsilon^2}/\varepsilon$.

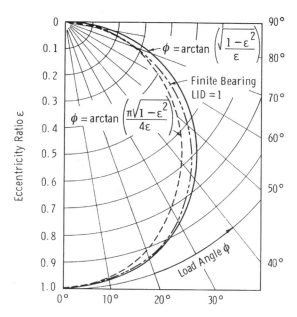

Figure 3.7. Journal loci: (−−−) short-bearing theory, Eq. (3.48); (− ·· −) finite bearing; (——) semicircle.

or in dimensionless form, after performing the indicated integration across the film with respect to y, by

$$q_s \equiv \frac{Q_s}{NRLC}$$
$$= -\frac{1}{6}\left(\frac{D}{L}\right)^2 \int_0^\pi H^3 \frac{\partial \bar{p}}{\partial \bar{z}}\bigg|_{\bar{z}=\pm 1} d\theta. \tag{3.51}$$

Substituting for $\partial \bar{p}/\partial z$, we have

$$q_s = 2\pi \varepsilon. \tag{3.52}$$

The predictions of short-bearing theory and accurate two-dimensional numerical solutions at $L/D = 1/4$ are compared in Table 3.1.

Although the length/diameter ratio of numerous industrial bearings might be small enough for their performance to be calculated on the basis of short-bearing theory, the latter theory cannot be used unless the bearing is of 180° (noncavitating film) or 360° (cavitating film) arc. To remove this constraint of the short-bearing theory, and also to increase the range of applicable L/D ratios, O'Donoghue, et al. (1970) proposed solution for pressure, see Eq. (3.39), in the form

$$\bar{p}(\theta, \bar{z}) = \bar{p}_c(\theta)(1 - \bar{z}^2). \tag{3.53}$$

Here $\bar{p}_c(\theta)$ is the center line pressure. Substitution of Eq. (3.53) into Eq. (3.35) yields the differential equation

$$\frac{d}{d\theta}\left(H^3 \frac{d\bar{p}_c}{d\theta}\right) - 2\left(\frac{D^2}{L}\right)H^3 \bar{p}_c = 12\pi \frac{\partial H}{\partial \theta}. \tag{3.54}$$

Table 3.1. *Performance prediction for a full journal bearing*
$(L/D = 1/4)$

Parameter	Short-bearing approximation, $\theta_2 = \pi$/exact solution[a]		
ε	0.1	0.6	0.9
S	15.84/16.20	1.00/1.07	0.053/0.074
ϕ	82.71/82.31	46.32/46.72	20.83/21.85
c_μ	314.2/322.1	24.67/26.73	2.40/3.50
q_s	0.628/0.621	3.77/3.72	5.65/5.59

[a]Computer solution of finite bearing obtained with the Swift-Stieber
boundary condition $\bar{p} = \partial \bar{p}/\partial\theta = 0$ at $\theta = \theta_2$.

Table 3.2. *Performance prediction for a full journal bearing*
$(L/D = 1/2)$

	Short-bearing[a] Eq. (3.37)	Modified short bearing[b] Eq. (3.54)	Finite bearing[b] Eq. (3.35)
$\varepsilon = 0.1$			
S	3.96	4.496	4.310
c_μ	78.56	88.703	85.6
$\varepsilon = 0.6$			
S	0.250	0.331	0.319
c_μ	6.169	8.399	8.10
$\varepsilon = 0.9$			
S	0.0133	0.0359	0.0313
c_μ	0.602	1.790	1.60

[a]Trailing edge boundary at $\theta = \pi$.
[b]Trailing edge boundary at $\bar{p} = \partial \bar{p}/\partial\theta = 0$.

Performance calculations based on Eq. (3.54) are displayed in Table 3.2. The boundary condition used here is the Swift-Stieber condition, $p = \partial p/\partial\theta = 0$ at $\theta = \theta_2$.

Boundary Conditions

When in pure form, liquids can withstand tensile stresses that are certainly of the order of tens or even hundreds of atmospheres (Temperly, 1975). If contaminated they will cavitate, however, when the pressure drops below the saturation pressure of the dissolved gases (gaseous cavitation). Vapor cavitation (boiling) of the liquid occurs when the pressure falls to the vapor pressure.

Under normal operating conditions a lubricant film of converging-diverging geometry is expected to cavitate within the diverging part of the clearance, where, on the assumption of a continuous lubricant film, theory predicts negative pressures. This much is clear. Still, the subjects of considerable discussion, however, are (1) the exact position of the film-cavity interface and (2) the boundary conditions that apply at that interface.

A typical pressure curve for the lubricant film of a journal bearing shows the pressure increasing from its value at inlet, which is located at say, $\theta = 0$, with the angular coordinate

θ, until it reaches a maximum somewhere still within the convergent part of the clearance space. Thereafter, the pressure decreases sharply to a small negative value (the subcavity pressure) just to rise again to the level of the cavity pressure p_{cav}. Dyer and Reason (1976) showed that if the journal eccentricity is smaller than some critical value, ε_{crit}, where ε_{crit} is inversely proportional to the bearing clearance, then a tensile stress greater than the oil vapor pressure may be developed in the film. They actually measured a tensile stress of 740 kPa in the oil film of a steadily loaded journal bearing. The cavity pressure is essentially constant and equals the saturation pressure of the lubricant. The saturation pressure, on the other hand, is equal to or is just below the ambient atmospheric pressure, as the lubricant is exposed to the ambient atmosphere for long periods of time under normal operating conditions.

The solid curve in Figure 3.6 represents the lubricant pressure as obtained under the so-called *Sommerfeld boundary condition*, which assumes the clearance space to be full of lubricant and allows for subambient pressures. This condition yields results that are physically unreasonable (e.g., shaft displacement is always at right angles to the applied load), except in special circumstances. For instance, by locating an oil groove at the position of minimum film pressure, Floberg (1961) demonstrated that a continuous, full Sommerfeld pressure curve can be maintained experimentally. The journal locus is represented by a straight line under such conditions, with a load angle of $\phi = \pi/2$. Raimondi and Boyd (1958) refer to the Sommerfeld condition as a *type I boundary condition* and find it useful for calculating bearings that operate under high ambient pressures.

The half-Sommerfeld or *Gümbel boundary condition*, although obtaining the pressure on the assumption of a continuous lubricant film, neglects the subambient pressure loop completely when calculating bearing performance. The short-bearing performance in Eqs. (3.45)–(3.52) was obtained under this condition. Although the Gümbel condition yields a pressure curve that is at variance with experimental data, it does give closed-form solution for bearing performance and is, therefore, still employed in theoretical work. A pressure profile obtained with the Gümbel condition is shown by the dashed curve in Figure 3.6.

H. W. Swift, on the basis of a stability argument, and W. Stieber, from considerations of flow continuity at the film-cavity interface, arrived at identical conditions, namely that

$$\frac{\partial \bar{p}}{\partial \theta} = 0 \qquad p = p_{cav} \tag{3.55}$$

at the cavitation boundary. The cavity pressure differs little from, and is usually taken to be equal to, atmospheric pressure. The *Swift-Stieber boundary condition*, as it is referred to in the literature, has been shown by Cameron and Wood to lead to both minimum potential energy and maximum load capacity of the bearing, and by Christopherson to yield minimum bearing friction (Cameron, 1966). The Swift-Stieber condition is unable to predict the subcavity pressures that occur just upstream from the cavitation boundary. It is nevertheless the most widely used boundary condition in numerical work. It leads to fairly good agreement with experimental data, particularly at large eccentricities, and is easy to incorporate into most numerical schemes. A typical pressure distribution obtained with the Swift-Stieber condition is shown by the dashed curve in Figure 3.8.

For a long bearing, the pressure gradient $d\bar{p}/d\theta$ can be obtained from Eq. (3.36) by integration

$$\frac{d\bar{p}}{d\theta} = \frac{12\pi}{H^2} + \frac{A'}{H^3}. \tag{3.56}$$

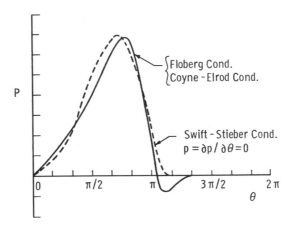

Figure 3.8. Circumferential pressure distribution according to various trailing edge boundary conditions.

Here A is an integration constant. When this expression for $d\bar{p}/d\theta$ together with $U_1 = 0$ is substituted into Eq. (2.59a), we obtain

$$\bar{u} = \frac{u}{R\omega}$$
$$= \left(3 + \frac{A''}{H}\right)(\bar{y}^2 - \bar{y}) + \bar{y}, \tag{3.57}$$

where $\bar{u}$ is the dimensionless circumferential velocity and $\bar{y} = y/h$ is the dimensionless coordinate across the film.

The constant A'' in Eq. (3.57) can be evaluated from continuity considerations, which demand that at the cavity-film interface where $H = H_{cav}$ we have

$$\int_0^l \bar{u}\,d\bar{y} = \frac{H_\infty}{H_{cav}} \equiv \hat{\alpha}. \tag{3.58}$$

Equation (3.58) assumes that the cavity-fluid interface is a straight line beneath which all the fluid flows to form a uniform layer of thickness $h_\infty = C H_\infty$. Back substitution into Eq. (3.56) yields

$$\frac{d\bar{p}}{d\theta} = \frac{12\pi}{H^2}(1 - 2\hat{\alpha}), \tag{3.59}$$

The pressure gradient at film separation would thus be completely specified by Eq. (3.59) had the film separated from the stationary surface at right angles. For this case, Coyne and Elrod (1970a, 1970b) determined the value of $\hat{\alpha}$ in terms of the group $\mu U/\sigma$, the surface tension parameter of Taylor (1964). Here σ is the surface tension of the lubricant.

The second boundary condition of Coyne and Elrod at the film-cavity interface is

$$p = -\frac{\sigma}{r} + \Delta p, \tag{3.60}$$

where r is the radius of curvature of the interface. The precise value of the transition pressure correction Δp, which is dependent on $\mu U/\sigma$, is of minor importance, particularly at small values of the surface tension parameter. Pressure profiles based on the Coyne-Elrod conditions show a subcavity pressure loop as indicated in Figure 3.8.

Smith (1975) sought to apply the Coyne-Elrod condition to bearings of finite width and found good agreement for all values of the parameter $\mu U/\sigma$ at high eccentricities. At moderate eccentricities the condition led to a contradiction, which precluded its applicability.

In contradiction to Coyne and Elrod, Floberg (1964) observed that in the cavitated region oil flow takes place in narrow strips, the quantity of lubricant adhering to the runner and passing under the cavities being negligible. Because of the low viscosity of the air or gases that occupy the cavities between the strips and because of the geometry of the strips, the pressure is essentially constant within the cavitated region and is equal to the saturation pressure of the dissolved gases. Under normal loading, subcavity pressures are negligible, and the lowest lubricant film pressure is equal to the cavitation pressure. At the end of the pressure buildup the oil flow leaving the continuous-film domain is

$$Q_{cav}^- = \frac{Uh}{2} - \frac{h^3}{12\mu}\frac{\partial p}{\partial x}, \tag{3.61}$$

while the flow entering the cavitation region is

$$Q_{cav}^+ = \tilde{\omega}\frac{Uh}{2}. \tag{3.62}$$

The symbol $\tilde{\omega}$ stands for the fractional width of the oil in the cavitation region.

From the equality $Q_{cav}^- = Q_{cav}^+$, we obtain

$$(1 - \tilde{\omega})\frac{Uh}{2} - \frac{h^3}{12\mu}\frac{\partial p}{\partial x} = 0. \tag{3.63}$$

By assumption $p \geq p_{cav}$ and $0 < \tilde{\omega} < 1$, thus Eq. (3.63) is satisfied only if

$$\frac{\partial p}{\partial x} = 0, \tag{3.64a}$$

$$\tilde{\omega} = 1. \tag{3.64b}$$

The second of these conditions means that oil will fill the whole width at the film-cavity interface.

If the average pressure in the film is low, then the subcavity pressure will have an influence on the position of the film-cavity interface and, according to Floberg (1965), should be taken into account. There is now a finite number of lubricant strips in the cavitated region. The assumption that no oil enters or leaves the gas-filled regions between the strips leads to the condition

$$\frac{\partial \bar{p}}{\partial \theta} - \frac{\partial \bar{p}}{\partial \bar{z}}\frac{\partial \theta}{\partial \bar{z}} = \frac{12\pi}{H^2}. \tag{3.65}$$

Condition (3.65) is applicable at both upstream and downstream film-cavity interfaces.

Both the Coyne-Elrod, Eqs. (3.59) and (3.60), and the Floberg, Eq. (3.65), conditions yield subcavity pressures upstream of the lubricant strip, in agreement with experiments. They are, however, difficult to implement in numerical schemes.

Savage (1977) considered the leading edge of the cavity and wrote an interface force balance in the form

$$p(c) + \frac{\sigma}{r} = 0, \tag{3.66}$$

where $p(c)$ is the fluid pressure at the cavitation boundary, σ is the surface tension of the lubricant, and r is the radius of curvature of the cavity-fluid interface. The interface $x = c$ is constantly subject to small disturbances. Let ξ be such a small disturbance, caused by fluctuation of fluid pressure, so that a point on the cavitation boundary originally at (c, y) is displaced to a new position $(c + \xi, y)$.

In its new, perturbed, position the force on the interface is given by the residue of Eq. (3.66) when the latter is written for that new position. Since $|\xi| < c$ we are permitted to write

$$F(c + \xi) = p(c + \xi) + \frac{\sigma}{r(c + \xi)}$$
$$= F(c) + F'(c)\xi + O(\xi^2) \tag{3.67}$$
$$= \frac{d}{dx}\left(p + \frac{\sigma}{r}\right)\xi + O(\xi^2).$$

The interface will return to its original position $x = c$ under the action of $F(c + \varepsilon)$, provided that F and ξ have opposite algebraic signs. Thus the criterion for the existence of a straight cavity-fluid interface is

$$\frac{d}{dx}\left(p + \frac{\sigma}{r}\right) < 0. \tag{3.68}$$

In writing Eq. (3.68) we neglected terms of order ξ^2.

Both the Sommerfeld condition and the Gümbel condition are easy to apply in analytical work but give results that are at variance with experimental data. The Swift-Stieber condition, although unable to reproduce the subcavity pressure loop, leads to acceptable results for bearing performance and is easy to implement in most numerical methods. The separation and Floberg conditions are difficult to implement in any numerical scheme and will not be considered further. The interested reader is referred to *Cavitation and Related Phenomena in Lubrication*, edited by Dowson et al. (1975). See also Dowson and Taylor (1979).

Long-Bearing Theory[7]

The pressure differential equation under the condition of vanishing axial flow, valid for long bearings, is given by Eq. (3.36)

$$\frac{d}{d\theta}\left(H^3 \frac{d\bar{p}}{d\theta}\right) = 12\pi \frac{dH}{d\theta}. \tag{3.36}$$

Integration twice with respect to θ yields

$$\bar{p}(\theta) = 12\pi \int_0^\theta \frac{H(\theta') - A}{H^3(\theta')} d\theta' + B. \tag{3.69}$$

The pressure distribution in Eq. (3.69) will be subjected to some of the simpler boundary conditions discussed previously.

Sommerfeld Condition

To determine the integration constants A and B in Eq. (3.69), we specify the Sommerfeld boundary conditions for the full (360° arc) bearing

$$\bar{p}(0) = \bar{p}_i$$
$$\bar{p}(2\pi) = \bar{p}(0). \tag{3.70}$$

---

[7] Caution is advised when employing the long-bearing theory; if the condition $(L/D) > 2$ is not satisfied, the results of the theory will be misleading.

Substituting $\bar{p}(\theta)$ from Eq. (3.69) into Eq. (3.70) we have

$$A = \frac{\displaystyle\int_0^{2\pi} (d\theta / H^2(\theta))}{\displaystyle\int_0^{2\pi} (d\theta / H^3(\theta))}, \qquad B = \bar{p}_i.$$

Using the Sommerfeld substitution, Eq. (3.42), we find that

$$\int \frac{d\theta}{H^3(\theta)} = \frac{1}{(1-\varepsilon^2)^{5/2}} \left(\psi - 2\varepsilon \sin \psi + \frac{\varepsilon^2 \psi}{2} + \frac{\varepsilon^2}{4} \sin 2\psi \right) \tag{3.71}$$

and

$$\int \frac{d\theta}{H^2(\theta)} = \frac{1}{(1-\varepsilon^2)^{3/2}} (\psi - \varepsilon \sin \psi). \tag{3.72}$$

Since the limits $\theta = 0, 2\pi$ correspond to the limits $\psi = 0, 2\pi$, we have for the Sommerfeld boundary condition

$$A = \frac{2(1-\varepsilon^2)}{2+\varepsilon^2}, \tag{3.73}$$

and the dimensionless pressure distribution is given by

$$\bar{p}(\theta) = \frac{12\pi\varepsilon \sin \psi}{(2+\varepsilon^2)(1-\varepsilon^2)^{3/2}} (2 - \varepsilon^2 - \varepsilon \cos \psi) + \bar{p}_i. \tag{3.74}$$

From Eq. (3.42) we have

$$\cos \psi = \frac{\varepsilon + \cos \theta}{1 + \varepsilon \cos \theta}, \tag{3.75a}$$

$$\sin \psi = \frac{\sqrt{1-\varepsilon^2} \sin \theta}{1 + \varepsilon \cos \theta}. \tag{3.75b}$$

Substituting Eq. (3.75) into Eq. (3.74), we obtain the pressure distribution in terms of the original variable θ:

$$\bar{p}(\theta) = \frac{12\pi\varepsilon \sin \theta (2 + \varepsilon \cos \theta)}{(2+\varepsilon^2)(1+\varepsilon \cos \theta)^2} + \bar{p}_i. \tag{3.76}$$

The function $\bar{p}(\theta)$ as given by Eq. (3.76) is 2π periodic in θ and is antisymmetric with respect to $\theta = \pi$. Thus the theory predicts negative values of the gauge pressure $\bar{p}(\theta) - \bar{p}_i$ of the same magnitude as its positive values.

The components of the oil film force are obtained when the pressure of Eq. (3.76) is substituted into Eq. (3.41). In dimensionless form we have

$$f_R = \frac{1}{2} \int_0^{\theta_2} \bar{p} \cos \theta \, d\theta, \tag{3.77a}$$

$$f_T = \frac{1}{2} \int_0^{\theta_2} \bar{p} \sin \theta \, d\theta. \tag{3.77b}$$

To evaluate Eq. (3.77) we need the following integrals:

$$\int_0^{\theta_2} \bar{p} \cos\theta \, d\theta = (\bar{p} \sin\theta)_0^{\theta_2} - \int_0^{\theta_2} \sin\theta \frac{d\bar{p}}{d\theta} d\theta$$

$$= -\int_0^{\theta_2} \sin\theta \frac{d\bar{p}}{d\theta} d\theta$$

$$= -12\pi \int_0^{\theta_2} \sin\theta \left[\frac{H(\theta) - A}{H^3(\theta)} \right] d\theta \tag{3.78}$$

$$= \frac{12\pi}{1 - \varepsilon^2} \left[\cos\psi + \frac{A}{1 - \varepsilon^2} \left(\frac{\varepsilon}{4} \cos 2\psi - \cos\psi \right) \right]_0^{\psi_2},$$

$$\int_0^{\theta_2} \bar{p} \sin\theta \, d\theta = (-\bar{p} \cos\theta)_0^{\theta_2} + \int_0^{\theta_2} \cos\theta \frac{d\bar{p}}{d\theta} d\theta$$

$$= 12\pi \int_0^{\theta_2} \cos\theta \left[\frac{H(\theta) - A}{H^3(\theta)} \right] d\theta$$

$$= \frac{12\pi}{(1 - \varepsilon^2)^{3/2}} \left\{ \sin\psi - \varepsilon\psi - \frac{A}{1 - \varepsilon^2} \left[(1 + \varepsilon^2) \sin\psi \right. \right. \tag{3.79}$$

$$\left. \left. - \varepsilon \left(\frac{3\psi}{2} + \frac{\sin 2\psi}{4} \right) \right] \right\}_0^{\psi_2}.$$

For the Sommerfeld condition we have

$$\theta_2 = \psi_2 = 2\pi \qquad A = \frac{2(1 - \varepsilon^2)}{2 + \varepsilon^2}$$

and find that

$$f_R = 0, \tag{3.80a}$$

$$f_T = \frac{12\pi^2 \varepsilon}{(2 + \varepsilon^2)(1 - \varepsilon^2)^{1/2}} = \frac{1}{S}. \tag{3.80b}$$

These equations show that under Sommerfeld condition the displacement of the journal is always at right angles to the applied load; that is, $\phi = \pi/2$. This most unsatisfactory result demonstrates the incorrectness of the Sommerfeld condition when applied to cavitating films.

Gümbel Condition
The Gümbel condition is $\theta_2 = \psi_2 = \pi$ and Eqs. (3.77) – (3.79) give

$$f_R = -\frac{12\pi\varepsilon^2}{(2 + \varepsilon^2)(1 - \varepsilon^2)}, \tag{3.81a}$$

$$f_T = \frac{6\pi^2 \varepsilon}{(2 + \varepsilon^2)(1 - \varepsilon^2)^{1/2}}. \tag{3.81b}$$

The journal locus is calculated from

$$\tan \phi = \left| \frac{f_T}{f_R} \right|.$$ (3.82a)

When substituting for the force components from Eq. (3.81), we obtain

$$\phi = \arctan \frac{\pi}{2} \frac{\sqrt{1 - \varepsilon^2}}{\varepsilon}.$$ (3.82b)

The bearing Sommerfeld number is calculated from Eq. (3.81) and has the value

$$S = \frac{(2 + \varepsilon^2)(1 - \varepsilon^2)}{6\pi \varepsilon \sqrt{4\varepsilon^2 + \pi^2(1 - \varepsilon^2)}}.$$ (3.83)

Swift-Stieber Conditions
For the long bearing, these conditions are represented by

$$\bar{p}(\theta_{\text{cav}}) = 0,$$ (3.84a)

$$\left. \frac{d\bar{p}}{d\theta} \right|_{\theta_{\text{cav}}} = 0,$$ (3.84b)

where θ_{cav} is the unknown angular position of the cavitation boundary. In general $\theta_{\text{cav}} = \theta_{\text{cav}}(\bar{z})$ when θ_{cav} is calculated from Eq. (3.84), but for the long bearing $\theta_{\text{cav}} = \text{const.}$ Substituting $\bar{p}(\theta)$ from Eq. (3.69) into Eq. (3.84b), we obtain

$$A = H(\theta_{\text{cav}}).$$ (3.85)

Back substitution into Eq. (3.69) yields the pressure distribution

$$\bar{p}(\theta) = 12\pi \int_0^\theta \frac{H(\theta) - H(\theta_{\text{cav}})}{H^3(\theta)} d\theta.$$ (3.86)

The value of θ_{cav} is as yet unknown but can be determined from the remaining boundary condition, Eq. (3.84a); thus substitution of $\bar{p}(\theta)$ yields the condition

$$\int_0^{\theta_{\text{cav}}} \frac{H(\theta) - H(\theta_{\text{cav}})}{H^3(\theta)} d\theta = 0.$$ (3.87)

When written in terms of the Sommerfeld angle ψ of Eq. (3.42a), Eq. (3.87) is equivalent to

$$\varepsilon(\sin \psi_{\text{cav}} \cos \psi_{\text{cav}} - \psi_{\text{cav}}) + 2(\sin \psi_{\text{cav}} - \psi_{\text{cav}} \cos \psi_{\text{cav}}) = 0.$$ (3.88)

Here ψ_{cav} corresponds to θ_{cav}, and we made use of

$$H(\theta_{\text{cav}}) = \frac{1 - \varepsilon^2}{1 - \varepsilon \cos \psi_{\text{cav}}}.$$ (3.89)

The values of ψ_{cav} and θ_{cav} are displayed in Table 3.3 (Szeri and Powers, 1967). The entries of Table 3.3 represent accurate solutions of Eq. (3.88).

Table 3.3. *Position of cavitation boundary in long bearings, as calculated from the Swift-Stieber condition*

ε	ψ_{cav} (rad)	θ_{cav} (rad)
0.1	4.44510	4.34974
0.2	4.39769	4.21195
0.3	4.35099	4.08021
0.4	4.30484	3.95451
0.5	4.25905	3.83438
0.6	4.21346	3.71892
0.7	4.16785	3.60645
0.8	4.12203	3.49369
0.9	4.07574	3.37195

The dimensionless force components are obtained from Eqs. (3.77), (3.86), and (3.89):

$$f_R = -\frac{3\pi\varepsilon(1 - \cos\psi_{cav})^2}{(1 - \varepsilon^2)(1 - \varepsilon\cos\psi_{cav})}, \tag{3.90a}$$

$$f_T = -\frac{6\pi(\psi_{cav}\cos\psi_{cav} - \sin\psi_{cav})}{(1 - \varepsilon^2)^{1/2}(1 - \varepsilon\cos\psi_{cav})}. \tag{3.90b}$$

When deriving Eq. (3.90), we made use of Eq. (3.87).

The bearing Sommerfeld number is obtained by substituting for f_R and f_T in Eq. (3.46):

$$\frac{1}{S} = \frac{3\pi}{(1 - \varepsilon^2)^{1/2}(1 - \varepsilon\cos\psi_{cav})}\left[\frac{\varepsilon^2(1 - \cos\psi_{cav})^4}{1 - \varepsilon^2} + 4(\psi_{cav}\cos\psi_{cav} - \sin\psi_{cav})^2\right]^{1/2}. \tag{3.91}$$

The journal-center locus is given by

$$\tan\phi = \frac{2(\sin\psi_{cav} - \psi_{cav}\cos\psi_{cav})}{(1 - \cos\psi_{cav})^2}\frac{\sqrt{1 - \varepsilon^2}}{\varepsilon}. \tag{3.92}$$

To calculate the frictional losses we observe that at $U_1 = 0$, $U_2 = U = R\omega$, Eq. (2.59a) gives

$$\begin{aligned}
\tau &= \mu\frac{\partial u}{\partial y} \\
&= \frac{1}{2}\frac{\partial p}{\partial x}(2y - h) + \frac{\mu}{h}U.
\end{aligned} \tag{3.93}$$

The shear stress on the journal is

$$\tau_0 = \frac{h}{2}\frac{\partial p}{\partial x} + R\omega\frac{\mu}{h}. \tag{3.94}$$

Table 3.4. *Long-bearing solutions*

ε	Gümble condition		Swift-Stieber condition	
	S, Eq. (3.83)	ϕ, Eq. (3.82)	S, Eq. (3.91)	ϕ, Eq. (3.92)
0.1	0.33704	86.339	0.24144	69.032
0.2	0.16736	82.596	0.12373	66.900
0.3	0.11004	78.679	0.08376	64.464
0.4	0.08053	74.472	0.06289	61.638
0.5	0.06177	69.819	0.04931	58.296
0.6	0.04795	64.477	0.03895	54.234
0.7	0.03639	58.035	0.02993	49.098
0.8	0.02549	49.675	0.02110	42.181
0.9	0.01392	37.263	0.01151	31.666

The total shear force on the journal is obtained by integrating τ_0 over the journal surface

$$
\begin{aligned}
F_\mu &= \int_{-L/2}^{L/2} \int_0^{\pi D} \left(\frac{h}{2} \frac{\partial p}{\partial x} + \frac{R \omega \mu}{h} \right) dx\, dz \\
&= \frac{1}{2} \pi \mu D L N \frac{R}{C} \int_{-1}^1 \int_0^{2\pi} \left(\frac{H}{4\pi} \frac{\partial \bar p}{\partial \theta} + \frac{1}{H} \right) d\theta\, d\bar z.
\end{aligned}
\tag{3.95}
$$

Integrating the first term by parts, we find that

$$
F_\mu = \pi \mu D L N \frac{R}{C} \left(\frac{\varepsilon f_T}{2\pi} + \frac{2\pi}{\sqrt{1 - \varepsilon^2}} \right).
\tag{3.96}
$$

The friction variable, c_μ, is obtained from Eq. (3.96) and the definition of the attitude angle ϕ:

$$
\begin{aligned}
c_\mu &\equiv \frac{R}{C} \frac{F_\mu}{W} \\
&= \frac{\varepsilon \sin \phi}{2} + \frac{2\pi^2 S}{\sqrt{1 - \varepsilon^2}}.
\end{aligned}
\tag{3.97}
$$

Of course, it would have been better to employ Eq. (3.94) only in the range $0 \le \theta \le \theta_{cav}$ and estimate the drag in the cavitated region from flow continuity considerations.

Table 3.4 contains long-bearing Sommerfeld numbers calculated by the Gümbel condition from Eq. (3.83) and by the Swift-Stieber condition from Eq. (3.91). The entries in Table 3.4 do not fully agree with the results usually quoted in the literature (Raimondi and Boyd, 1958; Cameron, 1966; Pinkus and Sternlicht, 1961), possibly because previous calculations relied on inaccurate values of θ_{cav}.

Finite Journal Bearings

The length/diameter ratio of industrial bearings is customarily in the range $0.25 < L/D < 1.5$; neither the short-bearing nor the long-bearing approximations apply to these bearings. Furthermore, the angular extent of large industrial bearings is rarely 360°; common ranges are $30° < \beta < 60°$ in pivoted-pad bearings and $120° < \beta < 160°$ in the "viscosity pump"

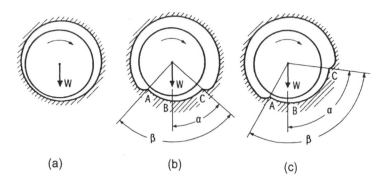

Figure 3.9. Fixed type journal bearings. (a) Full bearing, $\beta = 360°$; (b) partial bearing, centrally loaded, $\alpha = \beta/2$; and (c) partial bearing, eccentrically loaded (offset), $\alpha \neq \beta/2$. (Reprinted with permission from Raimondi, A. A. and Szeri, A. Z. Journal and thrust bearings. In Booser E. R., *CRC Handbook of Lubrication*. Copyright CRC Press, Boca Raton, Florida, © 1984.)

bearings of large rotating machinery. If the pad is centrally loaded (Figure 3.9), the ratio $\alpha/\beta = 1/2$, where the angle α is measured from the pad leading edge to the load line. For offset loading, $\alpha/\beta \neq 1/2$.

Over a finite bearing pad of diameter D, axial length L, and arc β, the lubricant pressure satisfies the equation

$$\frac{\partial}{\partial\theta}\left(H^3\frac{\partial\bar{p}}{\partial\theta}\right) + \left(\frac{D}{L}\right)^2\frac{\partial}{\partial\bar{z}}\left(H^3\frac{\partial\bar{p}}{\partial\bar{z}}\right) = 12\pi\frac{\partial H}{\partial\theta}. \qquad (3.35)$$

Solutions of Eq. (3.35) are usually sought subject to the Swift-Stieber boundary conditions

$$\bar{p} = 0 \quad \text{at} \quad \bar{z} = \pm 1, \qquad (3.98\text{a})$$

$$\bar{p} = 0 \quad \text{at} \quad \theta = \theta_1, \theta_1 + \beta, \qquad (3.98\text{b})$$

$$\bar{p} = \frac{\partial\bar{p}}{\partial\theta} \quad \text{at} \quad \theta = \theta_{\text{cav}}(\bar{z}). \qquad (3.98\text{c})$$

As the dimensionless film thickness is given by

$$\begin{aligned} H &= 1 + \varepsilon\cos\theta \\ &= 1 + \varepsilon\cos\left(\theta_1 + \frac{x}{R}\right), \end{aligned} \qquad (3.99)$$

where θ_1 is the angular coordinate of the pad leading edge $x = 0$, the pressure differential equation and its boundary conditions contain four dimensionless parameters in all:

$$\{L/D, \ \beta, \ \varepsilon, \ \theta_1\}. \qquad (3.100)$$

Two of these parameters, L/D and β, describe bearing geometry and remain fixed for a particular bearing. The other two parameters, ε and θ_1, specify the position of the rotating shaft within the bearing and are therefore dependent on loading conditions and lubricant viscosity.

Loading conditions and lubricant viscosity of an isothermal bearing can be characterized with the aid of two parameters, the Sommerfeld number S, which is the inverse of the dimensionless lubricant force Eq. (3.46), and the ratio α/β, where α is the angular position of the pad trailing edge relative to the load line (Figure 3.9). Specifying the couple (ε, θ_1) in Eqs. (3.35) and (3.98) and solving these yields a pair of values $(S, \alpha/\beta)$ from the formulas

$$S = \left(f_R^2 + f_T^2 \right)^{-1/2}, \tag{3.46}$$

$$\frac{\alpha}{\beta} = \frac{1}{\beta}[\pi - (\theta_1 + \phi)]. \tag{3.101}$$

The dimensionless force components f_R and f_T are given by

$$f_R = \frac{1}{2} \int_0^1 \int_{\theta_1}^{\theta_1+\beta} \bar{p} \cos\theta \, d\theta \, d\bar{z}, \tag{3.102a}$$

$$f_T = \frac{1}{2} \int_0^1 \int_{\theta_1}^{\theta_1+\beta} \bar{p} \sin\theta \, d\theta \, d\bar{z}, \tag{3.102b}$$

and

$$\phi = \arctan \left| \frac{f_T}{f_R} \right|. \tag{3.102c}$$

Thus, corresponding to each ordered pair (ε, θ_1) there exists another ordered pair $(S, \alpha/\beta)$; the mapping of the (ε, θ_1) plane into the $(S, \alpha/\beta)$ plane is defined by the pressure differential equation (3.35) and its boundary conditions (3.98), Eq. (3.46), and Eq. (3.101). This mapping is one-to-one and invertible. In a physical experiment, one specifies the load vector, the speed of rotation, and the lubricant viscosity – that is, the couple $(S, \alpha/\beta)$ – and permits the journal to select its own equilibrium position (ε, θ_1).

We thus have two equivalent parametric representations of journal bearing operations. The parameters in Eq. (3.100) are the natural set of parameters to use in numerical work, while the set

$$\{L/D, \ \beta, \ S, \ \alpha/\beta\} \tag{3.103}$$

is the obvious one to employ in a physical experiment.

The designer employs this latter set of parameters (3.103), and the task of the numerical analyst is to find the (ε, θ_1) couple that corresponds to the designer's $(S, \alpha/\beta)$ couple. This necessarily leads to solution of a nonlinear problem, which can be written in the symbolic form

$$S - \Omega_1(\varepsilon, \theta_1) = 0, \tag{3.104a}$$

$$\frac{\alpha}{\beta} - \Omega_2(\varepsilon, \theta_1) = 0. \tag{3.104b}$$

Here the functions Ω_1 and Ω_2 represent integrals of the pressure differential equation and the integrals involved in Eqs. (3.46) and (3.101).

The set of nonlinear equations (3.104) is conveniently solved by Newton's method. The nth iterated solution can be obtained from

$$\begin{bmatrix} \dfrac{\partial\Omega_1}{\partial\varepsilon} & \dfrac{\partial\Omega_1}{\partial\theta_1} \\[2.5ex] \dfrac{\partial\Omega_2}{\partial\varepsilon} & \dfrac{\partial\Omega_2}{\partial\theta_1} \end{bmatrix} \begin{bmatrix} \varepsilon^{(n)} - \varepsilon^{(n-1)} \\[1ex] \theta_1^{(n)} - \theta_1^{(n-1)} \end{bmatrix} = \begin{bmatrix} \Omega_1 - S \\[1ex] \Omega_2 - \dfrac{\alpha}{\beta} \end{bmatrix}, \qquad n = 1, 2, 3, \ldots. \tag{3.105}$$

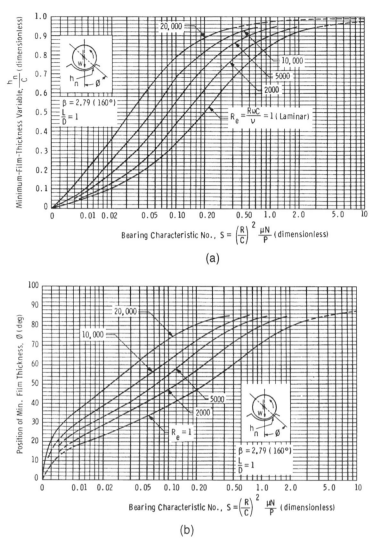

Figure 3.10. Performance curves for a centrally loaded, fixed partial journal bearing: (a) minimum film thickness, (b) attitude angle, (c) friction variable, and (d) inlet flow variable. (Reprinted with permission from Raimondi, A. A. and Szeri, A. Z. Journal and thrust bearings. In Booser, E. R., *CRC Handbook of Lubrication*. Copyright CRC Press, Boca Raton, Florida. © 1984.)

An extensive set of solutions of centrally loaded, isothermal bearings was published by Raimondi and Boyd (1958). Solutions for eccentrically loaded partial arc bearings were compiled by Pinkus and Sternlicht (1961).

The performance curves in Figures 3.10, taken from Raimondi and Szeri (1984), are for a centrally loaded fixed-pad partial bearing of $L/D = 1$, $\beta = 160°$ and various values of the Reynolds number. (Some of the curves in the figures are for turbulent flow conditions. Lubrication in the turbulent regime is discussed in Chapter 6, where we will again make reference to Figure 3.10.)

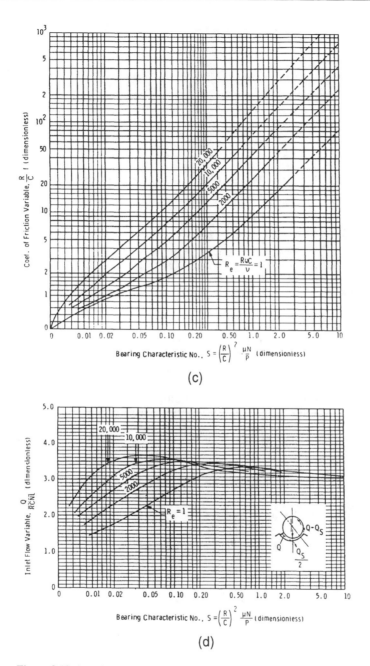

Figure 3.10 *(cont.)*

Figure 3.11 illustrates the effect of shifting the load position (i.e., varying the value of α/β, where α is measured from the pad leading edge to the load vector, and β is the pad angle) on the position of the journal center.

It can be seen from Figure 3.11 that if $\alpha/\beta > 0.665$ for $L/D = 1$ and $\beta = 120°$ the journal can be expected to rub the cap of the bearing.

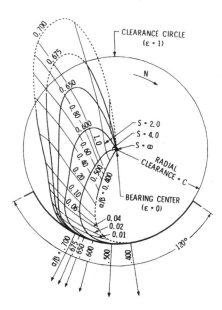

Figure 3.11 Locus of journal center for offset loads, $L/D = 1$, $\beta = 120°$. (Reprinted with permission from Raimondi, A. A. Theoretical study of the effect of offset loads on the performance of a 120° partial journal bearing. *ASLE Trans.*, **2**, 147–157, 1959.)

Pivoted-pad bearings, in which the pad is free to rotate so that it may choose its own orientation relative to the load line, are often used in practice. For these, the value of α is dictated by the pivot position, as the resultant torque on the pad must vanish. Pivoted pad journal bearings are discussed in Chapter 4, within the context of lubricant film dynamic properties.

Cavitation Algorithm

In this section we discuss an algorithm, due originally to Elrod and Adams (Elrod and Adams, 1974; Elrod, 1981) and modified subsequently by Vijayaraghavan and Keith (1989), that is designed to automatically handle cavitation in liquid-lubricated bearings.

We begin the analysis with the time dependent form of the Reynolds equation valid for compressible flow, Eq. (11.6)

$$\frac{\partial(\rho h)}{\partial t} + \frac{\partial}{\partial x}\left(\frac{\rho h U}{2} - \frac{\rho h^3}{12\mu}\frac{\partial p}{\partial x}\right) - \frac{\partial}{\partial z}\left(\frac{\rho h^3}{12\mu}\frac{\partial p}{\partial z}\right) = 0. \tag{3.106}$$

The pressure-density relationship that is required by Eq. (3.106) is given through the definition of the *bulk modulus* β

$$p = p_c + \beta \ln\left(\frac{\rho}{\rho_c}\right), \tag{3.107}$$

where p_c and ρ_c are the pressure and density, respectively, in the cavitated region.[8]

---

[8] The cavitation pressure is approximately constant and the lubricant flow in the cavitation zone is due to shear.

In the region of cavitation the principle of mass conservation leads to the equation

$$\frac{\partial(\rho h)}{\partial t} + \frac{\partial}{\partial x}\left(\frac{\rho h U}{2}\right) = 0. \tag{3.108}$$

In an effort to combine Eqs. (3.106) and (3.108) into a single equation, Elrod and Adams introduced a switching function, g, defined by

$$g = \begin{cases} 1, & \text{in full film region} \\ 0, & \text{in cavitated region.} \end{cases} \tag{3.109}$$

Introducing g from Eq. (3.109), using $p = p_c + g\beta \ln \phi$ from Eq. (3.107) and employing the notation $\rho = \rho_c \phi$, we obtain from Eq. (3.108)

$$\frac{\partial(\phi h)}{\partial t} + \frac{\partial}{\partial x}\left(\frac{U h \phi}{2} - \frac{\beta h^3}{12\mu}g\frac{\partial \phi}{\partial x}\right) - \frac{\partial}{\partial z}\left(\frac{\beta h^3}{12\mu}g\frac{\partial \phi}{\partial z}\right) = 0. \tag{3.110}$$

The variable ϕ has dual meaning: it may be interpreted as the ratio of densities in the full film region and as the volume fraction of the lubricant in the cavitated region (Vijayaraghavan and Keith, 1990a).

In the full film region where $g = 1$, Eq. (3.110) is an elliptic partial differential equation. In the cavitated region $g = 0$ and Eq. (3.110) yields

$$\frac{\partial \Phi}{\partial t} + \frac{U}{2}\frac{\partial \Phi}{\partial x} = 0 \tag{3.111}$$

in terms of the new variable $\Phi = \phi h$. Differentiating Eq. (3.111) twice, first with respect to t then with respect to x, we obtain

$$\frac{\partial^2 \Phi}{\partial t^2} + \frac{U}{2}\frac{\partial^2 \Phi}{\partial t \partial x} = 0,$$
$$\frac{\partial^2 \Phi}{\partial t \partial x} + \frac{U}{2}\frac{\partial^2 \Phi}{\partial x^2} = 0. \tag{3.112}$$

Eliminating the mixed derivatives in Eqs. (3.112), we have

$$\frac{\partial^2 \Phi}{\partial t^2} - \left(\frac{U}{2}\right)^2\frac{\partial^2 \Phi}{\partial x^2} = 0. \tag{3.113}$$

As this is a hyperbolic equation, we have a change of type of the PDE at the cavitation boundary; the numerical methods designed for solving Eq. (3.105) with $g = 1$ for full film and Eq. (3.113) for cavitating film, must take this change into consideration (Vijayaraghavan and Keith, 1991).

In the cavitated region Vijayaraghavan and Keith (1990a) recommend the use of second-order upwind-differencing, while central-differencing suffices in the full film region. They also follow Jameson (1975) by adding high-order artificial viscosity terms (Anderson et al., 1984) to the shear flow term in the cavitated region

$$\frac{\partial}{\partial x}(\phi h) \equiv \frac{\partial \Phi}{\partial x}$$

$$\approx \frac{\partial}{\partial x}\left[\Phi - (1 - g)\left(\frac{\partial^2 \Phi}{\partial x^2}\frac{\Delta x^2}{2} - \frac{\partial^3 \Phi}{\partial x^3}\frac{\Delta x^3}{8}\right)\right], \tag{3.114}$$

where Δx is the mesh spacing. With the addition of the artificial viscosity terms, the whole equation can now be centrally differenced with the consequences that (1) in the full film region, central differencing remains in effect on substituting $g = 1$, and (2) in the cavitated region, the central-differencing automatically switches to second-order upwind-differencing on account of the switch function being zero there.

For numerical work Eq. (3.110) is nondimensionalized

$$\frac{\partial(\phi H)}{\partial \bar{t}} + \frac{1}{4\pi} \frac{\partial}{\partial \bar{x}} (\phi H) = \frac{\bar{\beta}}{48\pi^2} \frac{\partial}{\partial \bar{x}} \left(H^3 g \frac{\partial \phi}{\partial \bar{x}} \right) + \frac{\bar{\beta}}{48(L/D)^2} \frac{\partial}{\partial \bar{z}} \left(H^3 g \frac{\partial \phi}{\partial \bar{z}} \right). \qquad (3.115)$$

Here we used

$$\bar{x} = x/2\pi R, \quad \bar{z} = z/L, \quad H = h/C, \quad \bar{\beta} = \frac{\beta}{\mu\omega}\left(\frac{C}{R}\right)^2, \quad \Phi = \phi H \quad \bar{t} = \omega t.$$

The finite differencing of Eq. (3.115) is explained in more detail in Vijayaraghavan and Keith, 1990b). Its final results are[9]

$$\left(\frac{\partial \bar{\Phi}}{\partial \bar{x}}\right)_i \approx \frac{1}{2\Delta \bar{x}} [g_{i+1/2}\bar{\Phi}_{i+1} + (2 - g_{i+1/2} - g_{i-1/2})\bar{\Phi}_i - (2 - g_{i-1/2})\bar{\Phi}_{i-1}]$$

$$+ \frac{1}{2\Delta x} [(1 - g_{i+1/2})\bar{\Phi}_i - (2 - g_{i+1/2} - g_{i-1/2})\bar{\Phi}_{i-1} + (1 - g_{i-1/2})\bar{\Phi}_{i-2}]$$

$$\qquad (3.116a)$$

for the shear flow term and

$$\left[\frac{\partial}{\partial \bar{x}}\left(-H^3 g \frac{\partial \phi}{\partial \bar{x}}\right)\right]_i = \left[\frac{\partial}{\partial \bar{x}}\left(-H^3 \frac{\partial g(\phi - 1)}{\partial \bar{x}}\right)\right]_i$$

$$\approx -\frac{1}{\Delta \bar{x}^2}\left[H_{i+1/2}^3 g_{i+1}(\phi_{i+1} - 1) - \left(H_{i+1/2}^3 + H_{i-1/2}^3\right)g_i(\phi_i - 1)\right.$$

$$\left. + H_{i-1/2}^3 g_{i-1}(\phi_{i-1} - 1)\right] \qquad (3.116b)$$

for the pressure term in $\bar{x}$. To obtain the $\bar{z}$ equation, replace i by j and $\bar{x}$ by $\bar{z}$ in Eq. (3.116b).

As relaxation methods require large computing times to convergence, Woods and Brewe (1989) incorporated a multigrid technique into the Elrod algorithm. Vijayaraghavan and Keith (1990b), on the other hand, advocate Newton iteration, coupled with an approximate factorization technique. The algorithm has been successfully applied to finite grooved bearings and flared misaligned bearings (Vijayaraghavan and Keith, 1990c). In context of coating flows see Gurfinkel and Patera (1997).

Effect of Surface Roughness
All previous developments were based on the highly unrealistic assumption of perfectly smooth bearing surfaces. In reality, however, engineering surfaces are covered with asperities. Even for a ground surface, asperities might reach 1.25 μm in height and ten times this value in lateral spacing; the latter distance we equate with the in-plane characteristic length

---

[9] Note that the shear flow term has been split to preserve tridiagonality of the coefficient matrix.

L_{xz}. The minimum film thickness in a journal bearing, say, of diameter $D = 25$ mm operating at eccentricity $\varepsilon = 0.5$ is $h_{\min} = 12.5$ μm; this minimum film thickness is selected here to represent the across-the-film characteristic length, L_y. Because the average asperity height is one order of magnitude smaller than the minimum film thickness, we might be tempted to ignore surface roughness altogether. However, the local characteristic lengths are of the same order of magnitude, $L_y = L_{xz} = 12.5$ μm, violating the thin film assumption of lubrication analysis, and it becomes questionable whether the Reynolds equation is at all valid. In cases when the lubrication approximation still holds even though the surfaces are rough, we are said to be dealing with *Reynolds roughness*. When there is significant pressure variation across the film due to surface roughness, to the extent that the lubrication approximation is no longer valid, Stokes equation instead of Reynolds equation must be employed; in this latter case we are dealing with *Stokes roughness* (Elrod, 1973). Just where the demarcation between these two roughness regimes lies, is not currently known. Compounding the difficulties is the fact that the asperity height distribution for most machined surfaces is random, and statistical methods must be applied when attempting to model lubrication between rough surfaces.[10]

Tzeng and Saibel (1967) were among the first to apply statistical methods to lubrication of rough surfaces. They investigated the inclined plane slider having one-dimensional roughness transverse to the direction of relative motion. This analysis was, almost immediately, extended by Christensen and Tonder (Christensen and Tonder, 1969; Tonder and Christensen, 1971), and there have been many other attempts since. The method employed by these investigators is based on statistical averaging of the Reynolds Equation. As remarked by Tripp (1983), the Reynolds equation has the property, unusual among equations of mathematical physics, that the boundary conditions are incorporated into the equation – it is this feature of the Reynolds equation that offers hope for ensemble averaging the equation itself.

Christensen and Tonder (1969) make two fundamental assumptions in their analysis: (1) the magnitude of the pressure ripples due to surface roughness is small and the variance of the pressure gradient in the roughness direction is negligible, and (2) the flow in the direction transverse to roughness direction has negligible variance.

The first step in the analysis is to average the Reynolds equation. Applying the expectation operator, E, to each of the terms of Eq. (2.68), we write

$$\frac{\partial}{\partial x} E\left(h^3 \frac{\partial p}{\partial x} \right) + \frac{\partial}{\partial z} E\left(h^3 \frac{\partial p}{\partial z} \right) = 6\mu U \frac{\partial E(h)}{\partial x}. \tag{3.117}$$

The film thickness, h, is separated into two parts $h = \bar{h} + \eta$, where $\bar{h}$ denotes the nominal, smooth, part of the film geometry while η is the variation due to surface roughness. For reasons not quite well explained, other than a reference to assumption (1), Christensen and Tonder assert that the expected value of the product of two stochastic variables equals the product of the expected values, and write

$$\frac{\partial}{\partial x} E\left(h^3 \frac{\partial p}{\partial x} \right) = \frac{\partial}{\partial x}\left[\frac{\partial \bar{p}}{\partial x} E(h^3) \right], \tag{3.118}$$

where $\bar{p} = E(p)$.

---

[10] There have been attempts to represent surface roughness by Fourier series. These efforts, however, should be considered as investigating waviness rather than roughness.

The flow in the z direction, assumed of having zero variance by assumption (2) above, can be written as

$$h^3 \frac{\partial p}{\partial z} = q_z.$$

On averaging, we have

$$\frac{\partial \bar{p}}{\partial z} = q_z E\left(\frac{1}{h^3}\right), \qquad q_z = \frac{1}{E(h^{-3})} \frac{\partial \bar{p}}{\partial z}. \tag{3.119}$$

By combining Eqs. (3.118) and (3.119), we obtain the ensemble averaged Reynolds equation

$$\frac{\partial}{\partial x}\left[\frac{\partial \bar{p}}{\partial x} E(h^3)\right] + \frac{\partial}{\partial z}\left[\frac{\partial \bar{p}}{\partial z} \frac{1}{E(h^{-3})}\right] = 6\mu U \frac{\partial}{\partial x} E(h), \tag{3.120a}$$

valid for a bearing with longitudinal, one-dimensional roughness (Christensen and Tonder, 1969). For transverse, one-dimensional roughness, Tonder and Christensen (1972) find

$$\frac{\partial}{\partial x}\left[\frac{\partial \bar{p}}{\partial x} \frac{1}{E(h^{-3})}\right] + \frac{\partial}{\partial z}\left[\frac{\partial \bar{p}}{\partial z} E(h^3)\right] = 6\mu U \frac{\partial}{\partial x}\left[\frac{E(h^{-2})}{E(h^{-3})}\right]. \tag{3.120b}$$

They have also attempted to extend these result to three-dimensional roughness, but only with limited success.

In 1978, Patir and Cheng (1978) published an analysis of three-dimensional surface roughness, resulting in flow factors that are to be included in the Reynolds equation. Their idea was to (1) characterize the surface by its autocorrelation function (see Peklenik, 1968), (2) computer-generate sets of rough surfaces that have the specified autocorrelation, (3) solve Reynolds equation for microbearings constructed with the computer-generated surfaces, (4) calculate flow factors for use with a Reynolds equation that employs the nominal film thickness and accounts for the surface roughness of the microbearings, and (5) average a large number of flow factors to obtain statistical representation.

For given Gaussian surface roughness distribution we can calculate the autocorrelation in a straightforward manner. If $\eta_{i,j}$ denotes the asperity amplitude at the nodal position $x_i = i\Delta x, z_j = j\Delta z$, the discrete autocorrelation function is defined by the $n \times m$ matrix

$$R_{p,q} = R(p\Delta x, q\Delta z) = E(\eta_{i,j}\eta_{i+p,j+q}) \qquad \begin{matrix} p = 0, 1, 2, \ldots, n-1 \\ q = 0, 1, 2, \ldots, m-1. \end{matrix} \tag{3.121}$$

Our goal is to invert Eq. (3.121), i.e., to generate an $N \times M$ matrix of roughness amplitudes having Gaussian height distribution that possesses a specified autocorrelation (Patir, 1978). To this end, using random number generator, we construct an $(N+n) \times (M+m)$ matrix $(\kappa_{i,j})$ whose elements are independent, identically distributed, Gaussian random numbers. The roughness heights are then obtained as the sum

$$\eta_{i,j} = \sum_{k=1}^{n}\sum_{\ell=1}^{m} a_{k\ell}\kappa_{i+k,j+l}, \qquad \begin{matrix} i = 1, 2, \ldots, N \\ j = 1, 2, \ldots, M \end{matrix} \tag{3.122}$$

where the $a_{k\ell}$ are the coefficients yet to be determined, so as to give the desired correletion matrix.

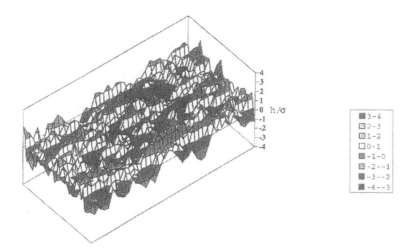

Figure 3.12 Computer-generated surface with isotropic roughness and specified autocorrelation function.

To find the coefficients a_{kl}, substitute Eq. (3.122) into Eq. (3.121), taking into account that the κ_{ij} are uncorrelated and have unit variance, i.e.,

$$E(\kappa_{ij}\kappa_{kl}) = \begin{cases} 1 & \text{if } i = k,\, j = l \\ 0 & \text{otherwise} \end{cases} \tag{3.123}$$

and obtain

$$R_{p,q} = \sum_{k=1}^{n-p}\sum_{l=1}^{m-q} a_{k,l}\, a_{k+p,l+q} \qquad \begin{array}{l} p = 0, 1, \ldots, n-1 \\ q = 0, 1, \ldots, m-1. \end{array} \tag{3.124}$$

This represents nm simultaneous, deterministic equations from which to calculate the set of nm coefficients a_{kl}. Once these coefficients are known, substitution into Eq. (3.122) yields the required asperity height distribution (Patir, 1978). Hu and Tonder (1992) further improved on Patir's scheme for generating rough surfaces of specified statistical properties by using Fast Fourier Transform (FFT) methods to commute between spectral and physical space.

It is now possible, either with the help of Eq. (3.122) or by utilizing Hu and Tonder's scheme, to generate surfaces with specified autocorrelation. Figure 3.12 shows such a computer-generated surface.

It is also possible to construct microbearings with, and to fit-finite-difference mesh to, the computer generated surfaces. Solution of Reynolds equation will then proceed in two phases. To calculate the pressure flow factors ϕ_x and ϕ_z, the microbearing is subjected to mean pressure gradients $(p_1 - p_2)/(x_2 - x_1)$ and $(p_1 - p_2)/(z_2 - z_1)$, respectively, where $(x_2 - x_1)(z_2 - z_1)$ is the dimension of the microbearing. The pressure flow factor ϕ_x, for example, is calculated from

$$\phi_x = E\left(h\frac{\partial p}{\partial x}\right) \Bigg/ \left[\bar{h}^3 \frac{(p_2 - p_1)}{(x_2 - x_1)}\right]$$

To calculate the shear flow factor ϕ_s, sliding is specified and the mean pressure gradient eliminated (Patir and Cheng, 1979).

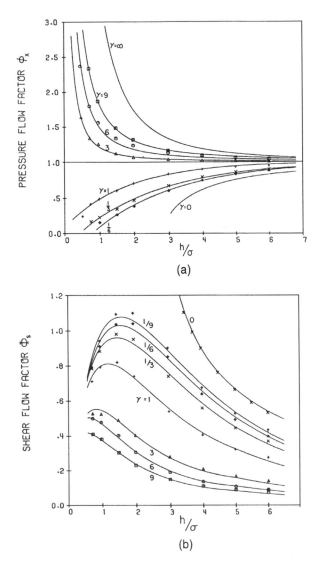

Figure 3.13 Flow factors for rough surfaces: (a) pressure flow factor, (b) shear flow factor; $\gamma = \lambda_{0.5x}/\lambda_{0.5z}$ (Eq. 1.3). (Reprinted with permission from Patir, N. and Cheng, H. S. An average flow model for determining effects of three-dimensional roughness on partial hydrodynamic lubrication. *ASME Journal of Lubrication Technology*, 100, 12–17, 1978; Patir, N. and Cheng, H. S. Application of average flow model to lubrication between rough sliding surfaces. *ASME Journal of Lubrication Technology*, 101, 220–230, 1979.)

In the next step, the flow factors are statistical averaged, i.e., they are calculated for a large number of microbearings and averaged. Reynolds equation now takes the form

$$\frac{\partial}{\partial x}\left(\phi_x \frac{h^3}{12\mu}\frac{\partial \bar{p}}{\partial x}\right) + \frac{\partial}{\partial z}\left(\phi_z \frac{h^3}{12\mu}\frac{\partial \bar{p}}{\partial z}\right) = \frac{U_1 + U_2}{2}\frac{\partial \bar{h}}{\partial x} + \frac{U_1 - U_2}{2}\sigma\frac{\partial \phi_s}{\partial x} + \frac{\partial \bar{h}}{\partial t}. \quad (3.125)$$

Part (a) of Figure 3.13 plots the pressure flow factor ϕ_x, and part (b) plots the shear flow factor ϕ_s, versus h/σ, where $\sigma^2 = R_{0,0}$ [see Eq. (1.9)]. Contact of opposing asperities is first made when $h/\sigma \approx 3$; for values less than this the lubrication approximation no longer holds (Stokes roughness regime) and the flow factors become increasingly inaccurate. In any case, the flow factors are very sensitive to the numerical scheme employed (Peeken et al. 1997).

The theory has been extended to EHL (Ai and Zheng, 1989; Chang and Webster, 1991). Others to discuss lubrication between rough surfaces were Elrod (1973, 1979), who used a two variable expansion procedure, Tripp (1983), Mitsuya and Fukui (1986), Bhushan and Tonder (1989), and Ai and Cheng (1994, 1996). The last two of these papers treats the prescribed surface profile as a deterministic function and employs the multigrid method (Lubrecht, ten Napel, and Bosma, 1986) for solving the Reynolds equation.

When the asperities touch, the lubrication approximation ceases to be valid (Sun and Chen, 1977) and one has to solve Stokes equations. To date, there has been little work done on Stokes roughness.

Thrust Bearings

Thrust bearings in their simplest form consist of two inclined plane surfaces that slide relative to one another. The geometry of the bearing surface is commonly rectangular or sector shaped, but other geometries are possible.

Plane Slider

The schematics of the fixed plane slider, the prototype of thrust bearings, is shown in Figure 3.14(a). Let B represent the dimension of the slider in the direction of relative motion; then the gradient m of the surfaces is calculated from

$$m = \frac{h_2 - h_1}{B}. \tag{3.126}$$

Here h_1 and h_2 represent the minimum value and the maximum value, respectively, of the film thickness.

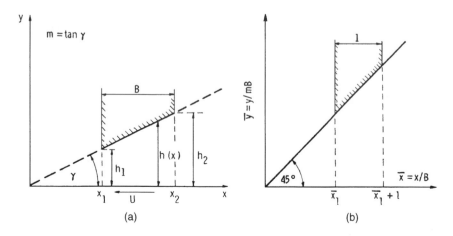

Figure 3.14 (a) The plane slider and (b) its nondimensional representation.

The applicable form of the Reynolds equation is given in Eq. (2.68). To render this equation dimensionless, we make the following transformation:

$$x = B\bar{x} \qquad y = mB\bar{y} \qquad z = \frac{L}{2}\bar{z} \qquad p = \frac{6\mu U}{Bm^2}\bar{p}. \tag{3.127}$$

On the nondimensional $\bar{x}, \bar{y}$ plane of Figure 3.14, part (b), the $\bar{x}$ axis represents the runner surface of all plane sliders. The bearing surface projects onto the 45° line through the origin. In this representation the dimensionless parameter $\bar{x}_1 = x_1/B$ serves to locate the bearing pad, which is now of unit length when measured along the $\bar{x}$ direction, relative to the origin of the $(\bar{x}, \bar{y})$ coordinate system. To specify bearing axial length and thereby to complete the geometric description of the lubricant film, we need one more parameter, say, the aspect ratio B/L.

On this basis, when substituting Eq. (3.127) into the Reynolds equation and its boundary conditions, we would expect the appearance of the two dimensionless parameters $\bar{x}_1$ and B/L. This is indeed the case, for we find that

$$\frac{\partial}{\partial\bar{x}}\left(\bar{x}^3\frac{\partial\bar{p}}{\partial\bar{x}}\right) + 4\left(\frac{B}{L}\right)^2\frac{\partial}{\partial\bar{z}}\left(\bar{x}^3\frac{\partial\bar{p}}{\partial\bar{z}}\right) = -1, \tag{3.128}$$

$$\bar{p}(\bar{x}, \pm 1) = 0, \tag{3.129a}$$

$$\bar{p}(\bar{x}_1, \bar{z}) = \bar{p}(\bar{x}_1 + 1, \bar{z}) = 0. \tag{3.129b}$$

The boundary conditions in Eq. (3.129) specify zero pressure on the pad boundaries, as we have $\bar{x}_1 = x_1/B$ and $\bar{x}_1 + 1 = (x_1 + B)/B = x_2/B$.

The distribution of the dimensionless pressure is identical in all bearings that have identical values of B/L and $\bar{x}_1$. The runner velocity U, the lubricant viscosity μ, and the slope m enter into the calculations only as constant multipliers when computing actual pressures.

The lubricant film force is given by

$$F = \int_{-L/2}^{L/2}\int_{x_1}^{x_2} p(x, z)\, dx\, dz$$

or in nondimensional form by[11]

$$f \equiv \frac{Fh_1^2}{\mu U L B^2} \\ = 6\bar{x}_1^2\int_0^1\int_{\bar{x}_1}^{\bar{x}_1+1} \bar{p}(\bar{x}, \bar{z})\, d\bar{x}\, d\bar{z}. \tag{3.130}$$

The x coordinate of the center of pressure may be found from

$$F x_p = \int_{-L/2}^{L/2}\int_{x_1}^{x_2} xp(x, z)\, dx\, dz. \tag{3.131}$$

Writing $x_p = x_1 + B\delta$, in terms of dimensionless variables, Eq. (3.131) reduces to

$$\delta = -\bar{x}_1 + \frac{6\bar{x}_1^2}{f}\int_0^1\int_{\bar{x}_1}^{\bar{x}_1+1} \bar{x}\,\bar{p}(\bar{x}, \bar{z})\, d\bar{x}\, d\bar{z}. \tag{3.132}$$

---

[11] In place of the dimensionless force as defined in Eq. (3.130), some authors, notably Raimondi and Boyd (1955), employ the *Kingsbury number* K_f. These two quantities are related through $K_f = \bar{x}_1^2/f$.

Here δ is the nondimensional distance between the center of pressure and the leading edge of the pad.

Because of the simplicity of the boundary condition Eq. (3.129), Eq. (3.128) has a straightforward analytical solution. Equation (3.128) is first made homogeneous by assuming the pressure to be of the form

$$\bar{p}(\bar{x}, \bar{z}) = \bar{p}_\infty(\bar{x}) - \bar{p}^*(\bar{x}, \bar{z}). \tag{3.133}$$

Here $\bar{p}_\infty(\bar{x})$ is the long-bearing solution and $\bar{p}^*$ represents the correction resulting from the finiteness of the bearing. The function $\bar{p}_\infty(\bar{x})$ satisfies

$$\frac{d}{d\bar{x}}\left(\bar{x}^3 \frac{d\bar{p}_\infty}{d\bar{x}}\right) = -1, \tag{3.134a}$$

$$\bar{p}_\infty(\bar{x}_1) = \bar{p}_\infty(\bar{x}_1 + 1) = 0, \tag{3.134b}$$

$$\bar{p}_\infty = \frac{(\bar{x} - \bar{x}_1)(\bar{x}_1 + 1 - \bar{x})}{(2\bar{x}_1 + 1)\bar{x}^2}. \tag{3.134c}$$

The boundary-value problem for $\bar{p}^*(\bar{x}, \bar{z})$ is

$$\frac{\partial}{\partial \bar{x}}\left(\bar{x}^3 \frac{\partial \bar{p}^*}{\partial \bar{x}}\right) + 4\left(\frac{B}{L}\right)^2 \frac{\partial}{\partial \bar{z}}\left(\bar{x}^3 \frac{\partial \bar{p}^*}{\partial \bar{z}}\right) = 0, \tag{3.135a}$$

$$\bar{p}^*(\bar{x}_1, \bar{z}) = \bar{p}^*(\bar{x}_1 + 1, \bar{z}) = 0, \tag{3.135b}$$

$$\bar{p}^*(\bar{x}, \pm 1) = \bar{p}_\infty(\bar{x}), \tag{3.135c}$$

and if the form $\bar{p}^* = \phi(\bar{x})\psi(\bar{z})$ is assumed, the functions ϕ and ψ satisfy

$$\frac{d}{d\bar{x}}\left(\bar{x}^3 \frac{d\phi}{d\bar{x}}\right) + 4\left(\frac{B}{L}\right)^2 \lambda^2 \bar{x}^3 \phi = 0, \tag{3.136a}$$

$$\phi(\bar{x}_1) = \phi(\bar{x}_1 + 1) = 0, \tag{3.136b}$$

$$\frac{d^2\psi}{d\bar{z}^2} - \lambda^2 \psi = 0. \tag{3.137}$$

The function ψ, normalized at the ends of its interval, is

$$\psi = \frac{\cosh \lambda \bar{z}}{\cosh \lambda}. \tag{3.138}$$

The Sturm-Liouville problem, Eq. (3.136), was solved by Hays (1958) in terms of Bessel functions but at the cost of some computational stability problems. Instead we interpret Eqs. (3.136a) and (3.136b) as the Euler condition for the isoperimetric problem of the minimization of

$$\int_{\bar{x}_1}^{\bar{x}_1+1} \bar{x}^3 \left(\frac{d\phi}{d\bar{x}}\right)^2 d\bar{x},$$

subject to the conditions

$$\int_{\bar{x}_1}^{\bar{x}_1+1} \bar{x}^3 \phi^2(\bar{x}) \, d\bar{x} = \text{const.}$$

and

$$\phi(\bar{x}_1) = \phi(\bar{x}_1 + 1) = 0.$$

Introducing λ as the Lagrange multiplier and changing the dependent variable to

$$v(\bar{x}) = \bar{x}^{3/2} \phi(\bar{x}),$$

the problem becomes that of minimizing (Szeri and Powers, 1970):

$$I = \int_{\bar{x}_1}^{\bar{x}_1+1} \left[(v')^2 + \frac{3}{4\bar{x}^2} v^2 - 4 \left(\frac{B}{L} \right)^2 \lambda^2 v^2 \right] d\bar{x}. \tag{3.139}$$

Adapting the Rayleigh-Ritz method (Hildebrand, 1965), we assume that $v(\bar{x})$ can be represented by a series of functions that satisfy the boundary conditions. If sines are chosen, $v(\bar{x})$ has the form

$$v_n(\bar{x}) = \sum_m x_{mn} \sin m\pi (\bar{x} - \bar{x}_1), \tag{3.140}$$

under the restriction that

$$\sum_m x_{mn}^2 = 1.$$

By the usual arguments of variational calculus, the problem in Eq. (3.139) is transformed into the matrix problem

$$AX = X\Lambda^2. \tag{3.141}$$

Here $X = [x_m]$, $\Lambda^2 = \text{diag}\{\lambda_1^2, \lambda_2^2, \ldots\}$, and $A = [a_{mn}]$, which is the Rayleigh-Ritz matrix whose elements are (δ_{mn} is the Kronecker delta)

$$a_{mn} = \frac{1}{4} \left(\frac{L}{B} \right)^2 (mn\pi^2 \delta_{mn} + q_{mn}).$$

The symbol q_{mn} stands for the definite integral

$$\frac{3}{2} \int_{\bar{x}_1}^{\bar{x}_1+1} \frac{\sin n\pi (\bar{x} - \bar{x}_1) \sin m\pi (\bar{x} - \bar{x}_2)}{\bar{x}^2} d\bar{x}.$$

Assuming that the problem can be solved, the expression for $\bar{p}^*(\bar{x}, \bar{z})$ is

$$\bar{p}^*(\bar{x}, \bar{z}) = \bar{x}^{-3/2} \sum_n a_n v_n(\bar{x}) \frac{\cosh \lambda_n \bar{z}}{\cosh \lambda_n}. \tag{3.142}$$

Here the coefficients a_n are chosen by orthogonality of the eigenfunctions to make $\bar{p}^*$ satisfy the boundary condition Eq. (3.135c)

$$a_n = 2 \int_{\bar{x}_1}^{\bar{x}_1+1} \bar{x}^{3/2} \bar{p}_\infty(\bar{x}) v_n(\bar{x}) \, d\bar{x}. \tag{3.143}$$

Often the lubricant pressure is of no interest. If that is the case, we can dispense with computing $\bar{p}$ from Eq. (3.133) and evaluate the load capacity, oil flow, and center of pressure (pivot position) directly (Szeri and Powers, 1970). The procedure will be demonstrated for the load capacity alone.

The nondimensional oil film force is written as

$$f = f_\infty - f^*, \tag{3.144}$$

where f_∞, the long-bearing force, is generated by $\bar{p}_\infty(\bar{x})$ and has the representation

$$f_\infty = 6\bar{x}_1^2 \left[\ln\left(1 + \frac{1}{\bar{x}_1}\right) - \frac{2}{2\bar{x}_1 + 1} \right]. \tag{3.145}$$

It follows that f^* is the force correction resulting from the finiteness of the bearing. Substitution of Eq. (3.143) into Eq. (3.142) yields

$$f^* = \int_0^1 \int_{\bar{x}_1}^{\bar{x}_1+1} \bar{p}^*(\bar{x}, \bar{z})\, d\bar{x}\, d\bar{z}$$

$$= 12\bar{x}_1^2 \sum_k \left(\sum_n b_n x_{nk} \right) \lambda_k^{-1} \tanh \lambda_k \left(\sum_m x_{mk} c_m \right). \tag{3.146a}$$

Our expression for f^* may be written in the compact form, using matrix notation

$$f^* = 12\bar{x}_1^2 b^T X \Lambda^{-1} \tanh \Lambda X^T c. \tag{3.146b}$$

Here

$$c_m = \int_{\bar{x}_1}^{\bar{x}_1+1} \frac{\sin m\pi(\bar{x} - \bar{x}_1)}{\bar{x}^{3/2}}\, d\bar{x},$$

$$b_m = \int_{\bar{x}_1}^{\bar{x}_1+1} \bar{x}^{3/2} \bar{p}_\infty(\bar{x}) \sin m\pi(\bar{x} - \bar{x}_1)\, d\bar{x},$$

and b and c are column vectors, having elements $b_1, b_2, b_3, \ldots$ and $c_1, c_2, c_3, \ldots$, respectively.

It is permissible to write, Eq. (3.146b),

$$X\Lambda^{-1} \tanh \Lambda X^T = X\Lambda^{-1} X^T (X \tanh \Lambda X^T) = A^{-1/2} \tanh A^{1/2}, \tag{3.147a}$$

as the following equalities hold

$$X\Lambda^n X^T = (X\Lambda X^T)^n = A^{n/2} \tag{3.147b}$$

for any rational number n, and

$$X \tanh \Lambda X^T = X \sinh \Lambda X^T X (\cosh \Lambda)^{-1} X^T.$$

In order to transform this last expression, we note that

$$X(\cosh \Lambda)^{-1} X^T = [(X^T)^{-1} \cosh \Lambda (X)^{-1}]^{-1}$$

$$= (X \cosh \Lambda X^T)^{-1}$$

$$= (\cosh X\Lambda X^T)^{-1}.$$

Here we took into account the orthogonality of the eigenvectors of the Sturm-Liouville problem and utilized the series expansion for $\cosh \Lambda$.

Our final expression for the dimensionless force Eq. (3.144) is

$$f = 6\bar{x}_1^2 \left[\ln\left(1 + \frac{1}{\bar{x}_1}\right) - \frac{2}{2\bar{x}_1 + 1} - 2b^T A^{-1/2} \tanh A^{1/2} c \right]. \tag{3.148}$$

The term $A^{-1/2} \tanh A^{1/2}$ can be calculated without recourse to the eigenvalue problem in Eq. (3.141) by making use of Padé approximations of the form

$$\frac{\tanh \lambda^{1/2}}{\lambda^{1/2}} \approx \frac{\sum_n e_n \lambda^n}{\sum_m d_m \lambda^m}. \tag{3.149}$$

Since hydrodynamic action requires convergence of the film shape in the direction of relative motion, fixed pad sliders cannot carry load when the direction of operation is reversed. If the bearing is to be operated in either the forward or the reverse direction, a combination of two pads with their surfaces sloping in opposite directions is required. Operation of such a configuration is possible only if the diverging lubricant film cavitates. If the ambient pressure is too high for the film to cavitate, then pivoted pad bearings are often used. Normally a flat pad will carry load efficiently only if the pivot is offset – that is, located between the center of the pad and the trailing edge – so that $0.5 < \delta < 1$. But with the pivot so placed, the pad can carry load effectively in only one direction. Szeri and Powers (1970) tabulate oil film force, oil flow, and pivot location for the parameter range $1/8 < \bar{x}_1 < 10$ and $1/8 < L/B < 16$.

When the pads are loaded, they deform, and the film shape is no longer given by $h = mx$. Crowning might also be machined into the pad deliberately. Curved pads were first investigated by Raimondi (1960).

Sector thrust bearing
The film shape for the sector thrust pad of Figure 3.15 is given by

$$h = h_2 - (h_2 - h_1)\frac{\theta}{\beta}, \tag{3.150}$$

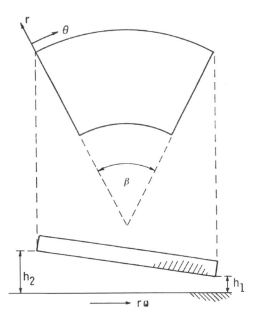

Figure 3.15 Sector thrust pad geometry.

where h_2 and h_1 are the maximum and minimum film thicknesses, respectively, and β is the angular span of the sector.

The Reynolds equation (2.68) in cylindrical polar coordinates, with V_0 set equal to zero, is

$$\frac{\partial}{\partial r}\left(rh^3\frac{\partial p}{\partial r}\right) + \frac{1}{r}\frac{\partial}{\partial \theta}\left(h^3\frac{\partial p}{\partial \theta}\right) = 6\mu\omega r\frac{\partial h}{\partial \theta}. \tag{3.151}$$

Introducing the new variable (Tao, 1959)

$$\phi = \frac{\beta}{h_2 - h_1}h$$

and the notation

$$K = \frac{6\mu\omega}{(h_2 - h_1)^2}\beta^2$$

into Eq. (3.131), the Reynolds equation reduces to

$$r\frac{\partial}{\partial r}\left(r\frac{\partial p}{\partial r}\right) + \frac{1}{\phi^3}\frac{\partial}{\partial \phi}\left(\phi^3\frac{\partial p}{\partial \phi}\right) = -Kr^2\phi^{-3}. \tag{3.152}$$

The solution of Eq. (3.152) is sought in the form

$$p(r, \phi) = p_1(r, \phi) + p_2(r, \phi),$$

where $p_1(r, \phi)$ satisfies the homogeneous equation and $p_2(r, \phi)$ is a particular solution of Eq. (3.152). The homogeneous equation is reduced to two ordinary differential equations:

$$r^2\zeta'' + r\zeta' - \lambda^2\zeta = 0, \tag{3.153a}$$

$$\xi'' + 3\phi^{-1}\xi' + \lambda^2\xi = 0, \tag{3.153b}$$

by introduction of

$$p_1(r, \phi) = \zeta(r)\xi(\phi).$$

Equation (3.153a) is of the Sturm-Liouville type when specifying zero boundary conditions for pressure and has solutions in terms of Bessel functions. Equation (3.133b), on the other hand, is amenable to direct integration. Details of the solution are given by Tao (1959).

Numerical solutions of the sector thrust bearing are given by Pinkus (1958) and Pinkus and Sternlicht (1961). The plot in Figure 3.16, showing the nondimensional load variable plotted against the radial slope parameter for various values of the tangential slope parameter, is taken from Raimondi and Szeri (1984). These solutions were obtained via Galerkin's method, employing global interpolating functions. Pivoted-pad thrust bearings employ supporting pivots, as shown in Figure 3.17, at the center of film pressure. While the performance of pivoted-pad bearings is theoretically identical to that of a fixed-pad bearing having the same slope, the pivoted type has the advantages of (1) self-aligning capability, (2) automatically adjusting pad inclination to optimally match the needs of varying speed and load, and (3) operation in either direction of rotation. Theoretically, the pivoted-pad can be optimized for all speeds and loads by judicious pivot positioning, whereas the

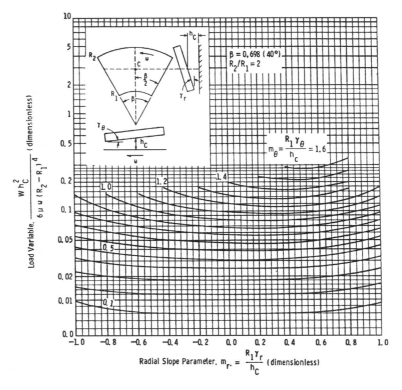

Figure 3.16 Load capacity chart for thrust pad sector. (Reprinted with permission from Raimondi, A. A. and Szeri, A. Z. Journal and thrust bearings. In Booser E. R., *CRC Handbook of Lubrication.* Copyright CRC Press, Boca Raton, Florida. © 1984.)

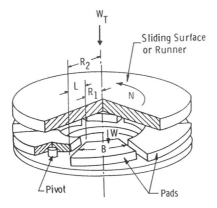

Figure 3.17 Pivoted-pad thrust bearing geometry. (Reprinted with permission from Raimondi, A. A. and Szeri, A. Z. Journal and thrust bearings. In Booser, E. R., *CRC Handbook of Lubrication.* Copyright CRC Press, Boca Raton, Florida. © 1984.)

fixed-pad bearing can be designed for optimum performance only for one operating condition. Although pivoted-pad bearings involve somewhat greater complexity, standard designs are readily available for medium to large size machines. Pivoted-pad thrust bearings are further discussed in Chapters 4 and 9, within the context of dynamic properties and thermal effects.

3.3 Nomenclature

A	area of hydrostatic pad
B	slider dimension (in direction of motion)
C	radial clearance
D	bearing diameter
F, F_R, F_T	oil film force, radial and tangential components
F_μ	friction force
H	dimensionless film thickness
H_p, H_f, H_T	pumping power, shear power, total power
K_f	Kingsbury number
L	bearing length
N	shaft speed
O_B, O_J	center of bearing, journal
P	lubricant force per projected bearing area
Q	rate of flow
Q_s	side leakage
$R = R_B, R_J$	radius of bearing, journal
R_1, R_2	hydrostatic pad inner and outer radii
R_B, R_C	resistance resulting from bearing, capillary
S	Sommerfeld number
T	temperature
U, V	surface velocity components
U_0	effective surface velocity
U_M	maximum surface velocity
V	surface velocity vector
V_0	squeeze velocity
W	external load
a_f	area factor
c	specific heat
c_μ	coefficient of friction
d	diameter of capillary restrictor
f, f_R, f_T	dimensionless lubricant force, radial and tangential components
h, h_1, h_2	film thickness, minimum and maximum values
h_f	friction factor
ℓ	length of capillary restrictor
m	slope of bearing surfaces
n	geometric constant
p, p_c	pressure, center line pressure
p_r, p_a, p_s	recess, ambient, supply pressures
p_i	inlet pressures

q_s	dimensionless side leakage
q_f	flow factor
t	time
$u_i(u, v, w)$	lubricant velocity components
v	lubricant velocity vector
$x_i(x, y, z)$	orthogonal Cartesian coordinates
x_p	pivot position
Γ_i, Γ_o	recess boundary, pad outside boundary
Ξ	angular coordinate
α	position of load relative to pad leading edge
β	pad angle
δ	dimensionless pivot position
ε	eccentricity ratio
θ	angular coordinate measured from line of centers
θ_1, θ_2	angular coordinates of pad leading edge, trailing edge
λ	dimensionless bearing stiffness
μ	lubricant viscosity
ξ	resistance ratio
ρ	lubricant density
σ	surface tension
τ	shear stress
τ_0	wall stress
ϕ	attitude angle
ω	shaft angular velocity
ω_w	angular frequency of applied load
$\bar{(\ )}$	dimensionless quantity
$(\ )_{\mathrm{cav}}$	evaluated at fluid-cavity interface

3.4   References

Ai, X. and Cheng, H. S. 1994. A transient EHL analysis for line contacts with measured surface roughness using multigrid technique. *ASME Journal of Tribology*, **116**, 549–558.

Ai, X. and Cheng, H. S. 1996. The effect of surface structure on EHL point contacts. *ASME Journal of Tribology*, **118**, 569–566.

Ai, X. and Zheng, L. 1989. A general model for microelastohydrodynamic lubrication and its full numerical solution. *ASME Journal of Tribology*, **111**, 569–576.

Anderson, D. A., Tannehil, J. C. and Pletcher, R. H. 1984. *Computational Fluid Mechanics and Heat Transfer*. Hemisphere Publishing Co., New York.

Bhushan, B. and Tonder, K. 1989. Roughness induced shear and squeeze film effects in magnetic recording – Part I: Analysis. *ASME Journal of Tribology*, **111**, 220–227.

Cameron, A. 1966. *The Principles of Lubrication*. Wiley, New York.

Chang, L. and Webster, M. N. 1991. A study of elastohydrodynamic lubrication of rough surfaces. *ASME Journal of Tribology*, **103**, 110–115.

Christensen, H. and Tonder, K. 1969. Tribology of rough surfaces: stochastic models of hydrodynamic lubrication. *Sintef Report*, No. 10/69-18. Trondheim, Norway.

Coyne, J. C. and Elrod, H. G. 1970a. Conditions for the rupture of a lubricating film, Part 1, Theoretical model. *ASME Trans. Ser. F*, **92**, 451–457.

Coyne, J. C. and Elrod, H. G. 1970b. Conditions for the rupture of a lubricating film, Part II, New boundary conditions for Reynolds' equation. *ASME Pap., 70-Lub-3.*

Dai, R. X., Dong, Q. and Szeri, A. Z. 1992. Approximations in lubrication theory. *ASME Journal of Tribology*, **114**, 14–25.

Dowson, D. and Taylor, C. M. 1979. Cavitation in bearings. *Ann. Rev. Fluid. Mech.*, **11**, 35–66.

Dowson, D., Godet, M. and Taylor, C. M. (eds.) 1975. *Cavitation and Related Phenomena in Lubrication.* Institute of Mechanical Engineers, London.

DuBois, G. B. and Ocvirk, F. W. 1955. Analytical derivation and experimental evaluation of short-bearing approximation for full journal bearings. *Natl. Advis. Comm. Aeronaut. Rep., 1157.*

Dyer, D. and Reason, B. R. 1976. A study of tensile stresses in a journal-bearing oil film. *J. Mech. Eng. Sci.*, **18**, 46–52.

Elrod, H. G. 1973. Thin film lubrication theory for Newtonian fluids with surfaces possessing striated roughness or grooving. *ASME Journal of Lubrication Technology*, **95**, 484–489.

Elrod, H. G. 1979. A general theory for laminar lubrication with Reynolds roughness. *ASME Journal of Lubrication Technology*, **101**, 8–14.

Elrod, H. G. 1981. A cavitation algorithm. *ASME Journal of Lubrication Technology*, **103**, 350–354.

Elrod, H. G. and Adams, M. L. 1974. A computer program for cavitation and starvation problems. *Leeds-Lyon Conference on Cavitation*, Leeds University, England.

Floberg, L. 1961. On hydrodynamic lubrication with special reference to cavitation in bearings. *Chalmers Tek. Hoegsk. Handl.*, Dissertation.

Floberg, L. 1964. Cavitation in lubricating oil films. In *Cavitation in Real Liquids*, R. Davies (ed.). American Elsevier, New York.

Floberg, L. 1965. On hydrodynamic lubrication with special reference to sub-cavity pressures and number of streamers in cavitation regions. *Acta Polytech. Scand. Mech. Eng. Ser.*, **19**, 3–35.

Gourley, W. E. 1977. *Laminar flow between closely spaced rotating disks with variable viscosity.* M.Sc. Thesis, Univ. of Pittsburgh.

Gross, W. A. 1962. *Gas Film Lubrication.* John Wiley & Sons, Inc.

Gurfinkel Castillo, M. E. and Patera, A. T. 1997. Three-dimensional ribbing instability in symmetric forward-roll film in coating. *J. Fluid Mech.*, **335**, 323–359.

Hays, D. F. 1958. Plane sliders of finite width. *ASLE Trans.*, **1**, 233–240.

Heller, S. and Shapiro, W. 1968. A numerical solution for the incompressible hybrid journal bearing with cavitation. *ASME Trans., Ser. F*, **90**, 508–515.

Hildebrand, F. B. 1965. *Methods of Applied Mathematics.* Prentice-Hall Inc.

Hu, Y. Z. and Tonder, K. 1992. Simulation of 3-D random surface by 2-D digital filter and Fourier analysis. *Int. J. Mach. Tools Manufact.*, **32**, 38–93.

Jameson, A. 1975. Transonic potential flow calculations in conservative form. *Proceedings of 2nd Computational Conference*, pp. 148–161, Hartford.

Lubrecht, A. A., ten Napel, W. E. and Bosma, R. 1986. Multigrid, an alternative method for calculating film thickness and pressure profile in elastohydrodynamic line contacts. *ASME Journal of Tribology*, **110**, 551–556.

Lund, J. W. 1964. Spring and damping coefficients for the tilting-pad journal bearing. *ASLE Trans.*, **7**, 342–352.

Mitsuya, Y. and Fukui, S. 1986. Stokes roughness effects on hydrodynamic lubrication. *ASME Journal of Tribology*, **108**, 151–158.

O'Donoghue, J. P., Koch, P. R. and Hooke, C. J. 1970. Approximate short bearing analysis and experimental results obtained using plastic bearing liners. *Proc. Inst. Mech. Eng.*, **184**, pt. 3L, 190–196.

Opitz, H. 1968. Pressure pad bearings. In *Lubrication and Wear: Fundamentals and Application to Design. Proc. Inst. Mech. Eng.*, **182**, pt. 3A.

Patir, N. 1978. A numerical procedure for random generation of rough surfaces. *WEAR*, **47**, 263–277.

Patir, N. and Cheng, H. S. 1978. An average flow model for determining effects of three-dimensional roughness on partial hydrodynamic lubrication. *ASME Journal of Lubrication Technology*, **100**, 12–17.

Patir, N. and Cheng, H. S. 1979. Application of average flow model to lubrication between rough sliding surfaces. *ASME Journal of Lubrication Technology*, **101**, 220–230.

Peeken, H. J., Knoll, G., Rienäcker, A., Lang, J. and Schönen, R. 1997. On the numerical determination of flow factors. *ASME Journal of Tribology*, **119**, 259–264.

Peklenik, J. 1968. New developments in surface characterization and measurements by means of random process analysis. *Proc. Inst. Mech. Engrs., London*, **182** (3K), 108–126.

Pinkus, O. 1958. Solutions of the tapered-land sector thrust bearing. *ASME Trans.*, **80**, 1510–1516.

Pinkus, O. and Sternlicht, B. 1961. *Theory of Hydrodynamic Lubrication*. McGraw-Hill, New York.

Raimondi, A. A. 1959. Theoretical study of the effect of offset loads on the performance of a 120° partial journal bearing. *ASLE Trans.*, **2**, 147–157.

Raimondi, A. A. 1960. The influence of longitudinal and transverse profile on the load capacity of pivoted pad bearings. *ASLE Trans.*, **3**, 265–276.

Raimondi, A. A. and Boyd, J. 1955. Applying bearing theory to the analysis and design of pad-type bearings. *ASME Trans.*, **77**, 287–309.

Raimondi, A. A. and Boyd, J. 1958. A solution for the finite journal bearing and its application to analysis and design. *ASLE Trans.*, **1**, 159–209.

Raimondi, A. A. and Szeri, A. Z. 1984. Journal and thrust bearings. In *Handbook of Lubrication*, E. R. Booser (ed.). Vol. II. CRC Press.

Reddi, M. M. 1970. Finite element solution of the incompressible lubrication problem. *ASME Trans. Ser. F*, **91**, 524–533.

Reynolds, O. 1886. On the theory of lubrication and its application to Mr. Beauchamp Tower's experiments. *Philos. Trans. Roy. Soc.*, **177**, pt. 1.

Rippel, H. C. 1963. *Cast Bronze Hydrostatic Bearing Manual*. Cast Bronze Bearing Institute, Cleveland.

Rowe, W. B. and O'Donoghue, J. P. 1971. A review of hydrostatic bearing design. In *Externally Pressurized Bearings*, pp. 157–187. Instit. Mech. Engrs., London.

Safar, Z. S. and Szeri, A. Z. 1972. A variational solution for the 120-degree partial journal bearing. *J. Mech. Eng. Sci.*, **14**, 221–223.

Savage, M. D. 1977. Cavitation in lubrication. *J. Fluid Mech.*, **80**, 743–767.

Smith, E. 1975. *A Study of Film Rupture in Hydrodynamic Lubrication*. Ph.D. Thesis, Univ. of Leeds.

Sommerfeld, A. 1904. Zür hydrodynamische theorie der Schmiermittelreibung. *Z. Math. Phys.*, **50**, 97–155.

Sun, D. C. and Chen, K. K. 1977. First effects of Stokes roughness of hydrodynamic lubrication. *ASME Journal of Lubrication Technology*, **99**, 2–9.

Szeri, A. Z. 1975. Hydrostatic bearing pads: a matrix iterative solution. *ASLE Trans.*, **19**, 72–78.

Szeri, A. Z. 1980. *Tribology: Friction, Lubrication, and Wear*. Hemisphere Publishing Corporation, New York.

Szeri, A. Z. and Adams, M. L. 1978. Laminar flow between closely spaced rotating disks. *J. Fluid Mech.*, **86**, 1–14.

Szeri, A. Z. and Philips, C. 1974. An iterative solution for the plane hybrid bearing. *ASLE Trans.*, **18**, 116–122.

Szeri, A. Z. and Powers, D. 1967. Full journal bearings in laminar and turbulent regimes. *J. Mech. Eng. Sci.*, **9**, 167–176.

Szeri, A. Z. and Powers, D. 1970. Pivoted plane pad bearings: a variational solution. *ASME Trans., Ser. F*, **92**, 466–472.

Tao, L. N. 1959. The hydrodynamic lubrication of sector thrust bearings. *6th Midwest. Conf. Fluid Mech.*, Austin, Tex., 406–416.

Taylor, G. I. 1964. Cavitation in hydrodynamic lubrication. In *Cavitation in Real Liquids*, R. Davies (ed.). American Elsevier, New York.

Temperley, H. N. V. 1975. The tensile strength of liquids. In *Cavitation and Related Phenomena in Lubrication*, pp. 11–13. Institution of Mechanical Engineers, London.

Tonder, K. and Christensen, H. 1972. Waviness and roughness in hydrodynamic lubrication. *Proc. Instn. Mech. Engrs.*, **186**, 807–812.

Tripp, J. H. 1983. Surface roughness effects in hydrodynamic lubrication: the flow factor method. *ASME Journal of Luburication Technology*, **105**, 458–465.

Tzeng, S. T. and Saibel, E. 1967. Surface roughness effect on slider lubrication. *ASLE Trans.*, **10**, 334–338.

Vijayaraghavan, D. and Keith, T. G. 1989. Development and evaluation of a cavitation algorithm. *ASLE Trans.*, **32**, 225–233.

Vijayaraghavan, D., and Keith, T. G. 1990a. An efficient, robust, and time accurate numerical scheme applied to a cavitation algorithm. *ASME Journal of Tribology*, **112**, 44–51.

Vijayaraghavan, D. and Keith, T. G. 1990b. Grid transformation and adaption techniques applied in the analysis of cavitated journal bearings. *ASME Journal of Tribology*, **112**, 52–59.

Vijayaraghavan, D. and Keith, T. G. 1990c. Analysis of a finite grooved misaligned journal bearing considering cavitation and starvation effects. *ASME Journal of Tribology*, **112**, 60–67.

Vijayaraghavan, D., Keith, T. G. and Brewe, D. E. 1991. Extension of transonic flow computational concepts in the analysis of cavitated bearings. *ASME Journal of Tribology*, **113**, 539–546.

Woods, C. M. and Brewe, D. E. 1988. The solution of the Elrod algorithm for a dynamically loaded journal bearing using multigrid techniques. *ASME/STLE Tribology Conference*, Baltimore, Paper JAT 8-10.

CHAPTER 4

Dynamic Properties of Lubricant Films

The behavior of rotors is strongly influenced by the characteristics of their supports. The forces generated on a journal by the lubricant film of its bearings are nonlinear functions of the position and velocity of the journal center.[1] Thus, to calculate the critical speeds and vibration amplitudes of rotors and to examine their stability against self-excited vibrations, knowledge of the response of the bearing lubricant film to journal displacements and velocities is essential.

In Figure 4.1 we schematize a rotor, of weight $2W$, and its support. Under steady load, the journal center is displaced from the bearing center to the steady operating position, O_{J_s}.

Rotor response to small excitation will be as shown in Figure 4.2. When the supporting bearings are rigid, shaft vibration amplitude varies with shaft speed, as indicated by the solid curve. This type of response would be expected when the rotor is running on rolling-contact bearings. Rigidly supported rotors cannot be operated at a *critical speed*[2] and can become "hung" on the critical when attempting to drive through.

We add spring and considerable damping to the system, in addition to the shaft spring and damping already present, when replacing the rigid supports with hydrodynamic bearings. This additional stiffness and damping will, as indicated in Figure 4.2, lower the critical speed below that calculated for rigid supports, and reduce the amplitude of vibration of the rotor.

The above example shows excitation occurring at running speed. In practice, however, excitation may occur at speeds other than running speed and might be caused by magnetic pulls, aerodynamic forces on turbine or compressor blades (Alford, 1965), gear impacts etc. Lubricant films themselves might originate destructive self-excited vibrations, which include *oil whip* at somewhat less than one-half running speed (Hagg, 1946). The self-excited vibration occurring at exactly one-half (or other exact submultiple) of the running speed is known as *subharmonic resonance* (Den Hartog, 1956).

In Figure 4.3 we reduce the rotor-bearing configuration of Figure 4.1 to a simple dynamical system consisting of a mass, springs, and dashpots. In this schematic, half the rotor mass, $M = W/g$, is concentrated at O_{J_s}, the steady running position of the journal. If excitation, F, at some frequency, Ω, is applied to this system, the mass center will respond by orbiting about O_{J_s}, its instantaneous orbital position denoted by O_J. It is implied here that the dynamic displacement $(O_{J_s} - O_J)$ is small relative to bearing clearance. For "large" dynamic displacements and velocities the behavior of lubricant films is strongly nonlinear, rendering the representation in Figure 4.3 very approximate.

---

[1] A more thorough examination of the problem reveals that the force on an orbiting journal is dependent on acceleration as well as on position and velocity (see e.g., Szeri, Raimondi, and Giron, 1983). For simplicity, however, we do not include acceleration, i.e., fluid inertia, effects here, and leave their analysis to Chapter 5.

[2] By critical speed we understand any rotor resonance that is excited by rotor unbalance. There can be, thus, several critical speeds.

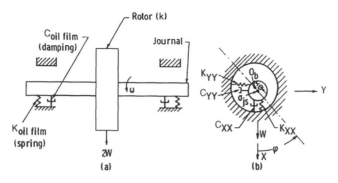

Figure 4.1. Dynamical elements of rotor-shaft configuration. (Reprinted with permission from Raimondi, A. A. and Szeri, A. Z. Journal and thrust bearings. In Booser, E. R., *CRC Handbook of Lubrication*. Copyright CRC Press, Boca Raton, Florida. © 1984.)

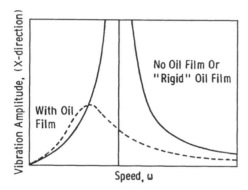

Figure 4.2. Effect of oil film on shaft response. (Reprinted with permission from Raimondi, A. A. and Szeri, A. Z. Journal and thrust bearings. In Booser, E. R., *CRC Handbook of Lubrication*. Copyright CRC Press, Boca Raton, Florida. © 1984.)

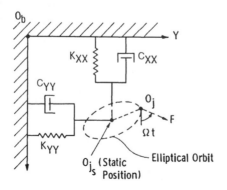

Figure 4.3. Representation of oil film as a simple dynamical system of springs and dampers (cross-film springs K_{xy}, K_{yx} and dampers C_{xy}, C_{yx} are not shown). (Reprinted with permission from Raimondi, A. A. and Szeri, A. Z. Journal and thrust bearings. In Booser, E. R., *CRC Handbook of Lubrication*. Copyright CRC Press, Boca Raton, Florida. © 1984.)

For small dynamic displacements and velocities of the journal the oil film forces may be linearized about their static equilibrium value, but even in this linear approximation the lubricant film cannot be simulated by a simple elastic-dissipative system. Cross-coupling stiffness and damping are needed to describe the relationship between the incremental oil-film forces and the journal displacements and velocities that cause them:

$$
\begin{pmatrix} dF_x \\ dF_y \end{pmatrix} = - \begin{pmatrix} K_{xx} & K_{xy} \\ K_{yx} & K_{yy} \end{pmatrix} \begin{pmatrix} x \\ y \end{pmatrix} - \begin{pmatrix} C_{xx} & C_{xy} \\ C_{yx} & C_{yy} \end{pmatrix} \begin{pmatrix} \dot{x} \\ \dot{y} \end{pmatrix}.
\tag{4.1}
$$

In this chapter, we will show how to evaluate the *linearized force coefficients* $K_{xx}, \ldots,$ C_{yy}. These coefficients define the rotor support for critical speed calculations.

For example, if a vibrating system, consisting of a rigid rotor of mass $2M$ running on hydrodynamic bearings, is excited by a force F at frequency Ω, its response is described by the equations of motion

$$
\begin{aligned}
M\ddot{x} + C_{xx}\dot{x} + C_{xy}\dot{y} + K_{xx}x + K_{xy}y &= F \cos \Omega t, \\
M\ddot{y} + C_{yy}\dot{y} + C_{yx}\dot{x} + K_{yy}y + K_{yx}x &= F \sin \Omega t.
\end{aligned}
\tag{4.2}
$$

By making the right-hand sides of these equations vanish, existence of any self-excited vibrations can also be investigated (Den Hartog, 1956). When the bearing constitutes an element in a complex dynamical system, it is usually incorporated in the system's equations of motion through its spring and damping coefficients. All eight oil film coefficients are required in order to make accurate dynamical analyses of rotor-shaft configurations.

4.1 Fixed Pad

Linearized Force Coefficients

Let O_B represent the center of the bearing in Figure 4.4. The static equilibrium position of the rotating shaft is O_{J_s}, the eccentricity is ε_0, and the attitude angle is ϕ_0. The components of the lubricant force, resolved along the fixed directions (R, T), are $(F_R)_0$ and $(F_T)_0$.[3]

If there is a small unbalanced force on the journal, the journal will orbit about its static equilibrium position (e_0, ϕ_0). At a particular instant it will occupy position O_J, a generic point on the orbit. In O_J the eccentricity is $e = e_0 + \Delta e$, the attitude angle is $\phi = \phi_0 + \Delta \phi$, and the journal possesses instantaneous velocities $\dot{e}$ and $e\dot{\phi}$. The instantaneous lubricant force now has components F_r and F_t, relative to the instantaneous radial and tangential coordinates r and t, respectively.

The instantaneous force will now be referred to the fixed (R, T) coordinate system,

$$
\begin{pmatrix} F_R \\ F_T \end{pmatrix} = \begin{pmatrix} \cos \Delta\phi & -\sin \Delta\phi \\ \sin \Delta\phi & \cos \Delta\phi \end{pmatrix} \begin{pmatrix} F_r \\ F_t \end{pmatrix}.
\tag{4.3}
$$

For small departures from equilibrium we have the approximations

$$
\cos \Delta\phi \approx 1, \qquad \sin \Delta\phi \approx \Delta\phi
$$

---

[3] Note that here $F_R > 0$ when pointing away from the origin of the (e, ϕ) coordinate system. This is contrary to the scheme adopted by some authors (Pinkus and Sternlicht, 1961; Lund, 1964), but follows accepted mathematical notation (see, e.g., Trumpler, 1966).

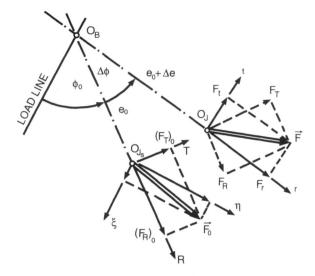

Figure 4.4. Force decomposition in journal bearings (for illustration only; not to scale).

and Eq. (4.3) can be written as

$$F_R = F_r - \Delta\phi F_t,$$
$$F_T = \Delta\phi F_r + F_t. \tag{4.4}$$

We are interested in evaluating the increase in the force components[4] over their equilibrium values, due to departure from the equilibrium

$$\Delta F_R = F_R - (F_R)_0,$$
$$\Delta F_T = F_T - (F_T)_0. \tag{4.5}$$

The force excess can be evaluated from Eq. (4.4) according to

$$\Delta F_R = F_r - (F_R)_0 - \Delta\phi F_t,$$
$$\Delta F_T = \Delta\phi F_r + F_t - (F_T)_0. \tag{4.6}$$

The scalar functions F_r and F_t are now expanded in Taylor series about the equilibrium position. For F_r we have, neglecting higher-order terms,

$$F_r = (F_R)_0 + \left(\frac{\partial F_R}{\partial e}\right)_0 de + \left(\frac{\partial F_R}{\partial \phi}\right)_0 d\phi + \left(\frac{\partial F_R}{\partial \dot{e}}\right)_0 d\dot{e} + \left(\frac{\partial F_R}{\partial \dot{\phi}}\right)_0 d\dot{\phi}, \tag{4.7}$$

where all derivatives are evaluated in the equilibrium position, as indicated by the zero subscript. The expansion is similar for F_t.

Substituting the Taylor expansions for F_r and F_t into Eq. (4.6) and, again, neglecting higher-order terms, we obtain (from here on we drop the zero subscript, remembering that

[4] The force increment is resolved relative to the fixed (R, T) axes in Eq. (4.5).

all derivatives are evaluated at equilibrium)

$$
\begin{pmatrix} dF_R \\ dF_T \end{pmatrix} = \begin{pmatrix} \dfrac{\partial F_R}{\partial \varepsilon} & \dfrac{\partial F_R}{\partial \phi} - F_T \\[2ex] \dfrac{\partial F_T}{\partial \varepsilon} & \dfrac{\partial F_T}{\partial \phi} + F_R \end{pmatrix} \begin{pmatrix} d\varepsilon \\ d\phi \end{pmatrix} + \begin{pmatrix} \dfrac{\partial F_R}{\partial \dot{\varepsilon}} & \dfrac{\partial F_R}{\partial \dot{\phi}} \\[2ex] \dfrac{\partial F_T}{\partial \dot{\varepsilon}} & \dfrac{\partial F_T}{\partial \dot{\phi}} \end{pmatrix} \begin{pmatrix} d\dot{\varepsilon} \\ d\dot{\phi} \end{pmatrix}.
$$
(4.8)

To calculate the force derivatives required in Eq. (4.8), we have to go to the Reynolds equation (3.33), which we nondimensionalize according to Eq. (3.34). However, under dynamic loading conditions it is appropriate to employ another nondimensional pressure $\hat{p}$ that is related to $\bar{p}$ of Eq. (3.34) through[5]

$$
\hat{p} = \frac{\bar{p}}{(1 - 2\dot{\phi}/\omega)}.
$$

Note that at static equilibrium ($\varepsilon = \varepsilon_0$, $\phi = \phi_0$, $\dot{\phi} = \dot{\varepsilon} = 0$) the pressure defined under dynamic loading, $\hat{p}$, reduces to the pressure defined under static loading, $\bar{p}$.

In terms of the dynamic pressure $\hat{p}$, as it will be referred to here, the nondimensional Reynolds equation has the form.

$$
\frac{\partial}{\partial \theta}\left(H^3 \frac{\partial \hat{p}}{\partial \theta} \right) + \left(\frac{D}{L} \right)^2 \frac{\partial}{\partial \bar{z}}\left(H^3 \frac{\partial \hat{p}}{\partial \bar{z}} \right) = -12\pi\varepsilon\sin\theta + 24\pi \frac{\dot{\varepsilon}/\omega}{\left(1 - 2\dfrac{\dot{\phi}}{\omega}\right)}\cos\theta.
$$
(4.9)

As noted from Eq. (4.9), the dynamic pressure $\hat{p}$ is a function of the variables ε, ϕ, $\dot{\varepsilon}/\omega$ and $\dot{\phi}/\omega$. Its dependence on the attitude angle ϕ enters through the definition of the film thickness

$$
H = 1 + \varepsilon\cos\theta
$$

$$
= 1 + \varepsilon\cos[\Xi - (\phi + \psi)],
$$

where $\Xi = 0$ is a fixed position (Figure 3.5).

We solve Eq. (4.9) subject to the Swift-Stieber boundary condition to obtain the force coefficients. The nondimensional force components are

$$
\bar{F}_R = \frac{F_R/LD}{\mu N(R/C)^2} = \left(1 - 2\frac{\dot{\phi}}{\omega}\right)\left[\frac{1}{2}\int_0^1\int_0^{\theta_2}\hat{p}\cos\theta\,d\theta\,d\bar{z}\right],
$$
(4.10a)

$$
\bar{F}_T = \frac{F_T/LD}{\mu N(R/C)^2} = \left(1 - 2\frac{\dot{\phi}}{\omega}\right)\left[\frac{1}{2}\int_0^1\int_0^{\theta_2}\hat{p}\sin\theta\,d\theta\,d\bar{z}\right].
$$
(4.10b)

Employing the notation

$$
f_R = \frac{1}{2}\int_0^1\int_0^{\theta_1}\hat{p}\cos\theta\,d\theta\,d\bar{z},
$$

$$
f_T = \frac{1}{2}\int_0^1\int_0^{\theta_1}\hat{p}\sin\theta\,d\theta\,d\bar{z},
$$

---

[5] Although Eq. (3.33) is linear in ω, $\dot{e}$, and $\dot{\phi}$, superposition of three separate solutions is not permitted, owing to the nonlinear condition $p \geq 0$, and to the fact that the $p = 0$ contour is, in general, dependent on all three parameters ω, $\dot{e}$, and, $\dot{\phi}$. However, at least two of the variables, ω and $\dot{\phi}$, enter the equation in the form ($\omega - 2\dot{\phi}$). It is this property of Eq. (3.34) that is being exploited here.

we can write Eq. (4.10) in the symbolic form

$$\bar{F}_R = \left(1 - 2\frac{\dot{\phi}}{\omega}\right) f_R\left[\varepsilon, \phi, \frac{\dot{\varepsilon}/\omega}{(1 - 2\dot{\phi}/\omega)}\right], \tag{4.11a}$$

$$\bar{F}_T = \left(1 - 2\frac{\dot{\phi}}{\omega}\right) f_T\left[\varepsilon, \phi, \frac{\dot{\varepsilon}/\omega}{(1 - 2\dot{\phi}/\omega)}\right]. \tag{4.11b}$$

The partial derivatives of the force components may now be evaluated:

$$\frac{\partial \bar{F}_R}{\partial \varepsilon} = (1 - 2\dot{\phi}/\omega)\frac{\partial f_R}{\partial \varepsilon},$$

$$\frac{\partial \bar{F}_R}{\partial \phi} = (1 - 2\dot{\phi}/\omega)\frac{\partial f_R}{\partial \phi},$$

$$\frac{\partial \bar{F}_R}{\partial (\dot{\varepsilon}/\omega)} = (1 - 2\dot{\phi}/\omega)\frac{\partial f_R}{\partial\left(\dfrac{\dot{\varepsilon}/\omega}{1 - 2\dot{\phi}/\omega}\right)}\frac{1}{(1 - 2\dot{\phi}/\omega)}, \tag{4.12}$$

$$\frac{\partial \bar{F}_R}{\partial (\dot{\phi}/\omega)} = -2 f_R + (1 - 2\dot{\phi}/\omega)\frac{\partial f_R}{\partial\left(\dfrac{\dot{\varepsilon}/\omega}{1 - 2\dot{\phi}/\omega}\right)}\frac{(2\dot{\varepsilon}/\omega)}{(1 - 2\dot{\phi}/\omega)^2}.$$

When Eq. (4.12) is evaluated at the equilibrium point $\varepsilon = \varepsilon_0, \phi = \phi_0, \dot{\varepsilon} = 0, \dot{\phi} = 0$, we obtain[6] (Szeri, 1966)

$$\begin{pmatrix} d\bar{F}_R \\ d\bar{F}_T \end{pmatrix} = \begin{pmatrix} \dfrac{\partial f_R}{\partial \varepsilon} & \dfrac{\partial f_R}{\varepsilon\partial \phi} - \dfrac{f_T}{\varepsilon} \\[3mm] \dfrac{\partial f_T}{\partial \varepsilon} & \dfrac{\partial f_T}{\varepsilon\partial \phi} + \dfrac{f_R}{\varepsilon} \end{pmatrix} \begin{pmatrix} d\varepsilon \\ \varepsilon d\phi \end{pmatrix} + \begin{pmatrix} \dfrac{\partial f_R}{\partial \dot{\varepsilon}/\omega} & -\dfrac{2 f_R}{\varepsilon} \\[3mm] \dfrac{\partial f_T}{\partial \dot{\varepsilon}/\omega} & -\dfrac{2 f_T}{\varepsilon} \end{pmatrix} \begin{pmatrix} d(\dot{\varepsilon}/\omega) \\ \varepsilon d(\dot{\phi}/\omega) \end{pmatrix}. \tag{4.13}$$

The first matrix on the right describes the oil film response to shaft displacement and is called the (nondimensional) stiffness matrix, $\bar{k}$. The second matrix describes response to velocities $\dot{\varepsilon}/\omega$ and $\varepsilon d(\dot{\phi}/\omega)$ and is called the (nondimensional) damping matrix, $\bar{c}$:

$$\bar{k} = \begin{pmatrix} \dfrac{\partial f_R}{\partial \varepsilon} & \dfrac{\partial f_R}{\varepsilon\partial \phi} - \dfrac{f_T}{\varepsilon} \\[3mm] \dfrac{\partial f_T}{\partial \varepsilon} & \dfrac{\partial f_T}{\varepsilon\partial \phi} + \dfrac{f_R}{\varepsilon} \end{pmatrix}, \tag{4.14a}$$

$$\bar{c} = \begin{pmatrix} \dfrac{\partial f_R}{\partial \dot{\varepsilon}/\omega} & -\dfrac{2 f_R}{\varepsilon} \\[3mm] \dfrac{\partial f_T}{\partial \dot{\varepsilon}/\omega} & -\dfrac{2 f_T}{\varepsilon} \end{pmatrix} \tag{4.14b}$$

(note that we dropped the zero suffix for convenience, remembering that all forces and force derivatives are to be evaluated under conditions of static equilibrium).

[6] We note that the rate of change of f_R with respect to its argument $(\dot{\varepsilon}/\omega)/(1 - 2\dot{\phi}/\omega)$ can be evaluated at arbitrary value of $\dot{\phi}$, say zero, varying $\dot{\varepsilon}/\omega$ only.

The nondimensional matrices $\bar{k}$ and $\bar{c}$ are related to their dimensional counterparts k and c, respectively, through

$$\bar{k} = \frac{C}{LD\mu N \left(\dfrac{R}{C}\right)^2}k, \qquad \bar{c} = \frac{C}{LD\mu N \left(\dfrac{R}{C}\right)^2}c. \tag{4.14c}$$

The elements k and c have dimensions force/length and force/velocity, respectively. The nondimensionalization

$$\bar{\bar{k}} = \frac{C}{W}k \quad \text{and} \quad \bar{\bar{c}} = \frac{C}{W}c, \tag{4.14d}$$

where W is the external load on the journal, is sometimes also employed in the literature. Then $\bar{\bar{k}} = S\bar{k}$ and $\bar{\bar{c}} = S\bar{c}$, where S is the Sommerfeld number.

Analytical Solutions

Long Bearings
Under dynamic conditions the pressure is given by

$$\hat{p} = 12\pi \left[\int_0^\theta \frac{d\theta}{H^2} - \frac{\displaystyle\int_0^{\theta_2} \frac{d\theta}{H^2}}{\displaystyle\int_0^{\theta_2} \frac{d\theta}{H^3}} \int_0^\theta \frac{d\theta}{H^3} \right]$$

$$+ 24\pi \left[\int_0^\theta \frac{\sin\theta}{H^2}d\theta - \frac{\displaystyle\int_0^{\theta_2} \frac{\sin\theta}{H^2}d\theta}{\displaystyle\int_0^{\theta_2} \frac{d\theta}{H^3}} \int_0^\theta \frac{d\theta}{H^3} \right] \frac{\dot{\varepsilon}/\omega}{(1 - 2\dot{\phi}/\omega)}. \tag{4.15}$$

Here $\theta = \theta_2$ is the position of the film-cavity interface. The integrals can be evaluated using the Sommerfeld substitution (3.42), in which the boundaries $\theta = 0$ and π, and $\theta = 0$ and 2π translate to $\psi = 0$ and π and $\psi = 0$ and 2π, respectively.

For the *Sommerfeld boundary condition* $\bar{p}(0) = \bar{p}(2\pi)$, Eq. (4.15) integrates to

$$\hat{p} = 12\pi \left\{ \frac{\varepsilon \sin\theta (2 + \varepsilon \cos\theta)}{(2 + \varepsilon^2)(1 + \varepsilon \cos\theta)^2} + \frac{1}{\varepsilon}\left[\frac{1}{(1 + \varepsilon \cos\theta)^2} - \frac{1}{(1 + \varepsilon)^2}\right] \frac{\dot{\varepsilon}/\omega}{(1 - 2\dot{\phi}/\omega)} \right\}. \tag{4.16}$$

[Under Gümbel boundary conditions, $p(0) = p(\pi)$, $p \geq 0$, there is an additional term on the right-hand side of Eq. (4.16) as shown by Trumpler (1966).]

The force components are calculated from Eq. (4.16) according to Eq. (3.78)

$$f_R = -\frac{12\pi^2}{(1 - \varepsilon^2)^{3/2}} \frac{\dot{\varepsilon}/\omega}{(1 - 2\dot{\phi}/\omega)},$$

$$f_T = \frac{12\pi^2 \varepsilon}{(2 + \varepsilon^2)(1 - \varepsilon^2)^{1/2}}. \tag{4.17}$$

There is no dependence on θ in full journal bearings so that $\partial f_R/\partial\theta = \partial f_T/\partial\theta = 0$, and on substituting Eq. (4.17) into Eq. (4.14) we obtain the $\bar{k}$ and $\bar{c}$ matrices, as shown in Table 4.1.

Table 4.1. *Analytical stiffness and damping coefficients*

| | Long bearing | | Short bearing |
	Sommerfeld BC.	Gümbel BC.	Gümbel BC.
$\bar{k}_{RR}$	0	$\dfrac{-24\pi\varepsilon(2+\varepsilon^4)}{(2+\varepsilon^2)^2(1-\varepsilon^2)^2}$	$-\dfrac{8\pi\varepsilon(1+\varepsilon^2)}{(1-\varepsilon^2)^3}\left(\dfrac{L}{D}\right)^2$
$\bar{k}_{RT}$	$\dfrac{-12\pi^2}{(2+\varepsilon^2)^2(1-\varepsilon^2)^{1/2}}$	$\dfrac{-6\pi^2}{(2+\varepsilon^2)^2(1-\varepsilon^2)^{1/2}}$	$\dfrac{-\pi^2}{(1-\varepsilon^2)^{3/2}}\left(\dfrac{L}{D}\right)^2$
$\bar{k}_{TR}$	$\dfrac{12\pi^2(2-\varepsilon^2+2\varepsilon^4)}{(2+\varepsilon^2)^2(1-\varepsilon^2)^{3/2}}$	$\dfrac{6\pi^2(2-\varepsilon^2+2\varepsilon^4)}{(2+\varepsilon^2)(1-\varepsilon^2)^{3/2}}$	$\dfrac{\pi^2(1+2\varepsilon^2)}{(1-\varepsilon^2)^{5/2}}\left(\dfrac{L}{D}\right)^2$
$\bar{k}_{TT}$	0	$\dfrac{-12\pi\varepsilon}{(2+\varepsilon^2)^2(1-\varepsilon^2)}$	$\dfrac{-4\pi\varepsilon}{(1-\varepsilon^2)^2}\left(\dfrac{L}{D}\right)^2$
$\bar{c}_{RR}$	$\dfrac{-12\pi^2}{(1-\varepsilon^2)^{3/2}}$	$\dfrac{12\pi}{(1-\varepsilon^2)^{3/2}}\left[\dfrac{\pi}{2}-\dfrac{8}{\pi(2+\varepsilon^2)}\right]$	$-\dfrac{2\pi^2(1+2\varepsilon^2)}{(1-\varepsilon^2)^{5/2}}\left(\dfrac{L}{D}\right)^2$
$\bar{c}_{RT}$	0	$\dfrac{24\pi\varepsilon}{(2+\varepsilon^2)(1-\varepsilon^2)}$	$\dfrac{8\pi\varepsilon}{(1-\varepsilon^2)^2}\left(\dfrac{L}{D}\right)^2$
$\bar{c}_{TR}$	0	$\dfrac{24\pi\varepsilon}{(1+\varepsilon)(1-\varepsilon^2)}$	$\dfrac{8\pi\varepsilon}{(1-\varepsilon^2)^2}\left(\dfrac{L}{D}\right)^2$
$\bar{c}_{II}$	$\dfrac{-24\pi^2}{(2+\varepsilon^2)(1-\varepsilon^2)^{1/2}}$	$\dfrac{-12\pi^2}{(2+\varepsilon^2)(1-\varepsilon^2)^{1/2}}$	$-\dfrac{2\pi^2}{(1-\varepsilon^2)^{3/2}}\left(\dfrac{L}{D}\right)^2$
S	$\dfrac{(2+\varepsilon^2)\sqrt{1-\varepsilon^2}}{12\pi^2\varepsilon}$	$\dfrac{(2+\varepsilon^2)(1-\varepsilon^2)}{6\pi\varepsilon\sqrt{4\varepsilon^2+\pi^2(1-\varepsilon^2)}}$	$\dfrac{(1-\varepsilon^2)^2(D/L)^2}{\pi\varepsilon\sqrt{\pi^2(1-\varepsilon^2)+16\varepsilon^2}}$

For the *Gümbel boundary conditions*, $\bar{p}(0)=\bar{p}(\pi)$, the force components are[7]

$$f_R = -\frac{12\pi\varepsilon^2}{(2+\varepsilon^2)(1-\varepsilon^2)}-\frac{12\pi}{(1-\varepsilon^2)^{3/2}}\left[\frac{\pi}{2}-\frac{8}{\pi(2+\varepsilon^2)}\right]\frac{\dot{\varepsilon}/\omega}{(1-2\dot{\phi}/\omega)}, \quad (4.18a)$$

$$f_T = \frac{6\pi^2\varepsilon}{(2+\varepsilon^2)(1-\varepsilon^2)^{1/2}}+\frac{24\pi\varepsilon}{(2+\varepsilon^2)(1-\varepsilon^2)}\frac{\dot{\varepsilon}/\omega}{(1-2\dot{\phi}/\omega)}. \quad (4.18b)$$

Short Bearings

Under the Gümbel condition, $p\geq 0$, the pressure distribution is

$$\left(\frac{D}{L}\right)^2\bar{p}=6\pi\frac{(1-\bar{z}^2)}{(1+\varepsilon\cos\theta)}\left[\varepsilon\sin\theta-2\frac{\dot{\varepsilon}/\omega}{(1-2\dot{\phi}/\omega)}\cos\theta\right] \quad (4.19)$$

[7] Note that if the pressure is evaluated under the Sommerfeld boundary conditions $p(0)=p(2\pi)=0$, and then integrated from $\theta=0$ to $\theta=\pi$ (i.e., $p\geq 0$), the dynamic part of f_R in Eq. (4.18a) will be different (Vance, 1988). Our analysis agrees with that of Hori (1959) and Trumpler (1966).

and the force components are

$$\left(\frac{D}{L}\right)^2 f_R = -\frac{4\pi\varepsilon^2}{(1-\varepsilon^2)^2} - \frac{2\pi^2(1+2\varepsilon^2)}{(1-\varepsilon^2)^{5/2}} \frac{\dot{\varepsilon}/\omega}{(1-2\dot{\phi}/\omega)},$$

$$\left(\frac{D}{L}\right)^2 f_T = \frac{\pi^2\varepsilon}{(1-\varepsilon^2)^{3/2}} + \frac{8\pi\varepsilon}{(1-\varepsilon^2)^2} \frac{\dot{\varepsilon}/\omega}{(1-2\dot{\phi}/\omega)}. \tag{4.20}$$

The linearized spring and damping coefficients derived from Eq. (4.20) are shown in Table 4.1, they are obtained by substituting Eq. (4.20) into Eq. (4.14).

Coordinate Transformations

In place of a single, fixed bearing pad, multiple pivoted-pads are often employed. The characteristics of pivoted-pad bearings are evaluated by combining the characteristics of single pads in suitable manner. With this in view, it is expedient to transform the characteristics of the single pad, i.e., the $\bar{k}$ and $\bar{c}$ matrices of Eq. (4.14), to a coordinate system (ξ, η) that is fixed relative to the pivot and is located in the static equilibrium position O_{JS}. The ξ axis passes through the pivot, and the (ξ, η) axes are related to the (R, T) axes by rotation through ϕ_0.

It is easily seen from Figure 4.4 that under the coordinate transformation

$$\begin{pmatrix} d\bar{\xi} \\ d\bar{\eta} \end{pmatrix} = Q\begin{pmatrix} d\varepsilon \\ \varepsilon d\phi \end{pmatrix}, \qquad Q = \begin{pmatrix} \cos\phi_0 & -\sin\phi_0 \\ \sin\phi_0 & \cos\phi_0 \end{pmatrix}, \tag{4.21}$$

where the angle ϕ_0 is measured from $\bar{\xi}$ to ε counterclockwise and $\bar{\xi} = \xi/C$, $\bar{\eta} = \eta/C$, the force components $(d\bar{F}_R, d\bar{F}_T)$ are transformed according to

$$\begin{pmatrix} d\bar{F}_\xi \\ d\bar{F}_\eta \end{pmatrix} = Q\begin{pmatrix} d\bar{F}_R \\ d\bar{F}_T \end{pmatrix}. \tag{4.22}$$

Substituting from Eqs. (4.21) and (4.14) into Eq. (4.22), we find that

$$\begin{pmatrix} d\bar{F}_\xi \\ d\bar{F}_\eta \end{pmatrix} = Q\bar{k}Q^T\begin{pmatrix} \bar{\xi} \\ \bar{\eta} \end{pmatrix} + Q\bar{c}Q^T\begin{pmatrix} \dot{\bar{\xi}} \\ \dot{\bar{\eta}} \end{pmatrix}. \tag{4.23}$$

Employing the notation[8]

$$\bar{K} = -Q\bar{k}Q^T, \bar{C} = -Q\bar{c}Q^T \tag{4.24}$$

and by substitution into Eq. (4.23),

$$\begin{pmatrix} d\bar{F}_\xi \\ d\bar{F}_\eta \end{pmatrix} = -\bar{K}\begin{pmatrix} \bar{\xi} \\ \bar{\eta} \end{pmatrix} - \bar{C}\begin{pmatrix} \dot{\bar{\xi}} \\ \dot{\bar{\eta}} \end{pmatrix}. \tag{4.25}$$

The component of the stiffness and damping matrices in Eq. (4.24) are given by (Lund,

---

[8] The transformation in Eq. (4.24) defines $\bar{k}$ and $\bar{c}$ as second-order Cartesian tensors.

1964; Szeri, 1966)

$$\bar{K}_{\xi\xi} = -\frac{\partial f_R}{\partial \varepsilon} \cos^2 \phi - \frac{\partial f_T}{\varepsilon \partial \phi} \sin^2 \phi + \left(\frac{\partial f_T}{\partial \varepsilon} + \frac{\partial f_R}{\varepsilon \partial \phi} \right) \frac{\sin 2\phi}{2} - \frac{f_\eta}{\varepsilon} \sin \phi,$$

$$\bar{K}_{\xi\eta} = -\frac{\partial f_R}{\varepsilon \partial \phi} \cos^2 \phi + \frac{\partial f_T}{\partial \varepsilon} \sin^2 \phi + \left(\frac{\partial f_T}{\varepsilon \partial \phi} - \frac{\partial f_R}{\partial \varepsilon} \right) \frac{\sin 2\phi}{2} + \frac{f_\eta}{\varepsilon} \cos \phi,$$

$$\bar{K}_{\eta\xi} = -\frac{\partial f_T}{\partial \varepsilon} \cos^2 \phi + \frac{\partial f_R}{\varepsilon \partial \phi} \sin^2 \phi + \left(\frac{\partial f_T}{\varepsilon \partial \phi} - \frac{\partial f_R}{\partial \varepsilon} \right) \frac{\sin 2\phi}{2} + \frac{f_\xi}{\varepsilon} \cos \phi,$$ $$\qquad (4.26a)$$

$$\bar{K}_{\eta\eta} = -\frac{\partial f_T}{\varepsilon \partial \phi} \cos^2 \phi - \frac{\partial f_R}{\partial \varepsilon} \sin^2 \phi - \left(\frac{\partial f_R}{\varepsilon \partial \phi} + \frac{\partial f_T}{\partial \varepsilon} \right) \frac{\sin 2\phi}{2} - \frac{f_\xi}{\varepsilon} \cos \phi,$$

$$\bar{C}_{\xi\xi} = -\frac{\partial f_R}{\partial \dot{\varepsilon}/\omega} \cos^2 \phi + \frac{\partial f_T}{\partial \dot{\varepsilon}/\omega} \frac{\sin 2\phi}{2} - \frac{2 f_\xi}{\varepsilon} \sin \phi,$$

$$\bar{C}_{\xi\eta} = \frac{\partial f_T}{\partial \dot{\varepsilon}/\omega} \sin^2 \phi - \frac{\partial f_R}{\partial \dot{\varepsilon}/\omega} \frac{\sin 2\phi}{2} + \frac{2 f_\xi}{\varepsilon} \cos \phi,$$

$$\bar{C}_{\eta\xi} = -\frac{\partial f_T}{\partial \dot{\varepsilon}/\omega} \cos^2 \phi - \frac{\partial f_R}{\partial \dot{\varepsilon}/\omega} \frac{\sin 2\phi}{2} - \frac{2 f_\eta}{\varepsilon} \sin \phi,$$ $$\qquad (4.26b)$$

$$\bar{C}_{\eta\eta} = -\frac{\partial f_R}{\partial \dot{\varepsilon}/\omega} \sin^2 \phi - \frac{\partial f_T}{\partial \dot{\varepsilon}/\omega} \frac{\sin 2\phi}{2} + \frac{2 f_\eta}{\varepsilon} \cos \phi.$$

Here we employed the notation

$$\begin{pmatrix} f_\xi \\ f_\eta \end{pmatrix} = Q \begin{pmatrix} f_R \\ f_T \end{pmatrix}. \qquad (4.26c)$$

and dropped the zero subscript on ϕ_0 to conform with accepted notation.

For a vertical load, $f_\eta = 0$ and $f_\xi = 1/S$; the Sommerfeld number S is defined in Eq. (3.46).

4.2 Stability of a Flexible Rotor

Consider a weightless elastic shaft supporting a disk of mass M at its midpoint. The shaft, in its turn, is supported by identical, single pad journal bearings at its end points. In Figure 4.5, the geometric center of a bearing pad is designated by O_B. Under the load $W = Mg$, the center of the rotating journal occupies its static equilibrium position O_{J_s}, while the mass center moves to O_M, due to the deflection of the shaft. We define an orthogonal Cartesian coordinate system (x, y) with origin in O_{J_s}, as shown in Figure 4.3, so O_M has coordinates (x_2, y_2).

If the rotor-bearing system is undisturbed, the rotor will remain in its static equilibrium position. If disturbed and the disturbances are small, the rotor will leave its equilibrium position and proceed along a closed orbit around it. This is what occurs in well designed, stable, rotor-bearing systems. Under other, unstable, conditions the rotor is unable to find a *limit cycle* and its path spirals outward until metal to metal contact between rotor and bearing occurs. There is great practical importance attached, therefore, to knowing the criteria that demarcates stable from unstable operation. To find this criteria we begin with the equations of motion:

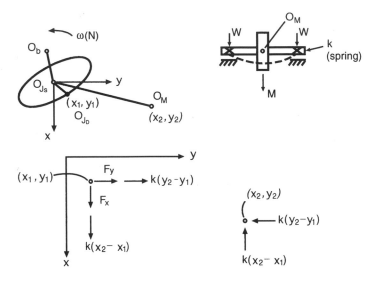

Figure 4.5. Schematic of flexible rotor supported on two identical journal bearings.

rotor:

$$-k(x_2 - x_1) = M\ddot{x}_2$$
$$-k(y_2 - y_1) = M\ddot{y}_2$$

(4.27)

bearing:

$$2dF_x + k(x_2 - x_1) = 0$$
$$2dF_y + k(y_2 - y_1) = 0$$

(4.28)

Here k is the shaft stiffness and dF_x, dF_y are the components of the lubricant force that is exerted on the shaft in excess of the equilibrium force $F_x = W$, $F_y = 0$. The force increments (dF_x, dF_y) are given by Eq. (4.1).

To solve Eqs. (4.27) and (4.28), we assume that both O_J and O_M undergo harmonic motion of the type

$$\begin{pmatrix} x_1 \\ y_1 \end{pmatrix} = \begin{bmatrix} X_1 \\ Y_1 \end{bmatrix} e^{vt}, \quad \begin{pmatrix} x_2 \\ y_2 \end{pmatrix} = \begin{bmatrix} X_2 \\ Y_2 \end{bmatrix} e^{vt}.$$

(4.29)

The eigenvalue v in Eq. (4.29) is , in general, complex,

$$v = \mathcal{R}(v) + i\mathcal{I}(v),$$

and we have

$$e^{vt} = e^{\mathcal{R}(v)t}[\cos \mathcal{I}(v)t + i \sin \mathcal{I}(v)t].$$

The real part of v is called the *damping exponent*. If $\mathcal{R}(v) < 0$, the motion is *stable*, and it is *unstable* if $\mathcal{R}(v) > 0$. The *neutral state* of stability (*threshold of stability*) is characterized by $\mathcal{R}(v) = 0$. The imaginary part of v is the *damped natural frequency* (the orbiting frequency).

Substituting Eq. (4.29) into Eqs. (4.27) and (4.28) and taking into account Eq. (4.1), we obtain

$$
\begin{bmatrix}
\dfrac{kMv^2}{Mv^2 + k} + 2K_{xx} + 2vC_{xx} & 2K_{xy} + 2vC_{xy} \\[4mm]
2K_{yx} + 2vC_{yx} & \dfrac{kMv^2}{Mv^2 + k} + 2K_{yy} + 2vC_{yy}
\end{bmatrix}
\begin{bmatrix} X_1 \\ Y_1 \end{bmatrix}
=
\begin{bmatrix} 0 \\ 0 \end{bmatrix}.
\tag{4.30}
$$

We nondimensionalize Eq. (4.30) according to

$$
\{K_{ij}, \omega C_{ij}, k\} = \frac{LD\mu N (R/C)^2}{C} \{\bar{K}_{ij}, \bar{C}_{ij}, \bar{k}\},
$$

$$
\{X_1, Y_1\} = C\{\bar{X}_1, \bar{Y}_1\}, \qquad v = \omega \bar{v}, \qquad \omega = \omega_N \bar{\omega}.
\tag{4.31}
$$

Here $\omega_N = \sqrt{k/M}$ is the natural frequency of the rotor and the overbar signifies, as before, nondimensional quantities.

Employing the rotation

$$
\alpha = \frac{\bar{k} \bar{v}^2 \bar{\omega}^2}{\bar{v}^2 \bar{\omega}^2 + 1} \qquad \text{(a real number)},
\tag{4.32}
$$

we obtain the nondimensional form of Eq. (4.30)

$$
\begin{bmatrix}
\alpha + 2\bar{K}_{xx} + 2\bar{v}\bar{C}_{xx} & 2\bar{K}_{xy} + 2\bar{v}\bar{C}_{xy} \\[2mm]
2\bar{K}_{yx} + 2\bar{v}\bar{C}_{yx} & \alpha + 2\bar{K}_{yy} + 2\bar{v}\bar{C}_{yy}
\end{bmatrix}
\begin{bmatrix} \bar{X}_1 \\ \bar{Y}_1 \end{bmatrix}
=
\begin{bmatrix} 0 \\ 0 \end{bmatrix}.
\tag{4.33}
$$

Equation (4.33) possesses a nontrivial solution if and only if the system determinant vanishes. By expanding the determinant and equating it to zero, we would obtain a fourth order polynomial in $\bar{v}$. Solving this so-called *frequency (characteristic) equation* would yield $\bar{v}$.

Instead of solving directly for $\bar{v}$, however, we will seek conditions for marginal stability, a state that is characterized by the vanishing of the real part of $\bar{v}$. To this end, we separate the determinant in Eq. (3.33) into its real and imaginary parts, which are then individually equated to zero.

The real part of the determinant of Eq. (3.33) is

$$
\begin{vmatrix}
\alpha + 2\bar{K}_{xx} & 2\bar{K}_{xy} \\[1mm]
2\bar{K}_{yx} & \alpha + 2\bar{K}_{yy}
\end{vmatrix}
+ \bar{v}^2
\begin{vmatrix}
2\bar{C}_{xx} & 2\bar{C}_{yx} \\[1mm]
2\bar{C}_{xy} & 2\bar{C}_{yy}
\end{vmatrix}
= 0
\tag{4.34}
$$

and its imaginary part is

$$
\begin{vmatrix}
2\bar{C}_{xx} & 2\bar{K}_{xy} \\[1mm]
2\bar{C}_{yx} & \alpha + 2\bar{K}_{yy}
\end{vmatrix}
\bar{v}
+
\begin{vmatrix}
\alpha + 2\bar{K}_{xx} & 2\bar{C}_{xy} \\[1mm]
2\bar{K}_{yx} & 2\bar{C}_{yy}
\end{vmatrix}
\bar{v} = 0.
\tag{4.35}
$$

Since $\bar{v} \neq 0$, we can divide Eq. (4.35) by $\bar{v}$, and on expanding the determinants obtain

$$
\alpha = \frac{2(\bar{K}_{xy}\bar{C}_{yx} + \bar{K}_{yx}\bar{C}_{xy} - \bar{K}_{yy}\bar{C}_{xx} - \bar{K}_{xx}\bar{C}_{yy})}{(\bar{C}_{xx} + \bar{C}_{yy})}.
\tag{4.36}
$$

Using the definition of α, we find the instability threshold value of the relative frequency as

$$
\bar{\omega}_c^2 = \frac{\alpha}{\bar{v}^2(\bar{k} - \alpha)}.
\tag{4.37}
$$

in terms of α, $\bar{k}$, and the yet undetermined whirl ratio $\bar{v}$.

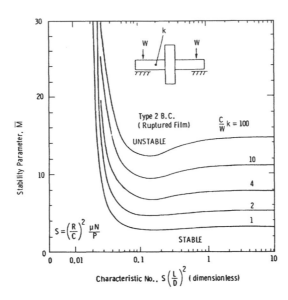

Figure 4.6. Stability of single mass rotor on full journal bearings, mounted on a rigid support. (Reprinted with permission from Raimondi, A. A. and Szeri, A. Z. Journal and thrust bearings. In Booser, E. R., *CRC Handbook of Lubrication*. Copyright CRC Press, Boca Raton, Florida. © 1984.)

If $\bar{\omega} > \bar{\omega}_c$, the system is unstable. At $\bar{\omega}_c = 1$, instability sets in at the system natural frequency, $\omega_N = \sqrt{k/M}$.

To render Eq. (4.37) useful, we need to determine the whirl ratio $\bar{\nu} = \nu/\omega$; this can be accomplished by expanding Eq. (4.34)

$$\bar{\nu}^2 = -\frac{\alpha^2 + 2(\bar{K}_{xx} + \bar{K}_{yy})\alpha + 4(\bar{K}_{xx}\bar{K}_{yy} - \bar{K}_{xy}\bar{K}_{yx})}{4(\bar{C}_{xx}\bar{C}_{yy} - \bar{C}_{yx}\bar{C}_{xy})}. \tag{4.38}$$

In the state of neutral stability $\bar{\nu}$ is purely imaginary:

$$\bar{\nu} = i\Im(\bar{\nu}) \tag{4.39}$$

and

$$\frac{\nu_{\text{whirl}}}{\omega} = \sqrt{-\bar{\nu}^2}, \tag{4.40}$$

where $\bar{\nu}^2$ is given by Eq. (4.38).

Note that both $\bar{\omega}_c$ and $\bar{\nu}$ are functions of the (dimensionless) shaft stiffness, $\bar{k}$, and the elements of the (dimensionless) bearing stiffness and damping matrices, $\bar{K}$ and $\bar{C}$; the latter two being evaluated at, and depending on, the static equilibrium position of the journal (ε_0, ϕ_0). Thus, for a given system, a stability chart can be prepared as a sole function of the static position of the journal, or, alternatively, as a sole function of the Sommerfeld number.

The stability characteristics of a flexible rotor carried on full journal bearings that are supported on a rigid foundation can be estimated from Figure 4.6 (Raimondi and Szeri, 1984).

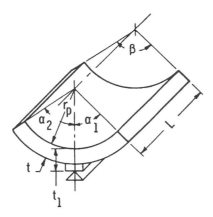

Figure 4.7. Pivoted-pad schematics. (Reprinted with permission from Raimondi, A. A. and Szeri, A. Z. Journal and thrust bearings. In Booser, E. R., *CRC Handbook of Lubrication*. Copyright CRC Press, Boca Raton, Florida. © 1984.)

In Figure 4.6 we plotted the stability parameter

$$\bar{M} = \frac{C}{W} M \omega^2 \qquad (4.41)$$

against the short-bearing Sommerfeld number.

4.3   Pivoted-Pad Journal Bearings

Tilting-pad journal bearings consist of a number of individually pivoted pads or shoes (Figure 4.7). Pivoting makes relatively high loading possible where shaft deflection or misalignment is a factor. The most important features of tilting-pad journal bearings are (1) small cross-coupling coefficients resulting in inherent stability, (2) availability of preloading to achieve relatively high stiffness (important with vertical rotors), and (3) operation with clearances smaller than considered desirable for fixed-pad journal bearings.

Normal practice is to construct all pads alike and space them uniformly around the circumference. When the number of pads is large, there is little difference in bearing performance between two alternatives: load line passing through a pivot or between pivots. When the number of pads is small, however, the load-between-pads orientation is preferred; in this case the load capacity is greater, the temperature rise is lower as the load is distributed more uniformly, and the lateral stiffness and damping are greater.

Load capacity is not unduly sensitive to pivot location when using oil lubricants. In these cases, the pivot is usually positioned at pad center to preserve independence from direction of journal rotation. When using low-viscosity fluids (water, liquid metals, and particularly gases), however, load capacity is sensitive to pivot location, and the pivot must be offset toward the trailing edge (Boyd and Raimondi, 1962).

Figure 4.8 illustrates a pad of angular extent β, which is machined to radius $R + C$ (position 1). In the absence of tilting, the film thickness is uniform (equal to C) and no hydrodynamic force can be developed. However, if the pad is now moved to position 2, by displacing the pivot radially inward a distance $(C - C')$, there results a nonuniform film thickness and, consequently, a hydrodynamic force that preloads the journal. Bearings of vertical machines

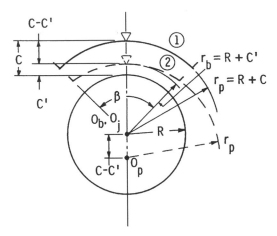

Figure 4.8. Preloading of pad. (Reprinted with permission from Raimondi, A. A. and Szeri, A. Z. Journal and thrust bearings. In Booser, E. R., *CRC Handbook of Lubrication.* Copyright CRC Press, Boca Raton, Florida. © 1984.)

operate almost concentric with their journal. To overcome the low radial stiffness and consequent *spragging*, vertical machines are often equipped with preloaded guide bearings.

Preload is characterized by the *preload coefficient* $m = (C - C')/C$, where $0 \leq m \leq 1$. Each pad is preloaded usually the same amount in vertical machines, while in horizontal machines often the top pads only are preloaded to prevent spragging.

A pivoted pad will track the orbiting journal by rocking about its pivot. The influence of pad inertia on the dynamic coefficients of the pad is negligible except when approaching pad resonance. At pad resonance, the journal and pad motions are $90°$ out of phase. The onset of pad resonance can be determined from the value of the *critical mass parameter* and requires calculation of the polar moment of inertia, I_p, of the pad.

Referring to Figure 4.7, we have

$$I_p = 2r_p^2 M_p \left\{ 1 + f_1 - \left[\frac{(\sin \alpha_1 + \sin \alpha_2)}{\beta} f_2 \right] \right\}, \qquad (4.42)$$

where

$$f_1 = \left[\left(\frac{t_1}{r_p} \right) + \frac{1}{2} \left(\frac{t_1}{r_p} \right)^2 + \frac{1}{2} \left(\frac{t}{r_p} \right) + \frac{1}{4} \left(\frac{t}{r_p} \right)^2 \right],$$

$$f_2 = \left[1 + \left(\frac{t_1}{r_p} \right) \right] \left[1 + \left(\frac{t}{r_p} \right) + \frac{1}{2} \left(\frac{t}{r_p} \right)^2 \right] \Big/ \left[1 + \frac{1}{2} \left(\frac{t}{r_p} \right) \right],$$

and M_p is the pad mass. Design data, such as shown in Figures 4.10 to 4.14, are often calculated on the (admittedly unrealistic) assumption that the pivot point is located on the pad surface; for this case $t_1/r_p = t/r_p = 0$ in Eq. (4.42).

The first researchers to compute the linearized spring and damping coefficients of a full journal bearing were Lund and Sternlicht (1962). Corresponding calculations for a partial bearing were made by Szeri (1966). Analysis of tilting pad bearings is more complicated. Lund (1964) presented a pad assembly method in which he assumed harmonic motion for both journal and pad. Shapiro and Colsher (1977) and later Allaire, Parsell, and Barrett

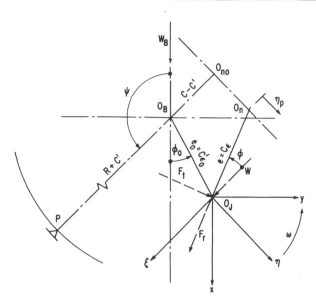

Figure 4.9. Coordinate systems for the pivoted-pad. (Reprinted with permission from Lund, J. W. Spring and damping coefficients for the tilting-pad journal bearing. *ASLE Trans.*, **7**, 342, 1964.)

(1981) published a pad perturbation method that calculates a complete coefficient matrix for the tilting pad bearing. In this latter scheme, the coefficient matrix is independent of pad motion; frequency dependence and inertial dependence enter the analysis of the rotor-bearing system only when a specific pad motion is assumed. In the sequel, we give details of the pad assembly method of Lund (1964), and refer to the pad perturbation method of Shapiro and Colsher (Allaire, Parsell, and Barrett, 1981).

Pad Assembly Method

Coefficient for Single Tilting Pad

In Figure 4.9, the symbol O_B marks the position of the bearing center and O_J the instantaneous position of the shaft center. The center of curvature of the pad is at O_{no} in the absence of load and at O_n when fully loaded and tilted. The bearing load vector, W_B, is in the vertical, and the pivot point, P, is located at an angle, ψ, relative to the load line. With respect to O_B, the eccentricity of the journal center, O_J, is $e_0 = c\varepsilon_0$ at the attitude angle ϕ_0. The eccentricity of the journal is $e = c\varepsilon$ relative to the instantaneous pad center, and the attitude angle, ϕ, is measured from the load line for the pad, which, by necessity, connects O_J and P. Representing the radius of the journal by R, the radius of the pad is $\overline{O_n P} = R + C$, while the pivot circle, centered at O_B, has radius $\overline{O_B P} = R + C'$. From geometric consideration, we have

$$
\begin{aligned}
\varepsilon \cos \phi &= 1 - \frac{C'}{C} - \varepsilon_0 \cos(\psi - \phi_0) \\
&= m - \varepsilon_0 \cos(\psi - \phi_0),
\end{aligned}
\tag{4.43}
$$

where m is the preload coefficient defined earlier.

If the position of the journal, O_J, relative to the bearing, O_B, is known, i.e., if the couple (ε_0, ϕ_0) is specified, Eq. (4.43) and the requirement that the load on the pad, W, passes through the pivot, P, are sufficient to determine the couple (ε, ϕ) for each pad. The components of the required lubricant force are obtained from the force equilibrium in Eq. (3.40b). Knowing (ε, ϕ) enables calculation of individual pad performance, which can then be summed (vectorially) to yield performance characteristics for the bearing.

Unfortunately (ε_0, ϕ_0) is not known a priori, and the best the designer can do is assume ε_0, and use the condition that W_B is purely vertical to calculate the corresponding ϕ_0. This procedure is, at least, tedious. If, however, the pivots are arranged symmetrically with respect to the load line, the pads are centrally pivoted and are identical, we have $\phi_0 \equiv 0$, and the journal moves along the load vector W_B.

In the following discussion of the pad assembly method, we assume that all conditions for $\phi_0 \equiv 0$ are met.

Referring to Figure 4.9, we note that under dynamic load the pad center oscillates about O_n. Denoting the amplitude of this oscillation of the pad center by η_p and representing the pad moment of inertia by I_p, Eq. (4.42), we can write the equation of motion of the pad as

$$I_p \frac{\ddot{\eta}_p}{R_p} = -R_p dF_\eta. \tag{4.44}$$

Setting $M_p = I_p / R_p^2$, Eq. (4.44) becomes

$$M_p \ddot{\eta}_p = K_{\eta\xi} \xi + C_{\eta\xi} \dot{\xi} + K_{\eta\eta}(\eta_p - \eta) + C_{\eta\eta}(\dot{\eta}_p - \dot{\eta}). \tag{4.45}$$

We can relate pad motion, η_p, to the motion of the journal, η, by assuming that both journal and pad execute synchronous motion according to

$$(\xi, \eta) = (\hat{\xi}, \hat{\eta})e^{i\omega t}, \qquad \eta_p = \hat{\eta}_p e^{i\omega t}. \tag{4.46}$$

Equation (4.46) is for the common case of unbalance excitation, when, necessarily, excitation occurs at the shaft running speed, ω.

Unlike fixed arc bearings, the spring and damping coefficients of pivoted-pad bearings are dependent upon the frequency, Ω, of the excitation force. For excitation at frequency $\Omega \neq \omega$, we should use

$$(\hat{\xi}, \hat{\eta})e^{i\Omega t}, \qquad \hat{\eta}_p e^{i\Omega t}$$

in Eq. (4.46). Bearing characteristics for non-synchronous excitation $\Omega/\omega \neq 1$ are presented by Raimondi and Szeri (1984). Here, for simplicity, we only treat synchronous excitation $\Omega/\omega = 1$ and use Eq. (4.46).

Substituting Eq. (4.46) into Eq. (4.45) yields

$$\hat{\eta} - \hat{\eta}_p = -[(K_{\eta\xi} + i\omega C_{\eta\xi})\xi + M_p \omega^2 \eta](p - iq), \tag{4.47}$$

where

$$(p, q) \equiv (K_{\eta\eta} - M_p \omega^2, \omega C_{\eta\eta}) / \left[\left(K_{\eta\eta}^2 - M_p \omega^2 \right)^2 + (\omega C_{\eta\eta})^2 \right].$$

Equation (4.47) can now be used to eliminate η_p from

$$\begin{pmatrix} dF_\xi \\ dF_\eta \end{pmatrix} = -K \begin{pmatrix} \xi \\ \eta - \eta_p \end{pmatrix} - C \begin{pmatrix} \dot{\xi} \\ \dot{\eta} - \dot{\eta}_p \end{pmatrix}, \tag{4.48}$$

which is the dimensional counterpart of Eq. (4.25), written for a pivoting pad. Thus, replacing $(\eta - \eta_p)$ in Eq. (4.48) from Eq. (4.47), we have

$$dF_\xi = -(K'_{\xi\xi} + i\omega C'_{\xi\xi})\hat{\xi} - (K'_{\xi\eta} + i\omega C'_{\xi\eta})\hat{\eta},$$
$$dF_\eta = -(K'_{\eta\xi} + i\omega C'_{\eta\xi})\hat{\xi} - (K'_{\eta\eta} + i\omega C'_{\eta\eta})\hat{\eta}. \tag{4.49a}$$

where $K'_{\xi\xi}, C'_{\xi\xi}$, etc. are the spring and damping coefficients for the tilting pad given by (Lund, 1964):

$$K'_{\xi\xi} = K_{\xi\xi} - (pK_{\xi\eta} + q\omega C_{\xi\eta})K_{\eta\xi} - (qK_{\xi\eta} - p\omega C_{\xi\eta})\omega C_{\eta\xi},$$
$$\omega C'_{\xi\xi} = \omega C_{\xi\xi} - (pK_{\xi\eta} + q\omega C_{\xi\eta})\omega C_{\eta\xi} + (qK_{\xi\eta} - p\omega C_{\xi\eta})K_{\eta\xi},$$
$$K'_{\xi\eta} = -M_p\omega^2(pK_{\xi\eta} + q\omega C_{\xi\eta}),$$
$$\omega C'_{\xi\eta} = M_p\omega^2(qK_{\xi\eta} - p\omega C_{\xi\eta}),$$
$$K'_{\eta\xi} = -M_p\omega^2(pK_{\eta\xi} + q\omega C_{\eta\xi}), \tag{4.49b}$$
$$\omega C'_{\eta\xi} = M_p\omega^2(qK_{\eta\xi} - p\omega C_{\eta\xi}),$$
$$K'_{\eta\eta} = -M_p\omega^2(pK_{\eta\eta} + q\omega C_{\eta\eta}) = -M_p\omega^2(1 + pM_p\omega^2),$$
$$\omega C'_{\eta\eta} = M_p\omega^2(qK_{\eta\eta} - p\omega C_{\eta\eta}) = (M_p\omega^2)^2q.$$

If the pad has no inertia, then $M_p = 0$ and only $K'_{\xi\xi}$ and $C'_{\xi\xi}$ are nonzero. This is on account of the pad freely tilting about the pivot.

The transformation from fixed, pad data to dynamic, pad data may be looked upon as allowing the pad to pitch so that the load will pass through the pivot.

Bearing Coefficients

In the previous section we obtained the spring and damping coefficients for the nth tilting pad relative to its own (local) coordinate system (ξ, η). It will be to our advantage during assembly if the coefficients for each of the pads are referred to a global coordinate system, (x, y) in Figure 4.9, of the bearing. The coordinate transformation from a local coordinate system (ξ, η) to the global system (x, y) is

$$\begin{pmatrix} x \\ y \end{pmatrix} = Q\begin{pmatrix} \xi \\ \eta \end{pmatrix}, \qquad Q = \begin{pmatrix} \cos(\pi - \psi_n) & \sin(\pi - \psi_n) \\ -\sin(\pi - \psi_n) & \cos(\pi - \psi_n) \end{pmatrix}, \tag{4.50a}$$

where ψ_n is the pivot angle of the nth pad (Figure 4.9), measured counterclockwise from the load line.

Under the coordinate transformation Eq. (4.50a), the components of the incremental force dF transform according to the formula

$$\begin{pmatrix} dF_x \\ dF_y \end{pmatrix} = Q\begin{pmatrix} dF_\xi \\ dF_\eta \end{pmatrix}. \tag{4.50b}$$

Combining Eqs. (4.49a), (4.50a), and (4.50b), we obtain

$$\begin{pmatrix} dF_x \\ dF_y \end{pmatrix} = -QK'Q^T\begin{pmatrix} x \\ y \end{pmatrix} - QC'Q^T\begin{pmatrix} \dot{x} \\ \dot{y} \end{pmatrix}. \tag{4.51}$$

Writing

$$K^{(n)} = QK'Q^T, \qquad C^{(n)} = QC'Q^T \tag{4.52}$$

for the stiffness and damping matrix of the nth pad, referred to the (x, y) global coordinate system, we have

$$\begin{pmatrix} dF_x^{(n)} \\ dF_y^{(n)} \end{pmatrix} = -\mathbf{K}^{(n)} \begin{pmatrix} x \\ y \end{pmatrix} - \mathbf{C}^{(n)} \begin{pmatrix} \dot{x} \\ \dot{y} \end{pmatrix}. \tag{4.53a}$$

The matrices in Eq. (4.53a) have components calculated by substituting Eq. (4.49) into Eq. (4.52):

$$
\begin{aligned}
K_{xx}^{(n)} &= K_{\xi\xi}' \cos^2 \psi_n + K_{\eta\eta}' \sin^2 \psi_n - \left(K_{\xi\eta}' + K_{\eta\xi}'\right) \cos \psi_n \sin \psi_n, \\
\omega C_{xx}^{(n)} &= \omega C_{\xi\xi}' \cos^2 \psi_n + \omega C_{\eta\eta}' \sin^2 \psi_n - \left(\omega C_{\xi\eta}' + \omega C_{\eta\xi}'\right) \cos \psi_n \sin \psi_n, \\
K_{xy}^{(n)} &= K_{\xi\eta}' \cos^2 \psi_n - K_{\eta\xi}' \sin^2 \psi_n + \left(K_{\xi\xi}' - K_{\eta\eta}'\right) \cos \psi_n \sin \psi_n, \\
\omega C_{xy}^{(n)} &= \omega C_{\xi\eta}' \cos^2 \psi_n - \omega C_{\eta\xi}' \sin^2 \psi_n + \left(\omega C_{\xi\xi}' - \omega C_{\eta\eta}'\right) \cos \psi_n \sin \psi_n, \\
K_{yx}^{(n)} &= K_{\eta\xi}' \cos^2 \psi_n - K_{\xi\eta}' \sin^2 \psi_n + \left(K_{\xi\xi}' - K_{\eta\eta}'\right) \cos \psi_n \sin \psi_n, \\
\omega C_{yx}^{(n)} &= \omega C_{\eta\xi}' \cos^2 \psi_n - \omega C_{\xi\eta}' \sin^2 \psi_n + \left(\omega C_{\xi\xi}' - \omega C_{\eta\eta}'\right) \cos \psi_n \sin \psi_n, \\
K_{yy}^{(n)} &= K_{\eta\eta}' \cos^2 \psi_n + K_{\xi\xi}' \sin^2 \psi_n + \left(K_{\xi\eta}' + K_{\eta\xi}'\right) \cos \psi_n \sin \psi_n, \\
\omega C_{yy}^{(n)} &= \omega C_{\eta\eta}' \cos^2 \psi_n + \omega C_{\xi\xi}' \sin^2 \psi_n + \left(\omega C_{\xi\eta}' + \omega C_{\eta\xi}'\right) \cos \psi_n \sin \psi_n.
\end{aligned}
\tag{4.53b}
$$

A summation over N pads that make up the bearing yields the bearing spring and damping coefficients:

$$\{\mathcal{K}, \mathcal{C}\} = \sum_{n=1}^{N_{\text{pad}}} \left\{ \mathbf{K}^{(n)}, \mathbf{C}^{(n)} \right\} \tag{4.54}$$

If the pads are assumed to have no inertia – and this assumption is good for conditions far from pad resonance – we may write

$$\mathbf{K}^{(n)} = K_{\xi\xi}' \mathbf{Y}_{(n)}, \quad \mathbf{C}^{(n)} = C_{\xi\xi}' \mathbf{Y}_{(n)}, \tag{4.55}$$

where

$$\mathbf{Y}_{(n)} = \begin{pmatrix} \cos^2 \psi_n & \dfrac{1}{2} \sin^2 \psi_n \\ \dfrac{1}{2} \sin^2 \psi_n & \sin^2 \psi_n \end{pmatrix}$$

for the nth pad.

At zero preload, symmetry about the x axis, and load-between-pads arrangement, Eq. (4.43) gives

$$\varepsilon_0 = \frac{\varepsilon \cos \phi}{-\cos \left[\dfrac{(N-1)\pi}{N} \right]}$$

where we put $\psi = (N-1)\pi/N$ for the lowermost pad. The maximum value of ε_0 is obtained with $\varepsilon = 1$ and $\phi = 0$

$$\max(\varepsilon_0) = \frac{1}{-\cos \left[\dfrac{(N-1)\pi}{N} \right]}.$$

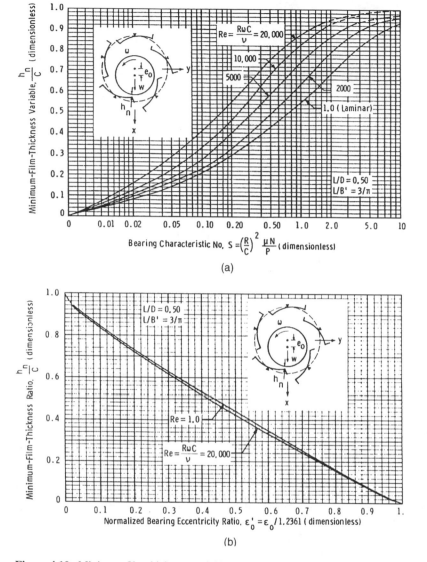

Figure 4.10. Minimum film thickness variable versus (a) bearing characteristic number and (b) normalized bearing eccentricity ratio: five 60° tilting pads, central loading, no preload, inertialess pad, $L/D = 0.5$. (Reprinted with permission from Raimondi, A. A. and Szeri, A. Z. Journal and thrust bearings. In Booser, E. R., *CRC Handbook of Lubrication*. Copyright CRC Press, Boca Raton, Florida. © 1984.)

The normalized bearing eccentricity $0 \leq \varepsilon_0' \leq 1$ is defined as

$$\varepsilon_0' = \frac{\varepsilon_0}{\max(\varepsilon_0)}.$$

For the five-padbearing of Figures 4.10 to 4.13, max $(\varepsilon_0) = 1.2361$.

Figures 4.10 to 4.13 plot performance characteristics for a five-pad pivoted-pad bearing of $L/D = 0.5$. The pads are identical and are arranged symmetrically with respect to the load line, hence the journal moves along the vertical, $\phi_0 = 0$. Figure 4.10 shows the

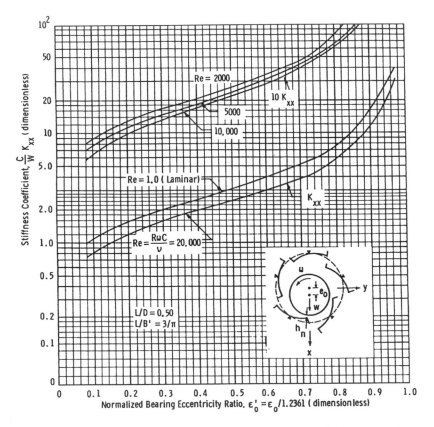

Figure 4.11. Bearing vertical stiffness: five 60° tilting pads, central loading, no pre-load, inertialess pad, $L/D = 0.5$. (Reprinted with permission from Raimondi, A. A. and Szeri, A. Z. Journal and thrust bearings. In Booser, E. R., *CRC Handbook of Lubrication*. Copyright CRC Press, Boca Raton, Florida. © 1984.)

variation of the minimum film thickness parameter $H_{\min}$ against (a) bearing characteristic number, and (b) against normalized bearing eccentricity ratio.

Figures 4.11 and 4.12 plot vertical and horizontal bearing stiffness, respectively, and Figure 4.13 contains information on bearing damping.

If the nth pad and the $(N - n + 1)$th pad are identical and are symmetrically placed relative to the load line, then

$$\psi_{(N-n+1)} = (2\pi - \psi_n),$$
$$\frac{1}{2} \sin \psi_{(N-n+1)} = -\frac{1}{2} \sin \psi_n,$$

hence, for symmetry about the load line, the cross-coupling terms in Eq. (4.53) vanish.

To investigate the motion of a pad, we write Eq. (4.47) in the form

$$\eta_p = \frac{(K_{\eta\xi} + i\omega C_{\eta\xi})\xi + (K_{\eta\eta} + i\omega C_{\eta\eta})\eta}{K_{\eta\eta} - M_p\omega^2 + i\omega C_{\eta\eta}}. \tag{4.56}$$

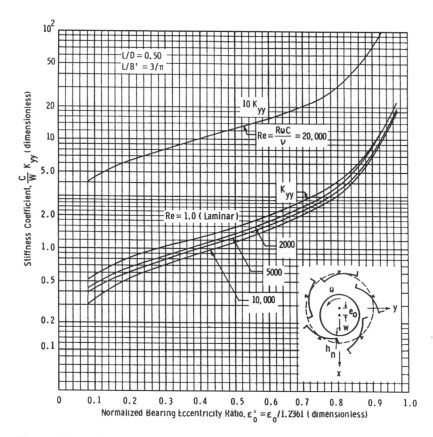

Figure 4.12. Bearing horizontal stiffness: five 60° tilting pads, central loading, no preload, inertialess pad, $L/D = 0.5$. (Reprinted with permission from Raimondi, A. A. and Szeri, A. Z. Journal and thrust bearings. In Booser, E. R., *CRC Handbook of Lubrication*. Copyright CRC Press, Boca Raton, Florida. © 1984.)

Let $\eta_p = \eta_0$ for zero pad inertia, $M_p = 0$

$$\eta_0 = \eta + \frac{K_{\eta\xi} + i\omega C_{\eta\xi}}{K_{\eta\eta} + i\omega C_{\eta\eta}}\xi. \tag{4.57}$$

The ratio η_p/η_0 is a complex quantity that can be written as

$$\left(\frac{\eta_p}{\eta_0}\right) = \left|\frac{\eta_p}{\eta_0}\right| \exp\{i[\arg(\eta_0) - \arg(\eta_p)]\},$$

where

$$\arg(\eta_0) - \arg(\eta_p) = \tan^{-1}\left\{\frac{\omega C_{\eta\eta} M_p \omega^2}{K_{\eta\eta}(K_{\eta\eta} - M_p \omega^2) + (\omega C_{\eta\eta})^2}\right\}. \tag{4.58}$$

The phase angle $[\arg(\eta_0) - \arg(\eta_p)]$ reaches $\pi/2$ when the denominator in Eq. (4.58) vanishes. At this juncture, the pad is said to possess critical mass, M_{CRIT}

$$\omega^2 M_{\text{CRIT}} = \frac{K_{\eta\eta}^2 + (\omega C_{\eta\eta})^2}{K_{\eta\eta}}. \tag{4.59a}$$

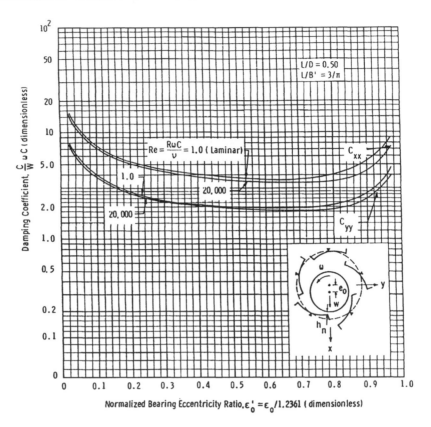

Figure 4.13. Bearing damping: five 60° tilting pads, central loading, no preload, inertialess pad, $L/D = 0.5$. (Reprinted with permission from Raimondi, A. A. and Szeri, A. Z. Journal and thrust bearings. In Booser, E. R., *CRC Handbook of Lubrication*. Copyright CRC Press, Boca Raton, Florida. © 1984.)

If the pad mass satisfies Eq. (4.59a), the conditions for pad resonance are satisfied. In dimensionless form, Eq. (4.59a) is often expressed as

$$\frac{C W_B M_{\text{CRIT}}}{\left[\mu D L \left(\dfrac{R}{C}\right)^2\right]^2} = \frac{1}{4\pi S^2} \frac{\left(\dfrac{C K_{\eta\eta}}{W_B}\right)^2 + \left(\dfrac{C \omega C_{\eta\eta}}{W_B}\right)^2}{\left(\dfrac{C K_{\eta\eta}}{W_B}\right)}, \tag{4.59b}$$

where W_B is the total load on the bearing's and S is the bearing's Sommerfeld number. The critical pad mass for the five-pad bearing is plotted in Figure 4.14.

The computational algorithm of tilting-pad bearings is as follows (Nicholas et al., 1979):

1. The fixed-pad stiffness and damping coefficients $\bar{K}$ and $\bar{C}$ are obtained by displacement and velocity perturbation about the equilibrium positions, Eq. (4.26).
2. The pad dynamic coefficients $\bar{K}'$ and $\bar{C}'$ are calculated using Eq. (4.49). These coefficients, along with the fixed-pad Sommerfeld number, S, are labeled with the dimensionless pivot film thickness, H_p, and stored.

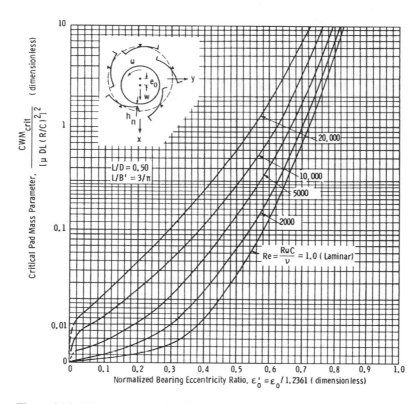

Figure 4.14. Pad critical mass: five 60° tilting pads, central loading, no preload, inertialess pad, $L/D = 0.5$. (Reprinted with permission from Raimondi, A. A. and Szeri, A. Z. Journal and thrust bearings. In Booser, E. R., *CRC Handbook of Lubrication*. Copyright CRC Press, Boca Raton, Florida. © 1984.)

3. A bearing eccentricity ratio ε_0 (as the shaft moves along the vertical, for centrally pivoted, symmetrically arranged pads, $\phi_0 \equiv 0$) is selected, and the pivot film thickness for the nth pad, H_{pn}, $n = 1, 2, 3, \ldots, N$, is calculated.
4. $\bar{K}', \bar{C}'$ at given H_{pi} are obtained by interpolation and transferred to the global (x, y) coordinate system, Eq. (4.52), to find $\bar{K}^{(n)}, \bar{C}^{(n)}, n = 1, 2, 3, \ldots, N$.
5. Use of Eq. (4.54) leads to the bearing dynamic coefficients $\mathcal{K}, \mathcal{C}$.
6. The dimensionless load on each pad is given by

$$\bar{F}_x^{(n)} = -\frac{\cos \psi_n}{S_n},$$

$$\bar{F}_y^{(n)} = -\frac{\sin \psi_n}{S_n}, \tag{4.60}$$

and for the complete bearing

$$F_x = \sum_{n=1}^{N} F_x^{(n)} = \frac{1}{S},$$

$$F_y = 0. \tag{4.61}$$

Pad Perturbation Method

We have seen that analysis of tilting-pad bearings, compared to fixed-pad bearings, is complicated by the fact that in addition to the degrees of freedom of the shaft, one has also to consider the degrees of freedom associated with the pivoting of the pads. Nevertheless, Lund's method does not provide the dynamical coefficients associated with the degrees of freedom of the pads. Instead, Lund assumes synchronous pad motion right from the start, thereby reducing pad data before interpolation and assembly.

The pad assembly method, on the other hand, calculates and stores the stiffness and damping coefficients associated with all degrees of freedom of a single pad over the whole range of eccentricities (pivot film thicknesses). The performance characteristics of the tilting-pad bearing are then calculated in the following steps:

(1) Fix the position of the journal ($\phi_0 = 0$ for symmetric arrangement of identical, centrally pivoted pads).
(2) From geometry, calculate the pivot film thickness for each pad and interpolate from stored pad data, to obtain pad characteristics at operating conditions.
(3) Obtain bearing characteristics by proper summation of individual pad characteristics.

Excitation frequency and pad inertia enter the analysis of the rotor-bearing system only when specific pad motion is assumed. Shapiro and Colsher (1977) describe this reduction of the results of the pad assembly method to "standard" 4×4 stiffness and damping matrices.

The rotor of the rotor-bearing system is assumed rigid and the bearings comprise N pads, each pad assuming its own orientation δ_i, $i = 1, \ldots, N$. The equations of motion for rotor (2 degrees of freedom) and pads (1 degree of freedom each) are

Rotor:

$$M\ddot{x} + K_{xx}x + K_{xy}y + C_{xx}\dot{x} + C_{xy}\dot{y} + \sum_{i=1}^{N}(K_{x\delta_i}\delta_i + C_{x\delta_i}\dot{\delta_i}) = 0 \qquad (4.62a)$$

$$M\ddot{y} + K_{yx}x + K_{yy}y + C_{yx}\dot{x} + C_{yy}\dot{y} + \sum_{i=1}^{N}(K_{y\delta_i}\delta_i + C_{y\delta_i}\dot{\delta_i}) = 0 \qquad (4.62b)$$

*i*th *pad*, $i = 1, \ldots, N$:

$$I_p\ddot{\delta_i} + K_{\delta_i\delta_i}\delta_i + C_{\delta_i\delta_i}\dot{\delta_i} + K_{\delta_i x}x + K_{\delta_i y}y + C_{\delta_i x}\dot{x} + C_{\delta_i y}\dot{y} = 0 \qquad (4.62c)$$

The coefficients in Eq. (4.62) have the definitions (Allaire, Parsell, and Barrett, 1981)

$$K_{xx} = -\frac{\Delta F_x}{\Delta x}, \ldots, C_{xx} = -\frac{\Delta F_x}{\Delta \dot{x}}, \ldots, K_{x\delta_i} = -\frac{\Delta F_x}{\Delta \delta_i}, \ldots C_{x\delta_i} = -\frac{\Delta F_x}{\Delta \dot{\delta_i}}, \ldots$$

$$K_{\delta_i\delta_i} = -\frac{\Delta M_i}{\Delta \delta_i}, \ldots, C_{\delta_i\delta_i} = -\frac{\Delta M_i}{\Delta \dot{\delta_i}}, \ldots, K_{\delta_i x} = -\frac{\Delta M_i}{\Delta x}, \ldots, C_{\delta_i x} = -\frac{\Delta M_i}{\Delta \dot{x}}, \ldots$$

The coefficients

$$K_{xx}, \ldots, K_{yy}$$

$$C_{xx}, \ldots, C_{yy}$$

$$K_{\delta_i x}, \ldots, K_{\delta_N y}$$

$$C_{\delta_1 x}, \ldots, C_{\delta_N y}$$

are obtained by perturbing the equilibrium state of the rotor while constraining the pads in their equilibrium position, while the coefficients

$$K_{x\delta_1}, \ldots, K_{y\delta_N}$$

$$C_{x\delta_1}, \ldots, C_{y\delta_N}$$

$$K_{\delta_1\delta_1}, \ldots, K_{\delta_N\delta_N}$$

$$C_{\delta_1\delta_1}, \ldots, C_{\delta_N\delta_N}$$

are obtained by constraining the rotor in its equilibrium state while perturbing the pitch angle of the ith pad, $i = 1, 2, \ldots, N$. There are a total of $8 + 10N$ coefficients.

The full stiffness matrix for a five-pad bearing, for example, is (Shapiro and Colsher, 1977; Allaire, Parsell, and Barrett, 1981)

$$
\begin{array}{c|ccccccc}
 & \Delta x & \Delta y & \Delta\delta_1 & \Delta\delta_2 & \Delta\delta_3 & \Delta\delta_4 & \Delta\delta_5 \\
\hline
-\Delta F_x & K_{xx} & K_{xy} & K_{x\delta_1} & K_{x\delta_2} & K_{x\delta_3} & K_{x\delta_4} & K_{x\delta_5} \\
-\Delta F_y & K_{yx} & K_{yy} & K_{y\delta_1} & K_{y\delta_2} & K_{y\delta_3} & K_{y\delta_4} & K_{y\delta_5} \\
-\Delta M_1 & K_{\delta_1 x} & K_{\delta_1 y} & K_{\delta_1\delta_1} & 0 & 0 & 0 & 0 \\
-\Delta M_2 & K_{\delta_2 x} & K_{\delta_2 y} & 0 & K_{\delta_2\delta_2} & 0 & 0 & 0 \\
-\Delta M_3 & K_{\delta_3 x} & K_{\delta_3 y} & 0 & 0 & K_{\delta_3\delta_3} & 0 & 0 \\
-\Delta M_4 & K_{\delta_4 x} & K_{\delta_4 y} & 0 & 0 & 0 & K_{\delta_4\delta_4} & 0 \\
-\Delta M_5 & K_{\delta_5 x} & K_{\delta_5 y} & 0 & 0 & 0 & 0 & K_{\delta_5\delta_5}
\end{array}
\qquad (4.63)
$$

and similarly for the damping matrix. Note that these matrices are $(N + 2) \times (N + 2)$, where N is the number of pads. Further details of the pad perturbation method of calculating tilting-pad bearing performance can be found in Shapiro and Colsher (1977) and Allaire et al. (1981). For an application of the pad perturbation method, see Section 4.4 on pivoted-pad thrust bearings.

Thermal effects on the stability characteristics of bearings can be considerable. Their importance can be gauged from Table 4.2, which is taken from Suganami and Szeri (1979).

Table 4.2. *Stiffness and damping coefficients and threshold speed of stability for single mass rigid rotor*

Regime	Model	K_{xx}	K_{xy}	K_{yx}	K_{yy}	C_{xx}	C_{xy}	C_{yx}	C_{yy}	$\bar{M}$
Laminar	Isothermal (ISO)	0.944	−0.076	4.31	5.83	0.598	1.25	1.25	9.25	1.79
	Adiabatic (ADI)	0.832	0.0038	3.85	5.89	0.492	0.940	0.934	7.95	3.08
	Thermohydrodynamic (THD), $D = 20$ cm	1.02	0.0415	4.69	6.14	0.565	1.13	1.12	8.74	2.31
Turbulent	Isothermal (ISO)	0.878	−0.0192	3.94	5.90	0.534	1.04	1.04	8.16	2.77
	Thermohydrodynamic (THD), $D = 60$ cm	0.792	0.052	3.6	5.94	0.455	0.768	0.769	7.01	14.5

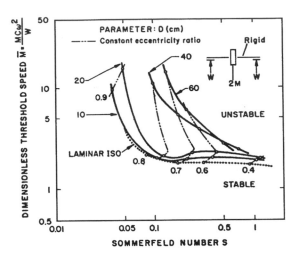

Figure 4.15. Stability of single mass rigid rotor ($N = 3600$ rpm, $C/R = 0.002$, $t = 50°$ C). (Reprinted with permission from Suganami, T. and Szeri, A. Z. A parametric study of journal bearing performance: the 80 degree partial arc bearing. *ASME Journal of Lubrication Technology*, **101**, 486–491. 1979.)

The domain of stable bearing operations changes considerably due to the viscosity nonuniformity that is caused by uneven viscous dissipation in the film. Figure 4.15 illustrates the size, i.e., the Peclet number, effect on stability boundaries.

The effects of inertia and turbulence on dynamic bearing characteristics and on the stability of rotor-bearing system have recently been discussed by Capone, Russo, and Russo (1991).

4.4   Pivoted-Pad Thrust Bearing

The bearing is constructed from N identical, equally spaced, sector–shaped pivoted pads. The pad angle is β, the pad radii are R_1, and $R_2 > R_1$, and the pad thickness is d. The pivot point is located at a distance $(d - \ell)$ from the pad working surface (see Figure 4.16). Because of this assumed uniformity of geometry, it is sufficient for us to consider performance of a single pad. Bearing performance will be obtained by suitable multiplication of single-pad performance.

The pad, pivoted in one point about which it is free to rotate (Figure 3.17), has three rotational degrees of freedom. During the motion of the runner about its equilibrium position, the pad will pivot in a manner that is determined by the dynamics of the whole system, viz., the runner, the oil film, and the pad.

We follow the work of Allaire, Parsell, and Barrett (1981), and evaluate 4×4 spring and damping matrices for each pad. The four degrees of freedom are the rectilinear motion of the runner and the rotational modes of the pad. The motion of the runner and the motion of the pad will be connected by means of Euler's rotational equations.

The inertial coordinate system (X, Y, Z) has its origin on the runner surface when the latter is in its undisturbed static equilibrium position, and its z axis is normal to the runner surface. When occupying its unperturbed static position, the pad is inclined to the runner at angles ψ_X, ψ_Y.

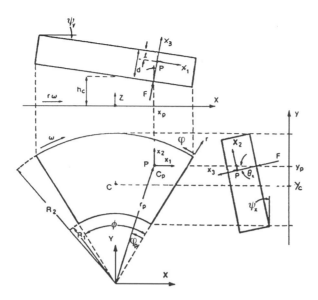

Figure 4.16. Pad schematics. (Reprinted with permission from Jeng, M. C. and Szeri, A. Z. A thermohydrodynamic solution of pivoted thrust pads: Part 3 – linearized force coefficients. *ASME Journal of Tribology*, **108**, 214–218, 1986.)

Let (x_1, x_2, x_3) be an inertial orthogonal Cartesian coordinate system, whose origin is located in the pivot point, at a distance $(d - \ell)$ from pad surface. The (x_1, x_2) plane is parallel to the pad surface in the unperturbed position and x_3 is normal to it.

We denote by $(\xi_1, \xi_2, \xi_3,)$ the body axes, with origin in the pivot point, that rotate with the pad. When the pad is occupying its static equilibrium position the ξ_i coordinate system coincides with the x_i system. The instantaneous (small) rotation of the ξ_i coordinate axes relative to the x_i coordinate axes during motion of the pad is measured by the rotation vector $\alpha = (\alpha_1, \alpha_2, \alpha_3)$.

Let F_0 and F represent the lubricant force in static equilibrium and during a small departure of the runner from the condition of static equilibrium, respectively. Then for small excursion of the runner $Z(t)$ about $Z = 0$, $\dot{Z} = 0$ we may write

$$dF = F - F_0 = \left[\frac{\partial F}{\partial Z}\right]_{Z=\dot{Z}=0} Z + \left[\frac{\partial F}{\partial \dot{Z}}\right]_{Z=\dot{Z}=0} \dot{Z}. \tag{4.64a}$$

The linearized force coefficients Eq. (4.64a) are the system stiffness

$$K = \left[\frac{\partial F}{\partial Z}\right]_{Z=\dot{Z}=0} \tag{4.64b}$$

and the system damping

$$C = \left[\frac{\partial F}{\partial \dot{Z}}\right]_{Z=\dot{Z}=0}. \tag{4.64c}$$

Had we fixed the bearing in any particular position, the linearized force coefficients K and C would be easy to calculate

$$-\frac{\partial F}{\partial Z} = \lim_{\Delta h_c \to 0} \frac{F(h_c + \Delta h_c, 0) - F(h_c, 0)}{\Delta h_c}, \qquad \Delta h_c = -Z,$$

$$-\frac{\partial F}{\partial \dot{Z}} = \lim_{\Delta \dot{h}_c \to 0} \frac{F(h_c, \Delta \dot{h}_c) - F(h_c, 0)}{\Delta \dot{h}_c}, \qquad \Delta \dot{h}_c = -\dot{Z}, \tag{4.65}$$

where h_c is the film thickness in the geometric center of the pad. If, however, the pad is free to pivot, the excursion $Z(t)$ of the runner introduces six unknowns to the problem: $\alpha_n(t)$ and $\dot{\alpha}_n(t)$, $n = 1, 2, 3$. Or, to state this differently, tilt of the pad not only causes a righting moment on the pad but introduces forces on the runner as well. The pad has three degrees of freedom, and thus the spring and damping matrices of the system are each 4×4, the four degrees of freedom being the Z motion of the runner and the rotational modes of the pad.

For a pivoting pad, we are thus forced to employ the more complicated Taylor expansion (see Shapiro and Colsher, 1977)

$$dF = \frac{\partial F}{\partial Z} Z + \frac{\partial F}{\partial \dot{Z}} \dot{Z} + \sum \left[\frac{\partial F}{\partial \alpha_n} \alpha_n + \frac{\partial F}{\partial \dot{\alpha}_n} \dot{\alpha}_n \right] \tag{4.66}$$

valid for Z, $\dot{Z}$, α_n and $\dot{\alpha}_n$ small.

In Eq. (4.66), we treat the instantaneous force F as a function of eight kinematic variables[9]

$$F = F(Z, \dot{Z}, \alpha_n, \dot{\alpha}_n), \qquad F_0 = F(0, 0, 0, 0), \qquad n = 1, 2, 3 \tag{4.67}$$

and, accordingly, the partial derivatives are evaluated keeping all but one variable constant, e.g.,

$$\frac{\partial F}{\partial Z} = \lim_{Z \to 0} \frac{F(Z, 0, 0, 0) - F(0, 0, 0, 0)}{Z},$$

$$\frac{\partial F}{\partial \alpha_2} = \lim_{\alpha_2 \to 0} \frac{F(0, 0, \alpha_2, 0) - F(0, 0, 0, 0)}{\alpha_2}. \tag{4.68}$$

Employing the notation

$$\frac{\partial F}{\partial Z} = K_{Z,Z}, \qquad \frac{\partial F}{\partial \dot{Z}} = C_{Z,Z},$$

$$\frac{\partial F}{\partial \alpha_j} = K_{Z,j}, \qquad \frac{\partial F}{\partial \dot{\alpha}_j} = C_{Z,j}, \qquad j = 1, 2, 3, \tag{4.69}$$

Equation (4.66) assumes the form

$$dF = K_{Z,Z} Z + C_{Z,Z} \dot{Z} + \sum_{n=1}^{3} [K_{Z,n} \alpha_n + C_{Z,n} \dot{\alpha}_n]. \tag{4.70}$$

The stiffness and damping coefficients in Eq. (4.70) are obtained via the pad perturbation method (Shapiro and Colsher, 1977).

_____

[9] We omit h_c, θ_x, and θ_y, which specify the equilibrium position of the pad, from the argument of F.

Equation (4.70) contains six unknowns, $\alpha_n, \dot{\alpha}_n, n = 1, 2, 3$, assuming that $Z, \dot{Z}$ are prescribed. Of course, the $\alpha_n, \dot{\alpha}_n$ are determined by the motion of the runner, i.e., they are dependent on $Z, \dot{Z}$ through the equations of motion for the pad. These equations (Euler's rotational equations) admit solutions with an exponential time factor and, therefore, can be reduced to a linear algebraic system in the amplitudes.

To make use of the above mentioned property of the Euler rotational equations or, ostensibly at this moment, to reduce the number of unknowns in Eq. (4.70), assume that the runner executes small harmonic motion with angular frequency Ω and amplitude ψ (Jeng and Szeri, 1986)

$$Z = \psi \exp(i\Omega t), \tag{4.71a}$$

$$\dot{Z} = i\Omega\psi \exp(i\Omega t). \tag{4.71b}$$

The motion of the pad, as induced by the motion of the runner, will also be harmonic, and we write[10]

$$\alpha_n = A_n \exp(i\Omega t), \tag{4.72a}$$

$$\dot{\alpha}_n = i\Omega A_n \exp(i\Omega t). \tag{4.72b}$$

With the aid of Eqs. (4.71) and (4.72), Eq. (4.70) takes the form

$$dF \exp(-i\Omega t) = [K_{Z,Z} + i\Omega C_{Z,Z}]\psi + \sum_{n=1}^{3}[K_{Z,n} + i\Omega C_{Z,n}]A_n. \tag{4.73}$$

In this form, the equation for dF contains only three unknowns, $A_n, n = 1, 2, 3$. These unknowns are related to the amplitude of the runner motion, ψ, through Euler's rotational equations (Goldstein, 1950)

$$I_{11}\ddot{\alpha}_1 - I_{12}\ddot{\alpha}_2 - I_{13}\ddot{\alpha}_3 = \tau_1,$$
$$-I_{21}\ddot{\alpha}_1 + I_{22}\ddot{\alpha}_2 - I_{23}\ddot{\alpha}_3 = \tau_2, \tag{4.74}$$
$$-I_{31}\ddot{\alpha}_1 - I_{32}\ddot{\alpha}_2 + I_{33}\ddot{\alpha}_3 = \tau_3,$$

where $I_{11}, I_{22}, \ldots, I_{33}$ are elements of the inertia matrix. These equations are written relative to the $\{x_1, x_2, x_3\}$ coordinate system, located in the pivot point P. Prior to writing Eq. (4.74), we made the assumption that both α and $\dot{\alpha}$ are small, so that products like $\dot{\alpha}_n, \dot{\alpha}_m; n, m = 1, 2, 3$, are neglected.

In Eq. (4.74), $\tau = (\tau_1, \tau_2, \tau_3)$ is the torque on the pad resulting from the applied forces and moments relative to $\{x_1, x_2, x_3\}$, $\alpha = (\alpha_1, \alpha_2, \alpha_3)$ is the rotation vector about the same axes, and $\alpha_n = A_n \exp(i\Omega t)$.

The torque components τ_1, τ_2, and τ_3 are dependent on the motion of both runner and pad. For small departures from equilibrium, we are permitted to terminate the Taylor expansion at first order and write

$$\tau_k = \frac{\partial \tau_k}{\partial Z}Z + \frac{\partial \tau_k}{\partial \dot{Z}}\dot{Z} + \sum_{n=1}^{3}\left[\frac{\partial \tau_k}{\partial \alpha_n}\alpha_n + \frac{\partial \tau_k}{\partial \dot{\alpha}_n}\dot{\alpha}_n\right], \quad k = 1, 2, 3. \tag{4.75}$$

---

[10] It would make no difference in the final outcome had we permitted a phase shift of pad motion relative to runner motion.

The definitions

$$\frac{\partial \tau_k}{\partial Z} = K_{k,Z}, \qquad \frac{\partial \tau_k}{\partial \dot{Z}} = C_{k,Z},$$
$$\frac{\partial \tau_k}{\partial \alpha_n} = K_{k,n}, \qquad \frac{\partial \tau_k}{\partial \dot{\alpha}_n} = C_{k,n}, \tag{4.76}$$

and Eqs. (4.71) and (4.72) enable us to write Eq. (4.75) in the form

$$\tau_k \exp(-i\Omega t) = [K_{k,Z} + i\Omega C_{k,Z}]\psi + \sum_{n=1}^{3}[K_{k,n} + i\Omega C_{k,n}]A_n, \quad k = 1, 2, 3. \tag{4.77}$$

Although we do not constrain the pad from motion about x_3, the coefficients $K_{k,3}$ and $C_{k,3}, k = 1, 2, 3$, will be neglected on account of being small (Mote, Shajer, and Telle, 1983). The torque component τ_3 is also found to be small and is neglected in the analysis.

Substituting Eq. (4.77) into Eq. (4.74), taking into account Eqs. (4.72), we obtain the following linear algebraic system

$$(M + N)A = R\psi. \tag{4.78}$$

Here R is a column vector of elements

$$R_k = K_{k,Z} + i\Omega C_{k,Z}, \tag{4.79}$$

and the matrices M and N have the definition

$$-M = (K_{k,m} + i\Omega C_{k,m} + \Omega^2 I_{k,m}), \tag{4.80}$$

$$N = \text{diag}(2\Omega^2 I_{11}, 2\Omega^2 I_{22}, 2\Omega^2 I_{33}). \tag{4.81}$$

The linear system, Eq. (4.78), yields

$$A = \psi(M + N)^{-1}R$$
$$= a\psi, \tag{4.82}$$

where

$$a = (M + N)^{-1}R.$$

Note that $a = \mathcal{R}(a) + i\mathcal{J}(a)$ is a complex vector, it includes both the amplitude and the phase angle of the pad motion about P.

The force response, dF, of Eq. (4.70) to the excitation Eqs. (4.71) and (4.72) is harmonic, and when Eq. (4.82) is substituted into Eq. (4.73), we obtain

$$dF \exp(-i\Omega t) = (\mathcal{K} + \mathcal{C}i\Omega)\psi. \tag{4.83}$$

From Eq. (4.83), the system stiffness and system damping are identified, respectively, as

$$\mathcal{K} = K_{Z,Z} + \sum_{n=1}^{3}\{K_{Z,n}\mathcal{R}(a_n) - \Omega C_{Z,n}\mathcal{J}(a_n)\} \tag{4.84}$$

and

$$\mathcal{C} = C_{Z,Z} + \sum_{n=1}^{3}\{C_{Z,n}\mathcal{R}(a_n) - K_{Z,n}\mathcal{J}(a_n)/\Omega\}. \tag{4.85}$$

The nondimensional inertia tensor

$$\bar{I} = \frac{\omega h_c}{\mu_*(\Delta R)^4} I \tag{4.86}$$

can be evaluated in closed form. It has components

$$\bar{I}_{11} = \bar{M}_p \left\{ \frac{1}{4}\left(1 + \frac{\sin\beta}{\beta}\right) + \frac{\bar{d}^2}{3(\bar{R}_2^2 + \bar{R}_1^2)}[\bar{\ell}^3 + (1-\bar{\ell})^3] + \frac{\bar{Y}_p(\bar{Y}_p - 2\bar{Y}_c)}{(\bar{R}_2^2 + \bar{R}_1^2)} \right\}, \tag{4.87a}$$

$$\bar{I}_{22} = \bar{M}_p \left\{ \frac{1}{4}\left(1 - \frac{\sin\beta}{\beta}\right) + \frac{\bar{d}^2}{3(\bar{R}_2^2 + \bar{R}_1^2)}[\bar{\ell}^3 + (1-\bar{\ell})^3] + \frac{\bar{X}_p^2}{(\bar{R}_2^2 + \bar{R}_1^2)} \right\}, \tag{4.87b}$$

$$\bar{I}_{33} = \bar{M}_p \left\{ \frac{1}{2} + \frac{[\bar{r}_p + \bar{R}_1]^2 - 2\bar{Y}_p\bar{Y}_c}{(\bar{R}_2^2 + \bar{R}_1^2)} \right\}, \tag{4.87c}$$

$$\bar{I}_{12} = \bar{M}_p \left\{ \frac{\bar{X}_p(\bar{Y}_p - \bar{Y}_c)}{(\bar{R}_2^2 + \bar{R}_1^2)} \right\}, \tag{4.87d}$$

$$\bar{I}_{13} = -\bar{M}_p \left\{ \frac{\bar{X}_p\bar{Z}_c}{(\bar{R}_2^2 + \bar{R}_1^2)} \right\}, \tag{4.87e}$$

$$\bar{I}_{23} = \bar{M}_p \left\{ \frac{2(\bar{R}_2^3 - \bar{R}_1^3)\bar{d}[\bar{\ell}^2 - (1-\bar{\ell})^2]}{3(\bar{R}_2^4 - \bar{R}_1^4)} \frac{\sin\beta/2}{\beta} - \frac{\bar{Y}_p\bar{Z}_c}{(\bar{R}_2^2 + \bar{R}_1^2)} \right\}, \tag{4.87f}$$

where

$$M_p = \frac{(R_2^2 - R_1^2)\beta t}{2}\rho_p = \frac{\mu_*(\Delta R)^4}{\omega h_c(R_2^2 + R_1^2)}\bar{M}_p \tag{4.88}$$

is the pad mass, ρ_p is the pad density, and d is the pad thickness.

In Eq. (4.87) we made use of the abbreviations

$$\bar{R}_1 = \frac{R_1}{\Delta R}, \quad \bar{d} = \frac{d}{\Delta R}, \quad \bar{\ell} = \frac{\ell}{d}, \quad \bar{R}_2 = \frac{R_2}{\Delta R}, \quad \bar{r} = \frac{r - R_1}{\Delta R}, \quad \Delta R = R_2 - R_1.$$
$$\tag{4.89}$$

The distance $(d - \ell)$ measures the separation between the center of pressure (r_p, φ_p) and the pivot location, P; if the pad is supported on its surface, then $d - \ell = 0$. The coordinates (X_p, Y_p, Z_p) and (X_c, Y_c, Z_c) are the coordinates of the pivot point, P, and the geometric center, C, of the pad, respectively, relative to the $\{X, Y, Z\}$ coordinate system (Figure 4.16). In normalized form, when normalization is achieved through division by ΔR, $\Delta R = R_2 - R_1$, they are calculated from the formulas below:

$$\bar{X}_p = [\bar{r}_p + \bar{R}_1]\sin\left[\beta\left(\frac{1}{2} - \bar{\varphi}_p\right)\right]$$

$$\bar{Y}_p = [\bar{r}_p + \bar{R}_1]\cos\left[\beta\left(\frac{1}{2} - \bar{\varphi}_p\right)\right] \tag{4.90}$$

$$\bar{Z}_p = 0,$$

$$\bar{X}_c = 0,$$

$$\bar{Y}_c = \frac{4}{3\beta}\left(\frac{\bar{R}_2^3 - \bar{R}_1^3}{\bar{R}_2^2 - \bar{R}_1^2}\right)\sin\frac{\beta}{2} \tag{4.91}$$

$$\bar{Z}_c = \frac{1}{2}\bar{d}(2\ell - 1).$$

Here $\bar{r}_p$ is the nondimensional radial coordinate of the center of pressure.

The foregoing equations may be nondimensionalized by specifying a force coefficient, c_F, and a torque coefficient, c_τ, through the formulas

$$c_F = \frac{\mu_*\omega(\Delta R)^4}{h_c^2},$$

$$c_\tau = \frac{\mu_*\omega(\Delta R)^4}{h_c}, \tag{4.92}$$

so that we may write

$$F = c_F\bar{F}, \qquad \tau = c_\tau\bar{\tau} \tag{4.93}$$

which serve as definition for the nondimensional quantities $\bar{F}$ and $\bar{\tau}$.

Defining, furthermore, nondimensional spring and damping coefficients through

$$K_{Z,j} = c_F\bar{K}_{z,j}; \qquad C_{Z,j} = \frac{c_F}{\omega}\bar{C}_{Z,j}, \tag{4.94a}$$

$$K_{Z,Z} = \frac{c_F}{h_c}\bar{K}_{Z,Z}; \qquad C_{Z,Z} = \frac{c_F}{h_c\omega}\bar{C}_{Z,Z}, \tag{4.94b}$$

$$K_{k,Z} = \frac{c_\tau}{h_c}\bar{K}_{k,Z}; \qquad C_{k,Z} = \frac{c_\tau}{\omega h_c}\bar{C}_{k,Z}, \tag{4.94c}$$

$$K_{m,n} = c_\tau\bar{K}_{m,n}; \qquad C_{m,n} = \frac{c_\tau}{\omega}\bar{C}_{m,n}, \tag{4.94d}$$

and setting

$$\bar{t} = \omega t, \qquad \bar{\psi} = \psi/h_c, \qquad \bar{\Omega} = \frac{\Omega}{\omega},$$

the dimensionless system stiffness and system damping, the counterpart of Eq. (4.84), is given by

$$\bar{\mathcal{K}} = \bar{K}_{Z,Z} + \sum_{n=1}^{3}\{\bar{K}_{Z,n}\mathcal{R}(a_n) - \bar{\Omega}\bar{C}_{Z,n}\mathcal{I}(a_n)\}, \tag{4.95a}$$

$$\bar{\mathcal{C}} = \bar{C}_{Z,Z} + \sum_{n=1}^{3}\{\bar{C}_{Z,n}\mathcal{R}(a_n) + \bar{K}_{Z,n}\mathcal{I}(a_n)/\bar{\Omega}\}, \tag{4.95b}$$

so that

$$\mathcal{K} = \frac{c_F}{h_c}\bar{\mathcal{K}}, \qquad \mathcal{C} = \frac{c_F}{h_c\omega}\bar{\mathcal{C}}. \tag{4.96}$$

The quantities $\mathcal{K}$ and $\mathcal{C}$ have physical dimensions of stiffness (= force/length) and damping (= force/velocity), respectively.

Table 4.3. *Linearized stiffness and damping coefficients*

	$\bar{\mathcal{K}}$		$\bar{\mathcal{C}}$	
$\bar{\Omega}$	$\bar{\mu} = 1.0$	$\bar{\mu} = \bar{\mu}(T)$	$\bar{\mu} = 1.0$	$\bar{\mu} = \bar{\mu}(T)$
0.1	0.4523	0.1717	0.3108	0.1407
0.5	0.4907	0.1746	0.3120	0.1393
1.0	0.4921	0.1747	0.3121	0.1392
5.0	0.4926	0.1747	0.3121	0.1392
10.0	0.4926	0.1747	0.3121	0.1392

$R_2/R_1 = 2.545$, $\beta = 24°$, $\bar{\varepsilon} = 0$, $m_x = 0.0$, $m_y = 1.0$, $\text{Re} = 500$. (Reprinted with permission from Jeng, M. C. and Szeri, A. Z. A thermohydrodynamic solution of pivoted thrust pads: Part 3 – linearized force coefficients. *ASME Journal of Tribology,* **108**, 214–218, 1986.)

Table 4.4. *Linearized stiffness and damping coefficients*

$\bar{M}_p$	$\bar{\mathcal{K}}$	$\bar{\mathcal{C}}$
479.82	0.4921	0.3121
47.982	0.4870	0.3119
4.7982	0.4453	0.3112
0.47982	0.4629	0.3219
0.047982	0.4925	0.3120

$R_2/R_1 = 2.545$, $\beta = 24°$, $\bar{\varepsilon} = 0$, $\bar{\Omega} = 1.0$, $m_x = 0.0$, $m_y = 1.0$, $\text{Re} = 500$. (Reprinted with permission from Jeng, M. C. and Szeri, A. Z. A thermohydrodynamic solution of pivoted thrust pads: Part 3 – linearized force coefficients. *ASME Journal of Tribology,* **108**, 214–218, 1986.)

Table 4.3 illustrates the effect of $\bar{\Omega}$, the ratio of excitational to rotational frequencies. At least for the geometry depicted, dependence of both the system stiffness $\bar{\mathcal{K}}$ and the system damping $\bar{\mathcal{C}}$ on $\bar{\Omega}$ appears to be insignificant. A stronger effect, that due to the temperature dependence of viscosity, is also depicted in this table; both the stiffness, $\bar{\mathcal{K}}$, and the damping, $\bar{\mathcal{C}}$, are affected by the temperature dependence of viscosity.

Table 4.4 tabulates $\bar{\mathcal{K}}$ and $\bar{\mathcal{C}}$ values for various magnitudes of the nondimensional pad mass, $\bar{M}_p$. The dependence of system stiffness and damping on $\bar{M}_p$ seems to be of little significance.

Figure 4.17 illustrates the dependence of $\bar{\mathcal{K}}$ and $\bar{\mathcal{C}}$ on the crowning of the pad. The pad deformation parameter, $\bar{\varepsilon}$, is defined by Jeng, Zhou, and Szeri (1986) as the ratio of the maximum deflection in the $y-z$ plane (Figure 4.16) to the pad center film thickness, h_c. This figure depicts calculations at $\bar{\Omega} = 1.0$ and $\bar{\Omega} = 0.1$, performed in an iterative manner so that the position of the pivot point remained fixed. This meant that the pad tilt parameter had to

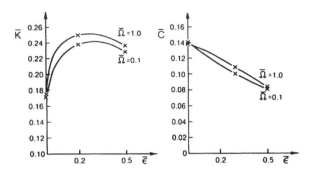

Figure 4.17. Variations of system stiffness $\bar{\mathcal{K}}$ and system damping $\bar{\mathcal{C}}$ with pad deformation parameter $\bar{e}$; $R_2/R_1 = 2.545$, $b = 24°$, $m_x = 0.0$, $m_y = 1.0$, Re = 500. (Reprinted with permission from Jeng, M. C. and Szeri, A. Z. A thermohydrodynamic solution of pivoted thrust pads: Part 3 – linearized force coefficients. *ASME Journal of Tribology*, **108**, 214–218, 1986.)

be adjusted from solution to solution and also from iteration to iteration between the pressure and the energy equations (Jeng, Zhou, and Szeri, 1986). Dependence of system stiffness and system damping on pad deformation seems to be strong. The solutions in Figure 4.17 were obtained from thermohydrodynamic bearing performance, i.e., the viscosity in this solution is dependent on temperature and thus has a three-dimensional distribution.

4.5   Nomenclature

C	pad clearance
$\mathcal{C}$	system damping
C'	pivot circle clearance
$C_{xx}, C_{xy}, C_{yx}, C_{yy}$	bearing damping coefficients
$C_{\xi\xi}, C_{\xi\eta}, C_{\eta\xi}, C_{\eta\eta}$	fixed-pad damping coefficients
$C'_{\xi\xi}, C'_{\xi\eta}, C'_{\eta\xi}, C'_{\eta\eta}$	tilting-pad damping coefficients
D	journal diameter
e	journal center eccentricity with respect to pad center
e_0	journal center eccentricity with respect to bearing center
W	load on pad
F_r, F_t	radial and tangential components of pressure-force
F_ξ, F_η	force components in ξ and η-directions
f_r, f_t	radial and tangential components of dimensionless pad force
f_ξ, f_η	compoents in ξ and η-directions of dimensionless pad fore
I	transverse mass moment of inertia of shoe around pivot
$\mathcal{K}$	system damping
$K_{xx}, K_{xy}, K_{yx}, K_{yy}$	bearing spring coefficients
$K_{\xi\xi}, K_{\xi\eta}, K_{\eta\xi}, K_{\eta\eta}$	fixed-pad spring coefficients
$K'_{\xi\xi}, K'_{\xi\eta}, K'_{\eta\xi}, K'_{\eta\eta}$	tilting-pad spring coefficients
L	bearing length, inches
M_{CRIT}	critical pad mass
R	journal radius
R_P	radius of pivot circle
S_B	bearing Sommerfeld number
S	pad Sommerfeld number

W	bearing load
x, y	coordinates of journal center with respect to the bearing
ε	eccentricity ratio with respect to the pad center
ε_0	eccentricity ratio with respect to the bearing center
η_p	amplitude for pad center motion
ξ, η	coordinates of journal center with respect to the pad
ϕ	attitude angle with respect to the pad load line
ϕ_0	attitude angle with respect to the bearing load line
n	angle from vertical (negative x-axis) to pad pivot point
ω	angular speed of shaft

Thrust Bearings

$\mathsf{A}(A_i)$	amplitude of angular perturbation
$C_{Z,Z}; C_{Z_n}$	damping
$C_{k,m}; C_{n,Z}$	damping
$\mathfrak{c}$	system damping
C	pad geometric center
C_p	center of pressure
f	lubricant force
$\mathsf{I}(I_{i,j})$	inertia tensor
$K_{Z,Z}; K_{Z,n}$	stiffness
$K_{k,n}; K_{n,Z}$	stiffness
$\mathcal{K}$	system stiffness
M_p	pad mass
R_1, R_2	inner, outer pad radius
Re	global Reynolds number ($= R_2\omega h_c/v_*$)
(X, Y, Z)	inertial coordinates
φ	amplitude of runner motion
Ω	frequency of runner motion
c_F, c_τ	force, torque coefficients
d	pad thickness
ℓ	depth of pivot point
m_x	radial tilt parameter ($= R_1\theta_x/h_c$)
m_y	azimuthal tilt parameter ($= R_1\theta_y/h_c$)
h_c	film thickness at C
θ_x, θ_y	pad tilt angles
t	time
(x_i)	
(x, y, z)	Cartesian coordinate systems
(ξ, η, ζ)	
$\alpha(\alpha_i)$	angular motion of pad
β	pad angle
$\bar{\rho}$	lubricant density
ω	runner angular velocity
v, μ	viscosity
$\tau, (\tau_i)$	torque
$(\dot{\ })$	time derivative

$(-)$ nondimensional quantity
$\mathcal{R}, \mathcal{J}$ real, imaginary part
$(\ )_*$ reference quantity

4.6   References

Alford, J. S. 1965. Protecting turbomachinery from self-excited rotor whirl. *ASME J. Eng. Power*, **87**, 333.

Allaire, P. E., Parsell, J. K. and Barrett, L. E. 1981. A pad perturbation method for the dynamic coefficients of tilting-pad journal bearings. *Wear*, **72**, 29–44.

Boyd, J. and Raimondi, A. A. 1962. Clearance considerations in pivoted pad journal bearings. *ASLE Trans.*, **5**, 418.

Capone, G., Russo, M. and Russo, R. 1991. Inertia and turbulence effects on dynamic characteristics and stability of rotor-bearings systems. *ASME Journal of Tribology*, **113**, 58–64.

Den Hartog, J. P. 1956. *Mechanical Vibrations*, 4th ed. McGraw-Hill, New York.

Goldstein, H. 1950. *Classical Mechanics*. Addison-Wesley.

Hagg, A. C. 1946. Influence of oil-film journal bearings on the stability of rotating machines. *ASME Trans. J. Appl. Mech.* **68**, A211.

Hori, Y. 1959. A theory of oil whip. *ASME J. Appl. Mech.*, **26**, 189–198.

Jeng, M. C. and Szeri, A. Z. 1986. A thermohydrodynamic solution of pivoted thrust pads: Part 3 – linearized force coefficients. *ASME Journal of Tribology*, **108**, 214–218.

Jeng, M. C., Zhou, G. R. and Szeri, A. Z. 1986. A thermohydrodynamic solution of pivoted thrust pads: Part 2 – static loading. *ASME Journal of Tribology*, **108**, 208–213.

Lund, J. W. 1964. Spring and damping coefficients for the tilting-pad journal bearing. *ASLE Trans.*, **7**, 342.

Lund, J. W. and Sternlicht, B. 1962. Rotor bearing dynamics with emphasis on attenuation. *ASME J. Basic Engr.*, **84**, 491–502.

Mote, C. D., Shajer, G. S. and Telle, L. I. 1983. Hydrodynamic sector bearings as circular saw guides. *ASME Journal of Lubrication Technology*, **105**, 67–76.

Nicholas, J. C., Gunter, E. J. and Allaire, P. E. 1979. Stiffness and damping properties for the five-pad tilting pad bearing. *ASLE Trans.*, **22**, 113–224.

Pinkus, O. and Sternlicht, B. 1961. *Theory of Hydrodynamic Lubrication*. McGraw-Hill, New York.

Raimondi, A. A. and Szeri, A. Z. 1984. Journal and thrust bearings. *CRC Handbook of Lubrication*, E. R. Booser (ed.), pp. 413–462.

Shapiro, W. and Colsher, R. 1977. Dynamic characteristics of fluid-film bearings. *Proc. 6th Turbomachinery Symp.*, Texas A&M University, 39–54.

Suganami, T. and Szeri, A. Z. 1979. A parametric study of journal bearing performance: the 80 degree partial arc bearing. *ASME Journal of Lubrication Technology*, **101**, 486–491.

Szeri, A. Z. 1966. Linearized force coefficients of a 110° partial journal bearing. *Proc. Inst. Mech. Engr.*, **181**, Pt. 3B, Paper No. 8.

Szeri, A. Z., Raimondi, A. A. and Giron, A. Linear force coefficients for squeeze-film damper. *ASME Journal of Lubrication Technology*, **105**, 326–334.

Trumpler, R. P. 1966. *Design of Film Bearings*. MacMillan, New York.

Vance, J. M. 1988. *Rotordynamics of Turbomachinery*. Wiley, New York.

CHAPTER 5

Effects of Fluid Inertia

The essence of lubrication theory is the recognition that the problem possesses two length scales (see Figure 2.7). Let the length scale in the "plane" of the film be denoted by L_{xz}, and let L_y be the length scale across the film; for conventional bearing geometries $(L_y/L_{xz}) = O(10^{-3})$. We utilize these length scales to normalize the equations of motion. To this end, define nondimensional coordinates, denoted by overbar, as follows

$$\{\bar{x}, \bar{y}, \bar{z}\} = \frac{1}{L_{xz}} \left\{ x, \left(\frac{L_{xz}}{L_y} \right) y, z \right\}. \tag{2.45}$$

Furthermore, let U_* represent the characteristic velocity in the plane of the film. The equation of continuity then requires $U_*(L_y/L_{xz})$ to be the velocity scale across the film, and we arrive at the following definition for normalized velocity:

$$\{\bar{u}, \bar{v}, \bar{w}\} = \frac{1}{U_*} \left\{ u, \left(\frac{L_{xz}}{L_y} \right) v, w \right\}, \tag{2.46}$$

where the overbar again denotes normalized, i.e., $O(1)$, nondimensional quantity. The nondimensional pressure and time are chosen to be

$$\bar{p} = \frac{p}{\rho U_*^2} \left(\frac{L_y}{L_{xz}} \right) \text{Re}, \qquad \bar{t} = \Omega t. \tag{2.47}$$

Here Ω is the characteristic frequency of the flow, and the Reynolds number has the definition $\text{Re} = L_y U_*/\nu$.

Substituting into the Navier-Stokes equations and neglecting terms of order $(L_y/L_{xz})^2$, we obtain

$$\Omega^* \frac{\partial \bar{u}}{\partial \bar{t}} + \text{Re}^* \bar{v} \cdot \nabla \bar{u} = -\frac{\partial \bar{p}}{\partial \bar{x}} + \frac{\partial^2 \bar{u}}{\partial \bar{y}^2}, \tag{2.53}$$

$$\Omega^* \frac{\partial \bar{w}}{\partial \bar{t}} + \text{Re}^* \bar{v} \cdot \nabla \bar{w} = -\frac{\partial \bar{p}}{\partial \bar{z}} + \frac{\partial^2 \bar{w}}{\partial \bar{y}^2}. \tag{2.54}$$

Here $\Omega^* = L_y^2 \Omega/\nu$ and $\text{Re}^* = (L_y/L_{xz})\text{Re}$ are the *reduced frequency* and the *reduced Reynolds number*, respectively, and $p = p(x, z)$ alone.

The continuity equation is, of course, form invariant under the transformation Eqs. (2.45) and (2.46):

$$\frac{\partial \bar{u}}{\partial \bar{x}} + \frac{\partial \bar{v}}{\partial \bar{y}} + \frac{\partial \bar{w}}{\partial \bar{z}} = 0. \tag{2.55}$$

According to Eqs. (2.53) and (2.54), lubricant inertia effects assume importance when $\Omega^* > 1$ and/or $\text{Re}^* > 1$. The three limiting cases that are instructive to consider here are: (1) *temporal inertia* limit, characterized by $\text{Re}^*/\Omega^* \to 0$, $\Omega^* > 1$; (2) the *convective inertia* limit, characterized by $\Omega^*/\text{Re}^* \to 0$, $\text{Re}^* > 1$; and (3) the *total inertia* limit, characterized by $\text{Re}^*/\Omega^* \to O(1)$, $\text{Re}^* > 1$.

5.1   Temporal Inertia Limit, $\mathrm{Re}^* \rightarrow 0, \Omega^* \geq 1$

When one of the bearing surfaces undergoes rapid, small-amplitude oscillation, the condition $\Omega^* \gg \mathrm{Re}^*$ is approximately satisfied. In this case, we retain the temporal inertia terms in Eqs. (2.53) and (2.54) but drop the terms representing convective inertia (Schlichting, 1968). Expressed in primitive variables, the equations of motion have now the reduced form

$$\rho \frac{\partial u}{\partial t} = -\frac{\partial p}{\partial x} + \mu \frac{\partial^2 u}{\partial y^2},$$

$$\rho \frac{\partial w}{\partial t} = -\frac{\partial p}{\partial z} + \mu \frac{\partial^2 w}{\partial y^2}. \tag{5.1}$$

We note that these equations are linear.

To accelerate a solid body through a fluid, a force must applied to accelerate the mass of the body itself. Additional force must also be applied to accelerate the mass of fluid that is being set in motion by the body. The added mass coefficient, C_M, is defined as the factor that multiplies the mass of the displaced fluid to give the mass of the accelerated fluid. Studies of added mass can be traced back to Stokes (Rosenhead, 1963). More recently, Chen et al. (1976) expressed Eq. (5.1) in terms of the stream function and found an exact solution to the resulting linear fourth-order partial differential equation in terms of Bessel functions. They were able to show that for a rod vibrating in a "large" cylinder filled with viscous fluid, both the *added mass coefficient*, C_M, and the *damping coefficient*, C_v, have their maximum at small Ω^* and decrease with $\Omega^* \uparrow$. This was also demonstrated by the analysis of Brennen (1976), who obtained a large Reynolds number limit for the added mass coefficient. Brennen found that for the ratio of C_M obtained at $\Omega^* \rightarrow 0$ to its value obtained at $\Omega^* \rightarrow \infty$ is ≈ 1.2. Tichy and Modest (1978) solved Eq. (5.1) for an arbitrary two-dimensional surface executing normal oscillation and found the classical lubrication solution for pressure to be in error, due to neglect of inertia forces. This analysis was extended later by Modest and Tichy (1979) to combined oscillation and sliding. Mulcahy (1980) was able to show that for $\Omega^* < 25$ the added mass coefficient, C_M, is independent of the reduced frequency and that in the range $0 < \Omega^* < 25$ the ratio C_M/C_v is linear in the reduced frequency according to the formula $C_M/C_v \approx \Omega^*/10$. For $\Omega^* > 25$ Mulcahy found that $C_M = C_M(\Omega^*)$ and that it is a decreasing function for $\Omega^* \uparrow$, as indicated by earlier investigators.

5.2   Convective Inertia Limit, $\Omega^* \rightarrow 0, \mathrm{Re}^* \geq 1$

In journal or thrust bearings in near steady state at high Reynolds number we have $\mathrm{Re}^* \gg \Omega^*$. This condition approximates to the limit $\Omega^*/\mathrm{Re}^* \rightarrow 0$, and we neglect temporal inertia in favor of convective inertia. When written in terms of primitive variables, this limit results in the equations

$$\rho \left(u \frac{\partial u}{\partial x} + v \frac{\partial u}{\partial y} + w \frac{\partial u}{\partial z} \right) = -\frac{\partial p}{\partial x} + \mu \frac{\partial^2 u}{\partial y^2},$$

$$\rho \left(u \frac{\partial w}{\partial x} + v \frac{\partial w}{\partial y} + w \frac{\partial w}{\partial z} \right) = -\frac{\partial p}{\partial z} + \mu \frac{\partial^2 w}{\partial y^2}. \tag{5.2}$$

Sestieri and Piva (1982) applied Eq. (5.2) to a plane slider in a numerical analysis. They showed significant inertia effects, up to 40% in load, for $Re^* = 8$. Care must be taken, however, when interpreting the results of Sestieri and Piva. As was pointed out by Constantinescu and Galetuse (1982), a reduced Reynolds number of $Re^* = 8$ for the slider geometry of Sestieri and Piva translates to $Re = 25,000$ in a conventional journal bearing. For long cylinders at $Re^* = 30$ San Andres and Szeri (1985) obtain, from numerical solutions of the exact equations, a 47% change in pressure due to inertia when $C/R = 1.0$. But as $(C/R) \downarrow$, the results from the exact equations decrease monotonically to the results of classical lubrication theory. Above $Re = 2,000$ the flow becomes turbulent, but even there inertial effects can be neglected, according to Constantinescu and Galetuse (1982).

The method of averaged inertia was employed for the steady problem by Osterle, Chou, and Saibel (1957) for a long bearing, and by Constantinescu (Constantinescu, 1970; Constantinescu and Galetuse, 1982). The more recent of these papers finds that in conventional journal bearings convective inertia effects can be neglected in the range $0 < Re^* < 10$. Although at the upper end of the range inertia effects do appear, they are still small in comparison with other nonlinear effects such as thermal distortion. Burton and Hsu (1974) find that at small downward vertical load the journal center might rise above the bearing center, as was shown experimentally by Black and Walton (1974). Others to use the method of averaged inertia were Launder and Leschziner (1978).

Journal Bearings

In one of the early publications on flow between eccentric rotating cylinders, Wannier (1950) discussed the problem without restricting the geometry; he used a complex variable technique to solve the biharmonic equation, satisfied by the stream function in Stokes flow. Wannier showed that the Reynolds equation of classical lubrication theory (Reynolds, 1886) constitutes the zero-order approximation to the Navier-Stokes equations when the stream function is expanded in powers of the film thickness. Wood (1957), using a modified bipolar coordinate system that reduces to polar coordinates when the eccentricity vanishes, analyzed the boundary layers that develop on the two cylinders at large Reynolds numbers. The small parameter of the perturbation analysis is the eccentricity ratio, and the solution is expressed in combinations of Bessel functions.

The effect of fluid inertia is estimated from a perturbation of the Stokes flow in Kamal's (1966) analysis. Kamal's inertial correction is incorrect, however, as was pointed out first by Ashino (1975) and later by Ballal and Rivlin (1976); even their solution of the Stokes problem shows disagreement with recent results. Another small perturbation analysis with the eccentricity ratio as the parameter was published by Kulinski and Ostrach (1967). Yamada (1968) neglected curvature effects and solved the boundary layer equations for the case of a rotating outer cylinder. Assuming a perturbation series in the clearance ratio, Yamada showed that the results of the unperturbed flow agree with those of lubrication theory. The importance of the inertial correction is found in the pressure distribution: the largest negative pressure is greater in magnitude than the largest positive pressure. This conclusion also received support from Sood and Elrod (1974).

DiPrima and Stuart (1972a) obtained inertial corrections to the linearized problem at small clearance ratios and at small values of the modified Reynolds number. Their zero-order approximation is identical to lubrication theory. Results presented by DiPrima et al. are in good agreement with those of Yamada. The perturbation analysis of Ballal and Rivlin

(1976) is one of the most complete analyses performed to date on the flow between eccentric rotating cylinders, but the solution is incorrect except for Stokes flow (San Andres and Szeri, 1985), as the boundary conditions are not satisfied for Re > 0. San Andres and Szeri (1985) worked out an accurate numerical solution of the exact equations and applied it to the wide-gap problem with arbitrary rotation of the cylinders. The analysis was recently extended to account for heat transfer in a fluid with temperature dependent viscosity by Dai, Dong, and Szeri (1991) and Kim and Szeri (1996) and to Rivlin-Ericksen fluids of third grade by Christie, Rajagopal, and Szeri (1987).

The questions we shall investigate here are:

(1)   What is *the effect of neglecting convective lubricant inertia*?
(2)   What is the *effect of neglecting curvature* of the lubricant film?

To answer these questions we shall examine three models for flow in a long journal bearing. The first model *retains both inertia and curvature* of the film and is based on the full Navier-Stokes equations. The second model *retains film curvature but neglects inertia* and results from application of the lubrication approximation in the natural coordinates that provide exact representation of the curved boundaries. The third model, classical lubrication theory, *neglects both inertia and film curvature* and results from application of the lubrication approximation in Cartesian coordinates.

Navier-Stokes Model

We employ a bipolar coordinate system for the representation of the flow field between infinite, rotating eccentric cylinders. The bipolar coordinate system $\{\hat{\alpha}, \hat{\beta}\}$ is related to the Cartesian coordinate system $\{X, Y\}$ through

$$\hat{\alpha} + i\hat{\beta} = -2\coth^{-1}\frac{(X + iY)}{a}, \tag{5.3}$$

where a is the separation between the pole and the origin of the $\{X, Y\}$ system (Figure 5.1). In the bipolar coordinate system, the cylinders of radii r_1 and r_2, $r_1 < r_2$, have the simple

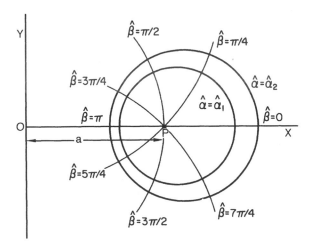

Figure 5.1. Geometry and coordinate systems for eccentric cylinders.

representation $\hat{\alpha} = \hat{\alpha}_1$ and $\hat{\alpha} = \alpha_2$, $\hat{\alpha}_1 < \alpha_2 < 0$, respectively. The scale factor, H, of the bipolar coordinate system (Ritchie, 1968) is

$$H = \frac{a}{(\cosh \hat{\alpha} - \cos \hat{\beta})}.$$

The equations of motion and continuity defining the two-dimensional flow field are first written relative to the bipolar coordinate system (Ritchie, 1968) and then nondimensionalized as follows:

$$\alpha = \Delta(\hat{\alpha} - \hat{\alpha}_1), \qquad \beta = \frac{\hat{\beta}}{2\pi}, \qquad \Delta = \frac{1}{\hat{\alpha}_2 - \hat{\alpha}_1},$$

$$\{U, V\} = \frac{r_1 \omega}{\sinh|\hat{\alpha}_1|}\{u, v\}, \qquad p = \frac{P}{\rho r_1^2 \omega^2}\sinh^2|\hat{\alpha}_1|, \qquad (5.4\text{a})$$

$$h = \frac{H}{a}, \qquad \mathrm{Re} = \frac{r_1 \omega C}{\nu}, \qquad C = r_2 - r_1, \qquad \eta = \frac{r_1}{r_2}.$$

Here P represents the pressure and $\{U, V\}$ are the physical components of velocity relative to the bipolar coordinate system. Dynamic conditions in the flow are represented by the Reynolds number, Re, calculated on the mean gap width. In terms of bipolar coordinates the eccentricity ratio is given by

$$\varepsilon = \frac{\sinh(\hat{\alpha}_1 - \hat{\alpha}_2)}{\sinh \hat{\alpha}_1 - \sinh \hat{\alpha}_2}. \qquad (5.4\text{b})$$

We intend to solve the steady-state problem. It is well documented, however, that for equal interpolation of velocity and pressure the mixed formulation of the steady-state Navier-Stokes equation yields a singular system. To circumvent this (*Babushka-Brezzi stability criteria*), Hughes, Franca, and Balestra (1986) employed an ingenious weighting procedure that resulted in a stable system. Though this system contains extra terms in the equation of mass conservation, in the limit its solutions are those of the Stokes problem. In 1991 de Sampiao obtained results similar to those of Hughes et al., by manipulating the steady-state equation for mass conservation and the time-discretized form of the momentum equations. Zienkiewicz and Woo (1991) generalized the procedure by considering the artificial compressibility formulation for the equation of mass conservation (Fletcher, 1991), instead of its steady-state form. We follow Zienkiewicz and Woo (1991) closely and apply their scheme to a Galerkin B-spline formulation of our nonzero Reynolds number flow.

Thus, although we are interested in the steady-state problem, we still write the equations of motion and the equation of mass conservation in their unsteady (nondimensional) form:

$$\frac{\partial u}{\partial t} = f^{(1)}(u, v, p),$$

$$\frac{\partial v}{\partial t} = f^{(2)}(u, v, p), \qquad (5.5)$$

$$\frac{1}{c^2}\frac{\partial p}{\partial t} + \mathrm{div}\, \boldsymbol{v} = 0.$$

Here we employed the notation

$$
f^{(1)}(u, v, p) = -\frac{\Delta}{h} u \frac{\partial u}{\partial \alpha} - \frac{1}{2\pi h} v \frac{\partial u}{\partial \beta} + uv \sin \hat{\beta} - v^2 \sinh \hat{\alpha}
$$
$$
- \frac{\Delta}{h} \frac{\partial p}{\partial \alpha} + \frac{(1 - \eta)}{\eta \, \mathrm{Re}} \left[\frac{1}{h^2} \left(\Delta^2 \frac{\partial^2 u}{\partial \alpha^2} + \frac{1}{(2\pi)^2} \frac{\partial^2 u}{\partial \beta^2} \right) \right.
$$
$$
\left. - \frac{2\Delta \sin \hat{\beta}}{h} \frac{\partial v}{\partial \alpha} + \frac{\sinh \hat{\alpha}}{\pi h} \frac{\partial v}{\partial \beta} - \frac{\cosh \hat{\alpha} + \cos \hat{\beta}}{h} u \right], \tag{5.6a}
$$

$$
f^{(2)}(u, v, p) = -\frac{\Delta}{h} u \frac{\partial v}{\partial \alpha} - \frac{1}{2\pi h} v \frac{\partial v}{\partial \beta} + uv \sinh \hat{\alpha} - u^2 \sinh \hat{\beta}
$$
$$
- \frac{1}{2\pi h} \frac{\partial p}{\partial \beta} + \frac{(1 - \eta)}{\eta \, \mathrm{Re}} \left[\frac{1}{h^2} \left(\Delta^2 \frac{\partial^2 v}{\partial \alpha^2} + \frac{1}{(2\pi)^2} \frac{\partial^2 v}{\partial \beta^2} \right) \right.
$$
$$
\left. + \frac{2\Delta \sin \hat{\beta}}{h} \frac{\partial u}{\partial \alpha} - \frac{\sinh \hat{\alpha}}{\pi h} \frac{\partial u}{\partial \beta} - \frac{\cosh \hat{\alpha} + \cos \hat{\beta}}{h} v \right]. \tag{5.6b}
$$

The divergence of the velocity field $v = (u, v)$ has the (nondimensional) form

$$
\mathrm{div}\, v = \frac{1}{h} \left(\Delta \frac{\partial u}{\partial \alpha} + \frac{1}{2\pi} \frac{\partial v}{\partial \beta} \right) - u \sinh \hat{\alpha} + v \sin \hat{\beta}. \tag{5.7}
$$

We could seek a steady-state solution to our problem by finding the time-asymptotic solution to Eq. (5.5), subject to no-slip conditions at the wall

$$
\begin{aligned}
u = 0, \; v = \sinh|\hat{\alpha}_1| \quad &\text{at} \quad \alpha = 0, \\
u = v = 0 \quad &\text{at} \quad \alpha = 1.
\end{aligned} \tag{5.8a}
$$

In addition to the boundary conditions, we must also ensure periodicity of the solution and its derivatives in β:

$$
\frac{\partial^{(n)} \varphi}{\partial \beta^n}(\alpha, 0) = \frac{\partial^{(n)} \varphi}{\partial \beta^n}(\alpha, 1) \qquad \begin{aligned} &n = 0, 1, 2, \ldots \\ &\varphi = u, v, p. \end{aligned} \tag{5.8b}
$$

Equations (5.5) and (5.6) define the pressure only within an arbitrary constant. We set this constant to zero by enforcing the condition $p(0, 0) = 0$. There can be no other conditions specified on pressure.

The condition $p(0, 0) = 0$ together with Eq. (5.8) and appropriate initial values could be used to obtain $\{u(\alpha, \beta, t), v(\alpha, \beta, t), p(\alpha, \beta, t)\}$ from Eq. (5.5). To this end, we would discretize the continuity equation according to

$$
\frac{p^{(n+1)} - p^{(n)}}{c^2 \delta t} = -\mathrm{div}\, v^{(n+1/2)}. \tag{5.9}
$$

The velocity components at the $n + 1/2$ time level can be obtained from the equations of motion (5.5):

$$
v^{(n+1/2)} = v^{(n)} + \frac{\delta t}{2} \left(f^{(1)}, f^{(2)} \right) \Big|_n \tag{5.10}
$$

so that now

$$\frac{p^{(n+1)} - p^{(n)}}{c^2 \delta t} = -\mathrm{div}\, \boldsymbol{v}^{(n)} - \frac{\delta t}{2}\mathrm{div}\big(f^{(1)}, f^{(2)}\big)\big|_n. \tag{5.11}$$

Steady state is characterized by $\partial u/\partial t = 0$, $\partial v/\partial t = 0$ in Eq. (5.5) and by $p^{(n+1)} = p^{(n)}$ in Eq. (5.11). In consequence, to arrive directly at steady state we solve the following system of equations:

$$f^{(1)}(u, v, p) = 0, \tag{5.12a}$$

$$f^{(2)}(u, v, p) = 0, \tag{5.12b}$$

$$\mathrm{div}\, \boldsymbol{v} + \frac{\delta t}{2}\mathrm{div}(f^{(1)}, f^{(2)}) = 0, \tag{5.12c}$$

subject to boundary conditions (5.8a) and periodicity condition (5.8b)

We intend to approximate the set of unknowns $\{u(\alpha, \beta), v(\alpha, \beta), p(\alpha, \beta)\}$ by piecewise polynomial functions. Thus, we partition the interval $[0, 1]$ for x, where x represents α or β, in turn, as

$$\pi: 0 = x_1 < x_2 < \cdots < x_\ell < x_{\ell+1} = 1.$$

Let $p_1(x), \ldots, p_\ell(x)$ be any sequence of ℓ polynomials, each of order k (i.e., of degree $<k$), and denote the collection of all piecewise polynomial functions $h(x)$ by $P_{k,\pi}$:

$$P_{k,\pi}\{h(x): h(x) = p_i(x) \quad \text{if} \quad x \in [x_i, x_{i+1}], 1 \leq i \leq \ell\}.$$

$P_{k,\pi}$ is a linear space, and since there are ℓ subintervals, the dimension of $P_{k,\pi}$ is $k\ell$.

Consider now subspaces $S_{k,\pi,v}$ of $P_{k,\pi}$ generated by imposing smoothness constraints on elements of $P_{k,\pi}$ at the interior breakpoints x_i, $2 \leq i \leq \ell$. Let $v = \{v_i\}_{i=2}^\ell$ be a nonnegative integer sequence, with $v_i \leq k$, all i, where v_i denotes the smoothness index of the piecewise polynomial subspace $S_{k,\pi,v}$ at the breakpoint x_i, so that $h^{(j)}(x_i^+) = h^{(j)}(x_i^-)$, $0 \leq j \leq v-1$. Then the dimension N of the subspace $S_{k,\pi,v}$ is given by

$$\dim S_{k,\pi,v} = k + \sum_{i=2}^{\ell}(k - v_i).$$

We now construct a basis for $S_{k,\pi,v}$ such that each element of the basis has local support and each element is nonnegative. To generate such a basis, we employ the recurrence relation (deBoor, 1978)

$$A_{i,k}(x) = \frac{x - t_i}{t_{i+k-1} - t_i}A_{i,k-1}(x) + \frac{t_{i+k} - x}{t_{i+k} - t_{i+k+1}}A_{i+1,k-1}(x)$$

$$A_{j,1} = \begin{cases} 1 & \text{for } x \in [t_j, t_{j+1}], \\ 0 & \text{otherwise.} \end{cases}$$

Here $t = \{t_i\}_{i=1}^{N+k}$ is any nondecreasing sequence such that

(1) $t_1 \leq t_2 \leq \cdots \leq t_k \leq x_\ell$ and $x_{\ell+1} \leq t_{N+1} \leq \cdots \leq t_{N+k}$;
(2) the number x_i, $2 \leq i \leq \ell$, occurs exactly $d_i = k - v_i$ times in t.

The sequence of $A_1, A_2, \ldots, A_N$ of B-splines of order k for the knot sequence t is a basis for $S_{k,\pi,\nu}$. The choice of t translates the desired amount of smoothness at breakpoints, and the Curry-Schoenberg theorem (deBoor, 1978) permits construction of a B-spline basis for any particular piecewise polynomial space $S_{k,\pi,\nu}$.

The B-splines thus defined provide a partition of unity:

$$\left.\begin{array}{l} A_i(x) \geq 0, 1 \leq i \leq N, \\ \displaystyle\sum_{i=1}^{N} A_i(x) = 1 \end{array}\right\} \quad x \in [x_1, x_{\ell+1}].$$

Other relevant properties of B-splines are

$$\left.\begin{array}{ll} A_1(x_1) = A_N(x_{\ell+1}) = 1, & A_j(x_1) = 0 \, (j > 1), \\ A_j(x_{\ell+1}) = 0 (j < N), & A_j(x) = 0 \, (x \notin [t_j, t_{j+k}]) \end{array}\right\}$$

$$\left.\begin{array}{ll} A_1'(x_1) = -A_2'(x_1) \neq 0, & A_j'(x_1) = 0 \, (j > 2), \\ A_N'(x_{\ell+1}) = -A_{N-1}'(x_{\ell+1}) \neq 0, & A_j'(x_{\ell+1}) = 0 \, (j < N - 1) \end{array}\right\}.$$

In the present calculations we employ quartic B-spline basis, i.e., $k = 5$, and write $\{A_i(\alpha): 1 \leq i \leq N_\beta\}$ for the set of normalized B-splines relative to $k_\alpha, \pi_\alpha, \nu_\alpha$.

In similar manner, we define the normalized B-splines $\{B_j(\beta): 1 \leq j \leq N_\beta\}$ on $0 \leq \beta \leq 1$. However, the requirement that the solution be periodic in β, Eq. (5.8b), suggest that we construct another basis $\{b_j(\beta): 1 \leq j \leq N_\beta - 3\}$ from the B_j that is periodic in β, and use the periodic $\{b_j\}$ in place of the nonperiodic $\{B_j\}$. Let $c = 2B_2''(0)/B_3''(0)$ and define matrices Φ and Ψ as

$$\Phi = \begin{bmatrix} I_3 & 0 & \Psi \\ 0 & I_{N_y-6} & 0 \end{bmatrix}, \qquad \Psi = \begin{bmatrix} c & 2 & 1 \\ -c & -1 & 0 \\ 1 & 0 & 0 \end{bmatrix},$$

where I_3 and I_{N_y-6} are unit matrices, then the new base vectors, $\{b_i\}_{i=1}^{N_y-3}$, are given by

$$b = \Phi B = (b_1, b_2, \ldots, b_{N_y-3})^T, \qquad B = (B_1, \ldots, B_{N_y})^T.$$

It can be verified that the sequence $\{b_i\}$ is a basis for a subspace of $S_{k,\pi,\nu}$ defined as

$$\Omega_{k,\pi,\nu} = \{\omega(z) \in S_{k,\pi,\nu}: \omega^{(n)}(0) = \omega^{(n)}(1), \qquad n = 0, 1, 2\}.$$

The expansions

$$u(\alpha, \beta) = \sum_{i=2}^{N_\alpha-1} \sum_{j=1}^{N_\beta-3} u_{ij} A_i(\alpha) b_j(\beta),$$

$$v(\alpha, \beta) = \sinh|\hat{\alpha}_1| A_1(\alpha) + \sum_{i=2}^{N_\alpha-1} \sum_{j=1}^{N_\beta-3} v_{ij} A_i(\alpha) b_j(\beta), \qquad (5.13)$$

$$p(\alpha, \beta) = \sum_{i=1}^{N_\alpha} \sum_{j=1}^{N_\beta-3} p_{ij} A_i(\alpha) b_j(\beta),$$

where the $A_i(\alpha)$, $i = 1, \ldots N_\alpha$ are normalized splines in α and the b_j, $j = 1, \ldots, N_\beta - 3$ are periodic splines in β, satisfy the boundary and periodicity conditions (5.8). The u_{ij}, v_{ij}, p_{ij} are unknown coefficients, to be determined later.

Part (a) of Figure 5.2 displays normalized B-splines $A_i(\alpha)$, $i = 1, \ldots, N_\alpha$, for $N_\alpha = 10$, $k_\alpha = 5$, $\nu_\alpha = 4$, and part (b) displays periodic B-splines b_j, $j = 1, \ldots, N_\beta - 3$, for

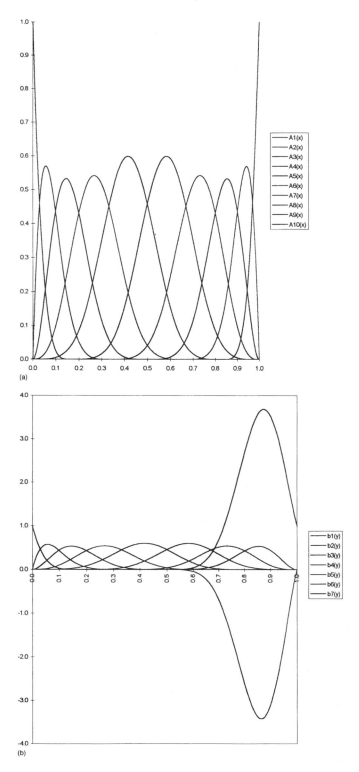

Figure 5.2. *B*-splines: (a) normalized splines $A_i(\alpha)$, (b) periodic splines $b_i(\beta)$. ($N_\alpha = N_\beta = 10$, $k_\alpha = k_\beta = 5$)

$N_\beta = 10, k_\beta = 5$ and $\nu_\beta = 4$. These splines possess three continuous derivatives at internal breakpoints.

Substituting expansions (5.13) into the equations of motion and continuity (5.12), multiplying through by $F_{rs}(\alpha, \beta)$, and integrating over the domain, in accordance with Galerkin's method, we obtain

$$\langle f^{(1)}(\alpha, \beta), F_{rs}(\alpha, \beta) \rangle = 0, \tag{5.14a}$$

$$\langle f^{(2)}(\alpha, \beta), F_{rs}(\alpha, \beta) \rangle = 0, \tag{5.14b}$$

$$\langle \mathrm{div}\, \boldsymbol{v}(\alpha, \beta), F_{rs}(\alpha, \beta) \rangle + \frac{\delta t}{2} \langle \mathrm{div}[f^{(1)}(\alpha, \beta), f^{(2)}(\alpha, \beta)], F_{rs}(\alpha, \beta) \rangle = 0. \tag{5.14c}$$

Here we used the symbol $F_{rs}(\alpha, \beta) \equiv A_r(\alpha) b_s(\beta)$ and the symbol $\langle \cdot, \cdot \rangle$ to represent the inner product of two functions calculated over the domain $0 \leq \alpha, \beta \leq 1$.

Our task can now be defined as solving the discretized system of equations (5.14) subject to the single condition $p(0, 0) = 0$, which takes the form

$$p_{1,1} = 0.$$

This formulation, which closely follows that of Zienkiewicz and Woo (1991), leads to a nonsingular algebraic system for the steady-state Navier-Stokes problem.

The system of nonlinear algebraic equations that result from the Galerkin discretization, Eq. (5.14), can be put into concise form:

$$\tilde{G}(\mu) = 0, \qquad \mu = (u, \lambda), \tag{5.15}$$

where u is the vector of state variables $\{u_{ij}, v_{ij}, p_{i,j}\}$ and λ is the vector of parameters $\Delta, \eta, \varepsilon$, and Re.

The set of nonlinear equations (5.15) is difficult to solve for parameter values outside a narrow range, and we employ parametric continuation in conjunction with Gauss-Newton iteration (Ortega and Rheinboldt, 1970). Parametric continuation is a scheme that allows for systematic determination of starting points for the iteration; using this technique one can trace out the solution in parameter space. The simplest continuation scheme would be to employ the solution obtained at, say, Re $= 1$ where the solution is easy to come by, as starting point for the iteration at, say, Re $= 10$, where the solution is more difficult to obtain.[1] There are, of course, efficient schemes to accomplish this. Details can be found, e.g., in Keller (1977) and Seydel (1988); the solution scheme detailed in Chapter 6 was taken from Szeri and Al-Sharif (1996).

Lubrication Approximation in Bipolar Coordinates

While in the previous section our aim was to solve the exact Navier-Stokes problem, in the present section the objective is to find approximate solutions for thin film but without neglecting film curvature. To discover the correct approximation, we make use of the geometry of the thin film by recognizing the existence of two different scale lengths that are three orders of magnitude apart. The length scales we adopt are $L_\beta = r_1$ along the principal dimension of the film and $L_\alpha = r_1(\alpha_2 - \alpha_1)$ across it. The continuity equation is normalized

---

[1] Parametric continuation is a particularly useful technique for mapping out solutions in case of solution multiplicity (Keller, 1977).

according to

$$\alpha = \Delta(\hat{\alpha} - \hat{\alpha}_1), \qquad \beta = \hat{\beta}/2\pi, \qquad V = r_1\omega v, \qquad \Delta = \frac{1}{(\hat{\alpha}_2 - \hat{\alpha}_1)} \qquad (5.16)$$

as the azimuthal velocity is of order $r_1\omega$. Then, to have the terms of the continuity equation balance, we must scale the $\hat{\alpha}$ component of the velocity as

$$U = \frac{1}{2\pi\Delta} r_1\omega u. \qquad (5.17)$$

The normalized pressure $\tilde{P}(\alpha, \beta)$ is defined through

$$\tilde{P} = 2\pi \frac{p}{\rho U_*^2} \left(\frac{L_\alpha}{L_\beta}\right) r_e,$$

$$U_* = r_1\omega, \qquad r_e = \frac{U_* L_\alpha}{\nu}, \qquad L_\alpha = r_1(\hat{\alpha}_2 - \hat{\alpha}_1), \qquad L_\beta = r_1. \qquad (5.18)$$

We are now ready to write the normalized equations of motion, neglecting terms of the order $(\hat{\alpha}_2 - \hat{\alpha}_1)^2$ or smaller. This order of approximation is consistent with lubrication approximation, as

$$\frac{r_2 - r_1}{r_1} = \frac{2\cosh\dfrac{\hat{\alpha}_2 + \hat{\alpha}_1}{2}\sinh\dfrac{\hat{\alpha}_2 - \hat{\alpha}_1}{2}}{\sinh\hat{\alpha}_1} \approx \frac{(\alpha_2 - \hat{\alpha}_1)}{\tanh\hat{\alpha}_1} \qquad (5.19)$$

and $\tanh(\hat{\alpha}_1) = O(1)$ for conventional bearing geometries.

The normalized equations of motion thus take the form

$$\frac{\partial\tilde{P}}{\partial\alpha} = 0$$

$$-\frac{1}{(2\pi)^2 H}\frac{\partial\tilde{P}}{\partial\beta} = \frac{1}{H^2\sinh\hat{\alpha}_1}\frac{\partial^2 v}{\partial\alpha^2} + r_e^*\left[\frac{1}{H}\left(u\frac{\partial v}{\partial\alpha} + v\frac{\partial v}{\partial\beta}\right)\right]. \qquad (5.20)$$

Note that $r_e^* \equiv \frac{1}{2\pi\Delta}r_e$ is the reduced Reynolds number of the problem, analogous to Re^* of Eq. (2.53). The condition $\mathrm{Re}^* \to 0$ of classical lubrication theory is equivalent, thus, to $r_e^* \to 0$ and, upon applying this limit, Eq. (5.20) reduces to

$$-\sinh\hat{\alpha}_1\frac{H}{(2\pi)^2}\frac{\partial\tilde{P}}{\partial\beta} = \frac{\partial^2 v}{\partial\alpha^2}, \qquad \frac{\partial\tilde{P}}{\partial\alpha} = 0, \qquad (5.21)$$

with boundary conditions

$$u = 0, v = 1 \quad \text{at } \alpha = 0, \qquad (5.22a)$$

$$u = v = 0 \quad \text{at } \alpha = 1. \qquad (5.22b)$$

To solve Eq. (5.21), we integrate twice with respect to α and obtain

$$v = \frac{1}{C}\int_0^\alpha\int_0^\phi H(\eta, \beta)\frac{d\tilde{P}(\beta)}{d\beta}d\eta\,d\phi + \alpha f(\beta) + F(\beta), \qquad C \equiv \frac{-(2\pi)^2}{\sinh\hat{\alpha}_1}. \qquad (5.23)$$

Here $f(\beta)$ and $F(\beta)$ are arbitrary functions of β and serve to make Eq. (5.23) satisfy the boundary conditions (5.22).

The complete solution of Eq. (5.21) that satisfies the boundary conditions (5.22) is

$$v = \frac{1}{C}\frac{d\tilde{P}(\beta)}{d\beta}\left[\int_0^\alpha\int_0^{\bar{\alpha}} H(\eta, \beta)\,d\eta\,d\bar{\alpha} - \alpha\int_0^1\int_0^{\bar{\alpha}} H(\eta, \beta)\,d\eta\,d\bar{\alpha}\right] + (1 - \alpha).$$

Table 5.1. *Convergence of solutions with* $(C/R) \to 0$

Theory	(C/R)	S	k_{RR}	k_{RT}	k_{TR}	k_{TT}
Navier-Strokes	0.002	0.16880	0.0	59.2402	−59.2492	0.0
	0.001	0.16883	0.0	59.2300	−59.2385	0.0
	0.0005	0.16885	0.0	59.2249	−59.2330	0.0
Reynolds	0.0	0.16886	0.0	59.2198	−59.2288	0.0
Bipolar lubrication	0.0005	0.16891	0.0	59.2037	−59.1951	0.0
	0.001	0.16898	0.0	59.1786	−59.1703	0.0
	0.002	0.16912	0.0	59.1285	−59.1208	0.0

$\varepsilon = 0.1, \mathrm{Re} = 0.0$ Sommerfeld condition. (From Dai, R. X., Dong, Q. M. and Szeri, A. Z. Approximations in hydrodynamic lubrication. *ASME Journal of Tribology*, **114**, 14–25, 1992.)

Substitution into the integrated (across the film) continuity equation yields

$$
\frac{\partial}{\partial \beta} \left\{ \frac{\partial \bar{P}}{\partial \beta} \int_0^1 H(\alpha, \beta) \left[\int_0^\alpha I(\bar{\alpha}, \beta) \, d\bar{\alpha} - \alpha \int_0^1 I(\bar{\alpha}, \beta) \, d\bar{\alpha} - C(1 - \alpha) \right] d\alpha \right\} = 0,
$$

$$
I(\alpha, \beta) \equiv \int_0^\alpha H(\phi, \beta) \, d\phi.
$$

(5.24)

The innermost integral of Eq. (5.24) was obtained analytically while the other integrals necessary to solve for $\tilde{P}$ were performed via Gaussian quadrature (Dai, Dong, and Szeri, 1992). The boundary conditions are $\tilde{P}(0) = \tilde{P}(1) = 0$.

To investigate the limit of the sequence of results obtained with sequentially smaller values of C/r_1, we compare at $\mathrm{Re} = 0$ solutions from the Navier-Stokes equations with those of the bipolar lubrication theory as $(\hat{\alpha}_2 - \hat{\alpha}_1) \to 0$. Table 5.1 displays results for Sommerfeld boundary conditions, while Table 5.2 was obtained for Gümbel conditions. These tables contain the Sommerfeld number (inverse of nondimensional force) and the nondimensional spring coefficients (4.14a). The data indicate that on decreasing the clearance ratio both the Navier-Stokes theory and the bipolar lubrication theory converge to a common limit, this limit being the classical lubrication theory of Reynolds.

The exact, zero Reynolds number solution of Ballal and Rivlin (1976) valid for arbitrary clearance ratio, can be employed to study film curvature effects. When this solution is expanded in powers of the clearance ratio, the first two terms correspond to the Myllerup and Hamrock (1994) solution, which employs regular perturbation:

$$
\bar{p} = \frac{12\pi \varepsilon \sin \theta (2 + \varepsilon \cos \theta)}{(2 + \varepsilon^2)(1 + \varepsilon \cos \theta)^2}
$$
$$
+ \left(\frac{C}{R} \right) \frac{4\pi \varepsilon \sin \theta (1 + 5\varepsilon^2 + 2\varepsilon(2 + \varepsilon^2) \cos \theta)}{(2 + \varepsilon^2)(1 + \varepsilon \cos \theta)^3} + O\left(\frac{C}{R} \right)^2.
$$

The first term is seen to be identical to the solution of the Reynolds equation under full film boundary conditions [cf. Eq. (3.76)], while the second term is the first order curvature correction

Table 5.2. *Convergence of solutions with* $(C/R) \rightarrow 0$

Theory	(C/R)	S	k_{RR}	k_{RT}	k_{TR}	k_{TT}
Navier-Stokes	0.002	0.33692	3.8022	29.6201	−29.6246	1.8943
	0.001	0.33698	3.8026	29.6150	−29.6192	1.8944
	0.0005	0.33701	3.8029	29.6125	−29.6165	1.8945
Reynolds	0.0	0.33704	3.8085	29.6099	−29.6144	1.8945
Bipolar lubrication	0.0005	0.33706	3.8080	29.6075	−29.6118	1.8943
	0.001	0.33732	3.8042	29.5851	−29.5892	1.8924
	0.002	0.33760	3.7999	29.5604	−29.5642	1.8903

$\varepsilon = 0.1$, $Re = 0.0$ Gümbel condition. (From Dai, R. X., Dong, Q. M. and Szeri, A. Z. Approximations in hydrodynamic lubrication. *ASME Journal of Tribology*, **114**, 14–25, 1992.)

Table 5.3. *Effect of film curvature on* $P/\mu N$

Boundary condition	(C/R)	$(P/\mu N) \times 10^{-5}$		
		Navier-Stokes	Reynolds	Bipolar lubrication
Sommerfeld	0.002	7.4202	7.4175	7.4052
	0.001	29.6754	29.6701	29.6454
	0.0005	118.6908	118.6802	118.6732
Gümbel	0.002	14.8104	14.8052	14.7824
	0.001	59.2311	59.2207	59.1786
	0.0005	236.8966	236.8826	236.8125

$\varepsilon = 0.1$, $Re = 0.0$. (From Dai, R. X., Dong, Q. M. and Szeri, A. Z. Approximations in hydrodynamic lubrication. *ASME Journal of Tribology*, **114**, 14–25, 1992.)

An alternative way of interpreting the results of Tables 5.1 and 5.2 is by calculating the nondimensional group $P/\mu N$. The results are shown in Table 5.3 at various values of (C/R). It may be concluded here that

$$P_{BP} < P_{RE} < P_{NS},$$

where P_{BP}, P_{RE}, and P_{NS} represent the specific pressure according to the bipolar lubrication theory, the lubrication theory of Reynolds, and the Navier-Stokes theory, respectively. Thus, neglect of the higher order terms in Eq. (2.40) underestimates the specific bearing pressure at all values of the clearance ratio (C/R), i.e., P_{BP}, $P_{RE} < P_{NS}$. Neglect of the curvature of the lubricant film, on the other hand, leads to an overestimate of the specific pressure, i.e., $P_{BP} < P_{RE}$. The net effect is shown in that the specific pressure of lubrication theory, P_{RE}, is bracketed by the specific pressures P_{BP} and P_{NS}. This convergence of both the bipolar lubrication theory and the Navier-Stokes theory to the classical lubrication theory shows that the latter constitutes the proper limit as $(C/R) \rightarrow 0$.

Table 5.4. *Effect of lubricant inertia* $(C/R = 0.002)$

ε	Boundary condition	Re	S	k_{RR}	k_{RT}	k_{TR}	k_{TT}
0.1	Sommerfeld	0.0	0.16880	7.3×10^{-6}	59.2402	-59.2492	7.6×10^{-6}
		50.0	0.16880	0.1836	59.2499	-59.2588	0.1944
		100.0	0.16879	0.3673	59.2525	-59.2603	0.3889
		500.0	0.16865	1.8374	59.2708	-59.2786	1.9447
		900.0	0.16836	3.3091	59.2923	-59.3151	3.5016
	Gümbel	0.0	0.33692	3.8122	29.6201	-29.6246	1.8943
		50.0	0.33565	3.9047	29.7259	-29.7068	1.9963
		100.0	0.33440	3.9977	29.8318	-29.7891	2.0946
		500.0	0.32451	4.7619	30.6823	-30.4498	2.9076
		900.0	0.31480	5.5616	31.5366	-31.1138	3.7663
0.5	Sommerfeld	0.0	0.03288	5.2×10^{-6}	60.8262	-67.7159	3.2×10^{-6}
		50.0	0.03288	0.1323	60.8262	-67.6976	0.0811
		100.0	0.03288	0.2750	60.8282	-67.6160	0.1491
		500.0	0.03288	1.3347	60.8296	-67.6094	0.7453
		900.0	0.03287	2.4734	60.8326	-67.5939	1.3415
	Gümbel	0.0	0.06174	27.2012	30.4131	-33.8579	11.1500
		50.0	0.06172	27.1428	30.4081	-33.7024	11.1937
		100.0	0.06169	27.1031	30.4077	-33.5617	11.2470
		500.0	0.06150	26.6237	30.4031	-32.4737	11.5515
		900.0	0.06129	26.0506	30.3997	-31.5835	11.8643
0.7	Sommerfeld	0.0	0.02143	1.1×10^{-7}	66.6671	-104.60	-2.8×10^{-8}
		50.0	0.02143	0.3630	66.6671	-104.60	-0.0173
		100.0	0.02143	0.7259	66.6671	-104.60	-0.0346
		500.0	0.02143	3.6294	66.6669	-104.60	-0.1730
		1100.0	0.02143	7.9823	66.6666	-104.61	-0.3798
	Gümbel	0.0	0.03636	73.4232	33.3335	-52.3011	-20.7935
		50.0	0.03632	73.6046	33.3782	-52.4787	-20.8022
		100.0	0.03628	73.7861	33.4228	-52.6563	-20.8110
		500.0	0.03597	75.2469	33.7795	-54.0772	-20.8839
		1100.0	0.03551	77.4519	34.3127	-56.2057	-21.0019

(From Dai, R. X., Dong, Q. M. and Szeri, A. Z. Approximations in hydrodynamic lubrication. *ASME Journal of Tribology*, **114**, 14–25, 1992.)

The Sommerfeld number, S, and the stiffness matrix, k, can now be calculated for $0 < \text{Re} < \text{Re}_{CR}$. Table 5.4 contains the results of these computations at $\varepsilon = 0.1, 0.5$, and 0.7. The data of Table 5.4 clearly demonstrate that the Sommerfeld number remains virtually constant for all changes $\text{Re} < \text{Re}_{CR}$. The largest change is encountered with Gümbel boundary conditions at $\varepsilon = 0.1$, amounting to -6.5% change in S when increasing the Reynolds number from zero to $\text{Re} = 900$, i.e., $\text{Re}^* = 1.8$. The critical Reynolds number here is $\text{Re}_{CR} = 932$. Looking at the data of Table 5.4, it does become obvious, however, that significant changes are encountered in the diagonal components of the stiffness matrix k_{RR}, k_{TT}, which increase apparently linearly with the Reynolds number. Changes in bearing stiffness, of course, alter the stability characteristics of the bearing

Table 5.5. *Comparison of numerical solution with small perturbation solution*

		$\bar{F}_R$	$\bar{F}_R$
ε	Re	Numerical solution	DiPrima and Stuart
0.1	10.0	0.0039	0.0038
	100.0	0.0389	0.0382
	500.0	0.1945	0.1910
	900.0	0.3502	0.3438
0.3	50.0	0.0446	0.0446
	100.0	0.0892	0.0892
	500.0	0.4461	0.4461
	900.0	0.8029	0.8030
0.5	50.0	0.0406	0.0372
	100.0	0.0746	0.0745
	500.0	0.3727	0.3724
	900.0	0.6708	0.6702
0.7	50.0	−0.0121	−0.0122
	100.0	−0.0242	−0.0244
	500.0	−0.1211	−0.1220
	1100.0	−0.2660	−0.2683
	1145.0	−0.2760	−0.2793

$C/R = 0.002$, Sommerfeld condition. (From Dai, R. X., Dong, Q. M. and Szeri, A. Z. Approximations in hydrodynamic lubrication. *ASME Journal of Tribology*, **114**, 14–25, 1992.)

for actual running conditions, as contrasted to the inertialess world of classical lubrication theory.

These calculations suggest a generalization of the conclusion offered by DiPrima and Stuart (1972b): "first order correction for inertia does not affect the vertical force or torque, but does introduce a horizontal force." Our extension is: *Correction for inertia does not affect the resultant force or torque nor the off-diagonal components of the stiffness matrix. It does, however, induce changes in the diagonal components, the changes being linear in the Reynolds number.* In fact, the conclusions of DiPrima and Stuart hold not only in the qualitative but also in the quantitative sense. The small perturbation theory calculates the radial force component, $\bar{F}_R$, for Sommerfeld boundary condition at ($C/R = 0.002$), from (DiPrima and Stuart, 1972b)

$$\bar{F}_R = \frac{2\pi^2}{35\varepsilon(2 + \varepsilon^2)} \left[-(7 + 8\varepsilon^2) + \frac{\sqrt{1 - \varepsilon^2}}{2 + \varepsilon}(14 + 44\varepsilon^2 + 5\varepsilon^4) \right] \mathrm{Re}^*. \qquad (5.25)$$

In Table 5.5, we compare our numerical results for $\bar{F}_R$ with that of Eq. (5.25) and show excellent agreement for Re $<$ Re$_{CR}$ and for all $\varepsilon \leq 0.7$ for conventional journal bearing geometry. This conclusion still holds for larger values of (C/R), as indicated in Table 5.6.

Table 5.6. *Effect of fluid inertia on solution*

Boundary conditions	Re	S	k_{RR}	k_{RT}	k_{TR}	k_{TT}	$(\bar{F}_R)_{\text{Numerical}}$	$(\bar{F}_R)_{\text{DiPrima}}$
Sommerfeld	0.0	0.03118	0.00	64.14	−74.73	0.00	0	0
	50.00	0.03089	13.17	64.26	−74.05	7.90	3.9483	3.7236
	100.0	0.03005	25.18	64.67	−72.13	15.74	7.8722	7.4472
	150.0	0.02880	35.23	65.39	−69.18	23.40	11.6990	11.1708
Gümbel	0.0	0.05883	28.32	32.07	−37.36	11.28	–	–
	50.0	0.05555	20.55	32.33	−24.60	15.85	–	–
	100.0	0.05088	11.71	32.92	−13.26	21.48	–	–
	150.0	0.04568	3.56	33.87	−4.35	27.75	–	–

$C/R = 0.02$, $\varepsilon = 0.5$, $\text{Re}_{CR} = 128.64$. (From Dai, R. X., Dong, Q. M. and Szeri, A. Z. Approximations in hydrodynamic lubrication. *ASME Journal of Tribology*, **114**, 14–25, 1992.)

The results obtained here demonstrate convincingly that lubricant inertia has negligible effect on load carrying capacity in noncavitating film, for isothermal laminar flow of the lubricant. The stability characteristics of the bearing are, however, affected by lubricant inertia. These conclusions apply to practical bearing operations directly, and assert that bearing load can be calculated from classical, i.e., noninertial, theory. To investigate stability, however, one must take lubricant inertia into account, even during laminar flow of the lubricant.

Solutions are obtained here under Sommerfeld and Gümbel boundary conditions, neither of which is particularly useful in practice. There is some evidence (You and Lu, 1987) that in journal bearings lubricant inertia has a tendency of stretching the film in the direction of rotation. This result of You and Lu was obtained by extrapolation from the small perturbation analysis of Reinhardt and Lund (1975). Ota et al., (1995) confirm the conclusions of You and Lu on fluid inertia caused film stretching.

Hydrostatic Bearings

For simplicity and because most of the published results on the effect of fluid inertia in hydrostatic lubrication are related to this bearing type, we consider here only the circular step bearing.

When both surfaces of a circular step bearing are stationary and parallel and when fluid inertia is neglected, theory requires a logarithmic pressure drop to maintain viscous dissipation. Experiments performed at low flow rates verify this theoretical result (Coombs and Dowson, 1965).

If the throughflow Reynolds number, defined by

$$R_Q = \frac{Q}{2\pi \nu h},$$

where Q is the volumetric rate of throughflow, is increased, convective inertia gains importance. Its effect is to increase the pressure in the radial direction rapidly at small radii and moderately at large radii. Convective inertia effects far outweigh viscous dissipation

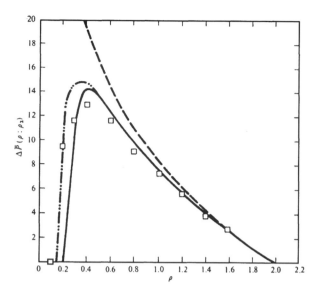

Figure 5.3. Flow between stationary parallel disks. Effect of convective inertia on pressure. ($\square$) Experimental values of Jackson and Symmons (1965). Theoretical curves: (- - - -) creeping flow, Eq. (3.26); ($—\cdot\cdot—$) Jackson and Symmons (1965); ($——$) Szeri and Adams. (Reprinted with the permission of Cambridge University Press from Szeri, A. Z. and Adams, M. L. Laminar throughflow between closely spaced rotating disks. *J. Fluid Mech.*, **86**, 1–14, 1978.)

at small values of the dimensionless radius, defined by

$$\rho = \frac{r}{h\sqrt{R_Q}},$$

whereas at large ρ the creeping flow solution remains essentially correct (Figure 5.3).[2]

Livesey (1960) argued that departure from parallel flow must be slight even at large flow rates and retained $u\,\partial u/\partial r$, where u is the creeping solution, as the significant inertia term. Livesey's analysis yields

$$\frac{\partial \bar{P}}{\partial \rho} = -\frac{12}{\rho} + \frac{2K}{\rho^3}. \tag{5.26}$$

In this equation

$$\bar{P}(\rho) = \frac{h}{\mu \nu R_Q} \int_0^h p(r, z)\, dz \tag{5.27}$$

is a dimensionless average pressure. The pressure gradient of creeping flow can be obtained from Eq. (3.6); in our present notion it is $d\bar{P}/d\rho = -12/\rho$. The value of K was given by Livesey as 0.6, subsequent authors gave $K = 0.72$ (Moller, 1963) and $K = 0.77143$ (Jackson and Symmons, 1965).

---

[2] This statement is in complete agreement with the conclusion that inertia becomes important for $\mathrm{Re}^* \geq 1$. In the present example the characteristic velocity is $U_* = Q/2\pi r$ and the characteristic dimension $L_y = h$. Then $\mathrm{Re} = (h/r)R_Q$ and $\mathrm{Re}^* = 1/\rho^2$; thus, inertia effects will be insignificant when $\rho > 1$.

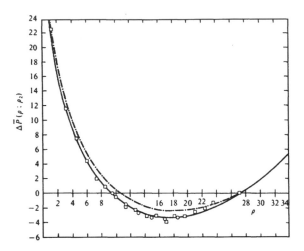

Figure 5.4. Flow between rotating parallel disks, $E = 2.9$. Effect of rotational inertia on pressure. Experimental values: ($\square$) Coombs and Dowson (1965), ($\bigcirc$) Nirmal (Szeri and Adams, 1978). Theoretical curves: ($-\cdots-$) Osterle and Hughes (1958), ($--$) Szeri and Adams. (Reprinted with the permission of Cambridge University Press from Szeri, A. Z. and Adams, M. L. Laminar flow between closely spaced rotating disks. *J. Fluid Mech.*, **86**, 1–14, 1978.)

There is further distortion of the creeping flow profile due to centrifugal inertia, if the runner surface is made to rotate (Figure 5.4). Osterle and Hughes (1958) demonstrated that this distortion can be large enough to diminish the load capacity of a hydrostatic bearing. Their solution was based on the assumption of negligible convective inertia and on a rotational inertia that was calculated from a linear circumferential velocity. The radial pressure gradient for this case was found to be

$$\frac{d\bar{P}}{d\rho} = -\frac{12}{\rho} + \frac{3\rho}{10E^2}. \tag{5.28}$$

Here $E = \nu/h^2\omega$ is the Ekman number and ω is the runner angular velocity. [The condition $Re^* \geq 1$ now assumes the form $\rho/E^2 \geq O(1)$.]

In a more thorough analysis of the problem, we consider two parallel disks of radius r_2 located at a distance h apart. The lower disk is stationary, and the upper disk rotates at the angular velocity ω. We write the equations of motion relative to cylindrical polar coordinates, assuming rotational symmetry, as

$$u_r \frac{\partial u_r}{\partial r} - \frac{u_\theta^2}{r} + u_z \frac{\partial u_r}{\partial z} = \nu \left(\nabla^2 u_r - \frac{u_r}{r^2} \right) - \frac{1}{\rho} \frac{\partial p}{\partial r}, \tag{5.29a}$$

$$\frac{u_r}{r} \frac{\partial}{\partial r}(r u_\theta) + u_z \frac{\partial u_\theta}{\partial z} = \nu \left(\nabla^2 u_\theta - \frac{u_\theta}{r^2} \right), \tag{5.29b}$$

$$u_r \frac{\partial u_z}{\partial r} + u_z \frac{\partial u_z}{\partial z} = \nu \nabla^2 u_z - \frac{1}{\rho} \frac{\partial p}{\partial z}, \tag{5.29c}$$

$$0 < r_1 < r < r_2 \qquad 0 < z < h.$$

Eliminating the pressure between the first and third equations by cross differentiation and writing the resulting equations in terms of the stream function Ψ, defined through

$$u_r = \frac{1}{r}\frac{\partial \Psi}{\partial z} \qquad u_z = -\frac{1}{r}\frac{\partial \Psi}{\partial r}, \tag{5.30}$$

we obtain the following system of equations (Szeri and Adams, 1978):

$$\frac{\partial \Psi}{\partial z}\frac{\partial (D^2\Psi)}{\partial r} - \frac{\partial \Psi}{\partial r}\frac{\partial (D^2\Psi)}{\partial z} - \frac{1}{r}\frac{\partial u_\theta^2}{\partial r} = \nu\left(r^3 D^2 + 4\frac{\partial}{\partial r}\right)D^2\Psi, \tag{5.31a}$$

$$\frac{\partial \Psi}{\partial z}\frac{\partial (r u_\theta)}{\partial r} - \frac{\partial \Psi}{\partial r}\frac{\partial (r u_\theta)}{\partial z} = \nu r^3 D^2(r u_\theta). \tag{5.31b}$$

The operator D^2 in Eqs. (5.31) has the definition

$$D^2 = \frac{1}{r^2}\left(\nabla^2 - \frac{2}{r}\frac{\partial}{\partial r}\right).$$

Equations (5.31) are supplemented with the boundary conditions

$$\frac{\partial \Psi}{\partial r} = \frac{\partial \Psi}{\partial z} = u_\theta = 0 \quad \text{at } z = 0,$$

$$\frac{\partial \Psi}{\partial r} = \frac{\partial \Psi}{\partial z} = 0, \qquad u_\theta = r\omega \quad \text{at } z = h, \tag{5.32}$$

$$\Psi(r, h) - \Psi(r, 0) = \frac{Q}{2\pi}.$$

The last condition of Eqs. (5.32) expresses global conservation of mass. In addition, we need boundary conditions on both Ψ and u_θ at $r = r_1$ (recess radius) and $r = r_2$.

It can be shown (Szeri, Schneider, Labbe, and Kaufman, 1983) that a minimum of four dimensionless parameters are required to characterize the problem as given by Eqs. (5.31) and (5.32). But if the spacing of the disks is narrow – that is, if $h/r_2 \ll 1$ – then the radial variation of the shear stress is negligible, and we have the approximation

$$D^2 \approx \frac{1}{r^2}\frac{\partial^2}{\partial z^2} \qquad \frac{h}{r_2} \ll 1. \tag{5.33}$$

Under the thin-film approximation, the problem can be characterized by a single dimensionless variable. To show this, we substitute

$$z = \zeta h, \qquad r = h\sqrt{R_Q}\,\rho$$
$$\Psi = h\nu R_Q \bar{\Psi}, \qquad u_\theta = h\omega\sqrt{R_Q}\,\rho\bar{\Omega} \tag{5.34}$$

into Eqs. (5.31) and obtain

$$\frac{d\bar{\Psi}}{\partial \zeta}\frac{\partial}{\partial \rho}\left(\frac{1}{\rho^2}\frac{\partial^2 \bar{\Psi}}{\partial \zeta^2}\right) - \frac{\partial \bar{\Psi}}{\partial \rho}\frac{\partial}{\partial \zeta}\left(\frac{1}{\rho^2}\frac{\partial^2 \bar{\Psi}}{\partial \zeta^2}\right) - \frac{1}{E^2}\frac{\partial (\rho\bar{\Omega}^2)}{\partial \zeta} = \frac{1}{\rho}\frac{\partial^4 \bar{\Psi}}{\partial \zeta^4}, \tag{5.35a}$$

$$\frac{\partial \bar{\Psi}}{\partial \zeta}\frac{\partial (\rho^2\bar{\Omega})}{\partial \rho} - \frac{\partial \bar{\Psi}}{\partial \rho}\frac{\partial (\rho^2\bar{\Omega})}{\partial \zeta} = \rho\frac{\partial^2 (\rho^2\bar{\Omega})}{\partial \zeta^2}. \tag{5.35b}$$

With the aid of Eq. (5.34), we also transform the boundary conditions into dimensionless form

$$\frac{\partial \bar{\Psi}}{\partial \rho} = \frac{\partial \bar{\Psi}}{\partial \zeta} = \bar{\Omega} = 0 \quad \text{at} \quad \zeta = 0,$$

$$\frac{\partial \bar{\Psi}}{\partial \rho} = \frac{\partial \bar{\Psi}}{\partial \zeta} = 0, \qquad \bar{\Omega} = 1 \quad \text{at} \quad \zeta = 1, \tag{5.36}$$

$$\Psi(\rho, 1) - \Psi(\rho, 0) = 1.$$

Equations (5.35a) and (5.35b) are now parabolic, and thus boundary conditions can no longer be prescribed at r_2. Furthermore, since the film is thin, the precise form of the upstream boundary (now initial) condition becomes unimportant; thus, in Eqs. (5.35) and (5.36) we have a one-parameter family of flows.

The parabolic system of Eqs. (5.35) and (5.36) has been solved, after reduction to ordinary differential equations by the Galerkin method, with a predictor-corrector equation solver. The solutions show strong interaction between circumferential and radial flows, leading to consistently lower torque on the stationary disk. The radial derivative of the dimensionless average pressure is given by (Szeri and Adams, 1978)

$$\frac{d\bar{P}}{d\rho} = -\frac{12}{\rho} + \frac{6}{5\rho^3} + \frac{\rho}{3E^2} + \sum_n \left[\frac{96}{\pi^2 \rho^2} X_n \left(f_n' - \frac{1}{\rho} f_n \right) - Z_n f_n \right.$$

$$\left. + \frac{\rho}{E^2} \left(\frac{1}{2} g_n^2 - \frac{2(-1)^n}{\pi n} g_n \right) \right] + \sum_{n,m} Y_{n,m} \left(\frac{\pi}{\rho^3} f_n f_m - \frac{2\pi}{\rho^2} f_n' f_m \right). \tag{5.37}$$

Here $f_n(\rho)$ and $g_n(\rho)$ are the coefficients in the Galerkin expansions for $\bar{\Psi}$ and $\bar{\Omega}$, respectively, and X_n, $Y_{n,m}$, and z_n are constants. The reader may wish to compare Eq. (5.37) to Eq. (5.28), obtained by simpler means.

At any radial position r,

$$P(r) = P(r_1) + \frac{\mu \nu R_Q}{h^2} \Delta \bar{P}(\rho; \rho_1), \tag{5.38}$$

where $\Delta \bar{P}(\rho; \rho_1) \equiv \bar{P}(\rho) - \bar{P}(\rho_1)$ and is calculated from Eq. (5.37). It can be shown that the initial conditions are washed out completely within a short distance, so that

$$\Delta \bar{P}(\rho_c; \rho_b) = \Delta \bar{P}(\rho_c; \rho_a) - \Delta \bar{P}(\rho_b; \rho_a) \qquad \rho_a < \rho_b < \rho_c \tag{5.39}$$

is independent of the initial conditions on $\bar{\Psi}$ and $\bar{\Omega}$ and is valid for all thin films without backflow, at points sufficiently far removed from the inlet.

In Figure 5.5, pressure profiles are plotted for various values of the Ekman number, i.e., for various rates of rotation at given viscosity, against the nondimensional radius. Let us say our recess boundary is located at $\rho_1 = 2$. In the absence of rotation, the profile labeled 'creeping flow' applies, and we see that no matter where the pad outer edge ρ_2, $\rho_2 > \rho_1$, might be located, the pressure decreases monotonically from the recess all the way to the pad outer edge. Now, since at the pad outer edge the pressure is ambient by supposition, the pressure will be above ambient everywhere within the clearance gap, yielding a force capable of supporting an external load. Conditions change drastically, however, when rotation is

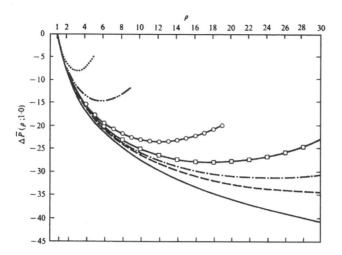

Figure 5.5. Dimensionless pressure profiles in fluid between rotating parallel disks for various values of the Ekman number: ...., $E = 0.5$; —··—, $E = 0.9$; —o—, $E = 2.0$; —□—, $E = 2.9$; —·—, $E = 3.9$; - - -, $E = 4.9$; ——, $E = 100.0$ (~creeping flow). (Reprinted with the permission of Cambridge University Press from Szeri, A. Z. and Adams, M. L. Laminar through flow between closely spaced rotating disks. *J. Fluid Mech.*, **86**, 1–14, 1978.)

added, i.e., when the Ekman number, is made finite. For finite values of the Ekman number, the pressure is no longer a monotonic decreasing function of ρ; for any nonzero rotation it is possible to select a pad outer radius such that a region of subambient pressure will exist within the clearance gap. If the region of subambient pressure is sufficiently large and the film does not cavitate, a negative pressure force might be developed that will tend to draw the runner towards the bearing. Of course the film is expected to cavitate when the pressure falls below ambient; in any case, and the hydrostatic bearing will loose its load carrying capacity. Subambient pressures and the resulting cavitation can be avoided by choosing $\rho_1 < \rho_2 < \rho_{\min}(E)$, where $\rho_{\min}$ is the location where $\Delta \bar{P}$ reaches its minimum for a given Ekman number.

The calculations above were performed for constant lubricant viscosity. It is possible to include the lubricant energy equation in Eqs. (5.29) and also to account for the temperature dependence of lubricant viscosity (Gourley, 1977).

5.3   Total Inertia limit, $\Omega^*/\text{Re}^* \to 1, \text{Re} \geq 1$

The conditions of limit (2) are by far the most demanding mathematically, as the equations are now time dependent as well as nonlinear:

$$\rho\left(\frac{\partial u}{\partial t} + u\frac{\partial u}{\partial x} + v\frac{\partial u}{\partial y} + w\frac{\partial u}{\partial z}\right) = -\frac{\partial p}{\partial x} + \mu\frac{\partial^2 u}{\partial y^2},$$

$$\rho\left(\frac{\partial w}{\partial t} + u\frac{\partial w}{\partial x} + v\frac{\partial w}{\partial y} + w\frac{\partial w}{\partial z}\right) = -\frac{\partial p}{\partial z} + \mu\frac{\partial^2 w}{\partial y^2}. \tag{5.40}$$

Equation (5.40) was solved by Reinhardt and Lund (1975) for journal bearings to first order in a perturbation series in Re^*, the zero-order solution representing the classical Reynolds theory. At small values of the Reynolds number regular perturbation in Re* will work, as was shown by Kuzma (1967), Tichy and Winer (1970), and Jones and Wilson (1974), but care must be exercised as the perturbation series is divergent (Grim, 1976). There are also numerical solutions available (Hamza, 1985).

Another approach is via the "method of averaged inertia" (Szeri, Raimondi, and Giron-Duarte, 1983), which was first employed in this country by Osterle, Chou, and Saibel (1975), but in connection with limit (3). For more recent work, see San Andres and Vance (1987).

The Method of Small Perturbations

Journal Bearings

We follow Reinhardt and Lund (1974) and apply small perturbation analysis to Eq. (5.40). We also put here $L_{xz} = R$ and $L_y = C$, where R is the radius and C the radial clearance, and employ Eqs. (2.45), (2.46), and (2.47), to nondimensionalize Eq. (5.40). Our small parameter is the reduced Reynolds number $\text{Re}^* \equiv (L_y/L_{xz})\text{Re} = C^2\omega/\nu$, and we write

$$\begin{Bmatrix} \bar{u} \\ \bar{v} \\ \bar{w} \\ \bar{p} \end{Bmatrix} = \begin{Bmatrix} \bar{u}^{(0)} \\ \bar{v}^{(0)} \\ \bar{w}^{(0)} \\ \bar{p}^{(0)} \end{Bmatrix} + \text{Re}^* \begin{Bmatrix} \bar{u}^{(1)} \\ \bar{v}^{(1)} \\ \bar{w}^{(1)} \\ \bar{p}^{(1)} \end{Bmatrix} + O(\text{Re}^{*2}). \tag{5.41}$$

Substituting these expansions into Eq. (5.40) and collecting like terms, we obtain

To zero order:

$$\begin{aligned} \frac{\partial \bar{p}^{(0)}}{\partial \bar{x}} &= \frac{\partial^2 \bar{u}^{(0)}}{\partial \bar{y}^2} \\ \frac{\partial \bar{p}^{(0)}}{\partial \bar{z}} &= \frac{\partial^2 \bar{w}^{(0)}}{\partial \bar{y}^2} \\ \frac{\partial \bar{u}^{(0)}}{\partial \bar{x}} + \frac{\partial \bar{v}^{(0)}}{\partial \bar{y}} + \frac{\partial \bar{w}^{(0)}}{\partial \bar{z}} &= 0. \end{aligned} \tag{5.42}$$

Following the procedure of Section 2.2, Eqs. (5.42) can be combined to yield a single equation in pressure,

$$\frac{\partial}{\partial \bar{x}}\left\{ \bar{h}^3 \frac{\partial \bar{p}^{(0)}}{\partial \bar{x}} \right\} + \frac{\partial}{\partial \bar{z}}\left\{ \bar{h}^3 \frac{\partial \bar{p}^{(0)}}{\partial \bar{z}} \right\} = 6\frac{\partial \bar{h}}{\partial \bar{x}} + 12\frac{\partial \bar{h}}{\partial \bar{t}}, \tag{5.43}$$

which we recognize as the Reynolds equation (2.68).

To first order:

$$\begin{aligned} \frac{\partial^2 \bar{u}^{(1)}}{\partial \bar{y}^2} &= \frac{\partial \bar{p}^{(1)}}{\partial \bar{x}} + \bar{u}^{(0)}\frac{\partial \bar{u}^{(0)}}{\partial \bar{x}} + \bar{v}^{(0)}\frac{\partial \bar{u}^{(0)}}{\partial \bar{y}} + \bar{w}^{(0)}\frac{\partial \bar{u}^{(0)}}{\partial \bar{z}} + \frac{\partial \bar{u}^{(0)}}{\partial \bar{t}} \\ \frac{\partial^2 \bar{w}^{(1)}}{\partial \bar{y}^2} &= \frac{\partial \bar{p}^{(1)}}{\partial \bar{z}} + \bar{u}^{(0)}\frac{\partial \bar{w}^{(0)}}{\partial \bar{x}} + \bar{v}^{(0)}\frac{\partial \bar{w}^{(0)}}{\partial \bar{y}} + \bar{w}^{(0)}\frac{\partial \bar{w}^{(0)}}{\partial \bar{z}} + \frac{\partial \bar{w}^{(0)}}{\partial \bar{t}} \\ \frac{\partial \bar{u}^{(1)}}{\partial \bar{x}} + \frac{\partial \bar{v}^{(1)}}{\partial \bar{y}} + \frac{\partial \bar{w}^{(1)}}{\partial \bar{z}} &= 0. \end{aligned} \tag{5.44}$$

It is, again, possible to eliminate the velocities and obtain a single equation in pressure:

$$
\frac{\partial}{\partial \bar{x}}\left\{\bar{h}^3 \frac{\partial \bar{p}^{(1)}}{\partial \bar{x}}\right\} + \frac{\partial}{\partial \bar{z}}\left\{\bar{h}^3 \frac{\partial \bar{p}^{(1)}}{\partial \bar{z}}\right\} = \frac{\partial}{\partial \bar{x}}\left\{-\frac{3\bar{h}^7}{560}\frac{\partial}{\partial \bar{x}}\left\{\left(\frac{\partial \bar{p}^{(0)}}{\partial \bar{x}}\right)^2 + \left(\frac{\partial \bar{p}^{(0)}}{\partial \bar{z}}\right)^2\right\}\right.
$$

$$
-\frac{3\bar{h}^6}{140}\frac{\partial \bar{h}}{\partial \bar{x}}\left(\frac{\partial \bar{p}^{(0)}}{\partial \bar{x}}\right)^2 + \frac{\bar{h}^5}{20}\frac{\partial^2 p^{(0)}}{\partial \bar{x}^2} + \frac{13\bar{h}^4}{140}\frac{\partial \bar{h}}{\partial \bar{x}}\frac{\partial \bar{p}^{(0)}}{\partial \bar{x}} - \frac{\bar{h}^2}{10}\frac{\partial \bar{h}}{\partial \bar{x}}
$$

$$
\left. -\frac{3\bar{h}^6}{140}\frac{\partial \bar{h}}{\partial \bar{z}}\frac{\partial \bar{p}^{(0)}}{\partial \bar{x}}\frac{\partial \bar{p}^{(0)}}{\partial \bar{z}} + \frac{13\bar{h}^4}{70}\frac{\partial \bar{p}^{(0)}}{\partial \bar{x}}\frac{\partial \bar{h}}{\partial \bar{t}} + \frac{\bar{h}^5}{10}\frac{\partial^2 \bar{p}^{(0)}}{\partial \bar{x}\,\partial \bar{t}}\right\}
$$

$$
+\frac{\partial}{\partial \bar{z}}\left\{-\frac{3\bar{h}^7}{560}\frac{\partial}{\partial \bar{z}}\left\{\left(\frac{\partial \bar{p}^{(0)}}{\partial \bar{x}}\right)^2 + \left(\frac{\partial \bar{p}^{(0)}}{\partial \bar{z}}\right)^2\right\}\right.
$$

$$
-\frac{3\bar{h}^6}{140}\frac{\partial \bar{h}}{\partial \bar{x}}\frac{\partial \bar{p}^{(0)}}{\partial \bar{x}}\frac{\partial \bar{p}^{(0)}}{\partial \bar{z}} + \frac{\bar{h}^5}{20}\frac{\partial^2 p^{(0)}}{\partial \bar{x}\,\partial \bar{z}} + \frac{13\bar{h}^4}{140}\frac{\partial \bar{h}}{\partial \bar{x}}\frac{\partial \bar{p}^{(0)}}{\partial \bar{z}}
$$

$$
\left. -\frac{3\bar{h}^6}{140}\frac{\partial \bar{h}}{\partial \bar{z}}\left(\frac{\partial \bar{p}^{(0)}}{\partial \bar{z}}\right)^2 + \frac{13\bar{h}^4}{70}\frac{\partial \bar{p}^{(0)}}{\partial \bar{z}}\frac{\partial \bar{h}}{\partial \bar{t}} + \frac{\bar{h}^5}{10}\frac{\partial^2 \bar{p}^{(0)}}{\partial \bar{z}\,\partial \bar{t}}\right\}. \tag{5.45}
$$

Thus, the zero-order pressure equation of the perturbation scheme is the classical Reynolds equation (5.40), while Eq. (5.42) gives the first-order pressure correction.

It is obvious from Eq. (5.40) that

$$
\bar{p}^{(0)} = \bar{p}^{(0)}(\bar{x}, \bar{z}; \bar{h}, \dot{\bar{h}}), \tag{5.46a}
$$

and, because the derivative $\partial p^{(0)}/\partial \bar{t}$ occurs in Eq. (5.42),

$$
\bar{p}^{(1)} = \bar{p}^{(1)}(\bar{x}, \bar{z}; \bar{h}, \dot{\bar{h}}, \ddot{\bar{h}}). \tag{5.46b}
$$

Let ψ be the angular coordinate measured from the load line that coincides with the vertical axis, as shown in Figure 5.6. Then $\psi = \theta + \phi$, where ϕ is the attitude angle, and

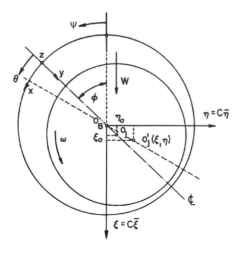

Figure 5.6. Journal bearing geometry for analysis of fluid inertia effects.

we have

$$\bar{h} = \frac{h}{C}$$

$$= 1 + \varepsilon \cos(\psi - \phi)$$

$$= 1 + \bar{\xi} \cos \psi + \bar{\eta} \sin \psi. \tag{5.47}$$

Here $\bar{\xi}$ and $\bar{\eta}$ are nondimensional Cartesian coordinates (Figure 5.6).

If we superimpose dynamic motions of small amplitudes, $\Delta \bar{\xi}$, and, $\Delta \bar{\eta}$, about the equilibrium position $(\bar{\xi}_0, \bar{\eta}_0)$, the film thickness will vary according to

$$\Delta \bar{h} = \Delta \bar{\xi} \cos \psi + \Delta \bar{\eta} \sin \psi. \tag{5.48}$$

From Eqs. (5.41) and (5.48), we have the first-order Taylor expansion about the equilibrium position $(\bar{\xi}_0, \bar{\eta}_0)$

$$\bar{p} = \bar{p}_0^{(0)} + \bar{p}_{\bar{\xi}}^{(0)} \Delta \bar{\xi} + \bar{p}_{\bar{\eta}}^{(0)} \Delta \bar{\eta} + \bar{p}_{\dot{\xi}}^{(0)} \Delta \dot{\bar{\xi}} + \bar{p}_{\dot{\eta}}^{(0)} \Delta \dot{\bar{\eta}}$$

$$+ \mathrm{Re}^* \Big[\bar{p}_0^{(1)} + \bar{p}_{\bar{\xi}}^{(1)} \Delta \bar{\xi} + \bar{p}_{\bar{\eta}}^{(1)} \Delta \bar{\eta} + \bar{p}_{\dot{\xi}}^{(1)} \Delta \dot{\bar{\xi}} + \bar{p}_{\dot{\eta}}^{(1)} \Delta \dot{\bar{\eta}} + \bar{p}_{\ddot{\xi}}^{(1)} \Delta \ddot{\bar{\xi}} + \bar{p}_{\ddot{\eta}}^{(1)} \Delta \ddot{\bar{\eta}} \Big]. \tag{5.49}$$

Equation (5.49) can now be substituted into Eqs. (5.43) and (5.48). Collecting again like terms, we end up with a set of 12 linear partial differential equations for the determination of

$$\left\{ \bar{p}_0^{(0)}, \bar{p}_{\bar{\xi}}^{(0)}, \ldots, \bar{p}_{\ddot{\eta}}^{(1)} \right\}. \tag{5.50}$$

In formulating these equations, we take into account that

$$\bar{h} = \bar{h}_0 + \Delta \bar{h},$$

where

$$\bar{h}_0 = 1 + \bar{\xi}_0 \cos \psi + \bar{\eta}_0 \sin \psi$$

and

$$(\bar{h})^n = (\bar{h}_0)^n + n(h_0)^{n-1} \Delta \bar{h} + O[(\Delta \bar{h})^2].$$

The vertical and horizontal components of the force perturbations corresponding to the various pressure perturbations, Eq. (5.49), are obtained from

$$\left\{ \begin{matrix} f_{\bar{\xi}, \bar{\omega}}^{(\alpha)} \\ f_{\bar{\eta}, \bar{\omega}}^{(\alpha)} \end{matrix} \right\} = \frac{\pi}{2} \left(\frac{D}{L} \right) \int_{-(L/D)}^{(L/D)} \int_{\bar{x}_1}^{\bar{x}_2} \bar{p}_{\bar{\omega}}^{(\alpha)} \left\{ \begin{matrix} \cos \psi \\ \sin \psi \end{matrix} \right\} d\psi \, d\bar{z} \qquad \begin{matrix} \alpha = 0, 1 \\ \bar{\omega} = \bar{\xi}, \dot{\bar{\xi}}, \ldots, \ddot{\bar{\eta}}. \end{matrix} \tag{5.51}$$

We arrange the force components obtained in Eq. (5.51) in matrix form:

$$\boldsymbol{K}^{(\alpha)} = - \begin{bmatrix} f_{\bar{\xi}, \bar{\xi}}^{(\alpha)} & f_{\bar{\xi}, \bar{\eta}}^{(\alpha)} \\ f_{\bar{\eta}, \bar{\xi}}^{(\alpha)} & f_{\bar{\eta}, \bar{\eta}}^{(\alpha)} \end{bmatrix}, \tag{5.52a}$$

$$\boldsymbol{C}^{(\alpha)} = - \begin{bmatrix} f_{\bar{\xi}, \dot{\bar{\xi}}}^{(\alpha)} & f_{\bar{\xi}, \dot{\bar{\eta}}}^{(\alpha)} \\ f_{\bar{\eta}, \dot{\bar{\xi}}}^{(\alpha)} & f_{\bar{\eta}, \dot{\bar{\eta}}}^{(\alpha)} \end{bmatrix}, \tag{5.52b}$$

$$\boldsymbol{D} = - \begin{bmatrix} f_{\bar{\xi}, \ddot{\bar{\xi}}}^{(1)} & f_{\bar{\xi}, \ddot{\bar{\eta}}}^{(1)} \\ f_{\bar{\eta}, \ddot{\bar{\xi}}}^{(1)} & f_{\bar{\eta}, \ddot{\bar{\eta}}}^{(1)} \end{bmatrix}. \tag{5.52c}$$

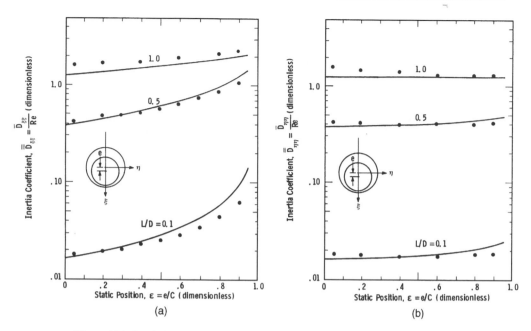

Figure 5.7. Inertial coefficients for journal bearing; (a) $D_{\xi\xi}$, (b) $D_{\eta\eta}$: — —, Szeri et al., (1983); • Reinhardt and Lund, (1974). (Reprinted with permission from Szeri, A. Z., Raimondi, A. A. and Giron-Duarte, A. Linear force coefficients for squeeze-film dampers. *ASME Journal of Lubrication Technology*, **105**, 326–334, 1983.)

Let $\Delta f = f - f_0$ signify the perturbation of the nondimensional oil-film force due to the small motion $(\Delta\bar\xi, \Delta\bar\eta)$ about the point of static equilibrium $(\bar\xi_0, \bar\eta_0)$; then, using the notation of Eq. (5.52), we can write

$$\Delta f = -\left(K^{(0)} + \mathrm{Re}^* K^{(1)}\right)\Delta\bar\xi - \left(C^{(0)} + \mathrm{Re}^* C^{(1)}\right)\Delta\dot{\bar\xi} - \mathrm{Re}^* D\,\Delta\ddot{\bar\xi}. \tag{5.53}$$

Here $\Delta\bar\xi = (\Delta\bar\xi, \Delta\bar\eta)$ is the (nondimensional) displacement vector. The diagonal elements of D, the inertia coefficient matrix, are plotted against ε in Figure 5.7.

It is obvious from Eq. (5.53) that the incremental force on the journal away from static equilibrium depends on acceleration as well as displacement and velocity of the journal, as was asserted earlier in Chapter 4.

Squeeze Flow Between Parallel Plates

Two distinct problems have been considered by researchers in this category: one in which the upper plate moves impulsively from rest with prescribed velocity toward the fixed lower plate (Kuzma, 1967; Tichy and Winer, 1970; Jones and Wilson, 1974; and Hamza and MacDonald, 1981), and another in which the upper plate moves under the action of a prescribed body force of constant magnitude (Weinbaum et al., 1985; and Yang and Leal, 1993).

We shall employ cylindrical polar coordinates (r, θ, z) and suppose that the fixed disk is at $z = 0$ and the moving disk is located at $z = h(t)$. Assuming the problem has axial symmetry, we may state it in terms of a Stokes stream function $\Psi(r, z)$

$$u_r = \frac{1}{r}\frac{\partial \Psi}{\partial z}, \qquad u_z = -\frac{1}{r}\frac{\partial \Psi}{\partial r}. \tag{5.54}$$

Let h_0 and R, the initial separation and the radius of the disks, be chosen as characteristic length in z and r, respectively. Also, let W_* be the characteristic velocity in z, the precise form of which is to be defined later. Then, the equation of continuity dictates that the characteristic velocity in r is

$$U_* = W_*\left(\frac{r_0}{h_0}\right).$$

We nondimensionalize the stream function and the coordinates according to

$$\Psi = R^2 W_* \bar{\Psi}, \qquad r = R\bar{r}, \qquad z = \bar{z}h_0.$$

The problem has two spatial dimensions r and z. We can eliminate dependence on r by making use of the von Karman (1921) similarity transformation

$$\bar{\Psi} = \bar{r}^2 \bar{G}(\bar{z}, \bar{t}), \qquad u_r = \bar{r}\frac{\partial \bar{G}(\bar{z}, \bar{t})}{\partial \bar{z}}, \qquad u_\theta = -2\bar{G}(\bar{z}, \bar{t}). \tag{5.55}$$

On substituting assumption (5.55) into the r component of the Navier-Stokes Eq. (5.29), we obtain (dropping the over bar)

$$\text{Re}[G_{zt} + (G_z)^2 - 2GG_{zz}] - G_{zzz} = -\frac{1}{r}\frac{\partial p}{\partial r}. \tag{5.56}$$

The z component of the equation simply serves to evaluate the rate of change of pressure across the film and need not be considered here.

To arrive at Eq. (5.56), we used

$$T_* = \frac{h_0}{W_*}, \qquad P_* = \frac{\mu W_*}{h_0}\left(\frac{R}{h_0}\right)^2, \qquad \text{Re} = \frac{h_0 W_*}{\nu}$$

for characteristic time, characteristic pressure, and squeeze Reynolds number, respectively, and dropped the bar that signifies nondimensionality.

The left side of Eq. (5.56) is at most a function of z and t, while the right side is at most a function of r and t, which prompts us to put

$$\frac{1}{r}\frac{\partial p}{\partial r} = A(t). \tag{5.57}$$

Here $A(t)$ is an instantaneous constant.

Constant Approach Velocity
This is characterized by the boundary conditions[3]

$$G = G_z = 0 \quad \text{at } z = 0, \tag{5.58a}$$

$$G = 1/2, G_z = 0 \quad \text{at } z = 1 - t, \tag{5.58b}$$

[3] Note that the instantaneous film thickness is given by $H(t) = \frac{h(t)}{h_0} = 1 - \frac{W}{h_0}t = 1 - \bar{t}$.

and the initial condition[4] (Jones and Wilson, 1974)

$$G = \frac{1}{2}z \quad \text{at} \ \ t = 0^+. \tag{5.59}$$

The characteristic velocity in the z-direction is chosen in this case to be the constant velocity of approach, $W_* = W$. To eliminate the pressure term, we differentiate Eq. (5.56) with respect to z. However, as the boundary condition (5.58a) is not easy to apply, we transform to a moving coordinate system (y, T) as follows:

$$G(z,t) \to F(y, T), \qquad 0 \le y = -z/T \le 1, \quad 0 \le T = t - 1 \le 1.$$

The boundaries are now fixed at $y = 0$ (stationary plate) and $y = 1$ (moving plate), and Eq. (5.59) takes the form

$$-yT F_{yyy} - 2T F_{yy} + T^2 F_{yyT} + 2T F F_{yyy} = \mathrm{Re}^{-1} F_{yyyy}. \tag{5.60}$$

The transformed boundary conditions are given by

$$F = F_y = 0, \quad \text{on} \ \ y = 0 \ \ \text{(fixed disk)} \tag{5.61a}$$

$$F = 1/2, \qquad F_y = 0, \quad \text{on} \ \ y = 1 \ \ \text{(moving disk)} \tag{5.61b}$$

and the initial condition by

$$F = y/2, \quad \text{at} \ \ T = -1. \tag{5.62}$$

Following Jones and Wilson (1974), we seek solution for small Reynolds number and assume that

$$F = \sum_{n=0} \varepsilon^n f_n(y), \qquad \varepsilon = T \, \mathrm{Re}. \tag{5.63}$$

Note that $|\varepsilon| = W_* h(t)/\nu$ is the instantaneous Reynolds number.

On substituting Eq. (5.63) into Eq. (5.60), we obtain

$$f_0'''' + \varepsilon \left[f_1'' - y f_0''' - 2 f_0'' + 2 f_0 f_0''' \right]$$
$$+ \varepsilon^2 \left[f_2'''' + f_2'' - y f_1''' - 2 f_1'' + 2 \left(f_0''' f_1 + f_0 f_1''' \right) \right] + \cdots = 0. \tag{5.64}$$

The leading-order term supplies the lubrication approximation

$$f_0 = \frac{3}{2}y^2 - y^3, \tag{5.65a}$$

and for the first-order correction we obtain

$$f_1 = \frac{1}{70}y^7 - \frac{1}{20}y^6 + \frac{3}{20}y^5 - \frac{1}{4}y^4 + \frac{5}{28}y^3 - \frac{3}{70}y^2. \tag{5.65b}$$

Though these functions satisfy the boundary conditions, they do not satisfy the initial condition and, therefore, cannot be valid for short times.

---

[4] To arrive at this initial condition, Jones and Wilson argue that vorticity is unchanged across $t = 0$ by the velocity impulse, and as the only vorticity component is rG_{zz}, we have $G_{zz} = 0$. This integrates to our initial condition.

From Eq. (5.59), the first three terms of the series (5.63) yield[5]

$$\frac{1}{r}\frac{\partial p}{\partial r} = 6T^3\left\{1 - \frac{5}{28}(T\,\mathrm{Re}) - \frac{277}{323,400}(T\,\mathrm{Re})^2 + \cdots\right\}. \tag{5.66a}$$

To study conditions at short times, Jones and Wilson scale the time variable with the Reynolds number and find that the leading inner solution term of Eq. (5.60) is

$$F_{yyT} = \mathrm{Re}\,F_{yyyy}.$$

Solution of this diffusion-type equation, found by separation of variables, leads to

$$\frac{1}{r}\frac{\partial p}{\partial r} = 6T^3\{1 + 4\exp[\lambda^2(T+1)/\mathrm{Re}] + \cdots\}, \tag{5.66b}$$

where $\lambda = 3\pi - 4/3\pi$ is an eigenvalue. From Eq. (5.66b), we can estimate the elapse of time after start for the initial condition to loose influence on the pressure gradient.

For large Reynolds number, the flow is of the boundary layer type and matching of inner and outer solutions is employed. Further details of the analysis are available in the paper by Jones and Wilson (1974).

Constant Applied Force

Following Yang and Leal (1993), for $\mathrm{Re} \ll 1$ we choose the characteristic velocity as

$$W_* = \frac{4}{\pi}\frac{fh_0^3}{\mu R^4}, \qquad f = mg,$$

then the characteristic time is

$$T_* = \frac{\pi}{4}\frac{\mu R^4}{fh_0^2}.$$

This choice of velocity and time scales is suggested by the equation of motion for the body encompassing the upper plate. For $\mathrm{Re} \ll 1$, we may neglect inertia of both fluid and body, and the equation of motion of the body reduces to a balance of viscous forces and the weight, f, of the body. The pressure term $A(t)$, Eq. (5.60), is then calculated from considering the equation of motion for the body containing the upper plate. We find in this manner that

$$A(t) = 1 + \mathrm{Re}\,\beta\frac{d^2H}{dt^2}, \qquad \beta = \frac{4h_0 m}{\pi\rho R^4}, \qquad \mathrm{Re} = \frac{4\rho f h_0^4}{\pi\mu^2 R^4} \ll 1. \tag{5.67}$$

Equation (5.60) now takes the form

$$\frac{\partial^2 G}{\partial z\,\partial t} + \left(\frac{\partial G}{\partial z}\right)^2 - 2G\frac{\partial^2 G}{\partial z^2} - \frac{1}{\mathrm{Re}}\frac{\partial^3 G}{\partial z^3} = \frac{1}{\mathrm{Re}} + \beta\frac{d^2H}{dt^2} \tag{5.68a}$$

with boundary conditions

$$G = -\frac{1}{2}\frac{dH}{dt}, \qquad \frac{\partial G}{\partial z} = 0 \quad \text{at } z = H,$$

$$G = \frac{\partial G}{\partial z} = 0 \quad \text{at } z = 0, \tag{5.68b}$$

$$G = 0, \qquad H = 1 \quad \text{at } t = 0.$$

---

[5] Note that the right-hand side of Eq. (5.66a) is to be divided by Re to obtain agreement with Eq. (13.2) of Jones and Wilson, as they scale the pressure with ρW^2 and the radial coordinate with d.

We seek solution of Eq. (5.68) in the form of a regular perturbation expansion:

$$G(z, t) = G^{(0)}(z, t) + Re\, G^{(1)}(z, t) + \cdots, \qquad (5.69a)$$

$$H(t) = H^{(0)}(t) + Re\, H^{(1)}(t) + \cdots. \qquad (5.69b)$$

The zero-order solution contains one unknown integration constant, c_0,

$$G^{(0)} = \frac{1}{4}\left(H^{(0)}z^2 - \frac{2}{3}z^3\right), \qquad H^{(0)} = \left(\frac{3}{t + c_0}\right)^{1/2}. \qquad (5.70)$$

The solution in Eq. (5.70) completely satisfies the boundary conditions but not the initial condition, confirming our earlier conclusion that regular perturbation in Re cannot solve Eq. (5.68) for short times. The reason is that the problem possesses two very different time scales, the classical signature of a singular perturbation problem. The "outer" time scale, $T_* = h_0/W_*$, and the expansions (5.69) characterize long-term behavior. Short-time behavior, on the other hand, is characterized by the "inner" time scale, $T_\nu = h_0^2/\nu$. The ratio of the time scales is $T_\nu/T_* = Re$, which we have assumed to be asymptotically small.

Equation (5.69) will suffice as the outer solution. To obtain the inner solution, we have to rescale the problem. The inner and outer (time) variables, $\tilde{t}$ and t respectively, are related by $\tilde{t} = t/Re$. Rescaling Eq. (5.68) accordingly, we obtain the equation valid for short times

$$\frac{\partial^2 \tilde{G}}{\partial z \partial \tilde{t}} - \frac{\partial^3 \tilde{G}}{\partial z^3} - 1 = Re\left[2\tilde{G}\frac{\partial^2 \tilde{G}}{\partial z^2} - \left(\frac{\partial \tilde{G}}{\partial z}\right)^2\right] + \frac{\beta}{Re}\frac{d^2\tilde{H}}{d\tilde{t}^2}. \qquad (5.71)$$

Here the tilde denotes the variables in the inner (short time) region.

To obtain the "inner" solution, i.e., to solve Eq. (5.71), we assume that

$$\tilde{G}(z, \tilde{t}) = \tilde{G}^{(0)}(z, \tilde{t}) + Re\, \tilde{G}^{(1)}(z, \tilde{t}) + \cdots, \qquad (5.72a)$$

$$\tilde{H}(\tilde{t}) = \tilde{H}^{(0)} + Re\, \tilde{H}^{(1)}(\tilde{t}) + \cdots, \qquad \tilde{H}^{(0)} = 1. \qquad (5.72b)$$

The governing equation and boundary and initial conditions for the leading-order terms $\tilde{G}^{(0)}$ and $\tilde{H}^{(1)}$ are

$$\frac{\partial^2 \tilde{G}^{(0)}}{\partial z \partial \tilde{t}} - \frac{\partial^3 \tilde{G}^{(0)}}{\partial z^3} = 1 + \beta\frac{d^2\tilde{H}^{(1)}}{d\tilde{t}^2}, \qquad (5.73a)$$

$$\tilde{G}^{(0)} = \frac{\partial \tilde{G}^{(0)}}{\partial z} = 0 \qquad \text{at } z = 0, \qquad (5.73b)$$

$$\frac{\partial \tilde{G}^{(0)}}{\partial z} = 0, \qquad \tilde{G}^{(0)} = -\frac{1}{2}\frac{d\tilde{H}^{(1)}}{d\tilde{t}} \qquad \text{at } z = 1, \qquad (5.73c)$$

$$\tilde{G}^{(0)} = 0, \qquad \tilde{H}^{(1)} = 0 \quad \text{at } \tilde{t} = 0, \qquad (5.73d)$$

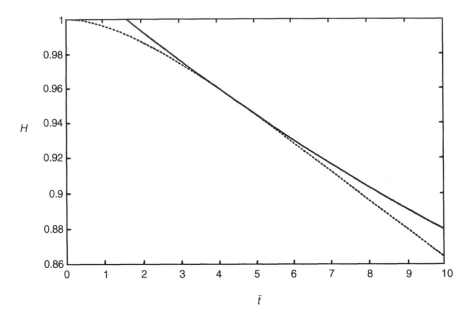

Figure 5.8. Composite uniform approximation to the film thickness as function of the small time variable $\tilde{t}$ at Re $= 0.1$; $——$, $\beta = 10$; $----$, $\beta = 0$. (Reprinted with permission from Yang, S.-M. and Leal, G. Thin fluid film squeezed with inertia between two parallel plane surfaces. *ASME Journal of Tribology*, **115**, 632–639, 1993.)

The system (5.73) is amenable to solution via separation of variables (Yang and Leal, 1953). Yang and Leal evaluate the first two terms in both inner and outer expansions, and from these construct H in the overlap region by matching. This is depicted in Figure 5.8.

The Method of Averaged Inertia

This method was employed for the steady state lubrication problem by Osterle, Chou, and Saibel (1957) and by Constantinescu (Constantinescu 1970; Constantinescu and Galetuse, 1982). Here we follow Szeri, Raimondi, and Giron (1983) and write Eq. (5.2) in the form

$$\frac{\partial u}{\partial t} + u\frac{\partial u}{\partial x} + v\frac{\partial u}{\partial y} + w\frac{\partial u}{\partial z} = -\frac{1}{\rho}\frac{\partial p}{\partial x} + \frac{1}{\rho}\frac{\partial T_{xy}}{\partial y},$$

$$\frac{\partial w}{\partial t} + u\frac{\partial w}{\partial x} + v\frac{\partial w}{\partial y} + w\frac{\partial w}{\partial z} = -\frac{1}{\rho}\frac{\partial p}{\partial z} + \frac{1}{\rho}\frac{\partial T_{zy}}{\partial y}. \tag{5.74}$$

As we are interested in average (across the film) quantities here, we integrate Eq. (5.74) across the film and on using the averaged continuity equation

$$\frac{\partial}{\partial x}\int_0^{h(x,t)} u\,dy + \frac{\partial}{\partial z}\int_0^{h(x,t)} w\,dy + \left(v - u\frac{\partial h}{\partial x}\right)\bigg|_{h(x,t)} = 0 \tag{5.75}$$

obtain

$$\frac{\partial}{\partial t}\int_0^h u\,dy + \frac{\partial}{\partial x}\int_0^h u^2\,dy + \frac{\partial}{\partial z}\int_0^h uw\,dy$$

$$-\left\{u\left(\frac{\partial}{\partial x}\int_0^h u\,dy + \frac{\partial}{\partial z}\int_0^h w\,dy\right)\right\}\bigg|_h = -\frac{h}{\rho}\frac{\partial p}{\partial x} + \frac{1}{\rho}\tau_{xy}\bigg|_0^h + u\bigg|_h\frac{\partial h}{\partial t},$$

(5.76a)

$$\frac{\partial}{\partial t}\int_0^h w\,dy + \frac{\partial}{\partial x}\int_0^h uw\,dy + \frac{\partial}{\partial z}\int_0^h w^2\,dy = -\frac{h}{\rho}\frac{\partial p}{\partial z} + \frac{1}{\rho}\tau_{zy}\bigg|_0^h. \qquad (5.76b)$$

Partial differentiation of Eq. (5.76a) with respect to x and of Eq. (5.76b) with respect to z and addition of the resulting equations yields

$$\frac{\partial}{\partial x}\left(\frac{h}{\rho}\frac{\partial p}{\partial x}\right) + \frac{\partial}{\partial z}\left(\frac{h}{\rho}\frac{\partial p}{\partial z}\right)$$

$$= \frac{\partial}{\partial x}\left(\frac{T_{xy}|_0^h}{\rho}\right) + \frac{\partial}{\partial z}\left(\frac{T_{zy}|_0^h}{\rho}\right) + U_P\frac{\partial^2 h}{\partial x\,\partial t} - \frac{\partial^2}{\partial x\,\partial t}\int_0^h u\,dy$$

$$- \frac{\partial^2}{\partial x^2}\int_0^h u^2\,dy - 2\frac{\partial^2}{\partial x\,\partial z}\int_0^h uw\,dy - \frac{\partial^2}{\partial z\,\partial t}\int_0^h w\,dy$$

$$- \frac{\partial^2}{\partial z^2}\int_0^h w^2\,dy + U_P\left(\frac{\partial^2}{\partial x^2}\int_0^h u\,dy + \frac{\partial^2}{\partial x\,\partial z}\int_0^h w\,dy\right) \qquad (5.77)$$

Here U_P is shaft surface velocity in rigid body translation.

We make our first significant assumption here: that the shape of the velocity profile is not strongly influenced by inertia, so the viscous terms are approximated by (Constantinescu, 1970)

$$T_{xy}|_0^h = -\frac{12\mu}{h}\left(U - \frac{1}{2}U_P\right); \qquad T_{zy}|_0^h = -\frac{12\mu}{h}W. \qquad (5.78)$$

Here

$$U = \frac{1}{h}\int_0^h u\,dy; \qquad W = \frac{1}{h}\int_0^h w\,dy \qquad (5.79)$$

are the components of the average velocity vector.

Our second significant assumption

$$\int_0^h u^2\,dy = hU^2; \qquad \int_0^h uw\,dy = hUW; \qquad \int_0^h w^2\,dy = hW^2 \qquad (5.80)$$

equates the average of a product to the product of the averages and is not possible to defend.[6] Nevertheless, it has been used by several researchers, beginning with Constantinescu (1970).

---

[6] Hashimoto (1994) appears to have overcome the necessity of having to assume Eq. (5.80). This, however, is purely illusory, as he makes the equally indefensible assumption that $\partial T_{xy}/\partial y = \text{const.}$ across the film, true only for zero inertia.

Assumptions (5.78) and (5.80) permit us to write (5.79) in the form

$$\frac{\partial}{\partial x}\left(\frac{h^3}{\mu}\frac{\partial p}{\partial x}\right) + \frac{\partial}{\partial z}\left(\frac{h^3}{\mu}\frac{\partial p}{\partial z}\right)$$

$$= 12\left\{v|_h - \frac{1}{2}U_P\frac{\partial h}{\partial x}\right\} + \frac{h^2}{\nu}\left\{U_P\frac{\partial^2 h}{\partial x\,\partial t} - \frac{\partial^2 (hU)}{\partial x\,\partial t} - \frac{\partial^2 (hU^2)}{\partial x^2}\right.$$

$$\left. - 2\frac{\partial^2 (hUW)}{\partial x\,\partial z} - \frac{\partial^2 (hW)}{\partial z\,\partial t} - \frac{\partial^2 (hW^2)}{\partial z^2} + U_P\frac{\partial^2 (hU)}{\partial x^2} + U_P\frac{\partial^2 (hW)}{\partial x\,\partial z}\right\}.$$

$$(5.81)$$

We will apply Eq. (5.81) to a squeeze film damper (Szeri et al., 1983). A schematic of the damper is shown in Figure 5.9.

To nondimensionalize Eq. (5.81), we apply the transformation

$$h = CH, \qquad t = \bar{t}/\omega, \qquad x = R\bar{x}, \qquad z = \frac{1}{2}\bar{z}, \qquad \bar{U} = U/\delta\omega,$$

$$(5.82)$$

$$\bar{W} = W/\delta\omega, \qquad \mathrm{Re} = \frac{\delta\omega C}{\nu}, \qquad p = 12\mu\omega\left(\frac{R}{C}\right)^2\bar{p}.$$

Here ω is the characteristic frequency, δ is the characteristic orbit radius, and $\delta\omega$ is the characteristic velocity of the system.

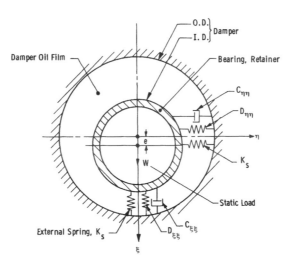

Figure 5.9. Basic elements (cross-coupling not shown) of squeeze-film damper. (Reprinted with permission from Szeri, A. Z., Raimondi, A. A. and Giron-Duarte, A. Linear force coefficients for squeeze-film dampers. *ASME Journal of Lubrication Technology*, **105**, 326–334, 1983).

The dimensionless form of Eq. (5.81) is

$$\frac{\partial}{\partial x}\left(H^3\frac{\partial \bar{p}}{\partial \bar{x}}\right) + \left(\frac{D}{L}\right)^2 \frac{\partial}{\partial \bar{z}}\left(H^3\frac{\partial \bar{p}}{\partial \bar{z}}\right) = \frac{\partial H}{\partial t} - \frac{1}{2}\left(\frac{\delta}{R}\right)U_P\frac{\partial H}{\partial \bar{x}} + \frac{\text{Re}}{12}\left(\frac{C}{R}\right)H^2$$

$$\times \left\{U_P\frac{\partial^2 H}{\partial \bar{x}\,\partial \bar{t}} - \frac{\partial^2 H\bar{U}}{\partial \bar{x}\,\partial \bar{t}} - \left(\frac{D}{L}\right)^2\frac{\partial^2 H\bar{W}}{\partial \bar{z}\,\partial \bar{t}} - \left(\frac{\delta}{R}\right)\left[\frac{\partial^2 H\bar{U}^2}{\partial \bar{x}^2} + 2\left(\frac{D}{L}\right)\right.\right.$$

$$\left.\left.\times\frac{\partial^2 H\bar{U}\,\bar{W}}{\partial \bar{z}\,\partial \bar{x}} + \left(\frac{D}{L}\right)^2\frac{\partial^2 H\bar{W}^2}{\partial \bar{z}^2} - U_P\frac{\partial^2 H\bar{U}}{\partial \bar{x}^2} - \left(\frac{D}{L}\right)\bar{U}_P\frac{\partial^2 H\bar{W}}{\partial \bar{x}\,\partial \bar{z}}\right]\right\}$$

$$(5.83)$$

Considering that $\delta = O(C)$ and therefore $(\delta/R) = O(10^{-3})$, it seems possible to simplify Eq. (5.83) by deleting all terms which are multiplied by (δ/R). We will not do this, however, as our intention is to introduce the short-bearing approximation at this stage via the equation of continuity Eq. (5.75).

Introducing the short bearing approximation $\bar{U} \cong \frac{1}{2}\bar{U}_P$ into Eq. (5.75), we have

$$\frac{\partial H\bar{W}}{\partial \bar{z}} \approx \left(\frac{L}{D}\right)\left(\frac{R}{\delta}\right)\left\{\frac{1}{2}\left(\frac{\delta}{R}\right)\bar{U}_P\frac{\partial H}{\partial \bar{x}} - \frac{\partial H}{\partial t}\right\} \approx -\left(\frac{L}{D}\right)\left(\frac{R}{\delta}\right)\frac{\partial H}{\partial \bar{t}}. \quad (5.84)$$

This leads to the following approximation of Eq. (5.83) when we neglect terms multiplied by (δ/R):

$$\frac{\partial}{\partial x}\left(H^3\frac{\partial \bar{p}}{\partial \bar{x}}\right) + \left(\frac{D}{L}\right)^2 \frac{\partial}{\partial \bar{z}}\left(H^3\frac{\partial \bar{p}}{\partial \bar{z}}\right) = \frac{\partial H}{\partial \bar{t}} + \frac{H^2}{12}\text{Re}^*\left\{\frac{\partial^2 H}{\partial \bar{t}^2} + \frac{2}{H}\left(\frac{\partial H}{\partial t}\right)^2\right\} \quad (5.85)$$

Here $\text{Re}^* = \text{Re}(C/\delta)$ is the reduced Reynolds number.

Further details can be found in Szeri, Raimondi, and Giron (1983). Here we show the diagonal terms of the inertia coefficient matrix D, i.e., the added mass tensor, in Figure 5.7. This figure also contains results from small perturbation analysis (Reinhardt and Lund, 1974). The analysis of Szeri, Raimondi, and Giron (1983), when cavitation is accounted for, yields results as shown in Figure 5.10.

Zhang, Ellis, and Roberts (1993) attempted to verify recent "averaged inertia" analyses, as applied to squeeze, film dampers. They found that in the cases they considered, the pressure field can be expressed as

$$p = 12\left(\frac{z^2}{2} - \frac{L^2}{8}\right) + 12\left\{\mu\left(\frac{1}{h^3}\frac{\partial h}{\partial t}\right) + \rho\left(\frac{k_1}{h}\frac{\partial^2 h}{\partial t^2}\right) - \rho\left[\frac{k_2}{h^2}\left(\frac{\partial h}{\partial t}\right)^2\right]\right\}. \quad (5.86)$$

The values of the numerical constants k_1 and k_2 in Eq. (5.86) for the various analyses are shown in Table 5.7. In the San Andres and Vance (1986) analysis, the k_1, k_2 values depend on the flow regime considered. At low Re, the analysis yields values identical to those of Tichy and Bou-Said (1991); at large Re the k_1, k_2 values agree with those of Szeri et al. (1985). At moderate Re values, San Andres and Vance propose k_1, k_2 between bounds identified in Table 5.7.

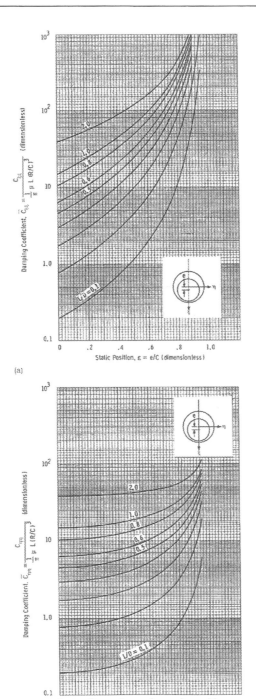

Figure 5.10. Inertia coefficients in presence of cavitation. (Reprinted with permission from Szeri, A. Z., Raimondi, A. A. and Giron-Duarte, A. Linear force coefficients for squeeze-film dampers. *ASME Journal of Lubrication Technology*, **105**, 326–334, 1983).

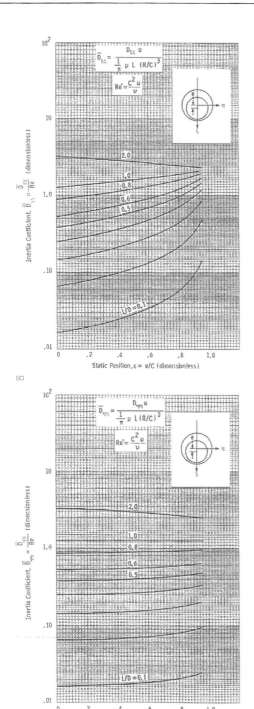

(c)

(d)

Figure 5.10. (*cont.*)

Table 5.7. *Numerical constants for inertia terms*

	Tichy and Bou-Said (1991) El-Shafei and Crandall (1991)	Szeri et al. (1983)	Trichy and Modest (1980) El-Shafei and Crandall (1991)
k_1	1/12	1/12	1/10
k_2	1/5	1/6	17/70

5.4   Nomenclature

A_i, B_i, b_i	B-splines
C	radial clearance
$\boldsymbol{C}, \boldsymbol{K}, \boldsymbol{D}$	linearized force coefficient matrices
E	Ekman number
F_R, F_T	radial, tangential force component
H	film thickness, dimensionless
L_y, L_{yz}	characteristic lengths
L_α, L_β	characteristic lengths
$\bar{P}$	average pressure, dimensionless
Q	flow rate
Re, R_Q, r_e	Reynolds numbers
R	radius
U, V, W	velocity components
W	load
S	Sommerfeld number
ϕ, ψ	angular coordinates
ψ	stream function
Ω	characteristic frequency, angular velocity
c	velocity of sound
h	film thickness
p	pressure
t	time
u, v, w	velocity components (Cartesian)
u_r, u_θ, u_z	velocity components (polar)
$(x, y), (\xi, \eta)$	Cartesian coordinates
(r, θ, z)	polar coordinates
$(\hat{\alpha}, \hat{\beta})$	bipolar coordinates
ρ, ζ	dimensionless coordinates
ρ, μ, ν	density, viscosity, kinematic viscosity
ω	angular velocity
$(\ )_*$	characteristic quantity
$(\bar{\ })$	dimensionless quantity
ε	eccentricity ratio

5.5 References

Ashino, I. 1975. Slow motion between eccentric rotating cylinders. *Bulletin, Japan Soc. Mech. Engrs.*, **18**, 280–285.

Ballal, B. and Rivlin, R. S. 1976. Flow of a Newtonian fluid between eccentric rotating cylinders. *Arch. Rational Mech. Anal.*, **62**, 237–274.

Black, H. F. and Walton, M. H. 1974. Theoretical and experimental investigations of a short 360° journal bearing in the transition superlaminar regime. *J. Mech. Eng. Sci.*, **16**, 286–297.

Brennen, C. 1976. On the flow in an annulus surrounding a whirling cylinder. *J. Fluid Mech.*, **75**, 173–191.

Burton, R. A. and Hsu, Y. C. 1974. The incompressible turbulent-thin-film short bearing with inertial effects. *ASME Trans., Ser. F*, **96**, 158–163.

Chen, S. S., Wambsganss, M. W. and Jendrzejczyk, J. 1976. Added mass and damping of a vibrating rod in confined viscous fluids. *ASME J. Appl. Mech.*, **43**, 325–329.

Christie, I., Rajagopal, K. R. and Szeri, A. Z. 1987. Flow of a non-Newtonian fluid between eccentric rotating cylinders. *Intl. J. Eng. Sci.*, **25**, 1029–1047.

Constantinescu, V. N. 1970. On the influence of inertia forces in turbulent and laminar self-acting films. *ASME Journal of Lubrication Technology*, **92**, 473–481.

Constantinescu, V. N. and Galetuse, S. 1982. Operating characteristics of journal bearing in turbulent inertial flow. *ASME Journal of Lubrication Technology*, **104**, 173–179.

Coombs, J. A. and Dowson, D. 1965. An experimental investigation of the effects of lubricant inertia in hydrostatic thrust bearings. *Proc. Inst. Mech. Eng.*, **179**, pt. 3, 96–108.

Dai, R. X., Dong, Q. M. and Szeri, A. Z. 1991. Flow of variable viscosity fluid between eccentric rotating cylinders. *Intl. J. Non-Linear Mech.*, **27**, 367–389.

Dai, R. X., Dong, Q. M. and Szeri, A. Z. 1992. Approximations in hydrodynamic lubrication. *ASME Journal of Tribology*, **114**, 14–25.

de Sampaio, P. A. B. 1991. Galerkin formulation for the incompressible Navier-Stokes equations using equal order interpolation for velocity and pressure. *Internat. J. Numer. Methods Engrg.*, **31**, 1134–1149.

DeBoor, C. 1978. *A Practical Guide to Splines*. Springer-Verlag, New York.

DiPrima, R. C. and Stuart, J. T. 1972a. Non-local effects in the stability of flow between eccentric rotating cylinders. *J. Fluid Mech.*, **54**, 393–415.

DiPrima, R. C. and Stuart, J. T. 1972b. Flow between eccentric rotating cylinders. *ASME Journal of Lubrication Technology*, **94**, 266–274.

El-Shafei, A. and Crandall, S. H. 1991. Fluid inertia forces in squeeze, film dampers in rotating machinery and vehicle dynamics. *ASME DE*, **35**, 219–228.

Fletcher, C. A. J. 1991. *Computational Techniques for Fluid Dynamics*. Springer-Verlag, New York.

Grim, R. J. 1976. Squeezing flows of Newtonian films. *App. Sci. Res.*, **32**, 149–166.

Hamza, E. A. 1985. A fluid film squeezed between two rotating parallel plane surfaces. *ASME Journal of Tribology*, **107**, 110–115.

Hashimoto, H. 1994. Viscoelastic squeeze film characteristics with inertia effects between two parallel circular plates under sinusoidal motion. *ASME Journal of Tribology*, **116**, 161–166.

Hughes, T. J .R., Franca, L. P. and Balestra, M. 1986. A new finite element formulation for computation fluid dynamics. *Comput. Methods Appl. Math. Engrg.*, **59**, 85–99.

Jackson, J. D. and Symmons, G. R. 1965. An investigation of laminar radial flow between two parallel disks. *Appl. Sci. Res. Sect. A*, **15**, 59–75.

Jones, A. F. and Wilson, S. D. R. 1974. On the failure of lubrication theory in squeezing flow. *ASME Journal of Lubrication Technology*, **97**, 101–104.

Kamal, M. M. 1966. Separation in the flow between eccentric rotating cylinders. *ASME J. of Basic Engineering*, **88**, 717–724.

Keller, H. B. 1977. Numerical solutions of bifurcation and non-linear eigenvalue problems. In *Applications of Bifurcation Theory*, P. Rabinowitz (ed.). Academic Press, New York.

Kim, E. and Szeri, A. Z. 1997. On the combined effects of lubricant inertia and viscous dissipation in long bearings. *ASME Journal of Tribology*, **119**, 76–84.

Kulinski, E. and Ostratch, S. 1967. Journal bearing velocity profiles for small eccentricity and moderate modified Reynolds number. *ASME J. Appl. Mech.*, **89**, 16–22.

Kuzma, D. 1967. Fluid inertia effects in squeeze films. *App. Sci. Res.*, **18**, 15–20.

Launder, B. E. and Leschziner, M. 1978. Flow in finite-width, thrust bearings including inertial effects, I and II. *ASME Trans., Ser. F*, **100**, 330–345.

Livesey, J. L. 1960. Inertia effects in viscous flows. *Int. J. Mech. Sci.*, **1**, 81–88.

Modest, M. F. and Tichy, J. A. 1978. Squeeze film flow in arbitrarily shaped journal bearings subject to oscillations. *ASME Journal of Lubrication Technology*, **100**, 323–329.

Mulcahy, T. M. 1980. Fluid forces on rods vibrating in finite length annular regions. *ASME J. Appl. Mech.*, **47**, 234–240.

Myllerup, C. M. and Hamrock, B. J. 1994. Perturbation approach to hydrodynamic lubrication theory. *ASME Journal of Tribology*, **116**, 110–118.

Ortega, J. M. and Rheinboldt, W. C. 1970. *Iterative Solution of Non-Linear Equations in Several Variables*. Academic Press, New York.

Osterle, J. F. and Hughes, W. F. 1958. Inertia induced cavitation in hydrostatic thrust bearings. *Wear*, **4**, 228–233.

Osterle, J. F., Chou, Y. T. and Saibel, E. 1975. The effect of lubricant inertia in journal bearing lubrication. *ASME Trans.*, **79**, Ser. F, 494–496.

Ota, T., Yoshikawa, H., Hamasuna, M., Motohashi, T. and Oi, S. 1995. Inertia effects on film rapture in hydrodynamic lubrication. *ASME Journal of Tribology*, **117**, 685–660.

Reinhardt, E. and Lund, J. W. 1975. The influence of fluid inertia on the dynamic properties of journal bearings. *ASME Trans., Ser. F*, **97**, 159–167.

Reynolds, O. 1986. On the theory of lubrication and its application to Mr. Beachamp Tower's experiments. *Phil. Trans. Roy. Soc.*, **177**, 157–234.

Ritchie, G. S. 1968. On the stability of viscous flow between eccentric rotating cylinders. *J. Fluid Mech.*, **32**, 131–144.

Rosenhead, L. 1963. *Laminar Boundary Layers*. Oxford University Press.

San Andres, A. and Szeri, A. Z. 1985. Flow between eccentric rotating cylinders. *ASME J. Appl. Mech.*, **51**, 869–878.

San Andres, A. and Vance, J. 1987. Force coefficients for open-ended squeeze-film dampers, executing small amplitude motions about an off-center equilibrium position. *ASLE Trans.*, **30**, 384–393.

Schlichting, H. 1968. *Boundary Layer Theory*, 6th ed. Pergamon, London.

Sestieri, A. and Piva, R. 1982. The influence of fluid inertia in unsteady lubrication films. *ASME Journal of Lubrication Technology*, **104**, 180–186.

Seydel, R. 1988. *From Equilibrium To Chaos*. Elsevier, New York.

Sood, D. R. and Elrod, H. G. 1974. Numerical solution of the incompressible Navier-Stokes equations in doubly-connected regions. *AIAA J.*, **12**, 636–641.

Szeri, A. Z. and Adams, M. L. 1978. Laminar through flow between closely spaced rotating disks. *J. Fluid Mech.*, **86**, 1–14.

Szeri, A. Z. and Al-Sharif, A. 1995. Flow between finite, steadily rotating eccentric cylinders. *Theoret. Comput. Fluid Dynamics*, **7**, 1–28.

Szeri, A. Z., Raimondi, A. A. and Giron-Duarte, A. 1983. Linear force coefficients for squeeze-film dampers. *ASME Journal of Lubrication Technology*, **105**, 326–334.

Szeri, A. Z., Schneider, S. J., Labbe, F. and Kaufman, H. N. 1983. Flow between rotating disks. Part 1: basic flow. *J. Fluid Mech.*, **134**, 103–131.

Tichy, J. and Bou-Said, B. 1991. Hydrodynamic lubrication and bearing behavior with impulsive loads. *STLE Tribology Transactions*, **34**, 505–512.

Tichy, J. and Modest, M. 1978. Squeeze film flow between arbitrary two-dimensional surfaces subject to normal oscillations. *ASME Journal of Lubrication Technology*, **100**, 316–322.

Tichy, J. and Winer, W. 1970. Inertial considerations in parallel circular squeeze-film bearings. *ASME Journal of Lubrication Technology*, 588–592.

Von Karman, T. 1921. Über laminare und turbuleute Reibung. *ZAMM*, **1**, 233–252.

Wannier, G. 1950. A contribution to the hydrodynamics of lubrication. *Quart. Appl. Math.*, **8**, 1–32.

Weinbaum, S., Lawrence, C. J. and Kuang, Y. 1985. The inertial drainage of a thin fluid layer between parallel plates with a constant normal force. Part 1. Analytical solutions: inviscid and small but finite-Reynolds-number limits. *J. Fluid Mechanics*, **121**, 315–343.

Wood, W. 1957. The asymptotic expansions at large Reynolds numbers for steady motion between non-coaxial rotating cylinders. *J. Fluid Mech.*, **3**, 159–175.

Yamada, Y. 1968. On the flow between eccentric cylinders when the outer cylinder rotates. *Japan Soc. Mech. Engrs.*, **45**, 455 462.

Yang, S.-M. and Leal, G. 1993. Thin fluid film squeezed with inertia between two parallel plane surfaces. *ASME Journal of Tribology*, **115**, 632–639.

You, H. L. and Lu, S. S. 1987. Inertia effects in hydrodynamic lubrication with film rupture. *ASME Journal of Tribology*, **109**, 86–90.

Zhang, J., Ellis, J. and Roberts, J. B. 1993. Observations on the nonlinear fluid forces in short cylindrical squeeze, film dampers. *ASME Journal of Tribology*, **115**, 692–698.

Zienkiewicz, O. C. and Woo, J. 1991. Incompressibility without tears: how to avoid restrictions of mixed formulation. *Internat. J. Numer. Methods Engrg.*, **32**, 1189–1203.

CHAPTER 6

Flow Stability and Transition

Classical lubrication theory is unable to predict the performance of large bearings accurately, particularly under conditions of heavy load and/or high rotational speed. The reason for this failure of the theory can be traced to two assumptions, (1) laminar flow and (2) constant, uniform viscosity. A third assumption, that of negligible fluid inertia, yields erroneous results only in special cases and, for the most part, can be left intact. The object of this and the next chapters is to investigate the shortcomings of classical theory due to the assumption of laminar flow. The effects of nonuniform viscosity will be discussed in Chapter 9, while Chapter 5 dealt with fluid inertia effects.

It will be shown in Chapter 7, Eq. (7.59), that under isothermal conditions the pressure generated in an infinitely long "turbulent" lubricant film of thickaness h has the same magnitude as the pressure that can be obtained in an otherwise identical "laminar" film of thickness

$$\frac{12}{12 + k_x(\mathrm{Re})} h.$$

The turbulence function $k_x(\mathrm{Re}) \geq 0$, thus a change from laminar to turbulent flow at fixed film geometry results in a change in load capacity (Figure 3.10). The rate of heat generation in the lubricant film also changes on passing from the laminar to the turbulent regime. It is thus essential to know at the design stage in which flow regime will the bearing operate.

> The actual conditions are far more complicated than alluded to in the previous paragraph. As suggested by Figure 9.7, the law governing energy dissipation changes abruptly on changing from the laminar mode of flow, in turbulent flow the rate of energy dissipation can be significantly higher. This higher rate of heat generations may lead to higher lubricant temperature, and, in consequence, to lower lubricant viscosity. Lower viscosity, in its turn, results in a loss of load capacity. There are thus two competing influences in effect; the apparent decrease in film thickness, which increases load capacity, and the decrease in lubricant viscosity, which tends to decrease it. The result of the ensuing competition under particular circumstances can only be predicted by detailed calculations (see Chapter 9).

There are two basic modes of flow in nature: laminar and turbulent. Under certain conditions, the laminar mode of flow is stable while under others, turbulence prevails. The rate of degradation of kinetic energy and the mechanism of diffusion are quite different in these two flow regimes, hence the ability to predict the appropriate flow regime in particular cases is of great technical importance. The process of moving from one flow regime to another is known as transition. In some cases, transition from the basic laminar flow is directly to turbulent flow, e.g., in a cylindrical pipe. In other cases, a sequence of secondary laminar flows separate the basic flow from turbulence, each member of the sequence being distinct from the others and each being stable in its own domain of parameter space. In Couette flow between concentric cylinders, Coles (1965) observed 74 such transitions, each appearing at a certain well-defined and repeatable speed.

Flow transition is preceded by flow instability. Instability of basic laminar flow occurs when a parameter, λ, defined by the ratio of destabilizing force to stabilizing force, reaches a critical value (Rosenhead, 1963). In isothermal flow, instability arises in one of two basic forms:

(1) *Centrifugal instability* occurs in flows with curved streamlines when the (destabilizing) centrifugal force exceeds in magnitude the (stabilizing) viscous force. The relevant parameter is the Taylor number:

$$\lambda \equiv T_a \sim \frac{\text{measure of centrifugal force}}{\text{measure of viscous force}}.$$

The instability is often characterized by a steady secondary laminar flow (Taylor-Görtler vortices).

(2) *Parallel flow instability* is characterized by propagating waves (Tollmien-Schlichting waves). The inertia force is destabilizing and the viscous force is stabilizing in this case, and the parameter is the Reynolds number:

$$\lambda \equiv \text{Re} \sim \frac{\text{measure of inertia force}}{\text{measure of viscous force}}.$$

Instability of this kind occurs in pipe flow and in the boundary layer.

In the following sections, we shall give more precise definition of the concept of stability, then discuss ways how the critical value of the parameter, λ_{CR}, that separates stable from unstable flows, may be calculated.

6.1   Stability

Let $\{U(x, t), P(x, t)\}$ and $\{u(x, t), p(x, t)\}$ represent the basic laminar flow and a perturbation of this flow, respectively. Furthermore, let

$$\{\hat{U}(x, t), \hat{P}(x, t)\} = \{U(x, t) + u(x, t), P(x, t) + p(x, t)\} \tag{6.1}$$

represent the perturbed flow.

If, with passage of time, the perturbed flow approaches the basic flow, we say that the basic flow is asymptotically stable. Otherwise, the basic flow is unstable. If the basic flow remains stable irrespective of the initial magnitude of the perturbation, the basic flow is said to be globally stable. If the flow is globally stable and is such that some norm of the perturbation never increases in time, not even instantaneously, the basic flow is said to be globally and monotonically stable.

In some cases, the basic flow is stable only if the initial value of the perturbation is suitably small. We then term the basic flow conditionally stable.

Stability Criteria

To give precise mathematical meaning to the stability criteria of the previous paragraph, define the average kinetic energy per unit mass, $\mathcal{E}(t)$, of the perturbation $u(x, t)$

$$\mathcal{E}(t) = \frac{1}{2V} \int_V |u|^2 \, dv. \tag{6.2}$$

We shall call the basic flow $\{U(x, t), P(x, t)\}$ *globally stable* if

$$\lim_{t \to \infty} \frac{\mathcal{E}(t)}{\mathcal{E}(0)} \to 0, \tag{6.3}$$

irrespective of the magnitude of $\mathcal{E}(0)$, the initial value of the kinetic energy of perturbation. The limiting value of the parameter λ for global stability is designated by the symbol λ_G. When $\lambda > \lambda_G$, a perturbation can always be found that destabilizes the basic flow.

We shall call the basic flow *globally and monotonically stable* if, in addition to Eq. (6.3), we also have

$$\lim_{t \to \infty} \frac{d\mathcal{E}(t)}{dt} < 0 \quad \text{for all } t > 0. \tag{6.4}$$

The greatest lower bound for global and monotonic stability, designated by λ_E and called the energy stability limit, represents a *sufficient condition for stability:* no matter how strong the perturbation may be, if $\lambda < \lambda_E$ the perturbation will die out monotonically. λ_E is called the energy stability limit because it is calculated by the energy method. If the basic flow is globally stable but not monotonically stable, a perturbation can always be found, the kinetic energy of which will initially increase with time before decaying to zero.

The flow is *conditionally stable* if

$$\lim_{t \to \infty} \frac{\mathcal{E}(t)}{\mathcal{E}(0)} \to 0, \quad \text{whenever } \mathcal{E}(0) < \delta. \tag{6.5}$$

δ is called the attracting radius of the stable flow. For global or unconditional stability $\delta \to \infty$. The linear limit of the parameter $\lambda = \lambda_L$ is defined by $\delta \to 0$. If $\lambda > \lambda_L$, the flow will be unstable, no matter how weak the perturbation may be. Therefore, the linear limit λ_L, so called because it is calculated from linear equations, represents a *sufficient condition for instability*.

It can be shown (Joseph, 1976) that $\lambda_L \geq \lambda_E$. In cases when these two stability limits coincide (e.g., in Bénard convection) an infinitesimal perturbation is just as dangerous as any finite perturbation. In other cases, the basic flow might be stable to infinitesimal disturbances but is unstable to finite disturbances. Pipe flow, by keeping out finite disturbances, can be kept laminar past Re = 100,000, yet, under ordinary circumstances, laminar flow is guaranteed only up to Re = 2,000 (Hinze, 1987). Figure 6.1, after Joseph (1976), is a schematic displaying the various stability regimes and defining criteria.

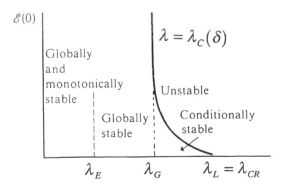

Figure 6.1. Stability limits for the basic flow.

Stability Analysis

To derive the equations that govern stability of the basic flow, consider a closed container of volume, $V(t)$, and surface, $S(t)$, completely filled with a viscous fluid. The fluid is brought into motion by external forces or by the motion of the boundary. The velocity and pressure of this motion, denoted by $\{U, P\}$ is governed by the Navier-Stokes (2.41) and continuity (2.44b) equations and appropriate boundary and initial conditions. In nondimensional form, we have

$$\frac{\partial U}{\partial t} + U \cdot \nabla U - \frac{1}{\mathrm{Re}}\nabla^2 U + \nabla P - f = 0, \tag{6.6a}$$

$$\mathrm{div}\, U = 0, \tag{6.6b}$$

$$\begin{aligned}&\text{B.C.:}\ U(x, t) = U_S(x, t), &&x \in S(t), \quad t \geq 0,\\ &\text{I.C.:}\ \ U(x, 0) = U_0(x), \quad \mathrm{div}\, U_0 = 0, &&x \in V(0).\end{aligned} \tag{6.6c}$$

Consider another motion of the fluid, defined by $\{\hat{U}, \hat{P}\} = \{U + u, P + p\}$, that is obtained by perturbing the initial condition U_0 to $\hat{U}_0 = U_0 + u_0$. The perturbed flow $\hat{U}$ satisfies the same boundary conditions as the basic flow U; it is also governed by the Navier-Stokes and continuity equations

$$\frac{\partial \hat{U}}{\partial t} + \hat{U} \cdot \nabla \hat{U} - \frac{1}{\mathrm{Re}}\nabla^2 \hat{U} + \nabla \hat{P} - f = 0, \tag{6.7a}$$

$$\mathrm{div}\, \hat{U} = 0. \tag{6.7b}$$

The evolution equation of the perturbation u is obtained by subtracting Eq. (6.6) from Eq. (6.7)

$$\frac{\partial u}{\partial t} + U \cdot \nabla u + u \cdot \nabla U + u \cdot \nabla u - \frac{1}{\mathrm{Re}}\nabla^2 u + \nabla p = 0, \tag{6.8a}$$

$$\mathrm{div}\, u = 0, \qquad u|_S = 0, \tag{6.8b}$$

$$u|_{t=0} = u_0(x), \qquad \mathrm{div}\, u_0 = 0. \tag{6.8c}$$

We are assured by the uniqueness property of the Navier-Stokes problem (Joseph, 1972) that the null solution $u(x, t) \equiv 0$ is the only possible solution to Eq. (6.8) for zero initial perturbation $u_0 \equiv 0$. But what if $u_0 \neq 0$? Will the perturbation vanish (stability) or will it increase (instability) in time?

The problem specified by Eq. (6.8) is a nonlinear initial boundary value problem. Though it does not lend itself to easy solution, it has been solved numerically in various cases, in particular for flow between concentric cylinders by Marcus (1984a,b). Such solutions trace the evolution in time of the initial perturbation u_0.

We may also use Eq. (6.8) to calculate the energy limit of stability $\lambda_E \equiv \mathrm{Re}_E$, to do this we first transform it into the evolution equation for the kinetic energy $\mathcal{E}(t)$ of the perturbation.

Energy Stability

To obtain the evolution equation for $\mathcal{E}(t)$, first form the scalar product of Eq. (6.8a) with the velocity and integrate the result over the volume $V(t)$. Application of the Reynolds

Transport Theorem[1] and the boundary condition $u|_S = 0$ leads to (Joseph, 1 1976)

$$\frac{d\mathcal{E}}{dt} = \mathcal{I} - \frac{1}{Re}\mathcal{D} = -\mathcal{D}\left(\frac{1}{Re} - \frac{\mathcal{I}}{\mathcal{D}}\right). \tag{6.9a}$$

Here

$$\mathcal{I} = -\int_V u \cdot \nabla U \cdot u \, dv \tag{6.9b}$$

is the production of kinetic energy of perturbation, and

$$\mathcal{D} = \int_V |\nabla u|^2 \, dv \tag{6.9c}$$

is its dissipation. Designating the maximum value[2] of the ratio $\mathcal{I}/\mathcal{D}$ in Eq. (6.9) by $1/Re_E$

$$\frac{1}{Re_E} = \max\left(\frac{\mathcal{I}}{\mathcal{D}}\right), \tag{6.10}$$

Eq. (6.9a) can be written in the form

$$\frac{d\mathcal{E}}{dt} = -\mathcal{D}\left(\frac{1}{Re} - \frac{1}{Re_E} + \frac{1}{Re_E} - \frac{\mathcal{I}}{\mathcal{D}}\right)$$

$$\leq -\mathcal{D}\left(\frac{1}{Re} - \frac{1}{Re_E}\right), \tag{6.11}$$

as

$$\frac{1}{Re_E} - \frac{\mathcal{I}}{\mathcal{D}} \geq 0.$$

Equation (6.11) shows that the flow is monotonically stable, Eq. (6.4), i.e., $d\mathcal{E}/dt \leq 0$, if $Re < Re_E$.

The energy stability limit Re_E can be calculated from Eq. (6.10) for given basic flow U by the following scheme: (1) for every admissible perturbation[3] u, calculate $\mathcal{I}$ from Eq. (6.9b) and $\mathcal{D}$ from Eq. (6.9c); (2) the particular perturbation that gives the largest value for the ratio $\mathcal{I}/\mathcal{D}$ is the critical perturbation, and this $\mathcal{I}/\mathcal{D}$ ratio is the inverse of the energy stability limit.

Rather than arbitrarily choosing perturbations in search of the critical one, as was done by Orr (1907), we find it more effective to pose Eq. (6.10) as a problem in the calculus

---

[1] The Reynolds Transport Theorem (Serrin, 1959a; White, 1991), $\frac{d}{dt}\int_{V(t)} \varphi \, dv = \int_{V(t)} (\frac{d\varphi}{dt} + \varphi \operatorname{div} v) \, dv$, is concerned with finding the time rate of change of the total property $\Phi(t) = \int_{V(t)} \phi \, dv$ associated with a material volume $V(t)$, i.e., a volume always containing the same fluid particles. Here φ represents property per unit volume and d/dt signifies material, or total, derivative (2–7).

[2] That the maximum exists in bounded regions follows from the Poincaré inequality (Joseph, 1976). The derivation is not valid when the region is unbounded, though a justification is available for infinite regions whenever the disturbances can be assumed spatially periodic at each time instant (Serrin, 1959b).

[3] A perturbation is admissible if it is divergence free and satisfies no-slip boundary conditions, Eq. (6.8c).

of variations and to calculate Re_E from a linear eigenvalue problem (Joseph, 1976). The energy stability limit for concentric, rotating cylinders (unloaded journal bearings) was first calculated by Serrin (1959b), but in this case, unlike that of Bénard convection, the linear limit and the energy limit are widely separated. Thus Re_E does not have practical significance for us in cylinder flows.

Linear Stability

There is another way we can make use of Eq. (6.8) for steady U. By assuming the perturbation $u(x, t)$ to be infinitesimal, we may neglect the quadratic term $u \cdot \nabla u$ and obtain

$$\frac{\partial u}{\partial t} + U \cdot \nabla u + u \cdot \nabla U - \frac{1}{\text{Re}} \nabla^2 u + \nabla p = 0. \tag{6.12}$$

As this equation is linear (the basic flow U is known), one can exploit superposition, i.e., solve Eq. (6.12) for each (Fourier) component of the perturbation (Chandrasekhar, 1961).

Equation (6.12) loses validity once the initially infinitesimal perturbation has grown too large, but for short times it will trace the evolution of the originally infinitesimal $u(x, t)$. As it contains time only through the first derivative, we look for solutions with an exponential time factor

$$u(x, t) = v(x)e^{-ict}, \quad p(x, t) = \pi(x)e^{-ict}. \tag{6.13}$$

By substituting Eq. (6.13) into Eq. (6.12), we find that solutions of the type Eq. (6.13) exist, provided there are numbers, c, for which the problem

$$-icv + U \cdot \nabla v + v \cdot \nabla U - \frac{1}{\text{Re}} \nabla^2 v + \nabla \pi = 0 \tag{6.14}$$

has nontrivial solutions. The particular numbers, c, for which Eq. (6.14) is amenable to solution, are the eigenvalues, and the corresponding solutions, $v(x)$, are the eigenfunctions.

Each of the infinity of eigenvalues is, in general, complex $c = c_r + ic_i$, and we write

$$e^{-ict} = e^{c_i t}(\cos c_r t - i \sin c_r t). \tag{6.15}$$

We call c_i the amplification factor. If $c_i < 0$ for all c, the disturbance will decay in time and the flow is stable.[4] If, on the other hand, $c_i > 0$ for at least one c, the disturbance will grow in time and the flow is unstable.[5] Neutral stability is characterized by $c_i = 0$ for the leading eigenvalue, while all other eigenvalues have negative imaginary parts; the lowest value of the Reynolds number at which this occurs is the linear limit of stability, indicated here by $R_{CR} \equiv (\text{Re})_L$ and referred to simply as the critical Reynolds number.

The term $(\cos c_r t - i \sin c_r t)$ in Eq. (6.15) represents a motion that is periodic in time, with period c_r. There are two ways in which neutral stability, $c_i = 0$, might be achieved. For certain basic flows, $c_r \neq 0$ at neutral stability, thus neutral stability is characterized by a time-periodic secondary (laminar) flow. If $c_r = 0$ when $c_i = 0$, the secondary (laminar) flow

[4] This is conditional stability, under the condition that initially the perturbation is infinitesimal. The flow might or might not be stable yet to finite disturbances.

[5] This is a sufficient condition for instability; if the flow is unstable to infinitesimal disturbance, it will certainly be unstable to finite disturbances.

appearing in the neutral stability state is steady, and we say that the basic flow exchanges its stability with (or transfers it to) the bifurcating flow (Chandrasekhar, 1960; Joseph, 1976). Which of these cases occurs in practice depends, of course, on the basic flow field $\{U, P\}$.

Taylor (1923) was the first to employ linear stability analysis in the study of flow between concentric cylinders, though the method had been applied half a century earlier to other flows (Kelvin, 1871). Flow between eccentric cylinders was investigated by DiPrima and Stuart (1972, 1975) among others. Cylinder flows will be discussed in detail in later sections.

Bifurcation Analysis

We saw in the previous section that the conditions for neutral stability signal the appearance of a new solution, $v(x)e^{-ict}$. This *bifurcating solution* might be steady or it might be unsteady, depending on whether $c_i = c_r = 0$ or $c_i = 0$, $c_r \neq 0$, respectively, at criticality.

For Re slightly larger than Re_{CR}, the bifurcating solution will grow exponentially, Eq. (6.13), the growth being dictated by the amplification factor, c_i. In some cases, this exponential growth remains unchecked, and we conclude that the bifurcating solution is unstable. In other cases, however, the perturbation will interact with the basic flow, modifying the rate of energy transfer, Eq. (6.9b), from basic flow to perturbation, so as to arrive at an equilibrium state. The equilibrated bifurcating flow is stable in the neighborhood of the critical point Re $= \text{Re}_{CR}$. Bifurcation analysis is concerned with stable bifurcating solutions (Iooss and Joseph, 1982).

Bifurcating solutions emanate from the basic flow at so-called *singular points* where the Jacobian matrix $(\partial v / \partial x)$ becomes singular (Seydel, 1988) – in fact, we locate such solutions by finding points on the solution curve where the Jacobian determinant vanishes. The Lyapunov conditional stability theorem (Joseph, 1976) tells us that the first bifurcation point found when increasing Re is identical to the linear limit of stability, Re_{CR}. In this sense, bifurcation analysis, which locates the first critical point of the basic flow, and linear stability analysis, which calculates the linear stability limit, are equivalent.

Where bifurcation has the advantage over linear stability analysis is that while the latter is valid only within an infinitesimal neighborhood of the critical point, bifurcation analysis permits the tracing of solution branches far away from critical points, as it employs the full Navier-Stokes and continuity equations. At the bifurcating point, we can switch to the new branch and continue the solution well into the supercritical, i.e., Re $> \text{Re}_{CR}$, range.

Bifurcating solutions that exist for values of Re $< \text{Re}_{CR}$ are called *subcritical*; those which exist for values of Re $> \text{Re}_{CR}$ are termed *supercritical* (Figure 6.2). It can be shown that subcritical bifurcating solutions are unstable and supercritical bifurcating solutions are stable (Iooss and Joseph, 1982).

We will illustrate two methods of calculating R_{CR} in the sequel. Bifurcation analysis will be employed to calculate the critical Reynolds number for eccentric rotating cylinders, while linear stability analysis will be used to study flow between rotating disks. Rotating disks are important in the computer industry. Rotating disk flows serve as an approximation to flows in annular thrust bearings. Eccentric cylinder flows, on the other hand, are closely related to flow in journal bearings.

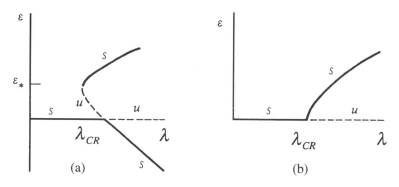

Figure 6.2. Simple bifurcating solutions: (a) subcritical, (b) supercritical bifurcation. The solid lines represent stable solutions, s, the dotted lines represent unstable solutions, u.

6.2   Flow between Concentric Cylinders

Centrifugal instability was first studied by Couette and by Lord Rayleigh. On the basis of energy considerations, Rayleigh derived a criterion of stability for inviscid fluids, viz., that if the square of the circulation increases outward, the flow is stable. An explanation of Rayleigh's criterion in terms of the centrifugal force field and pressure gradient was later given by von Karman (Lin, 1967).

To illustrate the argument constructed by von Karman, we consider a fluid ring located at $r = r_1$ and concentric with the axis of rotation. In steady state, the centrifugal force, $\rho v_1^2 / r_1$, on the element of the ring must be balanced by pressure forces; this serves us in estimating the steady-state pressure field. Consider now displacing the ring to $r = r_2$. The angular momentum (rv) will be conserved during this displacement in accordance with Kelvin's theorem,[6] so in its new surrounding the centrifugal force acting on the ring is given by

$$\rho \frac{v^2}{r} = \rho \left(\frac{r_1 v_1}{r_2} \right)^2 \frac{1}{r_2} = \rho \frac{(r_1 v_1)^2}{r_2^3}.$$

Because at $r = r_2$ the prevailing pressure force equals $\rho v_2^2 / r_2$, the fluid ring will continue in its motion outward, provided that

$$\rho \frac{(r_1 v_1)^2}{r_2^3} > \rho \frac{v_2^2}{r_2}.$$

This simplifies to the Rayleigh stability criterion: In the absence of viscosity a necessary and sufficient condition for a distribution of angular velocity to be stable is that

$$\frac{d}{dr}(r^2 \omega)^2 > 0 \tag{6.16}$$

everywhere in the interval.

When applied to flow between concentric cylinders, Rayleigh's criterion means that the flow is potentially unstable if the inner cylinder is rotating and the outer cylinder is

---

[6] Kelvin's circulation theorem states that in an inviscid fluid the angular momentum per unit mass of a fluid element (rv) remains constant [note that $\Gamma = 2\pi rv$ is the circulation round the circle $r = \text{const}$. (Milne-Thomson, 1968)].

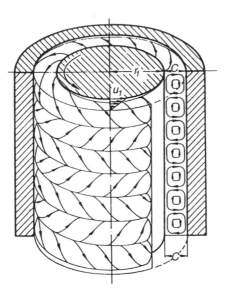

Figure 6.3. Taylor vortices between concentric cylinders, with the inner cylinder rotating and the outer cylinder stationary. (Reprinted with permission from Schlichting, H. *Boundary Layer Theory* 6th ed. Pergamon, London. Copyright 1968.)

stationary, but it is not susceptible to instability of the centrifugal type if the inner cylinder is at rest and the outer cylinder is rotating.[7] However, the flow in the latter case does become unstable at higher rotation (Coles, 1965), the shear flow instability occurring at Re $\sim$ 2,000.

When the inner cylinder of radius r_1 is rotated at the angular speed ω and the outer cylinder of radius r_2, $r_2 = r_1 + C$, $C > 0$ is kept stationary, the circumferential velocity (the only nonzero velocity component for long cylinders) is given by

$$v(r) = \frac{r_1^2 \omega}{r_2^2 - r_1^2}\left[\frac{r_2^2}{r} - r\right]. \tag{6.17}$$

On increasing the rate of rotation, this flow becomes unstable. Taylor (1923) showed that, when instability occurs, the disturbance assumes the form of cellular, toroidal vortices that are equally spaced along the axis of the cylinder (Figure 6.3). The appearance of this so-called *Taylor vortex flow* has a strong influence on the magnitude of the torque required to rotate the inner cylinder. The stability parameter of the problem is the Taylor number, which is proportional to the ratio of the centrifugal force to the viscous force. The definition we use is that of Drazin and Reid (1984) for the "average" Taylor number:

$$T = \frac{2(1 - \eta)}{(1 + \eta)}\text{Re}^2, \quad \text{Re} = \frac{r_1 \omega C}{v}, \quad \eta = \frac{r_1}{r_2} < 1,$$

where Re is the Reynolds number.

By assuming that the neutral (marginal) state of stability is stationary and that the radial clearance C is small[8] when compared to the radius r_1, Taylor was able to calculate the first

---

[7] Synge (Lin, 1967) showed Rayleigh's criterion to be sufficient for viscous fluids.
[8] The formula $T^{1/2} = (C/r_1)^{1/2}$ Re, valid for $\eta \to 1$, follows from the approximation $(1 - \eta)/(1 + \eta) \approx C/2r_1$.

critical value of T. The currently accepted minimum value of the critical Taylor number, T_{CR} is (Drazin and Reid, 1984)

$$T_{CR} = 1694.95, \quad \text{Re}_{CR} = 41.2 \left(\frac{r_1}{C}\right)^{1/2}. \tag{6.18}$$

This is within 1% of the value calculated by Taylor in 1923.

As the Taylor number is increased above its critical value, the axisymmetric Taylor vortices become unstable to nonaxisymmetric disturbances. Pairs of vortex cells that are symmetrical, $T = T_{CR}$, now become distorted (Mobbs and Younes, 1974). The boundaries between adjacent vortex cells assume a wavy form, with the waves traveling azimuthally at the average velocity of the basic Couette flow (Coles, 1965). The number of azimuthal waves increases on further increasing the Taylor number, until turbulence eventually makes its appearance. Concentric cylinder flows are discussed in detail by Koshmieder (1993).

Nonuniqueness of the flow was first demonstrated by Coles (1965), who documented as many as 26 distinct time-dependent states at a given Reynolds number. Coles found the different states by approaching the final Reynolds number along different paths in parameter space. Earlier, Landau and Lifshitz (1959) suggested that transition to turbulence may occur as an infinite sequence of supercritical bifurcations, each contributing a new degree of freedom to the motion. Although attractive, this conjecture had to be abandoned in light of theoretical (Ruelle and Takens, 1971; Newhouse, Ruelle, and Takens, 1978) and experimental (Gollub and Swinney, 1975; Gollub and Benson, 1980) research, which indicates that chaotic behavior occurs after a small number of transitions.

The value of the critical Taylor number depends on the clearance ratio C/r_1 (Coles, 1965) but not on the aspect ratio L/C (Cole, 1976), where L is the length of the cylinders. Only the Taylor number for the appearance for waviness is strongly dependent on L/C. Cole observed formation of a single counterrotating vortex pair in an open-ended annulus of length $L = 5C$. Schwartz et al. (1964) found that a superimposed axial flow has a stabilizing influence. The sole effect of slow axial flow is to translate the Taylor cells axially, but with higher axial flow the cell pattern assumes a spiral form. DiPrima (1960) also showed that superimposed axial flow has a stabilizing influence.

Couette flow is not a solution of the finite length problem. The presence of endplates, located a distance L apart, will force all three velocity components to vary with position. A corkscrew motion with axial structure, known as Ekman pumping, emerges even at small values of the Reynolds number. The experimental observations of Benjamin and Mullin (1981, 1982), performed in an apparatus having a relatively small aspect ratio, $\Gamma = L/C$, indicate at higher Reynolds numbers the existence of a number of distinct steady flows that are supported by identical boundary conditions. More importantly, these experiments reveal "an essentially continuous process: namely as Re is gradually raised through a narrow quasi-critical range, arrays of axisymmetric, counter-rotating cells spread from the end and finally link-up and become ordered prominently at the center of the Taylor apparatus" (Benjamin and Mullin, 1982).

Benjamin and Mullin (1982) experimented with a Taylor apparatus of $\eta = 0.6$ and $\Gamma = 12.6$ and found that the "primary" cellular flow, reached by gradual increases in Re, was comprised of 12 cells. This primary flow was stable for a range of Reynolds number values below the threshold for the onset of traveling waves. Benjamin and Mullin also found flows other than the primary flow: "normal" flows, exhibiting both inward flow at the stationary endplates and an even number of cells, and "anomalous" flows comprising an odd number of cells and/or exhibiting anomalous rotation. They observed a total of 20

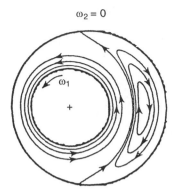

$\omega_2 = 0$

ω_1

+

Figure 6.4. Subcritical Couette flow with recirculation. (Reprinted with permission from San Andres, A. and Szeri, A. Z. Flow between eccentric rotating cylinders. *ASME J. Appl. Mech.*, **51**, 869–878, 1984.)

different flows, and, by applying degree theory, Benjamin (1978) predicted that the problem has at least 39 solutions, 19 of which are unstable.[9]

Theoretical analyses of infinite cylinder flows at slightly supercritical Reynolds numbers show a continuum of periodic solutions of wavelength λ, which are possibly stable when λ is located within some interval $(\lambda_c^-, \lambda_c^+)$ centered around the critical value, λ_c, for strict bifurcation of the Couette flow. Estimates for this λ interval were first made by Chandrasekhar (1961) and Kogelman and DiPrima (1976), and were subsequently improved by Nakaya (1975). The experiments of Burkhalter and Koschmieder (1977a) are not in contradiction with this theory, once the part of the flow that seems to be "directly influenced" by the endplates is neglected, and the observed wavelengths fall within the narrowest of the three $(\lambda_c^-, \lambda_c^+)$ intervals. It has been suggested (Benjamin and Mullin, 1982) that this "continuum" of cellular flows of the infinite geometry is the result of taking the limit $\Gamma \to \infty$, thereby increasing the multiplicity of steady flows, each of which represents the primary flow of its particular Γ interval.[10]

6.3   Flow between Eccentric Cylinders

For $R < R_{CR}$ and $\varepsilon < 0.3$, the flow is a Couette flow. On increasing the eccentricity ratio above $\varepsilon \approx 0.3$, a recirculation cell makes its appearance (Kamal, 1966), this flow being stable up to Re $<$ Re$_{CR}$. Such a recirculating flow pattern is shown in Figure 6.4 for Stokes flow at $\eta = 0.5$ and $\varepsilon = 0.5$.

On further increasing the Reynolds number past its critical value, Re = Re$_{CR}$, the two-dimensional basic laminar flow loses its stability to a new flow with three- dimensional structure, similar to the case of concentric cylinders (Dai, Dong, and Szeri, 1992). Wilcock (1950) and Smith and Fuller (1956) were the first to recognize that flow transition also takes place between eccentric cylinders. Analytically, the problem was first treated by DiPrima (1963). DiPrima's local theory shows the flow to be least stable at the position of maximum film thickness, along a vector extending from the center of the inner cylinder to the center of the outer cylinder. The critical Taylor number, T_{CR}, calculated at this position first decreases

---

[9] The Leray-Schauder degree theory is concerned with the topology of certain Banach-space mappings. In its application to fluid mechanics, the theory ascertains that, at any value of the parameter, the number of stable solution branches exceeds the number of unstable branches by one.

[10] Dai and Szeri (1990) were able to follow a branch comprised of 28 normal cells, and another branch of 30 anomalous cells, to $\Gamma = 200$. Note, that $\Gamma = 500$ for a short bearing of $L/D = 0.1$ and $r_1/C = 500$.

as the eccentricity ratio, $\varepsilon = e/C$ (where e is the eccentricity of the cylinders), increases from zero and remains below its concentric value in the range $0 < \varepsilon < 0.6$. For $\varepsilon > 0.6$, T_{CR} increases rapidly with increasing ε. This finding is, however, not in agreement with experimental results. Experimental data (e.g., Vohr, 1968) indicate that the eccentricity has a stabilizing effect over its whole range. Vohr also finds the maximum intensity of vortex motion not at the point of maximum film thickness, as predicted by local theory, but at the 50° position downstream from it.

DiPrima's local theory is based on the parallel flow assumption, i.e., on neglecting the effect of the azimuthal variation of the tangential velocity, and is no longer thought to represent the physics of the problem. In a second attempt to explain the dependence of T_{CR} on ε, DiPrima and Stuart (1972) examined the whole flow field (nonlocal theory), considering the tangential velocity to be a function of both r and θ. The most dramatic result of the newer theory is that it no longer places maximum vortex intensity at the position of maximum instability but rather at 90° downstream of it. This theory also shows $T_{CR}(\varepsilon)$ to be a monotonically increasing function, described by

$$T_{CR} = 1695 \left(1 + 1.162 \frac{C}{r_1} \right) (1 + 2.624\varepsilon^2). \tag{6.19}$$

Agreement with experiment is acceptable, however, only in the range $0 < \varepsilon < 0.3$. In a more recent nonlinear theory, DiPrima and Stuart (1975) found that the position of maximum vortex activity is at $\Theta_{max} = 90°$ only if the supercritical Taylor number vanished and $\Theta_{max} \to 0°$ as $T_{SC} \to \infty$. [The supercritical Taylor number is defined as $T_{SC} = (T - T_{CR})/\varepsilon$.] Although they found agreement in one specific case with Vohr's observation of $\Theta_{max} = 50°$ for maximum vortex activity, they raised the question of the applicability of their small perturbation solution under Vohr's experimental conditions.

Castle and Mobbs (1968) found two kinds of instabilities of flow between eccentric cylinders. The type of cellular flow that occurs at low speeds does not extend all the way to the stationary (outer) cylinder, while the second type of instability occurs at higher rotational speeds and exhibits vortices that straddle the clearance gap completely. According to Mobbs and Younes (1974), it is the latter group of vortices that are usually detected by experiments, as their existence is revealed by torque measurements on the stationary cylinder. The first, lower mode of instability lies close to that predicted by the local theories of Ritchie (1968) and DiPrima (1963). According to DiPrima and Stuart (1972), "it seems possible, therefore, that this incipient mode may be a manifestation of local instabilities." But they also admit the possibility that the slower mode represents another instability that cannot be accounted for by the nonlocal theory. The slower incipient mode was also detected by Versteegen and Jankowski (1969) and by and Frêne and Godet (1974). Koschmieder (1976), on the other hand, was unable to find instability of the first kind in his apparatus, nor could he detect maximal vortex action downstream from the position of maximum gap width. Nevertheless, he found $T_{CR}(\varepsilon)$ to be a monotonic increasing function, somewhat as predicted by nonlocal theories. It seems that the nonlocal theory correlates better with instability of the second type, while instability of the first type is predicted by local theories (Figure 6.4).

Li (1977) investigated the onset of instability in nonisothermal flows between rotating cylinders and found that a positive radial temperature gradient, as might exist in journal bearings, is strongly destabilizing, particularly at high Prandtl numbers. Li also suggested

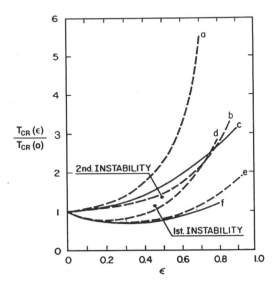

Figure 6.5. Effect of eccentricity ratio on critical speed for instability. (—) theory, (−−) experiment, (a) Castle and Mobbs (1968), (b) Vohr (1968), (c) DiPrima and Stuart (1972), (d) Castle and Mobbs (1968), (e) Frêne and Godet (1974), and (f) DiPrima (1963).

that under certain conditions the neutral (marginal) stability is oscillatory. Li's conclusion that a positive temperature gradient is destabilizing seems to be verified by the experiments of Gardner and Ulschmid (1980).[11]

Though there have been numerous calculations of concentric cylinder flows [Weinstein, 1977a,b; Frank and Meyer-Spasche, 1981; Marcus, 1984a,b; Cliff et al., 1985, 1986; Meyer-Spasche and Keller 1978 and 1980], the numerical treatment of eccentric cylinder flows has been hindered in the past by two circumstances: (1) the partial differential equations that model the basic flow contain the Reynolds number and the eccentricity ratio, as parameters and numerical solutions are difficult to obtain at large values of these parameters and (2) classical stability analysis leads to an eigenvalue problem for partial, rather than for ordinary, differential equations, increasing numerical complexity considerably. Nevertheless a linear stability analysis has been published recently by Oikawa, Karasndani, and Funakoshni (1989a,b), who use Chebyshev-Fourier expansion and the pseudo-spectral method (Gottlieb and Orszag, 1977).

In the next section, we investigate the effect of cylinder eccentricity on Couette-Taylor transition between infinite rotating cylinders, using bifurcation analysis. To do this, we employ Galerkin's method with B-spline basis functions to the system of equations that govern the motion of the fluid and locate critical points by detecting the vanishing of the Jacobian determinant during parametric continuation. The first bifurcation from Couette flow provides us with the linear stability limit (Joseph, 1976). At the bifurcation point, we switch to the new branch by locating its tangent in parameter space and continue the solution into the supercritical range; this allows investigation of the bifurcating flow and calculation of the torque which the bifurcating flow exerts on the cylinders.

---

[11] According to their data (Figure 9.9), turbulence must have set in at value $\text{Re} = 1{,}100$ of the global Reynolds number, $\text{Re} = R\omega C/\nu$. The value of the local Reynolds number $\text{Re}_h = R\omega h/\nu$ is even smaller at transition to turbulence; it is in the range $360 < \text{Re}_h < 560$.

Critical Reynolds Number

We employ a bipolar coordinate system for the representation of the flow field between infinite, rotating eccentric cylinders. The bipolar coordinate system $\{\hat{\alpha}, \hat{\beta}, \hat{\gamma}\}$ is related to the Cartesian coordinate system $\{X, Y, Z\}$ through

$$\hat{\alpha} + i\hat{\beta} = -2\coth^{-1}\frac{(X + iY)}{a}, \quad \hat{\gamma} = Z, \tag{6.20}$$

where a is the separation between the pole and the origin of the $\{X, Y, Z\}$ system (Figure 6.1). In the bipolar coordinate system, the cylinders of radii r_1 and r_2, $r_1 < r_2$, have the simple representation $\hat{\alpha} = \hat{\alpha}_1$ and $\hat{\alpha} = \hat{\alpha}_2$, $\hat{\alpha}_1 < \alpha_2 < 0$, respectively. The scale factor of the bipolar coordinate system is given by (Ritchie, 1968)

$$H = \frac{a}{(\cosh\hat{\alpha} - \cos\hat{\beta})}.$$

The nondimensional equations of motion and mass conservation are

$$\frac{\Delta}{h}u\frac{\partial u}{\partial\alpha} + \frac{1}{2\pi h}v\frac{\partial u}{\partial\beta} + w\frac{\partial u}{\partial\gamma} - uv\sin\hat{\beta} + v^2\sinh\hat{\alpha}$$

$$= -\frac{\Delta}{h}\frac{\partial p}{\partial\alpha} + \frac{(1-\eta)}{\eta\mathrm{Re}}\left[\frac{1}{h^2}\left(\Delta^2\frac{\partial^2 u}{\partial\alpha^2} + \frac{1}{(2\pi)^2}\frac{\partial^2 u}{\partial\beta^2}\right) + \vartheta^2\frac{\partial^2 u}{\partial\gamma^2}\right.$$

$$\left. - \frac{2\Delta\sin\hat{\beta}}{h}\frac{\partial v}{\partial\alpha} + \frac{\sinh\hat{\alpha}}{\pi h}\frac{\partial v}{\partial\beta} - \frac{\cosh\hat{\alpha} \mid \cos\hat{\beta}}{h}u\right], \tag{6.21a}$$

$$\frac{\Delta}{h}u\frac{\partial v}{\partial\alpha} + \frac{1}{2\pi h}v\frac{\partial v}{\partial\beta} + w\frac{\partial v}{\partial\gamma} - uv\sinh\hat{\alpha} + u^2\sin\hat{\beta}$$

$$= -\frac{1}{2\pi h}\frac{\partial p}{\partial\beta} + \frac{(1-\eta)}{\eta\mathrm{Re}}\left[\frac{1}{h^2}\left(\Delta^2\frac{\partial^2 v}{\partial\alpha^2} + \frac{1}{(2\pi)^2}\frac{\partial^2 v}{\partial\beta^2}\right) + \vartheta^2\frac{\partial^2 v}{\partial\gamma^2}\right.$$

$$\left. + \frac{2\Delta\sin\hat{\beta}}{h}\frac{\partial u}{\partial\alpha} - \frac{\sinh\hat{\alpha}}{\pi h}\frac{\partial u}{\partial\beta} - \frac{\cosh\hat{\alpha} + \cos\hat{\beta}}{h}v\right], \tag{6.21b}$$

$$\frac{\Delta}{h}u\frac{\partial w}{\partial\alpha} + \frac{1}{2\pi h}v\frac{\partial w}{\partial\beta} + w\frac{\partial w}{\partial\gamma}$$

$$= -\vartheta^2\frac{\partial p}{\partial\gamma} + \frac{(1-\eta)}{\eta\mathrm{Re}}\left[\frac{1}{h^2}\left(\Delta^2\frac{\partial^2 w}{\partial\alpha^2} + \frac{1}{(2\pi)^2}\frac{\partial^2 w}{\partial\beta^2}\right) + \vartheta^2\frac{\partial^2 w}{\partial\gamma^2}\right]. \tag{6.21c}$$

The divergence of the velocity field $v = (u, v, w)$ has the (nondimensional) form

$$\mathrm{div}\,v = \frac{1}{h}\left(\Delta\frac{\partial u}{\partial\alpha} + \frac{1}{2\pi}\frac{\partial v}{\partial\beta} + h\frac{\partial w}{\partial\gamma}\right) - u\sinh\hat{\alpha} + v\sin\hat{\beta}. \tag{6.22}$$

Equations (6.21) and (6.22) are nondimensional. They were nondimensionalized according to

$$\alpha = \Delta(\hat{\alpha} - \hat{\alpha}_1), \quad \beta = \frac{\hat{\beta}}{2\pi}, \quad \gamma = \frac{\hat{\gamma}}{r_1}, \quad \Delta = \frac{1}{\hat{\alpha}_2 - \hat{\alpha}_1},$$

$$\{U, V, W\} = \frac{r_1\omega}{\sinh|\hat{\alpha}_1|}\left\{u, v, \frac{1}{\vartheta}w\right\}, \quad \vartheta = \frac{a}{r_1}, \quad C = r_2 - r_1, \qquad (6.23)$$

$$p = \frac{P}{\rho r_1^2\omega^2}\sinh^2|\hat{\alpha}_1|, \qquad h = \frac{H}{a}, \quad \mathrm{Re} = \frac{r_1\omega C}{\nu}.$$

Here P represents the pressure and $\{U, V, W\}$ represents the physical components of velocity relative to the bipolar coordinate system. Dynamic conditions in the flow are represented by the Reynolds number, Re, calculated on the mean gap width.

We anticipate that although at small values of the Reynolds number fluid motion proceeds in $z = $ const. planes, at some higher rotation this basic laminar flow will give up its stability to a more complex laminar flow that possesses periodic structure in γ for cylinders of infinite length. Accordingly, we Fourier analyze the flow field as follows (Meyer-Spasche and Keller, 1980):

$$\{u, v, p\} = \sum_{j=0}^{N_z}\{u_j, v_j, p_j\}\cos j\kappa\gamma$$
$$w = \sum_{j=1}^{N_z} w_j \sin j\kappa\gamma \qquad (6.24)$$

and investigate the individual Fourier components, obtained by substituting Eqs. (6.24) into the equations of motion and continuity (Dai, Dong, and Szeri, 1992).

Let λ represent the length of a cell in the Z direction (not yet fixed). We shall restrict attention to

$$Z \in \left[-\frac{\lambda}{2}, \frac{\lambda}{2}\right],$$

thus the solution domain of Eqs. (6.20) is defined by

$$0 \le \alpha \le 1, \quad 0 \le \beta \le 1, \quad -\frac{\pi}{\kappa} \le \gamma \le \frac{\pi}{\kappa}, \qquad (6.25)$$

where $\kappa = 2\pi r_1/\lambda$ is the nondimensional wave number. The wavelength λ, and thus the wave number κ, is a parameter of the problem, and we obtain the critical Reynolds number, Re_{CR}, from computing $d\,\mathrm{Re}(\kappa)/d\kappa = 0$.

The boundary conditions accompanying Eqs. (6.21) are no slip on the walls

$$u(0, \beta, \gamma) = w(0, \beta, \gamma) = 0, v(0, \beta, \gamma) = \sinh|\hat{\alpha}_1|,$$
$$u(1, \beta, \gamma) = v(1, \beta, \gamma) = w(1, \beta, \gamma) = 0. \qquad (6.26a)$$

We also require periodicity of the solution and its derivatives in β

$$\phi^{(n)}(\alpha, 0, \gamma) = \phi^{(n)}(\alpha, 1, \gamma), \quad \phi = u, v, w, p \quad n = 0, 1, 2, \dots. \qquad (6.26b)$$

Equations (6.21) define pressure only within an arbitrary constant. We set this constant to zero by requiring that $p(0, 0, 0) = 0$.

At this stage, Dai et al., (1992) eliminated the pressure, p_j, $j \geq 1$, from the component equations so as to keep the discretized system nonsingular.[12]

The expansions

$$u_k(\alpha, \beta) = \sum_{i=2}^{N_\alpha-1} \sum_{j=1}^{N_\beta-3} u_{ijk} A_i(\alpha) b_j(\beta), \quad 0 \leq k \leq N_\gamma,$$

$$v_0(\alpha, \beta) = \sinh |\hat{\alpha}_1| A_1(\alpha) + \sum_{i=2}^{N_\alpha-1} \sum_{j=1}^{N_\beta-3} v_{ij0} A_i(\alpha) b_j(\beta),$$

$$v_k(\alpha, \beta) = \sum_{i=2}^{N_\alpha-1} \sum_{j=1}^{N_\beta-3} v_{ijk} A_i(\alpha) b_j(\beta), \quad 1 \leq k \leq N_\gamma, \quad (6.27)$$

$$w_k(\alpha, \beta) = \sum_{i=2}^{N_\alpha-1} \sum_{j=1}^{N_\beta-3} w_{ijk} A_i(\alpha) b_j(\beta), \quad 1 \leq k \leq N_\gamma,$$

$$p_0(\alpha, \beta) = \sum_{i=1}^{N_\alpha} \sum_{j=1}^{N_\beta-3} p_{ij} A_i(\alpha) b_j(\beta),$$

where the $A_i(\alpha)$, $i = 1, \ldots N_\alpha$ are normalized splines in α and the b_j, $j = 1, \ldots, N_\beta - 3$, are periodic splines in β, satisfy the boundary conditions (6.26a) and the periodicity condition (6.26b).

Next, we substitute Eqs. (6.27) into the component equations, obtained by Fourier decomposition of the equations of motion and continuity, and apply Galerkin's method (Dai et al., 1992). The discretized system of equations are then solved subject to the single condition $p(0, 0, 0) = 0$, which takes the form $p_{1,1} = 0$.

The system of algebraic equations that result from Galerkin's method can be written in the form

$$F(u, \sigma) = 0. \quad (6.28)$$

Here $F : U \oplus \Sigma \subset R^n \to R^m$, dim $U = m$, dim$\Sigma = n - m$. $u \in U$ is the vector of state variables and $\sigma \in \Sigma$ is the vector of parameters η, ν, Re, ε.

In the computational scheme, we fix three of the parameters, say η, ϑ, and ε, and vary the Reynolds number, Re; thus, $n - m = 1$ and the regular manifold of Eq. (6.28) is a path.

The objective here is to solve the nonlinear algebraic system Eq. (6.28) for various values of σ, now signifying a single parameter, the Reynolds number, $\sigma \equiv$ Re. The solution $u^* = u(\sigma^*)$ is guaranteed by the implicit function theorem in the neighborhood of the points (u, σ), where $F_u(u, \sigma)$, the Fréchet derivative of $F(u, \sigma)$, is nonsingular. The computational scheme for obtaining $u^* = u(\sigma^*)$ is then straightforward: (1) first locate a point (u^0, σ^0) in $(n + 1)$ space that lies within the neighborhood of attraction[13] of (u^*, σ^*), then (2) employ a suitable iterative scheme that guarantees convergence $(u^0, \sigma^0) \to (u^*, \sigma^*)$. Phase (1), i.e., the *predictor*, is simplest when (u^0, σ^0) is

[12] The pressure may be retained, at significant savings on algebra, if the scheme of Eq. (5.12) due to Zienkiewicz and Woo (1991) is employed. An extension to the full Navier–Stokes problem can be found in Szeri and Al-Sharif (1996).

[13] Attraction in terms of the numerical scheme, not in the sense of fluid dynamics.

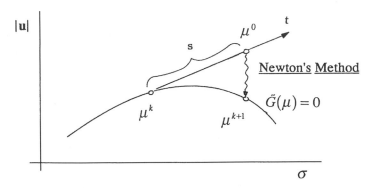

Figure 6.6. Schematics of the iterative solution of Eq. (6.29).

obtained by stepping along the tangent to the manifold from a known point (Figure 6.6). The step size, s, is constrained by the requirement that (u^0, σ^0) lie in the neighborhood of attraction of (u^*, σ^*).

For convenience of notation, we rewrite Eq. (6.28) in the form

$$\tilde{G}(\mu) = 0, \quad \mu = (u, \sigma). \tag{6.29}$$

Local iteration

We employ the Gauss-Newton method (Ortega and Rheinboldt, 1970)

$$D\tilde{G}(\mu^k)(\mu^k - \mu^{k+1}) = \tilde{G}(\mu^k) \quad k = 0, 1, 2, \ldots \tag{6.30}$$

to solve Eq. (6.29) by iteration, starting from a suitable initial point μ^0. Here

$$D\tilde{G}(x^k) = (D_u\tilde{G}(\mu^k), D_\sigma\tilde{G}(\mu^k)) \in R^{m \times n}$$

is the Jacobian of $\tilde{G}(\mu)$ evaluated at $\mu = \mu^k$. (Note that the Jacobian is calculated analytically.)

To solve Eq. (6.30) for μ^{k+1}, perform the Q-R factorization

$$D\tilde{G}(\mu^k)^T = Q\begin{pmatrix} \Re \\ 0 \end{pmatrix}, \tag{6.31}$$

where $Q \in R^{n \times n}$ is orthogonal and $\Re \in R^{m \times m}$ is upper triangular, and observe that

$$(\Re^T, 0)Q^T Q\begin{pmatrix} \Re^{-T} \\ 0 \end{pmatrix} = I$$

and

$$D\tilde{G}(x^k) = (\Re^T, 0)Q^T.$$

This implies that Eq. (6.30) has the solution

$$\mu^{k+1} = \mu^k - Q\begin{pmatrix} \Re^{-T} \\ 0 \end{pmatrix}\tilde{G}(\mu^k) \quad k = 0, 1, 2, \ldots, \tag{6.32}$$

where $\Re^{-T}$ is the inverse of $\Re$ transpose.

Local iteration of the solution proceeds thus along the following steps:

(1) Select[14] a starting point, $\mu = \mu^0$,
(2) For $k = 0, 1, 2, \ldots$ until convergence
   (a) Solve the triangular system: $\mathfrak{R}^T \varpi = \tilde{G}(\mu^k)$,
   (b) Compute the next iterate: $\mu^{k+1} = \mu^k - Q\binom{\varpi}{0}$

Continuation of the solution
The Q-R decomposition

$$D\tilde{G}(\mu^k)^T = Q\begin{pmatrix} \mathfrak{R} \\ 0 \end{pmatrix}$$

indicates that the last column of Q is tangential to the solution manifold, so that the tangent vector at μ^k is given by

$$t = Qe_n,$$

where $e_n = (0, 0, \ldots, 1)^T$.

The simplest way to find a starting point, μ^0, for the Gauss-Newton iteration is by computing

$$\mu^0 = \mu + st,$$

where s is a variable step and μ is a known point (Figure 6.5).

The iteration scheme depicted by Eq. (6.30) breaks down at critical points, μ_c, where the Jacobian $D_u\tilde{G}(\mu_c)$ is singular. If, as in the present example,

$$\text{rank}[D_u\tilde{G}(\mu_c)] = n - 1 \tag{6.33}$$

and $D_\sigma\tilde{G}(\mu_c) \in \text{range}[D_u\tilde{G}(\mu_c)]$, the point μ_c is a simple bifurcation point where exactly two branches with two distinct tangents intersect. [There is also a condition on the second derivatives. For technical details the reader is referred to Iooss and Joseph (1976) and to Keller (1977).]

Although a simple bifurcation point is characterized by the singularity of the augmented Jacobian $D\tilde{G}$, computationally we are looking for the change of sign of

$$\det\left[D\tilde{G}(\mu^k)^T, t\right]$$

for continuously varying tangent vector, t. This is achieved by examining the test function

$$\tau = \text{sgn}(r_{11} \times r_{22} \times \cdots \times r_{mm}),$$

where the r_{ii}, $1 \leq i \leq m$, are the diagonal elements of $\mathfrak{R}$.

In order to continue along the new branch at bifurcation, $\mu = \mu_c$, we need the tangent, q, to the bifurcating branch. Aided by the fact that the bifurcating branches emanate

---

[14] To obtain the solution at low Reynolds numbers, often $\mu^0 = 0$ will suffice, i.e., all variables may be set equal to zero prior to starting the iteration. At high Reynolds numbers this scheme, in general, will not work. Thus, when seeking solution at high Reynolds numbers, μ^0 must be a point (i.e., solution) that was obtained from low Reynolds number solutions by some type of extrapolation.

symmetrically at the bifurcation point, $\mu = \mu_c$, we are able to find a good approximation to the tangent of the bifurcating branch by simply calculating

$$q = Qe_m, \quad e_m = (0, \ldots, 0, 1, 0)^T.$$

Since $|r_{mm}| = \min(r_{ii}) \approx 0, 1 \leq i \leq m$, the corresponding row vector in $D\tilde{G}$ is closest to being a linear combination of the other rows of $D\tilde{G}$ and

$$q \in \text{null}[D\tilde{G}(\mu_c), t].$$

Evaluation and decomposition of the Jacobian consumes most of the computational effort. With a system of 1,452 equations and $R < R_{CR}$, Dai et al., (1992) required 120 sec of CPU time on the Cray Y-MP/832 for one Newtonian iteration, and three iterations to converge to an error $<10^{-6}$. For supercritical conditions, 300 sec per iteration was required, but convergence was achieved, again, in three steps. For continuation in Re, fixed ε, 20 solutions were necessary to reach the critical point, starting from a low Reynolds number solution that was obtained with $\mu^0 = 0$.

Figure 6.7 illustrates the variation of the normalized critical Reynolds number with eccentricity ratio, at $\eta = 0.912$. The bifurcation calculations were performed with $N_x = N_y = 16$ and $N_z = 3$. The figure also contains data from small perturbation solutions by DiPrima and Stuart (1972), with correction from DiPrima and Stuart (1975). The experimental data was obtained from torque measurements at $\eta = 0.91$ by Vohr (1968). The bifurcation results are in good agreement with the experimental data of Vohr; they also agree with DiPrima and Stuart as $\varepsilon \to 0$, where the DiPrima and Stuart small perturbation analysis is valid, but diverge from the latter as ε increases.

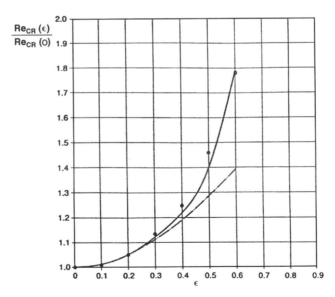

Figure 6.7. Variation of critical Reynolds number with eccentricity ratio, $\eta = 0.912$: —··—, DiPrima and Stuart (1972); $\circ$, Vohr (1968), experimental; —, Dai, Dong, and Szeri (1992). (Reprinted from *International Journal of Engineering Science*, Vol. 30, Dai, R. X., Dong, Q. M. and Szeri, A. Z. Flow between eccentric rotating cylinders: bifurcation and stability, pp. 1323–1340, Copyright (1990), with kind permission from Elsevier Science Ltd, The Boulevard, Langford Lane, Kidlington 0X5 1GB, UK.)

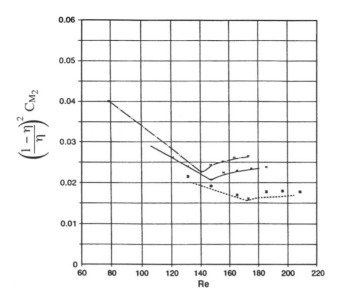

Figure 6.8. Comparison of numerical (Dai et al., 1990) and experimental (Castle and Mobbs, 1968) torque data at $\eta = 0.912$ ($-\cdot-\cdot-$, $\varepsilon = 0.0$; $-$ $\varepsilon = 0.2$; $-\,-\,-$ $\varepsilon = 0.4$;). (Reprinted from *International Journal of Engineering Science,* Vol. 30, Dai, R. X., Dong, Q. M. and Szeri, A. Z. Flow between eccentric rotating cylinders: bifurcation and stability, pp. 1323–1340, Copyright (1990), with kind permission from Elsevier Science Ltd, The Boulevard, Langford Lane, Kidlington 0X5 1GB, UK.)

Let M_1 and M_2 represent the torque on the inner and the outer cylinder, respectively, and define the torque coefficient C_{M_i} (Schlichting, 1960) through

$$C_{M_i} = \frac{2M_i}{\pi \rho \omega^2 L r_1^4}, \quad i = 1, 2. \tag{6.34}$$

Figure 6.8 compares torque results calculated on the outer cylinder at $\eta = 0.912$ with experimental data by Castle and Mobbs (1968) by plotting

$$C_{M_2} \frac{(1 - \eta)^2}{\eta^2},$$

indicating supercritical bifurcation at Re_{CR}. There is good agreement especially at small ε.

Torque Measurements

Torque measurements for the concentric case have been made by Taylor (1923), Wendt (1933), Donelly (1958), and Vohr (1968) and were reviewed by Bilgen and Boulos (1973). It is a simple matter to show that if the friction force, F_μ, depends on $\rho, \mu, \omega, r_1, C,$ and L, where C is the radial clearance and L is the length of the cylinders, the dependence must be of the form

$$\frac{F_\mu}{\rho \omega^2 r_1^3 L} = f\left(\frac{C}{r_1}, \frac{\rho \omega r_1 C}{\mu} \right) \tag{6.35}$$

on dimensional grounds.

Defining the torque coefficient C_M as in Eq. (6.34), Eq. (6.35) can be written in the approximate form (Bilgen and Boulos, 1973)

$$C_M = \lambda \left(\frac{C}{r_1}\right)^\alpha \text{Re}^\beta, \qquad (6.36)$$

where the Reynolds number is defined by $\text{Re} = \rho r_1 \omega C / \mu$.

By analyzing available experimental data, Bilgen and Boulos achieved the best fit to Eq. (6.36) with $\alpha \approx 0.3$. The value of the exponent β is dependent not only on the Reynolds number in the laminar or the turbulent regimes but on both the Reynolds number and the clearance ratio C/r_1 in the vortex flow regime.

Laminar regime

$\text{Re} \le 64$

$$C_M = 10 \left(\frac{C}{r_1}\right)^{0.3} \text{Re}^{-1.0} \qquad (6.37a)$$

Transition regime

$64 < \text{Re} \le 500$

$$C_M = 2 \left(\frac{C}{r_1}\right)^{0.3} \text{Re}^{-0.6} \qquad \text{if } \frac{C}{r_1} \ge 0.07 \qquad (6.37b)$$

$$C_M = 2 \left(2 + \frac{C}{r_1}\right) \left\{1 + 1.45\left[1 - \left(\frac{T_c}{T_a}\right)\right]\right\} \text{Re}^{-1.0} \qquad \text{if } \frac{C}{r_1} < 0.07 \qquad (6.37c)$$

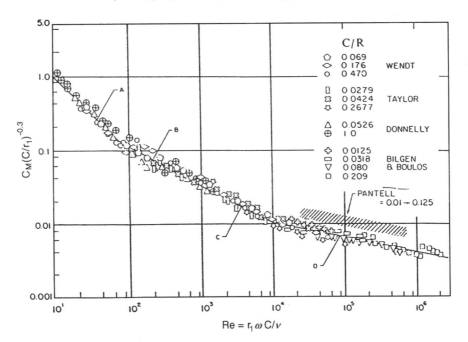

Figure 6.9. Variation of modified torque coefficient with Couette Reynolds number: (A) Eq. (6.37a); (B) Eq. (6.37b); (C) Eq. (6.37d); (D) Eq. (6.37e). (Reprinted with permission from Bilgen, E. and Boulos, R. Functional dependence of torque coefficient of coaxial cylinders on gap width and Reynolds numbers. *ASME Trans., Ser. G*, **95**, 122–126, 1973.)

Turbulent regime

Re > 500

$$C_M = 1.03 \left(\frac{C}{r_1}\right)^{0.3} \text{Re}^{-0.5} \qquad 500 < \text{Re} \le 10^4 \tag{6.37d}$$

$$C_M = 0.065 \left(\frac{C}{r_1}\right)^{0.3} \text{Re}^{-0.20} \qquad \text{Re} > 10^4 \tag{6.37e}$$

The maximum mean deviation of the experimental data from these equations is ±5.8, ±10.4, and $\pm8.35\%$ in the laminar, transition, and turbulent regimes, respectively (Bilgen and Boulos, 1973), as shown in Figure 6.9.

6.4   Rotating Disk Flows

Thorough stability analysis of thrust bearing flows has not yet been accomplished. Among the flows that have been analyzed, flow between parallel rotating disks is the flow closest to thrust bearing flows. Disk flows also warrant consideration due to their relevance to magnetic disk storage systems, and to flow over certain hydrostatic pads (Section 6.2).

The first systematic study of rotating disk flows was made by von Karman (1923), who assumed that the axial velocity is independent of the radial coordinate. This assumption led to a similarity transformation that was shown by Batchelor (1951) to be applicable even when the fluid is bounded by two infinite rotating disks. Based on the examination of the governing equations, Batchelor predicted that at high Reynolds numbers a thin boundary layer will develop on each disk, with the main body of the fluid rotating at a constant rate intermediate between disk velocities. This prediction was challenged by Stewartson (1953), who reasoned that at high Reynolds number the flow outside the boundary layers would be purely axial, thus inaugurating one of the longstanding controversies of fluid mechanics.

The similarity transformation, available when the disks are infinite, reduces the number of spatial dimensions of the problem to one. Although it is questionable whether the reduced model approximates to the physical problem, the equations resulting from the similarity transformation have been subject to intense analytical and numerical probing. An excellent review of the work on infinite disk flows can be found in Zandbergen and Dijkstra (1987).

Flow between finite, parallel disks was first investigated both experimentally and numerically by Szeri and Adams (1978) and by Szeri, Schneider, Labbe, and Kaufman (1983). Dijkstra and van Heijst (1983) found the finite disk solution to be unique for all values of the parameters considered: of the Batchelor type for weak and of the Stewartson type for strong counterrotation. Brady and Durlofsky (1987) investigated the relationship of the axisymmetric flow between large but finite coaxial rotating disks to the Karman similarity solution. They combined asymptotics with numerical analysis and showed that the finite disk solution and the similarity solution coincide over a decreasing portion of the flow domain as the Reynolds number is increased. Although this conclusion might seem counterintuitive and perhaps defies old wisdom, it does reinforce the assertion that finite disk and infinite disk flows are qualitatively different (Szeri, Giron, Schneider, and Kaufman, 1983). Despite the availability of solutions for finite disks, the computation of disk flow is far from elementary. Much of recent work on finite disk flows finds relevance in magnetic disk storage systems (Tzeng and Fromm, 1990; Tzeng and Humphrey, 1991; Humphrey, Schuler, and Iglesias, 1992; Radel and Szeri, 1997). In these applications nonuniformities in flow can affect disk rotational stability, with attendant drastic effects on read-write performance; the current read-write head to disk separation is on the order of 100 nm or less.

Turning now to stability of rotating disk flows, we find that most investigations are of flows bounded by a single rotating disk. Gregory, Stuart, and Walker (1955) examined stability at infinite Reynolds number and found good agreement with experimental data on the direction of wave propagation but overestimated the number of vortices that appear at criticality. Brown (1961) extended their analysis by considering stability at finite Reynolds number. The basic flow equations contain curvature and Coriolis terms in the work of Kobayashi, Kohama, and Takamadate (1980). Szeri and Giron (1984) retain the axial velocity in addition to the terms considered by Kobayashi et al. Their result shows favorable agreement with experimental data.

Experimentally, Faller (1963) found two types of instabilities. The waves of each of these form a series of horizontal roll vortices, whose spacing is related to the thickness of the boundary layer. Others who studied these instabilities include Faller and Kaylor (1966), Tatro and Mollo-Christensen (1967), Caldwell and Van Atta (1970), and Weidman and Redekopp (1976). The Ekman velocity profile exhibits numerous inflection points, each of which might give rise to instabilities. Faller and Kaylor found that the location of the type I vortices, which are stationary or nearly so, coincides with the first inflection point of the radial velocity. The analysis of Gregory, Stuart, and Walker (1955) of inviscid instabilities of the flow on a single, infinite, rotating disk confirms this.

Linear Stability Analysis

The flow field is bounded by two parallel disks of finite radii, located at $Z = 0$ and $Z = s$, with respect to the inertial cylindrical coordinate system (R, Θ, Z). The lower disk rotates with angular velocity Ω relative to the upper disk. Experimental evidence shows that the instabilities that may occur take the form of spirals, with their radius vector decreasing in the direction of rotation of the disks. Based on this evidence, and on the assertion that this type of instability should be a general feature of all rotating boundary layers (Greenspan, 1968), it is logical to select, as reference, an orthogonal, curvilinear coordinate system (x^1, x^2, x^3) that is related to the local geometry of the spiral vortices (Figure 6.10). The direction of the x^2 axis is chosen to coincide with axis of the spiral vortices (inclined locally at the yet unknown angle ε to the R direction). The x^1 axis intersects the x^2 axis perpendicularly, so that the x^1 axis is in the direction of propagation of the spiral vortices.

The origin of the (x^1, x^2, x^3) coordinate system is located on the lower disk at $R = r$, some Θ, via the transformation (Gregory, Stuart and Walker, 1955)

$$
T: \begin{cases} x^1 = r\left[\ln\left(\frac{R}{r}\right)\cos\varepsilon - (\Theta + \Omega t)\sin\varepsilon \right] \\[2mm] x^2 = r\left[\ln\left(\frac{R}{r}\right)\sin\varepsilon + (\Theta + \Omega t)\cos\varepsilon \right]. \\[2mm] x^3 = Z \end{cases}
\tag{6.38}
$$

The scale factors of the $\{x^i\}$ coordinate system are

$$
h_1 = h_2 = (R/r) = h, \quad h_3 = 1.
$$

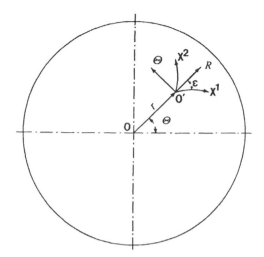

Figure 6.10. Coordinate systems employed in the linear stability analysis of disk flows.

We next calculate the physical components $\{V_1, V_2, V_3\}$ of the basic flow velocity relative to $\{x^i\}$, which are given by

$$\begin{cases} V_1 = \cos\varepsilon(U_R) - \sin\varepsilon(r\Omega - V_\Theta) \\ V_2 = \sin\varepsilon(U_R) + \cos\varepsilon(r\Omega - V_\Theta) \\ V_3 = W_z \end{cases} \tag{6.39}$$

where $\{U_R, V_\Theta, W_Z\}$ is the flow velocity in the cylindrical polar coordinate system $\{R, \Theta, Z\}$.

In accordance with the linear theory of stability, the basic flow is perturbed by an infinitesimal wave. We look for instability to perturbation which propagates in the x^1 direction, the direction of least stability, with speed $(\beta/\alpha)_r$, the real part of (β/α), relative to $\{x^i\}$ and a wavelength of $2\pi/\alpha$. Let $\{u; p\}$ represent the perturbation, then

$$\{u(x); p(x)\} = \{v(x^3); \pi(x^3)\}e^{i(\alpha x^1 - \beta t)}. \tag{6.40}$$

The linearized equations that govern the evolution of the perturbation [see Eq. (6.14)] are given by Greenspan (1968)

$$-i\lambda v + V \cdot \operatorname{grad} v + v.\operatorname{grad} V + 2\Omega \times v = -\operatorname{grad}\frac{\pi}{\rho} + v\nabla^2 v \tag{6.41}$$

$$\operatorname{div} v = 0.$$

Using the notation

$$\vartheta = \frac{s}{r}, \quad \delta = \frac{s}{hr}, \quad m_i = \frac{s}{h^2}\frac{\partial h}{\partial x^i}, \quad \Gamma_{ij} = \frac{s^2}{h^3}\frac{\partial^2 h}{\partial x^i \partial x^j}, \quad i, j = 1, 2,$$

Eq. (6.41) is nondimensionalized in accordance with

$$\{x^i\} = r\{x, y, \vartheta z\}, \quad \{V_1, V_2, V_3\} = V_0\{U, V, \vartheta W\}$$

$$\operatorname{Re} = \frac{V_0 s}{v}, \quad \sigma = \frac{\alpha s}{h}, \quad c = \frac{\beta h}{\alpha V_0}, \tag{6.42}$$

where $V_0 = r\Omega$ is the local surface velocity. Upon decomposing Eq. (6.41) along $\{x^i\}$ and substituting in Eq. (6.42), the following set of equations results:

$$-i\sigma cu - (m_1 v + i\sigma v - m_2 u)V + \vartheta\left(\frac{du}{dz} - i\sigma w\right)W$$

$$-\left(m_1 V + \delta\frac{\partial V}{\partial x} - m_2 U - \delta\frac{\partial U}{\partial y}\right)v + \left(\frac{\partial U}{\partial z} - \delta\vartheta\frac{\partial W}{\partial x}\right)w + 2\vartheta v \qquad (6.43a)$$

$$= -\frac{1}{V_0}\delta\frac{\partial\bar{\pi}}{\partial x} - \frac{1}{Re}\left[-\frac{d^2 u}{dz^2} + i\sigma\frac{dw}{dz} - (2m_1 m_2 - \Gamma_{12} + i\sigma m_2)v + \left(2m_2^2 - \Gamma_{22}\right)u\right],$$

$$-i\sigma cv + \vartheta\frac{dv}{dz}W + (m_1 v + i\sigma v - m_2 u)U + \frac{\partial V}{\partial z}w + \left(m_1 V + \delta\frac{\partial V}{\partial x} - m_2 U\right)u - 2\vartheta u$$

$$= -\delta\frac{\partial V}{\partial y}v - \frac{1}{Re}\left[-\frac{d^2 v}{dz^2} + (i\sigma m_2 - 2m_1 m_2 + \Gamma_{12})u + \left(2m_1^2 - \Gamma_{11} + \sigma^2\right)v\right],$$

$$(6.43b)$$

$$-i\sigma cw - \left(\frac{du}{dz} - i\sigma w\right)U - \frac{\partial v}{\partial z}V - \left(\frac{\partial U}{\partial z} - \delta\vartheta\frac{\partial W}{\partial x}\right)u + \left(\delta\vartheta\frac{\partial W}{\partial y} - \frac{\partial V}{\partial z}\right)v$$

$$= -\frac{1}{V_0}\frac{\partial\bar{\pi}}{\partial z} - \frac{1}{Re}\left[i\sigma\frac{du}{dz} + i\sigma m_1 w + \sigma^2 w + m_1\left(\frac{du}{dz} - i\sigma w\right) + m_1\frac{dv}{dz}\right],$$

$$(6.43c)$$

$$(m_1 + i\sigma)u + m_2 v + \frac{dw}{dz} = 0. \qquad (6.44)$$

Cross differentiation eliminates the pressure, and substitution from the equation of continuity (6.44) eliminates the velocity component u. We further simplify these equations by utilizing the approximations

$$\delta \approx \vartheta, \quad m_1 \approx \delta\cos\varepsilon, \quad m_2 \approx \delta\sin\varepsilon$$

$$\Gamma_{11} \approx \delta^2\cos^2\varepsilon, \quad \Gamma_{12} \approx \delta^2\sin\varepsilon\cos\varepsilon, \quad \Gamma_{22} \approx \delta^2\sin^2\varepsilon. \qquad (6.45)$$

The algebra is tedious, and we refer the reader here to Giron (1982) and Szeri, Giron, Schneider, and Kaufman (1983).

We seek solutions in the weak form

$$v(z) = \sum_{i=2}^{N-1} v_i B_i(z),$$

$$w(z) = \sum_{j=3}^{N-2} w_j B_j(z). \qquad (6.46)$$

Here the $B_i(z), 1 \leq i \leq N$ are B-splines defined over a partition in $z \in [0, 1]$. The expansions in Eqs. (6.46) satisfy the boundary conditions

$$v = w = \frac{dw}{dz} = 0 \quad (z = 0, 1). \qquad (6.47)$$

The condition $dw/dz = 0$ at $z = 0, 1$ is a derived boundary conditions; it is obtained from the equation of continuity (6.44).

Expansions (6.46), together with spline expansions for the (numerically or experimentally) known basic flow, are now substituted into the reduced equations of motion in an application of Galerkin's method. Multiplication through by the elements of the test functions $\{B_i(z)\}_{i=1}^{N}$ and integration over the domain $0 \leq z \leq 1$ lead to the complex algebraic eigenvalue problem

$$|X - cY| = 0, \tag{6.48}$$

where, in general, $c = c_r + i c_i$.

The flow is marginally or neutrally stable if there is one eigenvalue with $c_i = 0$ and $c_i > 0$ for all other eigenvalues. It is unstable if at least one eigenvalue exists with $c_i < 0$. The marginal state is steady if $c_r = 0$.

We now illustrate some results for parallel, infinite, disk flows at a fixed value, say $E^{-1} = 100$, of the Ekman number. To identify the most dangerous direction for wave propagation, the basic flow velocity is resolved along various directions as characterized by the angle ε (Figure 6.11). The neutral stability curve of each of the resolved velocity profiles is then calculated from the eigenvalue problem (6.48), by identifying (σ, Re) couples that yield the neutrality condition $c_i = 0$ (Figure 6.12). For each of the orientations, ε, the lowest achievable Reynolds number yielding $c_i = 0$ is now identified and plotted against ε in Figure 6.13. This last Figure shows that the perturbation that propagates at $\varepsilon \approx 17°$ to the radius is the most dangerous one, giving the critical Reynolds of $\text{Re}_{CR} = 5000$. Therefore, a basic flow characterized by $E^{-1} = 100$ is unstable to infinitesimal disturbances whenever $r \gtrsim 50s$.

The scheme described here will yield results even if the basic flow is available only experimentally (Szeri, Giron, Schneider, and Kaufman, 1983).

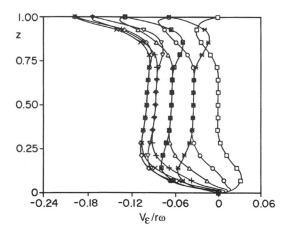

Figure 6.11. Basic flow velocity, resolved along direction ε : $\square$, $\varepsilon = 0°$; $\bigcirc$, $20°$; $\triangle$, $40°$; $+$, $60°$; $\times$, $80°$; $\diamond$, $100°$; ∇, $120°$; $\boxtimes$, $140°$; $*$, $160°$. (Reprinted with the permission of Cambridge University Press, from Szeri, A. Z., Giron, A., Schneider, S. J. and Kaufman, H. N. Flow between rotating disks. Part 2. Stability. *J. Fluid Mech.*, **134**, 133–154, 1983.)

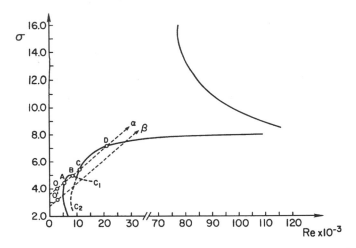

Figure 6.12. Stability diagram at $E^{-1} = 100$, $\varepsilon = 17°$. (Reprinted with the permission of Cambridge University Press from Szeri, A. Z., Giron, A., Schneider, S. J. and Kaufman, H. N. Flow between rotating disks. Part 2. Stability. *J. Fluid Mech.*, **134**, 133–154, 1983.)

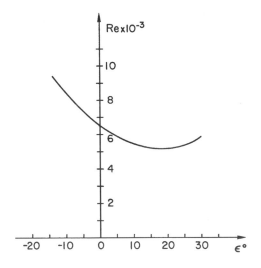

Figure 6.13. Variation of the Reynolds number for neutral stability with orientation ε. (Reprinted with the permission of Cambridge University Press from Szeri, A. Z., Giron, A., Schneider, S. J. and Kaufman, H. N. Flow between rotating disks. Part 2. Stability. *J. Fluid Mech.*, **134**, 133–154, 1983.)

6.5   Nomenclature

A_i	B-spline
C	radial clearance
C_M	coefficient of torque
$\mathcal{D}$	rate of energy dissipation
$\mathcal{E}$	kinetic energy of perturbation
H	scale factor

$\mathfrak{I}$ rate of energy production
P pressure
Re   Reynolds number
U basic flow velocity
b_i periodic spline
c eigenvalue
p pressure
r_1, r_2 inner, outer radii
$\boldsymbol{u}, \boldsymbol{v}$ perturbation
α, β, γ bipolar coordinates
R, Θ, Z cylindrical polar coordinates
$\{x^i\}$ curvilinear coordinates
ε coordinate orientation
U_R, V_Θ, W_Z basic flow velocity
V_i basic flow velocity
$(\ )_r.(\ )_i$ real, imaginary part
$(\ )_0$ initial value

6.6   References

Batchelor, G. K. 1951. Note on a class of solutions of the Navier-Stokes equations representing rotationally symmetric flow. *Q. J. Mech. Appl. Maths.*, **4**, 29.

Benjamin, T. B. 1978. Bifurcation phenomena in steady flows of a viscous liquid. *Proc. Roy. Soc. London*, **A 359**, 1–43.

Benjamin, T. B. and Mullin, T. 1981. Anomalous modes in the Taylor experiment. *Proc. Roy. Soc. London* **A 377**, 221–249.

Benjamin, T. B. and Mullin, T. 1982. Notes on the multiplicity of flows in the Taylor experiment. *J. Fluid Mech.*, **121**, 219–230.

Bilgen, E. and Boulos, R. 1973. Functional dependence of torque coefficient of coaxial cylinders on gap width and Reynolds numbers. *ASME Trans., Ser. G*, **95**, 122–126.

Brady, J. F. and Durlofsky, L. 1987. On rotating disk flow. *J. Fluid Mech.*, **175**, 363–394.

Brown, W. B. 1961. A stability criterion for three-dimensional boundary layers. In *Boundary Layer and Flow Control* (G. V. Lachman, ed.), Vol. 2, pp. 913–923. Pergamon Press.

Burkhalter, J. E. and Koschmieder, E. L. 1974. Steady supercritical Taylor vortices after sudden starts. *Phys. Fluids*, **17**, 1929–1935.

Caldwell, D. R. and Van Atta, C. W. 1970. Characteristics of Ekman boundary layer instabilities. *J. Fluid Mech.*, **44**, 79–95.

Castle, P. and Mobbs, F. S. 1968. Hydrodynamic stability of flow between eccentric rotating cylinders: visual observations and torque measurements. *Proc. Inst. Mech. Eng.*, **182**, 41–52.

Chandrasekhar, S. 1961. *Hydrodynamic and Hydromagnetic Stability*. Oxford University Press, Oxford.

Cliff, K. A. and Mullin, T. 1985. A numerical and experimental study of anomalous modes in the Taylor experiment. *J. Fluid Mech.*, **153**, 243–258.

Cliff, K. A. and Spence, A. 1986. Numerical calculations of bifurcations in the finite Taylor problem. In *Numerical Methods for Fluid Dynamics* (K. W. Morton and M. J. Baines, eds.), pp. 155–176. Clarendon Press, Oxford.

Cole, J. A. 1976. Taylor-vortex instability and annulus length effects. *J. Fluid Mech.*, **75**, 1–15.

Coles, D. 1965. Transition in circular Couette flow. *J. Fluid Mech.*, **21**, 385–425.

Dai, R. X. and Szeri, A. Z. 1990. A numerical study of finite Taylor flows. *Internat. J. Nonlinear Mech.*, **25**, 45–60.

Dai, R. X., Dong, Q. M. and Szeri, A. Z. 1992. Flow between eccentric rotating cylinders: bifurcation and stability. *Internat. J. Engrg. Sci.*, **30**, 1323–1340.

Dijkstra, D. and van Heijst, G. J. F. 1983. The flow between two finite rotating disks enclosed by a cylinder. *J. Fluid Mech.*, **128**, 123–254.

DiPrima, R. C. 1960. The stability of a viscous fluid between rotating cylinders with axial flow. *J. Fluid Mech.*, **9**, 621–631.

DiPrima, R. C. 1963. A note on the stability of flow in loaded journal bearings. *ASLE Trans.*, **6**, 249–253.

DiPrima, R. C. and Stuart, J. T. 1972. Non-local effects in the stability of flow between eccentric rotating cylinders. *J. Fluid Mech.*, **54**, 393–415.

DiPrima, R. C. and Stuart, J. T. 1975. The nonlinear calculation of Taylor vortex flow between eccentric rotating cylinders. *J. Fluid Mech.*, **67**, 85–111.

Donelly, R. J. 1958. Experiments on the stability of viscous flow between rotating cylinders. *Proc. Roy. Soc. London*, **A 266**, 312–325.

Drazin, P. G. and Reid, W. H. 1984. *Hydrodynamic Stability*. Cambridge University Press, Cambridge.

Faller, A. J. 1963. An experimental study of the instability of the laminar Ekman boundary layer. *J. Fluid Mech.*, **15**, 560–576.

Faller, A. J. and Kaylor, R. E. 1966. Investigations of stability and transition in rotating boundary layers. In *Dynamics of Fluids and Plasmas* (S. I. Pai, ed.), pp. 239–255. Academic.

Frank, G. and Meyer-Spasche, R. 1981. Computation of transitions in Taylor vortex flow. *Z. Agnew. Math. Phys.*, **32**, 710–720.

Frêne, J. and Godet, M. 1974. Flow transition criteria in a journal bearing. *ASME Journal of Lubrication Technology*, **96**, 135–140.

Gardner, W. W. and Ulschmid, J. G. 1974. Turbulence effects in two journal bearing applications. *ASME Trans., Ser. F*, **96**, 15–21.

Giron, A. 1982. *Stability of Rotating Disk Flows*. Ph.D. Dissertation, University of Pittsburgh.

Gollub, J. P. and Benson, S. V. 1980. Many roads to turbulent convection. *J. Fluid Mech.*, **100**, 449–470.

Gollub, J. P. and Swinney, H. L. 1975. Onset of turbulence in rotating fluids. *Phys. Rev. Let.*, **35**, 927–930.

Gottlieb, D. and Orszag, S. A. 1977. *Numerical Analysis of Spectral Methods: Theory and Applications*. SIAM, Philadelphia.

Greenspan, H. P. 1968. *The Theory of Rotating Fluids*. Cambridge University Press.

Gregory, N., Stewart, J. T. and Walker, W. S. 1955. On the stability of three-dimensional boundary layers with application to the flow due to a rotating disk. *Philos. Trans. Roy. Soc. London, Ser. A*, **248**, 155–199.

Hinze, J. O. 1974. *Turbulence*, 2d ed. McGraw-Hill, New York.

Humphrey, J. A. C., Schuler, C. A. and Iglesias, I. 1992. Analysis of viscous dissipation in disk storage systems and similar flow configurations. *Phys. Fluids A 4*, **7**, 1415–1427.

Iooss, G. and Joseph, D. D. 1980. *Elementary Stability and Bifurcation Theory*. Springer-Verlag, New York.

Joseph, D. D. 1976. *Stability of Fluid Motions*. Springer-Verlag, New York.

Kamal, M. M. 1966. Separation in the flow between eccentric rotating cylinders. *ASME J. Basic Engineering*, **88**, 717–724.

Karman, T. von. 1921. Über laminare und turbulente Reibung. *Z. Angew. Math. Mech.*, **1**, 233–252.

Keller, H. B. 1977. Numerical solutions of bifurcation and nonlinear eigenvalue problems. In *Applications of Bifurcation Theory* (P. Rabinowitz, ed.). Academic Press, New York.

Kelvin, Lord. 1871. Hydrokinetic solutions and observations. *Phil. Mag.*, **42**, 362–377.

Kobayashi, R., Kohama, Y. and Takamadate, C. 1980. Spiral vortices in boundary layer transition regime on a rotating disk. *Acta. Mech.*, **35**, 71–82.

Kogelman, S. and DiPrima, R. C. 1970. Stability of spatially periodic supercritical flows in hydrodynamics. *Phys. Fluids*, **13**, 1–11.

Koschmieder, E. L. 1976. Taylor vortices between eccentric cylinders. *Phys. Fluids*, **19**, 1–4.

Koschmieder, E. L. 1993. *Bénard cells and Taylor Vortices*. Cambridge University Press. Cambridge.

Landau, L. D. and Lifshitz, E. M. 1959. *Fluid Mechanics*. Pergamon Press, Oxford.

Li, C. H. 1977. The effect of thermal diffusion on flow stability between two rotating cylinders. *ASME Trans., Ser. F*, **99**, 318–322.

Lin, C. C. 1967. *The Theory of Hydrodynamic Stability*. Cambridge University Press, Cambridge.

Marcus, P. S. 1984a. Simulation of Taylor-Couette flow, I – numerical methods and comparison with experiment. *J. Fluid Mech.*, **146**, 45–64.

Marcus, P. S. 1984b. Simulation of Taylor-Couette flow, II – numerical results for wavy vortex flow with one traveling wave. *J. Fluid Mech.*, **146**, 65–113.

Meyer-Spasche, R. and Keller, H. B. 1978. *Numerical study of Taylor-vortex flows between rotating cylinders*. California Institute of Technology.

Meyer-Spasche, R. and Keller, H. B. 1980. Computations of the axisymmetric flow between rotating cylinders. *J. Comput. Phys.*, **35**, 100–109.

Milne-Thomson, L. M. 1968. *Theoretical Hydrodynamics*. Macmillan, New York.

Mobbs, F. R. and Younes, M. A. 1974. The Taylor vortex regime in the flow between eccentric rotating cylinders. *ASME Trans., Ser. F*, **96**, 127–134.

Nakaya, C. 1974. Domain of stable periodic vortex flows in a viscous fluid between concentric circular cylinders. *J. Phys. Soc. Japan*, **26**, 1146–1173.

Newhouse, S., Ruelle, D. and Takens, F. 1978. Occurrence of strange attractors near quasiperiodic flows on T^m, $m \geq 3$. *Common. Math. Phys.*, **64**, 35–40.

Oikawa, M., Karasndani, T. and Funakoshni, M. 1989a. Stability of flow between eccentric rotating cylinders. *J. Phys. Soc. Japan*, **58**, 2355–2364.

Oikawa, M., Karasndani, T. and Funakoshni, M. 1989b. Stability of flow between eccentric rotating cylinders with wide gap. *J. Phys. Soc. Japan*, **58**, 2209–2210.

Orr, W. M. F. 1907. The stability or instability of the steady motions of a perfect liquid and of a viscous liquid. *Proc. Roy. Irish Acad.*, **27A**, 9–138.

Ortega, J. M. and Rheinboldt, W. C. 1970. *Iterative Solution of Nonlinear Equations in Several Variables*. Academic Press, New York.

Radel, V. and Szeri, A. Z. 1997. Symmetry braking bifurcation in finite disk flow. *Phys. Fluids*, **9**(6), 1–7.

Ritchie, G. S. 1968. On the stability of viscous flow between eccentric rotating cylinders. *J. Fluid Mech.*, **32**, 131–144.

Rosenhead, L. 1963. *Laminar Boundary Layers*. Oxford University Press.

Ruelle, D. and Takens, F. 1971. On the nature of turbulence. *Commun. Math. Phys.*, **20**, 167–192.

San Andres, A. and Szeri, A. Z. 1984. Flow between eccentric rotating cylinders. *ASME J. Appl. Mech.*, **51**, 869–878.

Schlichting, H. 1968. *Boundary Layer Theory*, 6th ed. Pergamon, London.

Schwartz, K. W., Springer, B. E. and Donelly, R. J. 1964. Modes of instability in spiral flow between rotating cylinders. *J. Fluid Mech.*, **20**, 281–289.

Serrin, J. 1959a. Mathematical Principles of Classical Fluid Mechanics. In *Handbuch der Physik*, *VIII/I*, (S. Flügge and C. Truesdell, eds.). Springer Verlag, Heidelberg.

Serrin, J. 1959b. On the stability of viscous motion. *Arch. Rational Mech. Anal.*, **3**, 1.

Seydel, R. 1988. *From Equilibrium to Chaos*. Elsevier, New York.

Smith, M. I. and Fuller, D. D. 1956. Journal bearing operations at superlaminar speeds. *ASME Trans., Ser. F*, **78**, 469–474.

Stewartson, K. 1953. On the flow between rotating co-axial disks. *Proc. Camb. Phil. Soc.*, **3**, 333–341.

Szeri, A. Z. and Adams, M. L. 1978. Laminar throughflow between closely spaced rotating disks. *J. Fluid Mech.*, **86**, 1–14.

Szeri, A. Z. and Al-Sharif, A. 1995. Flow between finite, steadily rotating eccentric cylinders. *Theoret. Comput. Fluid Dyn.*, **7**, 1–28.

Szeri, A. Z. and Giron, A. 1984. Stability of flow above a rotating disk. *Int. J. Num. Methods.* **4**, 989–996.

Szeri, A. Z., Schneider, S. J., Labbe, F. and Kaufman, H. N. 1983. Flow between rotating disks. Part 1. Basic flow. *J. Fluid Mech.*, **134**, 103–132.

Szeri, A. Z., Giron, A., Schneider, S. J. and Kaufman, H. N. 1983. Flow between rotating disks. Part 2. Stability. *J. Fluid Mech.*, **134**, 133–154.

Tatro, P. R. and Mollo-Christensen, E. L. 1967. Experiments on Ekman layer stability. *J. Fluid Mech.*, **77**, 531–543.

Taylor, G. I. 1923. Stability of a viscous liquid contained between two rotating cylinders. *Philos. Trans. Roy., Soc.*, **A 223**, 289–343.

Tzeng, H. M. and Fromm, J. E. 1990. Air flow study in a cylindrical enclosure containing multiple co-rotating disks. *ISROMAC-3*, Hawaii.

Tzeng, H.-M. and Humphrey, J. A. C. 1991. Co-rotating disk flow in an axisymmetric enclosure with and without a bluff body. *Int. J. Heat and Fluid Flow*, **12**, 3, 194–201.

Versteegen, P. L. and Jankowski, D. F. 1969. Experiments in the stability of viscous flow between eccentric rotating cylinders. *Phys. Fluids*, **12**, 1138–1143.

Vohr, J. H. 1968. An experimental study of Taylor vortices and turbulence in flow between eccentric, rotating cylinders. *ASME Trans.*, **90**, 285–296.

Weidman, P. D. and Redekopp, L. G. 1975. On the motion of a rotating fluid in the presence of an infinite rotating disk. In *Proc. 12th Biennal Fluid Dyn. Symp.*, Bialowicza, Poland.

Weinstein, M. 1977a. Wavy vortices in the flow between long eccentric cylinders, I – linear theory. *Proc. Roy. Soc. London*, **A 354**, 441–457.

Weinstein, M. 1977b. Wavy vortices in the flow between long eccentric cylinders, II – nonlinear theory. *Proc. Roy. Soc. London*, **A 354**, 459–489.

Wendt, F. 1933. Turbulente strömungen zwischen zwei rotierenden konaxialen Zylindern. *Ing. Arch.*, **4**, 577–595.

White, F. M. 1991. *Fluid Mechanics*. McGraw-Hill, New York.

Wilcox, D. E. 1950. Turbulence in high speed journal bearings. *ASME Trans., Ser. F*, **72**, 825–834.

Zandbergen, P. J. and Dijkstra, D. 1987. Von Karman swirling flows. *Annual Rev. Fluid Mech.*, **19**, 465–492.

Zienkiewicz, O. C. and Woo, J. 1991. Incompressibility without tears: how to avoid restrictions of mixed formulation. *Internat. J. Numer. Methods Eng.*, **32**, 1189–1203.

CHAPTER 7

Turbulence

Instability of laminar flow leads to flow transition and, on further increase of the Reynolds number, to eventual turbulence. The object of this chapter is to investigate bearing performance under turbulent conditions.

In journal bearings, turbulence makes its first appearance at Re $\approx$ 2000 (DiPrima, 1963).[1] Opinions on the minimum value of the Reynolds number for turbulence in thrust bearings are somewhat divided, but here again the value Re $\approx$ 2000 seems acceptable (Frêne, 1977). Once turbulence has set in, the importance of the precise mechanism of the instability and of the transition that resulted in turbulence diminishes. Nevertheless, some authorities maintain that if turbulence was obtained by spectral evolution, then the cellular flow pattern will persist into turbulence and affect the velocity profile-shear stress relationship (Burton and Carper, 1967). Existing theories of turbulence do not account for such occurrences; nevertheless, they show substantial agreement with experimental data from near-isothermal bearing experiments.

7.1 Equations of Turbulent Motion

Turbulence is an irregular fluid motion in which the various flow properties, such as velocity and pressure, show random variation with time and with position. Because of this randomness the instantaneous value of a flow property has little practical significance, it is the average value of that property that is of engineering interest.

To make our ideas precise, we represent a dependent variable by the sum of its average (denoted by an uppercase letter and an overbar) and its fluctuating component (denoted by a lowercase letter and a prime), so that

$$p = \bar{P} + p' \qquad v_i = \bar{V}_i + v'_i \qquad i = 1, 2, 3. \tag{7.1}$$

The average value, say $\bar{U} \equiv \bar{V}_1$ of the velocity component $u \equiv v_1$, is interpreted as a time average and is calculated according to

$$\bar{U}(T) = \frac{1}{T} \int_0^T u(t + \tau)\, d\tau \qquad T_1 \leq T \leq T_2. \tag{7.2a}$$

Here T_1 is the time scale of turbulence (time scale of largest eddies in the flow) and T_2 is the time scale of "slow" variations of the flow that, because of their relative slowness, do not appropriately belong to turbulence.

In turbulence theory (Monin and Yaglom, 1973), $\bar{U}$ represents the stochastic average of u; that is, the average of a large number of ostensibly identical experiments. If $u(t)$ is

---

[1] This value seems to hold for isothermal bearing operations only. Strong thermal effects may lower the critical value of the Reynolds number considerably (Li, 1977; Gardner and Ulschmid, 1974).

a stationary random function of time, i.e., its mean value is constant and its correlation function depends only on the difference $\tau = (t_2 - t_1)$, then it can be proved that

$$\bar{U} = \lim_{T \to \infty} \frac{1}{T} \int_0^T u(t)\, dt,$$

provided that

$$\lim_{T \to \infty} \frac{1}{T} \int_0^T B(\tau)\, d\tau = 0.$$

This last condition on the integral of the correlation function $B(\tau) = \overline{u'(t)u'(t + \tau)}$ is easily satisfied in turbulence, since the velocities at distant (in either time or space) points are uncorrelated, leading to $B(\tau) \to 0$ as $\tau \to \infty$.

If the turbulence is homogeneous, say in the x direction, then we have correspondence between the stochastic average and the space average taken along x, according to

$$U = \lim_{X \to \infty} \frac{1}{X} \int_0^X u(x)\, dx. \tag{7.2b}$$

Although actual turbulent flows are in general, neither stationary nor homogeneous, averaging is defined according to the approximate formulas of Eq. (7.2).

When Eq. (7.1) is substituted into the Navier-Stokes equation [Eqs. 2.41b], we obtain

$$\rho \left[\frac{\partial(\bar{V}_i + v'_i)}{\partial t} + (\bar{V}_k + v'_k) \frac{\partial(\bar{V}_i + v'_i)}{\partial x_k} \right] = -\frac{\partial(\bar{P} + p')}{\partial x_i} + \frac{\partial}{\partial x_j}[2\mu(\bar{D}_{ij} + d'_{ij})]. \tag{7.3}$$

Here $\bar{D}_{ij} = \frac{1}{2}(\partial \bar{V}_i/\partial x_j + \partial \bar{V}_j/\partial x_i)$ and $d'_{ij} = \frac{1}{2}(\partial v'_i/\partial x_j + \partial v'_j/\partial x_i)$ are the stretching tensors for the mean motion and the fluctuation, respectively.

Our aim is to derive equations of motion for mean values. To this end, we average Eq. (7.3) according to the rules

$$\overline{f + g} = \bar{f} + \bar{g}, \tag{7.4a}$$

$$\overline{af} = a\bar{f} \quad a = \text{const.,} \tag{7.4b}$$

$$\overline{\lim_{n \to \infty} f_n} = \lim_{n \to \infty} \bar{f}_n, \tag{7.4c}$$

$$\overline{\bar{f}g} = \bar{f}\bar{g} \tag{7.4d}$$

first established by Reynolds. Averaging leads to

$$\rho \left[\frac{\partial \bar{V}_i}{\partial t} + \bar{U}_k \frac{\partial \bar{V}_i}{\partial x_k} \right] = -\frac{\partial \bar{P}}{\partial x_i} + \frac{\partial}{\partial x_j}(2\mu \bar{D}_{ij}) - \overline{\rho v'_j \frac{\partial v'_i}{\partial x_j}}. \tag{7.5}$$

The averaged equation of motion, Eq. (7.5), can be put into a more convenient form with the aid of the continuity equation for the fluctuation, which is derived below.

Substitution into the equation of continuity [Eq. (2.44b)] in terms of mean and fluctuating components gives

$$\frac{\partial \bar{V}_i}{\partial x_i} + \frac{\partial v_i'}{\partial x_i} = 0. \tag{7.6}$$

Taking the average value of this equation in accordance with Eqs. (7.4) leads to the equation of continuity for the mean flow. By subtracting the mean flow continuity equation from Eq. (7.6), we obtain the continuity equation for the fluctuation. That is, for the mean motion and the fluctuation we have, respectively,

$$\frac{\partial \bar{V}_i}{\partial x_i} = 0, \tag{7.7a}$$

$$\frac{\partial v_i'}{\partial x_i} = 0. \tag{7.7b}$$

Equation (7.7b) is now used to put the last term of Eq. (7.5) into the desired form:

$$\overline{v_j' \frac{\partial v_i'}{\partial x_j}} = \frac{\partial}{\partial x_j}\overline{(v_i' v_j')},$$

so that Eq. (7.5) now reads

$$\rho \left[\frac{\partial \bar{V}_i}{\partial t} + \bar{V}_k \frac{\partial \bar{V}_i}{\partial x_k} \right] = \frac{\partial}{\partial x_j}(-\bar{P}\delta_{ij} + 2\mu \bar{D}_{ij} - \overline{\rho v_i' v_j'}). \tag{7.8}$$

The term in parentheses on the right-hand side is the average stress tensor in turbulent flow,

$$\bar{T}_{ij} = -\bar{P}\delta_{ij} + 2\mu \bar{D}_{ij} - \overline{\rho v_i' v_j'}. \tag{7.9}$$

The stress tensor $\bar{T}_{ij}$ is, thus, the sum of contributions from the mean flow $-\bar{P}\delta_{ij} + 2\mu \bar{D}_{ij}$ and contributions from the turbulence fluctuation,

$$\tau_{ij} = -\overline{\rho v_i' v_j'}. \tag{7.10}$$

The latter is called the apparent (or Reynolds) stress tensor.[2] Its components are unknown variables in Eq. (7.8). For physical origin of τ_{ij} see Schlichting (1968).

The system consisting of the equations of motion [Eqs. (2.41b)], the equation of continuity [Eq. (2.44b)], and the appropriate boundary and initial conditions defines a mathematically well-posed problem for laminar flow, which, at least in theory, can be solved to obtain the four unknowns u_1, u_2, u_3, and p. The number of equations in the system available to characterize the mean flow remains the same as the flow becomes turbulent, the system now consisting of the mean flow equations of motion, Eq. (7.8), and of continuity, Eq. (7.7a), and the boundary and initial conditions; yet the number of unknowns has increased to 10. (The Reynolds stress tensor is symmetric and has, therefore, only six independent components.)

---

[2] In laminar flow $\bar{P} = p$ and $\bar{U}_i = u_i$ so that $-\overline{\rho v_i' v_j'} \equiv 0$, and we recover from Eq. (7.9) $T_{ij} = -p\delta_{ij} + 2\mu D_{ij}$, the constitutive equation for a Newtonian fluid.

In turbulent flow, we thus have only four equations from which to determine 10 unknowns. As there does not seem to be any possibility of deriving additional equations on purely theoretical grounds, we are forced into (1) making assumptions concerning the character of the flow and (2) considering experimental data, when wishing to close the turbulence problem. This predicament gives rise to the so-called phenomenological or semi-empirical models of turbulence.

It is, of course, possible to derive transport equations for components of the Reynolds stress tensor (Hinze, 1975). These transport equations have the form

$$
\frac{D}{Dt}(\overline{v_i' v_j'}) = -\left(\overline{v_j' v_k'}\frac{\partial \bar{V}_i}{\partial x_k} + \overline{v_i' v_k'}\frac{\partial \bar{V}_j}{\partial x_k}\right) - 2\nu \overline{\frac{\partial v_i'}{\partial x_k}\frac{\partial v_j'}{\partial x_k}} + \overline{\frac{p'}{\rho}\left(\frac{\partial v_i'}{\partial x_j} + \frac{\partial v_j'}{\partial x_i}\right)}
$$
$$
- \frac{\partial}{\partial x_k}\left(\overline{v_i' v_j' v_k'} - \nu\frac{\overline{\partial v_i' v_j'}}{\partial x_k} + \overline{\frac{p'}{\rho}(\delta_{jk}v_i' + \delta_{ik}v_j')}\right).
$$

(7.11)

Inclusion of the six additional equations in our system of governing equations is of no help with the closure problem as we have acquired 10 additional unknowns in the process: the 10 independent components of the third-order correlation tensor $T_{ijk} = \overline{v_i' v_j' v_k'}$. The closure problem is not peculiar to turbulence, it is common to all nonlinear stochastic processes.

A number of models of turbulence have been proposed, each one designed to supply additional equations. They range from the simple, such as the constant eddy viscosity hypothesis of Boussinesq, to the sophisticated, exemplified by the 28-equation model of Kolovandin (Ng and Spalding, 1972).

Before we discuss the various mathematical models of turbulence and examine how they may be applied to the turbulent flow of a lubricant in the clearance space of a bearing, we reduce the equations of motion by taking into account the simplifying features of the lubricant film geometry.

Let L_{xz}, L_y represent the characteristic dimensions of the film as in Figure 2.8, and let U_* be the characteristic velocity in the "plane" of the film. Then the continuity equation demands that the across-the-film characteristic velocity is $(L_y/L_{xz})U_*$. Normalized variables are defined as follows:

$$
(\xi, \eta, \zeta) = \frac{1}{L_{xz}}\left(x, \left(\frac{L_{xz}}{L_y}\right)y, z\right), \qquad \tau = \left(\frac{U_*}{L_{xz}}\right)t
$$
$$
(U, V, W) = \frac{1}{U_*}\left(\bar{U}, \left(\frac{L_{xz}}{L_y}\right)\bar{V}, \bar{W}\right)
$$

(7.12a)

$$
P = \frac{\bar{P}}{\rho U_*^2}\mathrm{Re}^*, \qquad \mathrm{Re}^* = \left(\frac{L_y}{L_{xz}}\right)\mathrm{Re} = \frac{L_y^2\omega}{\nu}.
$$

We assume that the three components (u', v', w') of the fluctuating velocity are each of order u_* and define nondimensional velocity fluctuations by

$$
(u, v, w) = \frac{1}{u_*}(u', v', w').
$$

(7.12b)

Substituting Eq. (7.12) into Eq. (7.8), rearranging, and setting $\kappa = (u_*/U_*)^2$ yields

$$\text{Re}^* \left[\frac{dU}{d\tau} + \kappa \left(\frac{\partial \overline{uu}}{\partial \xi} + \frac{\partial \overline{uw}}{\partial \zeta} \right) \right] + \kappa \, \text{Re} \frac{\partial \overline{uv}}{\partial \eta}$$
$$= -\frac{\partial P}{\partial \xi} + \frac{\partial^2 U}{\partial \eta^2} + \left(\frac{L_y}{L_{xz}} \right)^2 \left(\frac{\partial^2 U}{\partial \xi^2} + \frac{\partial^2 U}{\partial \zeta^2} \right),$$

(7.13a)

$$\left(\frac{L_y}{L_x} \right)^2 \left[\text{Re}^* \frac{dV}{d\tau} - \frac{\partial^2 V}{\partial y^2} - \left(\frac{L_y}{L_{xz}} \right)^2 \left(\frac{\partial^2 V}{\partial \xi^2} + \frac{\partial^2 V}{\partial \zeta^2} \right) \right]$$
$$= -\frac{\partial P}{\partial \eta} - \kappa \, \text{Re}^* \left[\frac{\partial \overline{vv}}{\partial \eta} + \left(\frac{L_y}{L_{xz}} \right) \left(\frac{\partial \overline{vu}}{\partial \xi} + \frac{\partial \overline{vw}}{\partial \zeta} \right) \right],$$

(7.13b)

$$\text{Re}^* \left[\frac{dW}{d\tau} + \kappa \left(\frac{\partial \overline{uw}}{\partial \xi} + \frac{\partial \overline{ww}}{\partial \zeta} \right) \right] + \kappa \, \text{Re} \frac{\partial \overline{vw}}{\partial \eta}$$
$$= -\frac{\partial P}{\partial \zeta} + \frac{\partial^2 W}{\partial \eta^2} + \left(\frac{L_y}{L_{xz}} \right)^2 \left(\frac{\partial^2 W}{\partial \xi^2} + \frac{\partial^2 W}{\partial \zeta^2} \right)$$

(7.13c)

where $\frac{d}{d\tau} = \frac{\partial}{\partial \tau} + U \frac{\partial}{\partial \xi} + V \frac{\partial}{\partial \eta} + W \frac{\partial}{\partial \zeta}$ is the (dimensionless) material derivative based on the mean velocity.

On the basis of film geometry, we delete the terms in Eqs. (7.13) that are multiplied by powers of (L_y/L_{xz}) and investigate these equations under two different assumptions:

(1) Assuming that $\kappa = (u_*/U_*)^2 = O(1)$, a condition that is said to exist in wake flow (Hinze, 1979), we have

$$\text{Re}^* \left[\frac{dU}{d\tau} + \frac{\partial \overline{uu}}{\partial \xi} + \frac{\partial \overline{uw}}{\partial \zeta} \right] + \text{Re} \frac{\partial \overline{uv}}{\partial \eta} = -\frac{\partial P}{\partial \xi} + \frac{\partial^2 U}{\partial \eta^2},$$

(7.14a)

$$\text{Re}^* \frac{\partial \overline{vv}}{\partial \eta} = -\frac{\partial P}{\partial \eta},$$

(7.14b)

$$\text{Re}^* \left[\frac{dW}{d\tau} + \frac{\partial \overline{uw}}{\partial \xi} + \frac{\partial \overline{ww}}{\partial \zeta} \right] + \text{Re} \frac{\partial \overline{vw}}{\partial \eta} = -\frac{\partial P}{\partial \zeta} + \frac{\partial^2 W}{\partial \eta^2}.$$

(7.14c)

Equations (7.14) can be combined into a single equation in the mean pressure if $\text{Re}^* \ll 1$; under this condition pressure variation across the film is negligible and Eqs. (7.14) become linear. Most existing turbulent lubrication theories rely on the assumptions,

$$\frac{L_y}{L_{xz}} \ll 1, \qquad \frac{u_*}{U_*} = O(1), \qquad \text{Re}^* \ll 1,$$

and lead to the equations (now in primitive variables)

$$\frac{\partial \bar{P}}{\partial x} = \frac{\partial}{\partial y} \left(\mu \frac{\partial \bar{U}}{\partial y} - \rho \overline{u'v'} \right),$$

(7.15a)

$$\frac{\partial \bar{P}}{\partial y} = 0,$$

(7.15b)

$$\frac{\partial \bar{P}}{\partial z} = \frac{\partial}{\partial y} \left(\mu \frac{\partial \bar{W}}{\partial y} - \rho \overline{v'w'} \right).$$

(7.15c)

However, we should realize that turbulent conditions are not observed unless $Re = O(10^3)$, yielding $Re^* \geq 1$ for common bearing geometries.

(2) Alternatively, if one assumes that $\kappa = (u_*/U_*)^2 = O(L_y/L_{xz})$, as is more appropriate for wall turbulence according to Hinze (1979), Eqs. (7.13) assume the form

$$Re^*\left(\frac{dU}{d\tau} + \frac{\partial \overline{uv}}{\partial \eta}\right) = -\frac{\partial P}{\partial \xi} + \frac{\partial^2 U}{\partial \eta^2}, \tag{7.16a}$$

$$0 = -\frac{\partial P}{\partial \eta}, \tag{7.16b}$$

$$Re^*\left(\frac{dW}{d\tau} + \frac{\partial \overline{vw}}{\partial \eta}\right) = -\frac{\partial P}{\partial \zeta} + \frac{\partial^2 W}{\partial \eta^2}. \tag{7.16c}$$

In this case, as easily seen from Eqs. (7.16), there is no turbulence in the zero Re^* limit. On the other hand, inertia effects must be included, as they are of the same order of magnitude as the surviving Reynolds stresses, in all turbulent, i.e., $Re^* > 0$, flows. And although the pressure remains constant across the film to order $(L_y/L_{xz})^2$, a Reynolds-type pressure equation is not possible to obtain on account of the nonlinearity of Eqs. (7.16).

In terms of primitive variables, Eq. (7.16) takes the form

$$\frac{\partial \bar{P}}{\partial x} = \frac{\partial}{\partial y}\left(\mu\frac{\partial \bar{U}}{\partial y} - \rho\overline{u'v'}\right) + \rho\frac{d\bar{U}}{dt},$$

$$\frac{\partial \bar{P}}{\partial y} = 0, \tag{7.17}$$

$$\frac{\partial \bar{P}}{\partial z} = \frac{\partial}{\partial y}\left(\mu\frac{\partial \bar{W}}{\partial y} - \rho\overline{v'w'}\right) + \rho\frac{d\bar{W}}{dt}.$$

According to this second analysis, mean-flow inertia is of the same order of magnitude as the surviving Reynolds stresses; in other words, there can be no turbulence without inertia.

In neither of the above two cases is the physics of turbulence well represented by a Reynolds-type equation. Nevertheless, we follow accepted practice in this chapter and consider (7.15) to be valid in thin films.

To calculate bearing performance under turbulent conditions, several researchers, (Constantinescu, 1962; Ng and Pan, 1965; Elrod and Ng, 1967) employ Eqs. (7.15), resorting to the more acceptable model of Eqs. (7.17) only when specifically investigating lubricant inertia (Constantinescu and Galetuse, 1979; Landau and Leschziner, 1978).

The earliest turbulent lubrication models employed equation Eqs. (7.15) and the ideas of (1) mixing length (Constantinescu, 1959), (2) eddy viscosity (Ng and Pan, 1965; Elrod and Ng, 1967), and (3) empirical drag laws (Hirs, 1973; Black and Walton, 1974). Closure is obtained in these models by relating the Reynolds stress to the mean flow characteristics. Such representations, when used in conjunction with Eqs. (7.15), can lead to relationships between the mean velocity and the mean pressure gradient. To obtain the governing equation for pressure, the mean velocity, now in terms of the mean pressure, is substituted into the equation of continuity in a development that parallels the laminar flow case. Several of the available models yield a formally identical turbulent Reynolds equation of the form

$$\frac{\partial}{\partial x}\left(\frac{h^3}{\mu k_x}\frac{\partial \bar{P}}{\partial x}\right) + \frac{\partial}{\partial z}\left(\frac{h^3}{\mu k_z}\frac{\partial \bar{P}}{\partial z}\right) = \frac{1}{2}U_0\frac{\partial h}{\partial x}. \tag{7.18}$$

Here $k_x = k_x(R_h)$ and $k_z = k_z(R_h)$, where $R_h = U_* h / \nu$ is the local Reynolds number, thus in these formulations turbulence is accounted for by weighting the film with some function of R_h. Since $k_x, k_z \geq 12$, the equality holding for laminar flow, the apparent film thickness, $h/k_x^{1/3}$, $h/k_z^{1/3}$, employed in performance calculations is smaller than the geometric film thickness, h, therefore, turbulent conditions yield higher pressures, for the same film geometry, than laminar conditions. Most turbulence models are in agreement with this conclusion, it is in the actual form of the turbulence functions $k_x(x, z; \text{Re})$, $k_z(x, z; \text{Re})$ where they differ (see Figure 7.4). But the difference in performance predictions between the various models is often insignificant, except perhaps at extreme conditions such as high eccentricity in a journal bearing.

7.2   Turbulence Models

Perhaps the first attempt to provide a mathematical model of turbulence was made by Boussinesq in 1877, when he proposed a relation between the Reynolds stresses and the mean velocity gradient in the form

$$\tau_{ij} = -\rho \overline{v_i' v_j'} = \rho \varepsilon_m \left(\frac{\partial \bar{V}_i}{\partial x_j} + \frac{\partial \bar{V}_j}{\partial x_i} \right), \tag{7.19}$$

so that the mean stress in the fluid is given by

$$\bar{T}_{ij} = -\bar{P} \delta_{ij} + \mu \left(1 + \frac{\varepsilon_m}{\nu} \right) \left(\frac{\partial \bar{V}_i}{\partial x_j} + \frac{\partial \bar{V}_j}{\partial x_i} \right). \tag{7.20}$$

Equation (7.20) is analogous to the constitutive equation of Stokes for laminar flow, except that, unlike the molecular viscosity, μ, the eddy viscosity, ε_m, is not a material constant.

The eddy viscosity, ε_m, depends on the structure of turbulence itself. Therefore, to complete the model of Boussinesq, ε_m has to be related to measurable quantities of the turbulent flow. Workers in the field of turbulence who use Boussinesq's model, Eq. (7.20), accomplish this in various ways.

Prandtl (1963) reasoned that the eddy viscosity is given by

$$\varepsilon_m = \ell^2 \left| \frac{\partial \bar{U}}{\partial y} \right| \tag{7.21}$$

when dealing with a two-dimensional mean flow along the solid wall, which is located at $y = 0$. In Prandtl's theory, the mixing length, ℓ, is analogous to the mean free path of the kinetic theory of gases. (It is shown in the kinetic theory of gases that if v_m is the rms. molecular velocity and ℓ is the mean free path between collisions, then $\mu \sim \rho v_m \ell$, where μ is the molecular viscosity.) To illustrate the analogy, without actually considering details, we follow Prandtl and consider a two-dimensional mean flow that is parallel to the $y = 0$ plane (a solid wall) so that

$$\bar{U} = \bar{U}(y) \qquad \bar{V} = \bar{W} = 0.$$

Although turbulence is three dimensional, we focus attention on only one Reynolds stress component $-\rho \overline{u' v'}$ and write, in accordance with Eq. (7.19), the approximation

$$\tau_{xy} \approx -\rho \overline{u' v'} = \rho \varepsilon_m \frac{d\bar{U}}{dy}. \tag{7.22}$$

Here we neglected the viscous stress, so Eq. (7.22) is not valid in regions where viscous effects are of the same order of magnitude as turbulence effects.

To have dimensional homogeneity in Eq. (7.22), the eddy viscosity, ε_m, must have the physical dimensions of (length^2)/(time). This requirement is satisfied when, in analogy to the kinetic theory, we choose ε_m to be represented by the product of a characteristic length and a characteristic velocity. The length parameter is selected in such a way that it permits the local friction velocity

$$v_* = \sqrt{|\tau_{xy}|/\rho}$$

to be used as the velocity parameter. Thus we have, by definition,

$$\varepsilon_m = \ell \times v_* \qquad \text{and} \qquad |\tau_{xy}| = \rho v_*^2.$$

Substitution for ε_m into Eq. (7.22) yields

$$|\tau_{xy}| = \rho v_*^2 = \rho \varepsilon_m \frac{d\bar{U}}{dy} = \rho(\ell \times v_*)\frac{d\bar{U}}{dy}.$$

By comparison of the second and last terms we arrive at Eq. (7.21), so that

$$\tau_{xy} = -\rho\overline{u'v'} = \rho\ell^2 \left|\frac{d\bar{U}}{dy}\right|\frac{d\bar{U}}{dy}. \tag{7.23}$$

[Note that the absolute sign in Eqs. (7.21) and (7.23) was introduced to ensure that $\varepsilon_m \geq 0$ whatever the sign of $d\bar{U}/dy$.]

In a simple application of the theory, where there is only one characteristic dimension associated with the problem (e.g., the distance from the wall in channel flow or the width of the turbulent mixing zone in jets and wakes), we may assume proportionality between this characteristic dimension and the mixing length ℓ.

The velocity profile that may be obtained from Eq. (7.23) by integration is not valid at the wall. There, in the so-called viscous sublayer, we have $\mu|d\bar{U}/dy| \gg \rho|\overline{u'v'}|$, and the velocity profile is determined as

$$\tau_w = \mu\frac{d\bar{U}}{dy}, \tag{7.24}$$

where τ_w is the wall stress.

In an application to boundary-layer flows, Prandtl integrated Eq. (7.23) within the constant stress layer; that is, in the layer outside the viscous sublayer that is still close enough to the wall so that the condition $\tau_{xy} = \tau_w$ is approximately satisfied. Within this layer, as the only length parameter is the distance from the wall, Prandtl assumed that

$$\ell = \kappa y, \tag{7.25}$$

where κ is a dimensionless constant that must be deduced from experiment.

When Eq. (7.25) and $\tau_{xy} \approx \tau_w$ are substituted into Eq. (7.23) and the latter is integrated, we obtain

$$\bar{U} = \frac{v_*}{\kappa} \ln y + C. \tag{7.26}$$

The integration constant C can be determined in either of two ways.

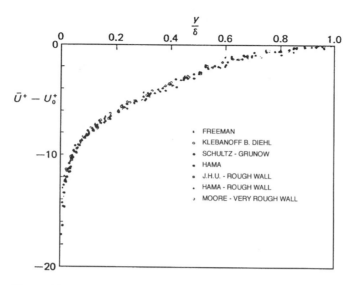

Figure 7.1. Experimental verification of the velocity defect law, Eq. (5.34). (Reprinted with permission from Clauser, F. H. The turbulent boundary layer. *Adv. Appl. Mech.*, **4**, 1–51, 1965.)

First, by fixing attention on the layer $y = y_0$ ($y_0 = h/2$ in channel flow and $y_0 = \delta$ in the boundary layer, where δ is the boundary-layer thickness), where the velocity, $\bar{U}$, is equal to its maximum value U_0, we have from Eq. (7.26)

$$U_0 = \frac{v_*}{\kappa} \ln y_0 + C. \tag{7.27}$$

The integration constant C can now be eliminated between the last two equations, and we obtain a particular form of the velocity defect law,

$$U_0^+ - \bar{U}^+ = \frac{1}{\kappa} \ln \frac{y_0}{y}, \tag{7.28}$$

that is valid within the constant stress layer. Here we used the notation $U_0^+ = U_0/v_*$ and $\bar{U}^+ = \bar{U}/v_*$. Experimental verification of the velocity defect law is given in Figure 7.1.

Alternatively, the constant C in Eq. (7.26) may be determined from the condition that the turbulent velocity distribution of Eq. (7.23) must join onto the velocity of the viscous sublayer of Eq. (7.24) somewhere in the vicinity of the wall. Thus, the condition on the velocity distribution in Eq. (7.24) is $\bar{U} = 0$ at $y = y_L$. The thickness of the viscous sublayer y_L is determined by conditions at the (smooth) wall, characterized by $v_* = \sqrt{\tau_w/\rho}$ and by the kinematic viscosity v. Then, by dimensional reasoning, we must have

$$y_L = \beta \frac{v}{v_*}, \qquad y_L^+ = \beta. \tag{7.29}$$

Substituting into Eq. (7.26), we find

$$\bar{U}^+ = \frac{1}{\kappa}(\ln y^+ - \ln \beta). \tag{7.30}$$

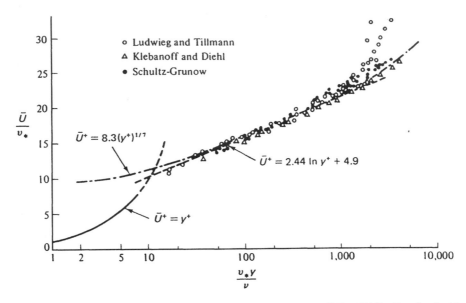

Figure 7.2. Experimental verification of the law of the wall, Eq. (5.32). (Reprinted with permission from Hinze, J. O. *Turbulence*, 2nd ed. McGraw-Hill, New York. Copyright 1975.)

This is Prandtl's universal logarithmic distribution. Here $y^+ = yv_*/\nu$ is the dimensionless distance from the wall, and the constant β depends on the nature of the wall surface.

In Eqs. (7.28) and (7.30), we found two universal velocity distribution laws. When these are written in the more general form,

$$U_0^+ - \bar{U}^+ = f(\hat{y}) \qquad \hat{y} = \frac{y}{y_0} \tag{7.31}$$

and

$$\bar{U}^+ = F(y^+) \qquad y^+ = \frac{yv_*}{\nu}, \tag{7.32}$$

they are known as the velocity defect law and the law of the wall, respectively.[3] Experimental verification of the law of the wall is shown in Figure 7.2.

The law of the wall, Eq. (7.32), is valid in the wall region, which comprises the viscous sublayer adjacent to the wall, the constant stress layer farther out in the fully turbulent region, and the buffer zone separating the two. The logarithmic velocity distribution of Prandtl, Eq. (7.30), which is Prandtl's simplified version of the law of the wall, does not account for the buffer zone but joins the velocity profile of the constant stress layer abruptly to the velocity profile of the viscous sublayer.

Much effort has been directed toward making Boussinesq's hypothesis, Eq. (7.20), applicable in the whole wall layer by providing a universal eddy viscosity profile there. It was

[3] The velocity defect law, Eq. (7.31), and the law of the wall, Eq. (7.32), can be arrived at on purely dimensional grounds. Then, the requirement that there must be a region of overlap where both equations are valid leads to the logarithmic laws, Eqs. (7.28) and (7.30), without use of the mixing length theory. For details of this more appealing approach see Townsend (1977) and Hinze (1975).

shown by Reichardt and later by Elrod (Hinze, 1975) that such an eddy viscosity distribution must satisfy two requirements:

$$\lim_{y^+ \to 0} \frac{\varepsilon_m}{\nu} = \text{const.} \times (y^+)^3,$$

$$\lim_{y^+ \to \infty} \frac{\varepsilon_m}{\nu} = \text{const.} \times (y^+). \tag{7.33}$$

The second condition is necessary to have agreement with Eq. (7.26), and the first follows from the equation of continuity when there is streamwise variation of the mean values.

One of the more successful eddy viscosity profiles was devised by Reichardt (Monin and Yaglom, 1973). His eddy viscosity

$$\frac{\varepsilon_m}{\nu} = k\left(y^+ - \delta_l^+ \tanh \frac{y^+}{\delta_\ell^+} \right), \tag{7.34}$$

where k and δ_ℓ^+ are constants, δ_ℓ^+ being related to the thickness of the viscous sublayer, satisfies the two conditions of Eq. (7.33) and is in good agreement with experimental data.

In the Kolmogoroff-Prandtl energy model, it is assumed that the eddy viscosity can be represented as the product of two characteristic quantities of turbulence

$$\varepsilon_m = A\left(\frac{\overline{q^2}}{2} \right)^{1/2} \Lambda, \tag{7.35}$$

where A is a numerical constant, $\overline{q^2}/2 = \overline{v_i' v_i'}/2$ is the kinetic energy of the turbulent fluctuation per unit mass, and Λ is an integral length scale.

The length scale Λ in Eq. (7.35) may be made proportional to the characteristic dimension of the problem, as suggested by Prandtl. In the analysis of flow near a wall, Wolfshtein (1969) employed the modified van Driest formula (Hinze, 1975)

$$\Lambda = y(1 - e^{-A_\eta \text{Re}_q}), \tag{7.36}$$

where A_η is a constant and $\text{Re}_q = y\sqrt{\overline{q^2}/2}\nu$ is a local Reynolds number. Ng and Spalding (1972) recommended that Λ be calculated from a transport equation.

The turbulent kinetic energy for use in Eq. (7.36) is given by a transport equation, which is obtained from Eq. (7.11) by contraction:

$$\frac{D}{Dt}\left(\frac{\overline{q^2}}{2} \right) = -\frac{\partial}{\partial x_i} \overline{v_i'\left(\frac{p'}{\rho} + \frac{q^2}{2} \right)} - \overline{v_i' v_j'} \frac{\partial \bar{V}_j}{\partial x_i} + 2\nu \frac{\partial}{\partial x_i} \overline{v_j' d_{ij}'} - \varepsilon. \tag{7.37}$$

The terms on the right-hand side of Eq. (7.37) represent (1) the convective diffusion by turbulence of the total turbulence energy, (2) the rate of production of turbulence, (3) the work by the viscous shear stresses of the turbulent motion, and (4) the rate of viscous dissipation of turbulent energy,

$$\varepsilon = 2\nu \overline{d_{ij}' d_{ij}'}$$

$$= \nu \overline{\left(\frac{\partial v_i'}{\partial x_j} + \frac{\partial v_j'}{\partial x_i} \right) \frac{\partial v_i'}{\partial x_j}}. \tag{7.38}$$

In internal flows subject to a strong favorable pressure gradient, Jones and Launder (1972) put $\varepsilon \sim (\overline{q^2})^{3/2}/\Lambda$ in Eq. (7.35), so that their eddy viscosity is given by

$$\varepsilon_m = C\frac{(\overline{q^2})^2}{4\varepsilon} \tag{7.39}$$

and ε is calculated from a transport equation (Hinze, 1975).

7.3   Constantinescu's Model

The approach in Constantinescu's turbulent lubrication model (Constantinescu, 1959) is based on the Prandtl mixing length hypothesis, Eq. (7.21). Following Prandtl's ideas, the mixing length is made to vanish at the walls and to vary linearly with the distance from the nearest wall:

$$\left.\begin{array}{ll} \ell = ky & 0 \leq y \leq \dfrac{h}{2} \\[2mm] \ell = ky' & 0 \leq y' \leq \dfrac{h}{2} \end{array}\right\}, \tag{7.40}$$

where $y' = h - y$. Substituting for ℓ from Eq. (7.40) and for $-\rho\overline{u'v'}$ from Eq. (7.23) into Eq. (7.15a), the equation of turbulent motion in a long bearing ($\partial\bar{P}/\partial z = 0$, $\bar{W} = 0$) is obtained. For $0 \leq y \leq h/2$ this equation has the nondimensional form

$$\frac{\partial}{\partial\hat{y}}\left(k^2\hat{y}^2\,\mathrm{Re}_h\left|\frac{\partial U}{\partial\hat{y}}\right|\frac{\partial U}{\partial\hat{y}} + \frac{\partial U}{\partial\hat{y}}\right) - \frac{h^2}{\mu U_*}\frac{\partial\bar{P}}{\partial x} = 0, \tag{7.41}$$

where $\mathrm{Re}_h = U_*h/\nu$ is the local Reynolds number and

$$\hat{y} = \frac{y}{h} \qquad U = \frac{\bar{U}}{U_*}.$$

$U_* = U_2$ is the velocity of the runner in the x direction relative to the stationary bearing surface ($U_1 = 0$).

Equation (7.41) may now be integrated,

$$A\hat{y}^2\frac{\partial U}{\partial\hat{y}}\left|\frac{\partial U}{\partial\hat{y}}\right| + \frac{\partial U}{\partial\hat{y}} + B\hat{y} - C = 0, \tag{7.42}$$

where we use the notation

$$A = k^2\,\mathrm{Re}_h, \qquad B = -\frac{h^2}{\mu U_*}\frac{\partial\bar{P}}{\partial x}.$$

The integration constant, C, in Eq. (7.42) corresponds to the dimensionless wall stress [as $\hat{y} \to 0$, Eq. (7.42) reduces to $C = (\partial U/\partial\hat{y})|_{\hat{y}=0} = h\tau_w/\mu U_*$]. It may therefore take on positive, zero, or negative values. In addition, in the lubricant film we encounter (1) pressure flow in the direction of motion ($B > 0$), (2) pure shear flow ($B = 0$), and (3) pressure flow opposing shear flow ($B < 0$). The velocity gradient, $\partial U/\partial\hat{y}$, might take on positive or negative values. Thus, since all possible flow situations must be accounted for, Eq. (7.42) represents a total of $3 \times 3 \times 2 = 18$ distinct cases.

We can show, however, that not all 18 cases yield real and therefore physically possible velocity profiles. As an example, consider the following combination of the parameters:

$$C > 0; \quad B = -B' < 0; \quad \frac{\partial U}{\partial \hat{y}} < 0. \tag{7.43}$$

Equation (7.42) now assumes the form

$$A\hat{y}^2 \left(\frac{\partial U}{\partial \hat{y}}\right)^2 - \frac{\partial U}{\partial \hat{y}} + B'\hat{y} + C = 0. \tag{7.44}$$

Solving formally for $\partial U/\partial \hat{y}$, we have

$$\frac{\partial U}{\partial \hat{y}} = \frac{1 \pm \sqrt{1 - 4A\hat{y}^2(B'\hat{y} + C)}}{2A\hat{y}^2}. \tag{7.45}$$

Equation (7.45) must satisfy two conditions simultaneously: (1) that $1 \geq 4A\hat{y}^2(B'\hat{y} + C)$, so that the solution is real, and (2) that $\partial U/\partial \hat{y} < 0$, so that the last condition in Eq. (7.43) is satisfied. This, of course, is not possible, leading to the conclusion that Eq. (7.43) does not represent a possible flow in the bearing clearance. Physically acceptable solutions of Eq. (7.42) are obtained only in the following 10 cases:

	1	2	3	4	5	6	7	8	9	10
B	−	−	−	−	0	0	+	+	+	+
C	−	−	0	+	+	−	−	0	+	+
$\partial U/\partial \hat{y}$	−	+	+	+	+	−	−	−	−	+

Instead of integrating Eq. (7.42) rigorously in these 10 cases, Constantinescu chose to follow Prandtl and divided the flow regime $0 \leq \bar{y} \leq 1/2$ into two layers.

In the viscous sublayer $0 \leq \hat{y} \leq \hat{y}_L$ the effect of the Reynolds stress is negligible, and here we have

$$\frac{\partial U}{\partial \hat{y}} + B\hat{y} - C = 0, \tag{7.46}$$

whereas in the turbulent core the effect of molecular viscosity is small, and we take

$$A\hat{y}^2 \frac{\partial U}{\partial \hat{y}} \left|\frac{\partial U}{\partial \hat{y}}\right| + B\hat{y} - C = 0. \tag{7.47}$$

Equations (7.46) and (7.47) are to be solved simultaneously, with similar equations for the other half of the channel. These calculations will now be illustrated in one case, at the position of maximum film pressure. There the following conditions and approximate differential equations apply:

Sliding surface ($\hat{y}' = 0$), *case 6*
$B = 0; \quad C = -C' < 0; \quad \partial U/\partial \hat{y}' < 0$

$$\frac{\partial U}{\partial \hat{y}'} = -C' \qquad \text{viscous sublayer} \tag{7.48a}$$

$$A\hat{y}'^2 \left(\frac{\partial U}{\partial \hat{y}'}\right)^2 + C' = 0 \qquad \text{turbulent core} \tag{7.48b}$$

Stationary surface ($\hat{y} = 0$)*, case 5*

$B = 0; \quad C > 0; \quad \partial U / \partial \hat{y} > 0$

$$\frac{\partial U}{\partial \hat{y}} = C \qquad \text{viscous sublayer} \tag{7.48c}$$

$$A\hat{y}^2 \left(\frac{\partial U}{\partial \hat{y}}\right)^2 - C = 0 \qquad \text{turbulent core.} \tag{7.48d}$$

Integrating Eqs. (7.48a) to (7.48d), we obtain the following four-segment velocity profile:

$$U = \begin{cases} -C'\hat{y}' + K' & 0 \le \hat{y}' \le \hat{y}'_L \\[2mm] C'_2 - \sqrt{\dfrac{C'}{A}} \ln \hat{y}' & \hat{y}'_L \le \hat{y}' \le 0.5 \\[2mm] C_2 + \sqrt{\dfrac{C}{A}} \ln \hat{y} & \hat{y}_L \le \hat{y} \le 0.5 \\[2mm] C\hat{y} + K & 0 \le \hat{y} \le \hat{y}_L. \end{cases} \tag{7.49}$$

The integration constants C, C', C_2, C'_2, K, and K' the boundary-layer thickness $\hat{y}_L$ and $\hat{y}'_L$ can be evaluated by imposing the following conditions:

(1) No slip at the solid walls
(2) Continuity of both the velocity and the velocity gradient at the edge of viscous sublayers
(3) Continuity of both the velocity and the velocity gradient in the center of the channel.

Imposing these conditions on Eq. (7.49), we find that $\hat{y}_L = \hat{y}'_L = (CA)^{-1/2}$; that is, $y_L^+ = 1/k$ and

$$U = \begin{cases} 1 - C\hat{y}' & 0 \le \hat{y}' \le (CA)^{-1/2} \\[2mm] 1 - \sqrt{\dfrac{C}{A}}[1 + \ln(\hat{y}'\sqrt{CA})] & (CA)^{-1/2} \le \hat{y}' \le 0.5 \\[2mm] \sqrt{\dfrac{C}{A}}[1 + \ln(\hat{y}\sqrt{CA})] & (CA)^{-1/2} \le \hat{y} \le 0.5 \\[2mm] C\hat{y} & 0 \le \hat{y} \le (CA)^{-1/2} \end{cases} \tag{7.50}$$

Here the constant, C, is given by the transcendental equation

$$1 - \sqrt{\frac{C}{A}}\left(2 + \ln \frac{CA}{4}\right) = 0. \tag{7.51}$$

Equation (7.51) has the solution $C = 37.8097$ at $A = 6400$, or Re $= 40,000$ and $k = 0.4$. When this value of C is substituted into Eq. (7.50), the velocity distribution in pure Couette flow at Re $= 40,000$ is obtained according to Constantinescu's model (Figure 7.3). The shear stress is uniform across the channel and has the value $\tau/\rho U_*^2 = 37.8097/ \text{Re}_h$.

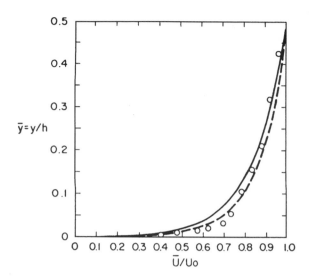

Figure 7.3. Velocity profile of plane Couette flow at Re = 40,000. Curves are theoretical predictions: (—) Constantinescu (1959), Eq. (7.50); (– – –) Elrod and Ng (1967). (o) experimental values of Robertson (1959).

If the flow was laminar throughout the channel, the velocity distribution would be represented by the single equation

$$U = \frac{1}{2}B_x(\hat{y} - \hat{y}^2) + \hat{y} \tag{7.52}$$

and the average velocity U_m by

$$U_m = \int_0^1 U\,d\hat{y} = U_{mp} + U_{ms}$$
$$= \frac{1}{12}B_x + \frac{1}{2}. \tag{7.53}$$

Here $U_{mp} = B_x/12$ is the average velocity of pressure flow and $U_{ms} = 1/2$ is the average velocity of shear flow. Thus, from Eq. (7.53) we find for laminar flow that

$$B_x = 12U_{mp}. \tag{7.54}$$

That is, in the laminar regime the average velocity of pressure-induced flow varies linearly with the pressure parameter B_x.

For turbulent flow $\bar{U}_{ms} = 0.5$ again, and when values of U_{mp} are plotted against B_x for different values of the Reynolds number, the resulting plots can be described by the approximate relation

$$B_x \equiv -\frac{h^2}{\mu U_*}\frac{\partial \bar{P}}{\partial x} = k_x(\text{Re}_h)U_{mp}. \tag{7.55}$$

An analysis, similar to the one above, yields the following relationship between axial pressure drop and average velocity of pressure flow in the same direction [the axial flow

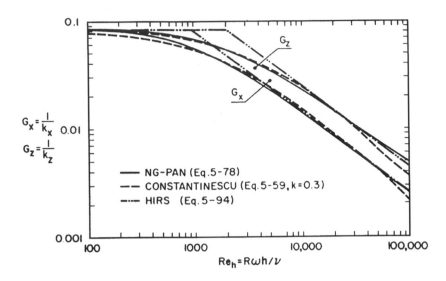

Figure 7.4. Variation of the turbulence functions G_x and G_z with Reynolds number. Comparison of various theories.

is decoupled from the circumferential flow in Constantinescu's analysis; for the Reynolds stress in Eq. (7.15c), he takes $-\rho \overline{v'w'} = \rho \ell^2 |\partial \overline{W}/\partial y| \partial \overline{W}/\partial y]$:

$$B_z \equiv -\frac{h^2}{\mu U_*}\frac{\partial \bar{P}}{\partial z} = k_z(\mathrm{Re}_h)W_{mp},\tag{7.56}$$

where W_{mp} is the dimensionless average velocity of the axial pressure flow induced by B_z.
Constantinescu's analysis yields through curve fitting

$$k_x = 12 + 0.53(k^2 \,\mathrm{Re}_h)^{0.725},\tag{7.57a}$$

$$k_z = 12 + 0.296(k^2 \,\mathrm{Re}_h)^{0.65}.\tag{7.57b}$$

Notice that for $\mathrm{Re} \to 0$, Eqs. (7.55) and (7.56) reduce to Eq. (7.54), i.e., laminar flow is included in Constantinescu's model. The variation of k_x and k_z with Re_h is shown in Figure 7.4.

Substituting $U_m = U_{mp} + U_{ms}$ and $W_m = W_{mp}$ into the once-integrated continuity equation,

$$\frac{\partial}{\partial x}(\bar{U}_m h) + \frac{\partial}{\partial z}(\bar{W}_m h) + \frac{1}{U_0}\frac{dh}{dt} = 0,\tag{7.58}$$

we obtain the differential equation that models the distribution of pressure in a turbulent lubricant film:

$$\frac{\partial}{\partial x}\left(\frac{h^3}{\mu k_x}\frac{\partial \bar{P}}{\partial x}\right) + \frac{\partial}{\partial z}\left(\frac{h^3}{\mu k_z}\frac{\partial \bar{P}}{\partial z}\right) = \frac{U_0}{2}\frac{\partial h}{\partial x} + V_0.\tag{7.59}$$

Here $-V_0 = V_1 - V_2$ is the squeeze velocity and $U_0 = U_2$, or U_1, depending on whether Eq. (7.59) is being used for a slider bearing or a journal bearing (see Section 2.2).

Predictions of Constantinescu's theory do not agree well with experimental data. Arwas and Sternlicht (1963) found that in order to correlate theory with experiment, the empirical constant k in Eq. (7.40) must be a function of the Reynolds number and the eccentricity ratio. In later publications, Constantinescu himself recommended $k = 0.125\mathrm{Re}_h^{0.07}$.

Constantinescu's turbulent lubrication theory may be criticized on the following grounds:

(1) The change from viscous sublayer to turbulent core is abrupt (discontinuity in shear stress), no account is taken of the buffer zone.

(2) The predicted thickness of the viscous sublayer does not agree with experimental findings. (For pure shear flow, we obtained $y_L^+ = 1/k \approx 2.5$, whereas measurements give twice this value.)

(3) The linearization $B_x = k_x(\mathrm{Re})U_{mp}$ is inaccurate at large B (i.e., the theory is not applicable for externally pressurized bearings or for self-acting bearings at large eccentricity ratios).

(4) Turbulent Couette flow has $U_{mp} = 0$; thus, its representation is not possible by Eq. (7.55).

(5) Orthogonal flows are decoupled.

This criticism is, of course, not intended to detract from Constantinescu's seminal work in turbulent lubrication.

7.4 Ng-Pan-Elrod Model

In an effort to construct a turbulent lubrication model that is consistent with channel flow data, Ng (1964) investigated the applicability of Reichardt's eddy diffusivity formulation, Eq. (7.34). For two-dimensional flow the Boussinesq hypothesis, Eq. (7.20), is

$$\frac{d\bar{U}^+}{dy^+} = \frac{1}{1 + \varepsilon_m/\nu}, \tag{7.60}$$

where the velocity has been made nondimensional with the shear velocity at the wall $v_* = (|\tau_w|/\rho)^{1/2}$. By integrating Eq. (7.60) with ε_m/ν as given by Reichardt's formula, and fitting the resulting velocity profile to experimental data in the $0 < y^+ < 1000$ range, Ng optimized Reichardt's constants and found the values

$$k = 0.4 \qquad \delta_\ell^+ = 10.7. \tag{7.61}$$

For $y^+ > 1000$, theoretical prediction and experimental data are at variance. Nevertheless, in his first paper Ng assumed that Eq. (7.34) applies not only in the constant-stress region but over the whole width of the channel. To take into account the presence of two boundaries, the flow is divided into two regions. In the upper layer, given by $y_1 < y < h$, the eddy viscosity is calculated with the upper wall shear stress τ_h. In the lower layer, defined by $0 < y < y_1$, the eddy viscosity is evaluated with the lower wall shear τ_0. It is required then that ε_m/ν be a unique function of $y^+ = yv_*/\nu$, where $v_* = (|\tau_h|/\rho)^{1/2}$ or $v_* = (|\tau_0|/\rho)^{1/2}$, depending on location, with the consequence that the nondimensional coordinate $\hat{y}_1 = y_1/h$ of the interface between upper and lower layers is given by

$$\hat{y}_1 = \frac{\sqrt{\tau_h}}{\sqrt{\tau_h} + \sqrt{\tau_0}}. \tag{7.62}$$

In a further development, the linearized theory of Ng and Pan (1965), Reichardt's eddy viscosity distribution is still assumed to hold over the whole channel width. Following a suggestion of Elrod's, however, ε_m/ν is now calculated with the local total shear, placing the boundary between upper and lower layers in the center plane of the channel. Thus, ε_m/ν remains isotropic and its derivative has a discontinuity at the channel center. Substituting

$$\tau_{xy} = \mu\frac{\partial \bar{U}}{\partial y} - \rho\overline{u'v'} = \mu\left(1 + \frac{\varepsilon_m}{\nu}\right)\frac{\partial \bar{U}}{\partial y} \tag{7.63a}$$

and

$$\tau_{zy} = \mu\frac{\partial \bar{W}}{\partial y} - \rho\overline{v'w'} = \mu\left(1 + \frac{\varepsilon_m}{\nu}\right)\frac{\partial \bar{W}}{\partial y} \tag{7.63b}$$

into Eq. (7.15), and integrating twice formally with respect to y, we obtain the velocity distribution

$$\bar{U} = \frac{1}{\mu}\left[\tau_{xy}\left(\frac{h}{2}\right)\int_0^y \frac{dy'}{1 + \varepsilon_m/\nu} + \frac{\partial \bar{P}}{\partial x}\int_0^y \frac{y' - h/2}{1 + \varepsilon_m/\nu}dy'\right], \tag{7.64a}$$

$$\bar{W} = \frac{1}{\mu}\left[\tau_{zy}\left(\frac{h}{2}\right)\int_0^y \frac{dy'}{1 + \varepsilon_m/\nu} + \frac{\partial \bar{P}}{\partial z}\int_0^y \frac{y' - h/2}{1 + \varepsilon_m/\nu}dy'\right]. \tag{7.64b}$$

Here $\tau_{xy}(h/2)$ and $\tau_{zy}(h/2)$ are integration constants to be determined from the boundary conditions.

Substituting Eqs. (7.64a) and (7.64b) into the continuity equation (7.7a) and integrating with respect to y would yield a differential equation in lubricant pressure. (A detailed derivation of this equation is given in Chapter 9 in a discussion on thermohydrodynamic theory.) Such an equation would be nonlinear, for to calculate $\bar{U}$ and $\bar{W}$ in Eqs. (7.64) one must already know the total shear $|\tau| = (\tau_{xy}^2 + \tau_{zy}^2)^{1/2}$. Thus, Eqs. (7.64) presuppose knowledge of the velocity field.

To avoid an iterative procedure, Ng and Pan made the assumption that the flow is a small perturbation of turbulent Couette flow

$$\tau_{xy} = \tau_c + \delta\tau_x \qquad \frac{\delta\tau_x}{\tau_c} \ll 1,$$

$$\tau_{zy} = \delta\tau_z \qquad \frac{\delta\tau_z}{\tau_c} \ll 1, \tag{7.65}$$

so that

$$|\tau| = \tau_c + \delta\tau_x + O\left(\delta\tau_x^2\right). \tag{7.66}$$

Here τ_c is the turbulent Couette stress and $\delta\tau_x$ and $\delta\tau_z$ are the perturbations of τ_c in the x and z directions, respectively.

But if τ is only a small perturbation of τ_c, the eddy viscosity that yields the shear stress τ is a small perturbation of the Couette flow eddy viscosity, and we may write

$$\frac{\varepsilon_m}{\nu}(\bar{y}; |\tau|) = \frac{\varepsilon_m}{\nu}(\bar{y}; \tau_c) + \frac{\partial(\varepsilon_m/\nu)}{\partial|\tau|}\bigg|_{|\tau|=\tau_c} \delta\tau_x + O(\delta\tau_x^2), \tag{7.67}$$

where $\bar{y} = y/h$ is the dimensionless normal coordinate. Using the notation

$$f_c(\bar{y}) = 1 + \frac{\varepsilon_m}{\nu}(\bar{y}; \tau_c)$$

$$= 1 + \kappa\left[\bar{y}h_c^+ - \delta_\ell^+ \tanh\left(\frac{\bar{y}h_c^+}{\delta_\ell^+}\right)\right], \tag{7.68}$$

where

$$h_c^+(x, z) = \frac{h}{\nu}\sqrt{\frac{|\tau_c|}{\rho}}$$

and

$$g_c(\bar{y}) = \tau_c \frac{\partial(\varepsilon_m/\nu)}{\partial|\tau|}\bigg|_{|\tau|=\tau_c}$$

$$= \frac{1}{2}\kappa\bar{y}h_c^+ \tanh^2\left(\frac{\bar{y}h_c^+}{\delta_\ell^+}\right), \tag{7.69}$$

Eq. (7.67) can be put in the abbreviated form

$$\frac{\varepsilon_m}{\nu}(\bar{y}; |\tau|) = f_c(\bar{y}) - 1 + g_c(\bar{y})\frac{\delta\tau_x}{\tau_c} + O(\delta\tau_x^2). \tag{7.70}$$

When

$$\left(1 + \frac{\varepsilon_m}{\nu}\right)^{-1} = \frac{1}{f_c(\bar{y})}\left(1 - \frac{g_c(\bar{y})}{f_c(\bar{y})}\frac{\delta\tau_x}{\tau_c}\right) + O(\delta\tau_x^2)$$

is substituted into Eqs. (7.64a) and (7.64b) and terms of order $(\delta\tau_x^2)$ are neglected, we obtain

$$U = \frac{\bar{U}}{U_*} = \frac{h\tau_c}{\mu U_*}\int_0^{\bar{y}}\frac{d\eta}{f_c(\eta)} + \left(\tau_{xy}\left(\frac{1}{2}\right) - \tau_c\right)\frac{h}{\mu U_*}\int_0^{\bar{y}}\frac{1}{f_c(\eta)}$$

$$\times\left(1 - \frac{g_c(\eta)}{f_c(\eta)}\right)d\eta + B_x\int_0^{\bar{y}}\frac{\frac{1}{2}-\eta}{f_c(\eta)}\left(1 - \frac{g_c(\eta)}{f_c(\eta)}\right)d\eta, \tag{7.71a}$$

$$W = \frac{\bar{W}}{U_*} = \frac{h\tau_{zy}(1/2)}{\mu U_*}\int_0^{\bar{y}}\frac{d\eta}{f_c(\eta)} + B_z\int_0^{\bar{y}}\frac{\frac{1}{2}-\eta}{f_c(\eta)}d\eta. \tag{7.71b}$$

Observing that in pure shear flow $B_x = 0$ and $\tau_{xy}(\frac{1}{2}) - \tau_c = 0$, we have

$$\frac{h\tau_c}{\mu}\int_0^1\frac{d\eta}{f_c(\eta)} = U_*. \tag{7.72}$$

Both $f_c(\bar{y})$ and $g_c(\bar{y})$ are symmetrical with respect to $\bar{y} = 1/2$. Thus, satisfaction of the remaining boundary conditions,

$$\bar{U} = U_0 \quad \text{at} \quad y = h, \qquad \bar{W} = 0 \quad \text{at} \quad y = h, \tag{7.73}$$

leads to

$$\tau_{xy}\left(\frac{1}{2}\right) = \tau_c, \qquad \tau_{zy}\left(\frac{1}{2}\right) = 0.$$

Equations (7.71a) and (7.71b) can be simplified to

$$U = \frac{1}{2} + \frac{(h_c^+)^2}{\mathrm{Re}_h} \int_{1/2}^{\bar{y}} \frac{d\eta}{f_c(\eta)} + B_x \int_0^{\bar{y}} \frac{\frac{1}{2} - \eta}{f_c(\eta)} \left(1 - \frac{g_c(\eta)}{f_c(\eta)}\right) d\eta, \tag{7.74a}$$

$$W = B_z \int_0^{\bar{y}} \frac{\frac{1}{2} - \eta}{f_c(\eta)} d\eta. \tag{7.74b}$$

Substituting for $\bar{U}$ and $\bar{W}$ in the continuity equation (7.7a) yields the linearized turbulent lubrication equation. This is formally identicalto Constantinescu's equation (7.59), but here the coefficients k_x and k_z are defined by

$$\frac{1}{k_x} \equiv G_x = \int_0^1 d\bar{y} \int_0^{\bar{y}} \frac{\frac{1}{2} - \eta}{f_c(\eta)} \left(1 - \frac{g_c(\eta)}{f_c(\eta)}\right) d\eta, \tag{7.75a}$$

$$\frac{1}{k_z} \equiv G_z = \int_0^1 d\bar{y} \int_0^{\bar{y}} \frac{\frac{1}{2} - \eta}{f_c(\eta)} d\eta. \tag{7.75b}$$

Both k_x and k_z depend on h_c^+ through Eqs. (7.68), (7.69), and (7.75). The local value of h_c^+, on the other hand, is dependent on the local film thickness through Eq. (7.72), which has the dimensionless form

$$\mathrm{Re}_h = \left(h_c^+\right)^2 \int_0^1 \frac{d\eta}{f_c(\eta)}, \tag{7.76}$$

The coefficients k_x and k_z depend, therefore, on the local Reynolds number. A good representation of this dependence was obtained by least-squares fitting of polynomials to Eqs. (7.75a) and (7.75b), with the results (Figure 7.4)

$$\frac{1}{k_x} = \begin{cases} \dfrac{1}{12} & \mathrm{Re}_h < 100 \\[2mm] \displaystyle\sum_n a_n (\log \mathrm{Re}_h)^{n-1} & 100 \le \mathrm{Re}_h < 10{,}000 \\[2mm] 0.014 - 0.0114(\log \mathrm{Re}_h - 4.0) & \mathrm{Re}_h > 10{,}000 \end{cases} \tag{7.77a}$$

$$\frac{1}{k_z} = \begin{cases} \dfrac{1}{12} & \mathrm{Re}_h < 100 \\[2mm] \displaystyle\sum_n b_n (\log \mathrm{Re}_h)^{n-1} & 100 \le \mathrm{Re}_h < 10{,}000 \\[2mm] 0.023 - 0.0182(\log \mathrm{Re}_h - 4.0) & \mathrm{Re}_h > 10{,}000 \end{cases} \tag{7.77b}$$

$$a_1 = -0.4489 \quad a_2 = 0.6703 \quad a_3 = -0.2904 \quad a_4 = 0.0502 \quad a_5 = -0.00306$$
$$b_1 = -0.3340 \quad b_2 = 0.4772 \quad b_3 = -0.1822 \quad b_4 = 0.02628 \quad b_5 = -0.001242.$$

To render the theory of Ng and Pan applicable even at large pressure gradients, Elrod and Ng substituted $\bar{U}$ and $\bar{W}$ from Eqs. (7.71a) and (7.71b) directly into the boundary condition Eqs. (7.73) and subsequently into the continuity equation (7.7a). They also removed the most objectionable component of the Ng and Pan model, namely, that Reichardt's eddy viscosity distribution is valid over the whole channel.

In the nonlinear model of Elrod and Ng (1967), Reichardt's formula is retained only in the constant stress region. In the core region the eddy, viscosity is assumed to be given by a constant value ε_c. This value is obtained from a generalized form of Clauser's formula (1965),

$$\varepsilon_c = \frac{1}{56} \int_0^{y_0} |U_0 - \bar{U}| \, dy, \tag{7.78}$$

where U_0 is the maximum value of $\bar{U}$ at the given $x = $ const. position.

Following the analysis of Elrod and Ng, we are again led to a pressure equation of the form of Eq. (7.59), but now $k_x = 1/G_x$ and $k_z = 1/G_z$ are nonlinear functions of the local values of the pressure gradient and the Reynolds number. For decreasing values of the pressure gradient, the functions G_x and G_z from nonlinear theory approach their respective values from the linearized theory.

The nonlinearity of the Elrod-Ng theory makes it applicable to flows at large values of the pressure gradient (externally pressurized bearings or self-acting bearings at large eccentricity), but because of the nonlinearity the solution must be iterative. This requirement renders the theory difficult to apply in comparison with the linearized theory of Ng and Pan.

7.5   Bulk Flow Model of Hirs

The bulk flow theory of turbulent lubrication by Hirs (1973) does not attempt to analyze turbulence in all its details. Instead it relies on an easily measured global characteristic of the flow, namely the relationship between average velocity and wall stress. In a hydrodynamic bearing, we encounter two basic flow types, pressure flow and shear flow. But more general flows, which result from the combined action of a pressure gradient and the sliding of one of the surfaces, are also encountered. All these flow types have to be considered by the theory [see the discussion following Eq. (7.42)].

Analyzing the then-available experimental data for pressure flow in a pipe, Blasius discovered a simple relationship between the wall stress and the Reynolds number that is calculated on the average velocity. This drag law, said to be valid for Re $\leq 10^5$, also applies to pressure flow between parallel plates separated by a distance h. In the latter case, it has the form

$$\frac{\tau_0}{\frac{1}{2}\rho U_a^2} = n_0(\text{Re})^{m_0}, \tag{7.79}$$

where Re $= U_a h/\nu$ is the Reynolds number based on the average velocity (Schlichting, 1968),

$$U_a = \frac{1}{h} \int_0^h \bar{U} \, dy.$$

In later years, while analyzing their own and also Couette's experimental results, Davies and White (1928) found a relationship between wall stress and average velocity of shear flow between two parallel surfaces. The drag law for Couette flow may be put into a form that is identical to the one obtained earlier by Blasius for pressure flow. Thus, following Hirs (1973), for shear flow between parallel plates a distance h apart we write

$$\frac{\tau_1}{\frac{1}{2}\rho U_a^2} = n_1 (\text{Re})^{m_1}. \tag{7.80}$$

In these equations, U_a represents the average value of the mean flow velocity relative to the surface on which the wall stress τ_1 is being evaluated. Thus, in Eq. (7.80) $U_a = U_*/2$, where U_* is the velocity of the sliding surface, when the equation is applied to the stationary surface, and $U_a = -U_*/2$ when applied to the sliding surface.[4]

For the particular case of Reynolds number equality between Eqs. (7.79) and (7.80), we have the following value for the ratio of Poiseuille wall stress to Couette wall stress

$$\frac{\tau_0}{\tau_1} = \frac{n_0}{n_1} (\text{Re})^{m_0 - m_1}$$

An exhaustive survey of available experimental data (Hirs, 1974) shows that $m_0 = m_1 = -0.25$ and $a = n_0/n_1 = 1.2$. Motivated by this relative insensitivity of the wall stress to the type of flow, Hirs assumed wall stress to be additive in the sense that if τ_1 is due to the relative velocity U_* of the surfaces and τ_0 is due to the pressure gradient $d\bar{P}/dx$, then the wall stress τ that is caused by the combined action of U_* and $d\bar{P}/dx$ can be calculated from $\tau = \tau_0 + a\tau_1$ on the stationary surface and $\tau = \tau_0 - a\tau_1$ on the sliding surfaces. Furthermore, the Couette shear τ_1 is approximately equal to $(n_1/n_0)\tau_0$.

A Couette flow of wall stress τ_1 is then equivalent, at least as far as Eq. (7.79) is concerned, to a pressure flow that is maintained by the fictitious pressure gradient

$$\frac{dP_1}{dx} = -\frac{2}{h} a\tau_1. \tag{7.81}$$

To arrive at this relationship, consider force equilibrium on a control volume of length Δx and height h under the action of pressure forces ph and $-(p + \Delta p)h$ and shear force $2a\tau_1$.

Utilizing the ideas above, Eqs. (7.79) and (7.80) may now be written for generalized channel flows. [Here and in what follows, we assume the bearing to be held stationary and put $U_1 = 0$. A slight generalization of ideas is required if the bearing surface is given a velocity different from zero. For details see Hirs (1973).]

Stationary surface

$$\frac{-h(d/dx)(\bar{P} + P_1)}{\rho U_a^2} = n_0 \left(\frac{U_a h}{\nu} \right)^{m_0} \tag{7.82a}$$

Sliding surface

$$\frac{-h(d/dx)(\bar{P} - P_1)}{\rho (U_a - U_*)^2} = n_0 \left(\frac{(U_a - U_*)h}{\nu} \right)^{m_0}. \tag{7.82b}$$

---

[4] The functions on the right-hand sides of Eqs. (7.79) and (7.80) must be odd functions. Alternatively, we may write Eqs. (7.79) and (7.80) in the more appropriate form: $2\tau_{0,1}/\rho U_a^2 = n_{0,1}(|\text{Re}|)^{m_{0,1}} \text{sgn}(U_a)$.

On eliminating the fictitious pressure gradient dP_1/dx between Eqs. (7.82a) and (7.82b), we obtain the actual pressure gradient in terms of the average velocity U_a and the relative velocity of sliding U:

$$\frac{d\bar{P}}{dx} = -\frac{n_0}{2}\left\{\frac{\rho U_a^2}{h}\left(\frac{U_a h}{\nu}\right)^{m_0} + \frac{\rho(U_a - U_*)^2}{h}\left[\frac{(U_a - U_*)h}{\nu}\right]^{m_0}\right\}. \tag{7.83}$$

It is worth noting that the magnitude of the weighing factor $a = n_0/n_1$ never enters Eq. (7.83) and thus has no effect on the pressure gradient.

Equation (7.83) is valid for unidirectional flow; that is, for flow in the direction of the representative pressure gradient. However, this direction need not coincide with the direction of relative velocity U [x direction in Eq. (7.83)], in which case there will be two component equations, one in the x direction and the other in the z direction.

The orthogonal components of Eq. (7.82a) are

$$\frac{-h(\partial/\partial x)(\bar{P} + P_1)}{\rho U_a S_a} = n_0\left(\frac{S_a h}{\nu}\right)^{m_0}, \tag{7.84a}$$

$$\frac{-h(\partial/\partial z)(\bar{P} + P_1)}{\rho W_a S_a} = n_0\left(\frac{S_a h}{\nu}\right)^{m_0}, \tag{7.84b}$$

where $S_a = (U_a^2 + W_a^2)^{1/2}$ is the magnitude of the average velocity vector $S_a = U_a i + W_a k$.

Raising all terms of Eqs. (7.84a) and (7.84b) to the second power and adding the results leads to the equation

$$\frac{-h(d/ds)(\bar{P} + P_1)}{\rho S_a^2} = n_0\left(\frac{S_a h}{\nu}\right)^{m_0}. \tag{7.85}$$

Equation (7.85) is identical, except for a slight difference in notation, to Eq. (7.82a), suggesting that our procedure for taking component equations (7.84a) and (7.84b) is correct.

When the orthogonal component equations are written for Eq. (7.82b) also, we have

$$\frac{-h(\partial/\partial x)(\bar{P} - P_1)}{\rho(U_a - U_*)S_b} = n_0\left(\frac{S_b h}{\nu}\right)^{m_0}, \tag{7.86a}$$

$$\frac{-h(\partial/\partial z)(\bar{P} - P_1)}{\rho W_a S_b} = n_0\left(\frac{S_b h}{\nu}\right)^{m_0}, \tag{7.86b}$$

where $S_b = [(U_a - U_*)^2 + W_a^2]^{1/2}$. If P_1 is eliminated between Eqs. (7.84a) and (7.86a) and between Eqs. (7.84b) and (7.86b), we obtain the pressure gradient in terms of the components of the average velocities, the velocity of the sliding surface, the film thickness, and the viscosity and density of the fluid.

The dimensionless pressure flow coefficients G_x and G_z can be obtained from

$$G_x = \frac{\frac{1}{2} - U_a/U_*}{(h^2/\mu U_*)(\partial\bar{P}/\partial x)} \qquad G_z = \frac{-W_a/U_*}{(h^2/\mu U_*)(\partial\bar{P}/\partial z)} \tag{7.87}$$

by substitution. If Couette flow predominates so that the following conditions apply,

$$G_x \frac{h^3}{\mu\nu} \frac{\partial \bar{P}}{\partial x} \ll \frac{1}{2}\mathrm{Re}_h \qquad G_z \frac{h^3}{\mu\nu} \frac{\partial \bar{P}}{\partial z} \ll \frac{1}{2}\mathrm{Re}_h,$$

then Eq. (7.87) reduces to

$$G_x = \frac{1}{2+m_0} G_z, \qquad G_z = \frac{2^{1+m_0}}{n_0} \mathrm{Re}_h^{-(1+m_0)}. \tag{7.88}$$

Hirs (1974) uses $n_0 = 0.066$ and $m_0 = -0.25$ for smooth surfaces and Re $\leq 10^5$. With these values, Eq. (7.88) gives the approximate formulas:

$$k_x \equiv \frac{1}{G_x} = 0.0687\,\mathrm{Re}_h^{0.75}, \tag{7.89a}$$

$$k_z \equiv \frac{1}{G_z} = 0.0392\,\mathrm{Re}_h^{0.75}. \tag{7.89b}$$

At high Reynolds numbers the predictions of Ng, Pan, and Elrod and those of Hirs are almost identical, and it is only in and near the transition regime that any significant discrepancy occurs. Neither of these models agrees completely in this region with the calculation of Ho and Vohr (1974), who used the Kolmogoroff-Prandtl energy model of turbulence, but Elrod and Ng come closest. Launder and Spalding (1972) caution against interpreting this as proof of the correctness of the Ng-Pan-Elrod model. According to these authors, consistently accurate prediction of internal flows under severe favorable pressure gradient can be achieved only if one calculates the length scale from a transport equation, rather than from an algebraic formula. In their calculations, Ho and Vohr (1974) estimated the length scale Λ in Eq. (7.70) from the van Driest formula, Eq. (7.71), as recommended by Wolfshtein (1969).

Of the theories discussed here, the Ng-Pan-Elrod theory and the bulk flow theory of Hirs are to be preferred over Constantinescu's model. There is, however, little to choose between the Ng-Pan-Elrod theory and the bulk flow theory. The latter may be extended to flow between rough or grooved surfaces as experimental data become available. Extension of the Ho-Vohr model to include fluid inertia is also possible. The effect of turbulence on journal bearing performance can be gauged from Figure 3.10 (Raimondi and Szeri, 1984). Thermal effects and turbulence are discussed in Chapter 9.

To illustrate the effect of turbulence, we calculate some of the performance characteristics of a $\beta = 160°$ fixed pad bearing, under the following conditions:

$$W = 10{,}000\,\mathrm{N}$$
$$N = 12{,}000\,\mathrm{rpm}$$
$$D = 0.1\,\mathrm{m}$$
$$L = 0.1\,\mathrm{m}$$
$$C = 0.0001\,\mathrm{m}$$
$$\mu = 8.2 \times 10^{-4}\,\mathrm{Pa\cdot s}$$
$$\nu = 8.23 \times 10^{-7}\,\mathrm{m^2/s}.$$

Table 7.1. *Bearing Performance Comparison*

	Laminar (Re $=$ 1)	Turbulent (Re $=$ 7,634)
Minimum film thickness (μm)	18.2	40.0
Attitude angle (deg.)	30	45
Viscous dissipation (W)	1520	4398

The specific pressure, Sommerfeld and Reynolds numbers are given by

$$P = \frac{W}{LD} = 10^6 \, \text{Pa}, \qquad S = \frac{\mu N}{P}\left(\frac{R}{C}\right)^2 = 0.041, \qquad \text{Re} = \frac{R\omega C}{\nu} = 7,634$$

(i)   Entering Figure 3.10 with the parameter values $S = 0.041$ and Re $= 1$, we obtain the column labeled "Laminar" in Table 7.1.

(ii) The actual parameter values $S = 0.041$ and Re $= 7,634$, on the other hand, yield the column labeled "Turbulent" in Table 7.1.

The real difference between turbulent and laminar predictions is not that, in reality, we require a 6 hp motor whereas laminar theory only asks for a 2 hp motor, but that most of the 4 hp difference between the two estimates is used up in raising the temperature of the lubricant; laminar theory is incapable of predicting this increased dissipation. For large bearings, such as employed in the power generation industry, as much as 600–800 hp is dissipated per bearing – if this is not taken into account at the design stage, bearing failure through overheating and seizure will result.

It was mentioned, when discussing laminar theory, that isothermal operation of bearings is a rarity and not the rule. This statement is even more true in the case of large bearings operating in the turbulent regime. In such cases there can be no excuse for designing on the assumption of isothermal operation, unless there is compelling evidence to support that assumption. Thermal effects in turbulent bearings are discussed in Chapter 9.

7.6   Nomenclature

A	modified local Reynolds number
B_x, B_z	components of dimensionless pressure gradient
C	radial clearance
D	bearing diameter
D_{ij}	stretching tensor
F_μ	friction force
G_x, G_z	turbulence functions
L	bearing axial length
N	shaft rotational speed
$\bar{P}$	mean turbulent pressure
P_1	fictitious pressure
R	journal radius

Re	global Reynolds number
Re_h	local Reynolds number
S	Sommerfeld number
S_A	average velocity
U_{mp}	component of mean pressure flow
U_{ms}	component of mean shear flow
a	shear–stress ratio
c_μ	friction coefficient
d'_{ij}	fluctuating component of stretching tensor
f_c, g_c	turbulent functions
h	film thickness
h_c^+	reduced film thickness
k	mixing length constant
k_x, k_z	resistance coefficients
ℓ	mixing length
m_0, m_1	exponents for pressure flow, shear flow
n_0, n_1	coefficients for pressure flow, shear flow
p	lubricant pressure
$q^2/2$	turbulent kinetic energy per unit mass
t	time
$u_i(u, v, w)$	velocity components
v_*	shear velocity
$x_i(x, y, z)$	ortogonal Cartesian coordinates
y_L	thickness of viscous sublayer
δ_{ij}	Kronecker delta
δ_ℓ^+	constant
ε	eccentricity ratio
ε	turbulent dissipation
$\varepsilon_m, \varepsilon_c$	eddy diffusivity, core eddy diffusivity
θ, η, ξ	orthogonal Cartesian coordinates
λ	coefficient
μ	dynamic viscosity
ν	kinematic viscosity
ρ	density
T_{ij}	stress tensor
τ_{ij}	apparent stress
τ_c	Couette stress
τ_w	wall stress
τ_0, τ_h	wall stress at $y = 0$, $y = h$
τ_0, τ_1	wall stress resulting from Poiseuille flow, Couette flow
$(\ )_1$	evaluated at surface 1 (reference surface)
$(\ )_2$	evaluated at surface 2
$(\ )_m$	average value
$(\bar{\ })$	mean turbulent quantities
$(\ )'$	fluctuating turbulent quantities

7.7   References

Arwas, E. B. and Sternlicht B. 1963. Analysis of plane cylindrical journal bearings in turbulent regime. *ASME Pap. 63-LUB-11*.

Black, H. F. and Walton, M. H. 1974. Theoretical and experimental investigations of a short 360° journal bearing in the transition superlaminar regime. *J. Mech. Eng. Sci.*, **16**, 287–297.

Burton, R. A. and Carper, H. J. 1967. An experimental study of annular flows with applications in turbulent film lubrication. *ASME Trans.*, **89**, 381–391.

Clauser, F. H. 1965. The turbulent boundary layer. *Adv. Appl. Mech.*, **4**, 1–51.

Constantinescu, V. N. 1959. On turbulent lubrication. *Proc. Inst. Mech. Eng.*, **173**, 881–889.

Constantinescu, V. N. 1962. Analysis of bearings operating in turbulent regime. *ASME Trans.*, **82**, 139–151.

Constantinescu, V. N. 1970. On the influence of inertia forces in turbulent and laminar self-acting films. *ASME Trans.*, **92**, 473–481.

Davies, S. T. and White, C. M. 1928. An experimental study of the flow of water in pipes of rectangular cross section. *Proc. Roy. Soc., London*, **A 92**, 119.

DiPrima, R. C. 1963. A note on the stability of flow in loaded journal bearings. *ASLE Trans.*, **6**, 249–253.

Elrod, H. G. and Ng, C. W. 1967. A theory for turbulent films and its application to bearings. *ASME Trans.*, **89**, 347–362.

Frene, J. 1977. Tapered land thrust bearing operating in both laminar and turbulent regimes. *ASLE Paper. 77-AM-8-4*.

Gardner, W. W. and Ulschmid, J. G. 1974. Turbulence effects in two journal bearing applications. *ASME Trans.*, **96**, 15–21.

Hinze, J. O. 1975. *Turbulence*, 2d ed. McGraw-Hill, New York.

Hirs, G. G. 1973. Bulk Flow Theory for Turbulence in Lubricant Films. *ASME Trans., Ser. F*, **95**, 137–146.

Hirs, G. G. 1974. A Systematic Study of Turbulent Film Flow. *ASME Trans., Ser. F*, **96**, 118–126.

Ho, M. K. and Vohr J. H. 1974. Application of Energy Model of Turbulence to Calculation of Lubricant Flows. *ASME Trans., Ser. F*, **96**, 95–102.

Jones, W. P. and Launder B. E. 1972. The prediction of laminarization with a two equation model of turbulence. *Int. J. Heat Mass Transfer*, **15**, 301–313.

Kettleborough, C. F. 1965. Turbulent and inertia flow in slider bearings. *ASLE Trans.*, **8**, 287–295.

Launder, B. E. and Spalding, D. B. 1972. *Lectures in Mathematical Models of Turbulence*. Academic Press, London.

Li, C. H. 1977. The effect of thermal diffusion on flow stability between two rotating cylinders. *ASME Trans.*, **99**, 318–322.

Monin, A. S. and Yaglom, A. M. 1973. *Statistical Fluid Mechanics*. Volume 1. MIT Press, Mass.

Ng, C. W. 1964. Fluid dynamic foundation of turbulent lubrication theory. *ASLE Trans.*, **7**, 311–321.

Ng, C. W. and Pan, C. H. T. 1965. A linearized turbulent Lubrication theory. *ASME Trans.*, **87**, 675–688.

Ng, K. H. and Spalding, D. B. 1972. Turbulence model for boundary layer near walls. *Phys. Fluids*, **15**, 20–30.

Prandtl, L. 1963. *The Mechanics of Viscous Fluids*. Division G, in Aerodynamic Theory (W. F. Durand, ed.). Vol. 3. Dover, New York.

Raimondi, A. A. and Szeri, A. Z. 1984. Journal and thrust bearings. In *Handbook of lubrication*, II (E. R. Booser, ed.). CRC Press, Boca Raton.

Robertson, J. M. 1959. On turbulent plane Couette flow. *Proc. 6th Midwestern Conf. Fluid Mech.* Austin, Texas, pp. 169–182.

Schlichting, H. 1968. *Boundary Layer Theory*, 6th ed. Pergamon, London.

Townsend, A. A. 1977. *The Structure of Turbulent Shear Flow*. 2d ed. Cambridge University Press, Cambridge.

Wolfshtein, M. 1969. The velocity and temperature distribution in one dimensional flow with turbulence augmentation and pressure gradient. *Int. J. Heat Mass Transfer*, **12**, 301–308.

CHAPTER 8

Elastohydrodynamic Lubrication

Elastohydrodynamic lubrication (EHL) is the name given to hydrodynamic lubrication when it is applied to solid surfaces of low geometric conformity that are capable of, and are subject to, elastic deformation. In bearings relying on EHL principles, the pressure and film thickness are of order 1 GP and 1 μm, respectively – under such conditions, conventional lubricants exhibit material behavior distinctly different from their bulk properties at normal pressure. In fact, without taking into account the viscosity-pressure characteristics of the liquid lubricant and the elastic deformation of the bounding solids, hydrodynamic theory is incapable of explaining the existence of continuous lubricant films in highly loaded gears and rolling-contact bearings. This is illustrated in the next section, by applying isoviscous lubrication theory to a rigid cylinder rolling on a plane.

When two convex, elastic bodies come into contact under zero load, they touch along a line (e.g., a cylinder and a plane or two parallel cylinders) or in a point (e.g., two spheres or two crossed cylinders). On increasing the normal contact load from zero, the bodies deform in the neighborhood of their initial contact and yield small, though finite, areas of contact; this deformation ensures that the surface stresses remain finite. For a *nominal line contact* the shape of the finite contact zone is an infinite strip, for a *nominal point contact* it is an ellipse. Nominal line contacts possess only one spatial dimension and are, therefore, easier to characterize than the two-dimensional nominal point contacts.

8.1 Rigid Cylinder Rolling on a Plane

The geometry of the cylinder-plane combination is shown in Figure 8.1. Let h_0 be the minimum separation between the infinitely long cylinder of radius R and the plane; then, at any angular position θ the film thickness is given by

$$h = -\frac{R}{n}(1 + n \cos \theta), \tag{8.1}$$

where $n = -R/(h_0 + R)$ is constant for given geometry. We note that Eq. (8.1) is of the same form as Eq. (3.29), which was derived for journal bearings, if only one puts $C = -R/n$ and $\varepsilon = n$. In consequence, many of the previously derived journal bearing formulas remain applicable.

For the (infinite) cylinder-plane geometry, the Reynolds equation (3.33) reduces to

$$\frac{d}{dx}\left(\frac{h^3}{\mu}\frac{dp}{dx}\right) = 6U_0\frac{dh}{dx}, \tag{8.2}$$

and its first integral is given by

$$\frac{dp}{dx} = 6\mu U_0 \frac{h - h_2}{h^3}. \tag{8.3}$$

267

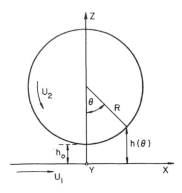

Figure 8.1. Cylinder-plane geometry and nomenclature.

We now introduce the transformation

$$\bar{p} = \frac{ph_0}{\mu U_0}, \qquad \bar{h} = \frac{hn}{R}, \qquad \theta = \sin^{-1}\left(\frac{x}{R}\right)$$

to obtain a nondimensional representation of the pressure gradient in Eq. (8.3),

$$\frac{d\bar{p}}{d\theta} = \frac{6h_0 n^2}{R}\left(\cos\theta\frac{\bar{h}(\theta) - \bar{h}(\theta_2)}{\bar{h}^3(\theta)}\right). \tag{8.4}$$

Implicit in this equation is the condition

$$\frac{d\bar{p}}{d\theta} = 0 \quad \text{at } \theta = \theta_2,$$

where θ_2 represents the as yet unknown position of the trailing edge liquid-cavity interface.
   The pressure distribution can be found formally from Eq. (8.4) by a second integration,

$$\bar{p} = \frac{6h_0 n^2}{R}\int_{\theta_1}^{\theta}\cos\theta\frac{\bar{h}(\theta) - \bar{h}(\theta_2)}{\bar{h}^3(\theta)}\,d\theta + B. \tag{8.5}$$

We subject this pressure distribution to the second of the Swift-Stieber boundary conditions, viz., $\bar{p}(\theta_2) = 0$, and find that θ_2 is determined by the condition

$$\int_{\theta_1}^{\theta_2}\cos\theta\frac{\bar{h}(\theta) - \bar{h}(\theta_2)}{\bar{h}^3(\theta)}\,d\theta = 0 \tag{8.6a}$$

or, when written in terms of the Sommerfeld angle ψ of Eq. (3.42), by the condition

$$\left[(\cos\psi_2 + n) - \sin\psi - \frac{1}{2}\sin\psi\cos\psi - \left(\frac{1}{2} + n\cos\psi_2\right)\psi\right]_{\psi_1}^{\psi_2} = 0, \tag{8.6b}$$

where, for a given ψ_1, Eq. (8.6b) serves to determine ψ_2.
   For simplicity, we assume that the continuous film commences at $x = -\infty$ and, therefore, put $\theta_1 = -\pi/2$. The corresponding $\psi_1 = \arccos(n)$ was found from Eq. (3.42a). Equation (8.6b) is now in the form $\psi_2 = \psi_2[-n/(n+1)]$, where $-n/(n+1) = R/h_0$. Table 8.1 lists ψ_2 for typical values of R/h_0. Once ψ_2 is known, $\bar{h}(\theta_2)$ can be calculated

Table 8.1. *Rigid cylinder on a plane*

R/h_0	ψ_2 (rad)	θ_2 (deg)	$\bar{w}'$
10	0.8520	11.3105	1.3985×10
10^2	0.8833	3.8200	2.2140×10^2
10^3	0.8868	1.2166	2.4103×10^3
10^4	0.8872	0.3850	2.4423×10^4
10^5	0.8887	0.1220	2.4444×10^5
10^6	0.8871	0.0385	2.4473×10^6

and substituted into Eq. (8.5) to obtain the pressure distribution. Having found the pressure distribution, we can evaluate the component of the lubricant force that acts normal to the plane. If w' represents this force per unit axial width, we have

$$w' = \int_{x_1}^{x_2} p \, dx = R \int_{\theta_1}^{\theta_2} p \cos \theta \, d\theta. \tag{8.7}$$

Substituting the dimensionless quantities defined above, integrating by parts as in Eqs. (3.78) and taking into account Eq. (8.4), Eq. (8.7) yields

$$\bar{w}' = -6n^2 \left[\int_{\theta_1}^{\theta_2} \frac{\sin\theta\cos\theta}{(1+n\cos\theta)^2} d\theta + \bar{h}(\theta_2) \int_{\theta_1}^{\theta_2} \frac{\sin\theta\cos\theta}{(1+n\cos\theta)^3} d\theta \right], \tag{8.8}$$

where we set $\bar{w}' = w'/\mu U_0$.

The integrals in Eq. (8.8) are easily evaluated when using either the Sommerfeld substitution, Eq. (3.42), or partial fractions. In terms of the Sommerfeld angle ψ, the result is

$$\bar{w}' = 6 \left[\frac{1 - n\cos\psi}{1-n^2} + \ln\frac{1-n^2}{1-n\cos\psi} + \frac{n^2}{2(1-n^2)^2} \frac{2n\cos\psi - \cos^2\psi}{1-n\cos\psi_2} \right]_{\psi_1}^{\psi_2}. \tag{8.9}$$

Table 8.1 contains $\bar{w}'$ as calculated from Eq. (8.9) at selected values of R/h_0. From the values in Table 8.1, we find that the approximate relationship

$$\frac{h_0}{R} = 2.44 \frac{\mu U_0}{w'} \tag{8.10}$$

holds over a wide range of R/h_0 and may serve as a means to calculate the minimum film thickness ratio for given mechanical input.

We now follow Dowson and Higginson (1977) and estimate the film thickness for a gear tooth contact. Typical values for the lubricant viscosity, surface speeds, and loads are

$$\mu = 0.075 \text{ Pa} \cdot \text{s}, \qquad\qquad U_1 = U_2 = 5.0 \text{ m/s},$$
$$U_0 = U_1 + U_2 = 10.0 \text{ m/s}, \qquad w' = 26.5 \text{ kN/cm}.$$

Substituting these values into Eq. (8.10), we find the dimensionless minimum film thickness to be

$$\frac{h_0}{R} = 0.69 \times 10^{-6}.$$

If $R = 2.5$ cm, assuming that in the neighborhood of contact the gear geometry is satis-
factorily represented by an "equivalent cylinder" of radius R rolling on a plane, we obtain

$$h_0 = 0.0172 \,\mu\text{m}. \tag{8.11}$$

This value is small in comparison with even the best surface finishes encountered in gear
manufacturing. It is almost two orders of magnitude smaller than the mean film thickness between two steel disks of $R = 3.81$ cm under a load of 1.96 kN/cm, as measured
by Crook (1958). Typical rms surface finish for ball bearings is 0.1 μm for the ball
and 0.25 μm for the raceway; for helical gears, ground and shaved, it is in the range
0.2–0.4 μm. Thus, our conclusion must be that classical hydrodynamic theory cannot
explain the existence of a continuous lubricant film in highly loaded contacts. We must,
therefore, look for appropriate extension of Reynolds' theory to explain continuous-film
lubrication of counter-formal contacts.

8.2   Elastohydrodynamic Theory

The above analysis of the cylinder-plane geometry was first performed by H. M.
Martin in 1916. Although Martin's negative conclusion on the existence of a continuous hydrodynamic film in highly stressed EHL contacts was completely in opposition
to experimental evidence, it, nevertheless, discouraged theoretical research on lubricated
counter-formal contacts for two decades.

The assumptions that limited Martin's analysis were (1) constant lubricant viscosity
and (2) rigid bounding surfaces. The first significant extension to classical hydrodynamic
theory came in 1936 when W. Peppler allowed the contacts to deform elastically; but Peppler
also put forth an erroneous proposition, viz., that the pressure cannot exceed the Hertzian
pressure of unlubricated contacts. The second extension, removal of the uniform, constant
viscosity constraint, came in 1945 when Gatcombe allowed the lubricant viscosity to change
with pressure. A good representation of the pressure dependence of viscosity of mineral
oils is given by the formula

$$\mu = \mu_0 \exp(\alpha p), \tag{8.12}$$

where μ_0 and α are material constants of the lubricant.[1]

Substituting for μ in Eq. (8.3) and integrating the resulting equation, we obtain

$$\frac{d\bar{q}}{d\theta} = \frac{6h_0 n^2}{R} \left(\frac{\bar{h} - \bar{h}_2}{\bar{h}^3} \cos\theta \right). \tag{8.13}$$

The (dimensionless) *reduced pressure*, defined by

$$\bar{q} = \frac{h_0}{\mu_0 U_0 \alpha}(1 - e^{-\alpha p}), \qquad q = \frac{\mu_0 U_0}{h_0}\bar{q}, \tag{8.14}$$

---

[1] The pressure-viscosity coefficient α of the Barus (1989) formula (8.12) is in the range $1.5 - 2.5 \times 10^{-8}$.
The lower end of the range is for "paraffinic" oils while the upper end for "naphtenic" oils (Jones et al.,
1975). However, this formula does not give good results at higher pressures, and application of the
Barus equation to pressures in excess of 0.5 MPa may lead to serious errors.

is thus shown to satisfy the same equation as the isoviscous pressure, Eq. (8.4). Any pressure distribution found for the constant viscosity case may, therefore, be readily used for pressure-dependent viscosity through the formula

$$
\begin{aligned}
p &= -\frac{1}{\alpha} \ln\left(1 - \bar{q}\,\frac{\mu_0 U_0 \alpha}{h_0}\right) \\
&= -\frac{1}{\alpha} \ln(1 - \alpha q).
\end{aligned}
\tag{8.15}
$$

Equation (8.15) yields infinitely large pressure at the position where $q = 1/\alpha$, as was first remarked by Blok (1950). The solid surfaces, of course, cannot sustain limitless pressures. They will deform elastically and our simple analysis breaks down; nevertheless, it points to the necessity of having to consider both the pressure dependence of the viscosity and the deformation of the surfaces when considering continuous film lubrication of highly loaded contacts. Figure 8.2 dramatizes the development of EHL theory. The individual contributions of elastic deformation of surfaces and of pressure dependence of lubricant viscosity are relatively modest. However, they combine nonlinearly to yield a much increased load capacity and a new film shape.

The first satisfactory solution to account for the effects of both elastic deformation and pressure dependence of viscosity was reported by A. N. Grubin in 1949. His work, and that

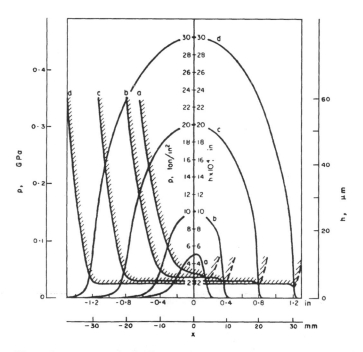

Figure 8.2. Pressure distribution and film shapes for same center line film thickness: (a) constant viscosity, rigid cylinders; (b) pressure-dependent viscosity, rigid cylinders; (c) constant viscosity, elastic cylinders; (d) pressure-dependent viscosity, elastic cylinders. (Reprinted by permission of the Council of the Institution of Mechanical Engineers from Dowson, D. and Higginson, G. R. A numerical solution to the elastohydrodynamic problem. *J. Mech. Eng. Sci.*, **1**, 6–15, 1959.)

of A. I. Petrushevich in 1951, established most of the essential properties of EHL solutions in concentrated contacts:

(1) The film is of almost uniform thickness over most of the contact zone. It displays, however, a typical and sudden decrease just upstream of the trailing edge.
(2) The pressure distribution curve follows the Hertzian ellipse over most of the contact zone.
(3) A sharp second pressure maximum exists (Figure 8.4), particularly at high speeds and light loads.

The first of these properties easily follows from Eq. (8.3), once it is acknowledged that $\mu = \mu(p)$. If μ is very large, and it might be several orders of magnitude larger than under atmospheric conditions, $(h - h_2)$ cannot deviate much from zero so as to constrain dp/dx from becoming excessive; this makes for an almost constant film thickness in the high-pressure zone. At the trailing edge, on the other hand, the pressure must rapidly drop to zero as here the Swift-Stieber conditions apply. To sustain the required large negative pressure gradient there, the film thickness must decrease sharply just upstream of the trailing edge.

The second and third properties of EHL contacts follow from the observation that at large load and small speed the film thickness is orders of magnitude smaller than the elastic deformation, thus the pressure distribution will be close to the Herzian distribution for dry contacts. At low load and high speed, on the other hand, the elastic deformation of the surfaces remains insignificant when compared to the film thickness, thus the pressure distribution will approach that of the rigid cylinder.

The physical parameters that are required to characterize the EHL problem can be conveniently combined into independent nondimensional parameters (groups); as the number of the resulting dimensionless groups is smaller than the number of physical parameters, this nondimensionalization of the problem facilitates presentation of the results. Assuming pure rolling and considering only nominal line contacts, Dowson and Higginson (1977) combined the physical variables of the isothermal EHL problem into four[2] convenient dimensionless parameters:

film thickness parameter $H = h_0/R$

load parameter $W = w/E'RL$

speed parameter $U = \mu_0\tilde{u}/E'R$

materials parameter $G = \alpha E'$

and through numerical solution of the appropriate equations, obtained the following relationship for the minimum film-thickness variable:

$$H = 1.6\frac{G^{0.6}U^{0.7}}{W^{0.13}}. \tag{8.16}$$

Here R is the effective radius at contact and L is its axial length, E' is the effective contact modulus, $\tilde{u} = U_0/2$ is the effective speed, and w is the load.

In practice the nondimensional parameters vary greatly, as $10^{-13} \leq U \leq 10^{-8}$, $3 \times 10^{-5} \leq W \leq 3 \times 10^{-4}$, and $2.5 \times 10^3 \leq G \leq 7.5 \times 10^3$.

---

[2] This set of nondimensional parameters must be augmented with the *ellipticity parameter*, $k = b/a$, when discussing nominal point contact. Here a and b are the semi-axes of the contact ellipse, expressible in terms of the principal curvatures of the contacting bodies. (The full definition of parameters for nominal point contact will be given later.)

Writing the Dowson-Higginson formula for minimum film thickness, Eq. (8.16), in the dimensional form, we obtain

$$h = 1.6 \frac{\alpha^{0.6} (\mu_0 \tilde{u})^{0.7} E'^{\,0.03} R^{0.43}}{w^{0.13}}. \tag{8.17}$$

Equation (8.17) indicates that the film thickness is virtually independent of the elastic modulus of the material and is only a weak function of the external load.

For complete solution of the isothermal EHL problem, we have to satisfy simultaneously the following equations:

(1)   The Reynolds equation,
(2)   The viscosity-pressure relationship,
(3)   The equations of elasticity.

It was discovered early on that the conventional iterative scheme represented by

PRESSURE

FILM SHAPE ← DEFORMATION

does not always converge. This is particularly true when there is large elastic deformation of the surfaces. To correct for this, Dowson and Higginson (1959) proposed the so-called *inverse hydrodynamic solution*. The essence of the inverse solution is calculation of the film shape that would produce a specified pressure distribution. This same pressure distribution is also employed to calculate surface deformations for plane strain. The remainder of the computation consists of systematic alteration of the pressure distribution to bring the film shape and surface deformation into agreement.

The inverse hydrodynamic method is essentially a one-dimensional method, but here it works well. As a starting point, we perform the indicated differentiation in Eq. (8.2) and obtain

$$h^3 \frac{d}{dx}\left(\frac{1}{\mu}\frac{dp}{dx}\right) - \frac{dh}{dx}\left(6U_0 - \frac{3h^2}{\mu}\frac{dp}{dx}\right) = 0.$$

The second term of this equation will vanish when

$$\frac{dh}{dx} = 0$$

or when

$$\frac{dp}{dx} = \frac{2\mu U_0}{h^2}$$

(the first condition defines the point of inflection $d^2 p/dx^2 = 0$ in an isoviscous lubricant).

Inspection of Figure 8.4 (below) will convince the reader that the first condition is satisfied at the point of minimum film thickness near the trailing edge, while the second condition is satisfied at a position $x = a$ within the inlet zone where the film thickness has the value

$$h_a = \sqrt{\frac{2\mu_a U_0}{(dp/dx)_a}}.$$

Substituting h_a into Eq. (8.3) yields

$$h_2 = \frac{2}{3}h_a. \tag{a}$$

The once-integrated Reynolds Eq. (8.3) may also be rearranged to yield

$$K\bar{h}^3 - \bar{h} + 1 = 0, \tag{b}$$

where

$$K = \frac{h_2^2}{6\mu U_0}\frac{dp}{dx}, \qquad \bar{h} = \frac{h}{h_2}.$$

Thus, knowing μ and U_0, the complete film shape can be found for an arbitrary dp/dx distribution from Eqs. (a) and (b). Details of the inverse hydrodynamic method are discussed by Dowson and Higginson (1959, 1977).

Figures 8.3 and 8.4 show pressure distributions and film shapes for two different material combinations, steel and mineral oil ($G = 5000$) and bronze and mineral oil ($G = 2500$).

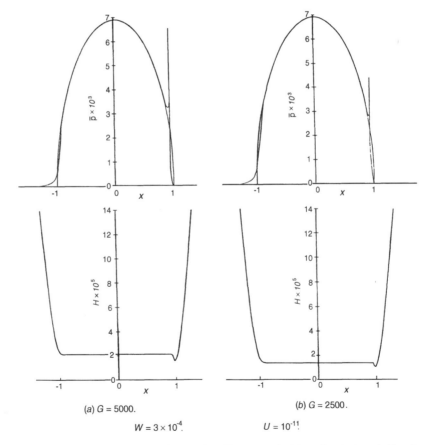

(a) $G = 5000$.   (b) $G = 2500$.

$W = 3 \times 10^{-4}$. $U = 10^{-11}$.

Figure 8.3. Pressure and film thickness in EHL contact for (a) steel and mineral oil and (b) bronze and mineral oil under high load, $\bar{p} = p/E'$, $H = h/R$. (Reprinted by permission of the Council of the Institution of Mechanical Engineers from Dowson, D. and Higginson, G. R. Effect of material properties on the lubrication of elastic rollers. *J. Mech. Eng. Sci.*, **2**, 188–194, 1960.)

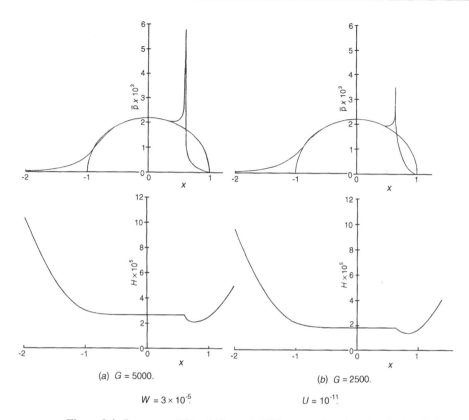

(a) $G = 5000$.   (b) $G = 2500$.

$W = 3 \times 10^{-5}$. $U = 10^{-11}$.

Figure 8.4. Pressure and film thickness in EHL contact for (a) steel and mineral oil and (b) bronze and mineral oil under low load, $\bar{p} = p/E'$, $H = h/R$. (Reprinted by permission of the Council of the Institution of Mechanical Engineers from Dowson, D. and Higginson, G. R. Effect of material properties on the lubrication of elastic rollers. *J. Mech. Eng. Sci.*, **2**, 188–194. 1960.)

These solutions were calculated by Dowson and Higginson (1960) using the inverse hydrodynamic method. The necking down of the film near the outlet, which might be as high as 25%, and the sharp second pressure maximum are easily noticeable in Figures 8.3 and 8.4. We note that the departure from the Hertzian pressure distribution is more pronounced at the lower load, consistent with our earlier reasoning.

The speed effect is well demonstrated by the curves of Figure 8.5, which were obtained for a liquid lubricant whose density was also allowed to change with pressure. The most significant effect of lubricant compressibility (not shown) is a slight lowering of the pressure peak and its displacement into the upstream direction from its constant density position.

A slight reduction of the pressure peak occurs also on introduction of thermal effects into the analysis, but only at higher speeds and moderate-to-high slip between rollers. At lower speeds, the pressure peak becomes more severe. The film thickness in the contact zone is not greatly influenced by temperature effects (Cheng and Sternlicht, 1965).

For pure rolling, Cheng (1965) and Cheng and Sternlicht (1965) reported no significant thermal effects on either the pressure level or the film thickness but found that the temperature has a major influence on the friction force. Kim and Sadeghi (1992), on the other hand, found that thermal effects can reduce film thickness by as much as 15%. Calculated and

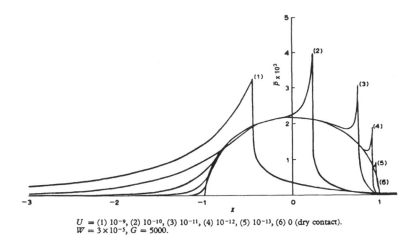

$U = (1)\ 10^{-9},\ (2)\ 10^{-10},\ (3)\ 10^{-11},\ (4)\ 10^{-12},\ (5)\ 10^{-13},\ (6)\ 0\ (\text{dry contact}).$
$W = 3 \times 10^{-5},\ G = 5000.$

Figure 8.5. Effect of surface velocity on pressure distribution in EHL contact, $\bar{p} = p/E'$. (Reprinted by permission of the Council of the Institution of Mechanical Engineers from Dowson, D., Higginson, G. R. and Whitaker, A. V. Elastohydrodynamic lubrication: A survey of isothermal solutions. *J. Mech. Eng. Sci.*, **4**, 121–126, 1962.)

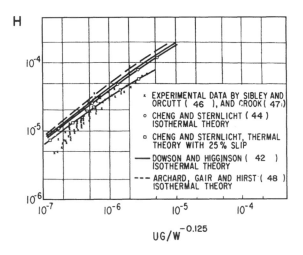

Figure 8.6. Comparison of predicted and measured film thickness variable in EHL contacts. (Reprinted with permission from Cheng, H. S. and Sternlicht, B. A numerical solution for the pressure, temperature and film thickness between two infinitely long lubricated rolling and sliding cylinders, under heavy loads. *ASME Trans., Ser. D*, **87**, 695–707, 1965.)

measured film thicknesses in EHL contacts are compared in Figure 8.6 (Dowson et al., 1959; Cheng and Sternlicht, 1965; Sibley and Orcutt, 1961; Crook, 1958; Archard et al., 1961).

We alluded previously to the fact that thin films and high contact pressures are not always found in, nor are essential to, elastohydrodynamic lubrication. The lubrication of natural and artificial biological joints (Tanner, 1966; Dowson, 1967), the viscous hydroplaning of automobile tires (Browne, Whicker, and Rohde, 1975), and the lubrication of compliant slider and journal bearings belong to this problem area.

Elastic deformation of surfaces can also cause significant effects in conformal bearings. Carl (1964) found that elastic distortion of journal bearings leads to increased film thickness and reduced peak pressure at low loads. Benjamin and Castelli (1971) were the first to treat finite compliant bearings analytically and found severe change in the journal locus, as compared to rigid bearings. Oh and Huebner (1973) applied the finite-element method to a finite bearing. Oh and Rohde (1977) extended finite-element analysis to compressible finite journal bearings. In another paper, Rohde and Oh (1975) presented the first complete thermoelastohydrodynamic analysis of a finite slider. An excellent review of pre-1978 work was given by Rohde (1978). For more recent work, see Mittwollen and Glienicke (1990), Bouchoule et al., (1996), and Monmousseau et al., (1996), as referenced in Chapter 9.

8.3   Contact Mechanics

Consider two convex bodies, $\bar{B}_1$ and $\bar{B}_2$, which make contact in a single point, O, under vanishing load. We intend to find out what happens at the contact as the load is increased. To facilitate this discussion, we replace our very general bodies $\bar{B}_1$, $\bar{B}_2$ by the ellipsoids, B_1, B_2, that are illustrated in Figure 8.7. The ellipsoid B_1 has the same principal curvatures, r_{1x} and r_{1y}, as the body $\bar{B}_1$ at O, while the ellipsoid B_2 possesses the same principal curvatures, r_{2x} and r_{2y}, as $\bar{B}_2$ at O. The principal curvatures, r_{1x} and r_{2x}, lie in the

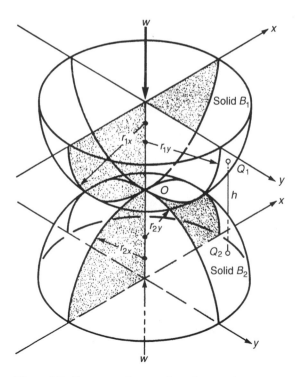

Figure 8.7. Geometry of counterformal contact between two convex bodies. (Hamrock, B. J. and Dowson, D. 1981. *Ball Bearing Lubrication.* Copyright John Wiley & Sons. © 1981. Reprinted by permission of John Wiley & Sons, Inc.)

same[3] plane $x = 0$ and r_{1y}, r_{2y} lie in the plane $y = 0$. The surfaces of the ellipsoids near the point of contact can be represented by the equations

$$
\begin{aligned}
z_1 &= A_1 x^2 + A_2 xy + A_3 y^2, \\
z_2 &= B_1 x^2 + B_2 xy + B_3 y^2.
\end{aligned}
\tag{8.18}
$$

The distance, h, between two points, such as Q_1 and Q_2 in Figure 8.7, is then

$$
h = z_1 + z_2 = (A_1 + B_1)x^2 + (A_2 + B_2)xy + (A_3 + B_3)y^2.
\tag{8.19}
$$

We can always choose directions for x and y such as to make the coefficient of the product xy in Eq. (8.19) vanish. This way we obtain

$$
h = z_1 + z_2 = Ax^2 + By^2 = \frac{1}{2R_x}x^2 + \frac{1}{2R_y}y^2.
\tag{8.20}
$$

Here R_x, R_y are the principal relative radii of curvature, defined by

$$
\frac{1}{R_x} = \left(\frac{1}{r_{1x}} + \frac{1}{r_{2x}} \right), \qquad \frac{1}{R_y} = \left(\frac{1}{r_{1y}} + \frac{1}{r_{2y}} \right).
$$

The contours of constant gap h between the undeformed surfaces are the ellipses

$$
\frac{x^2}{(\sqrt{2hR_x})^2} + \frac{y^2}{(\sqrt{2hR_y})^2} = 1,
\tag{8.21}
$$

the ratio of whose axes are given by $(R_y/R_x)^{1/2}$.

If now we press the bodies together, they deform in the neighborhood of the point of their first contact, O, and touch over a finite area. The contact area also has the shape of an ellipse, the ratio of whose semi-axes, b/a, depends on $(R_y/R_x)^{1/2}$ alone. But $\kappa = b/a$ equals $(R_y/R_x)^{1/2}$ only in the limit $R_y/R_x \to 1$ (Johnson, 1992).[4] The ratio of the semi-axes,

$$
\kappa = \frac{b}{a},
$$

is called the ellipticity parameter.[5]

Two additional indicators of contact geometry that will be used in the sequel are
Curvature sum:

$$
S_c = \frac{1}{\mathcal{R}} = \frac{1}{R_x} + \frac{1}{R_y}
\tag{8.22}
$$

Curvature difference:

$$
D_c = \frac{1}{S_c}\left[\left(\frac{1}{r_{1x}} - \frac{1}{r_{1y}} \right) + \left(\frac{1}{r_{2x}} - \frac{1}{r_{2y}} \right) \right]
$$

---

[3] For the general case when the principal radii of curvature are not aligned, see Timoshenko and Goodier (1951).

[4] The contact ellipse is somewhat skinnier than the ellipse $h = $ const., Eq. (8.21).

[5] We select x as the direction of relative motion, so $2b$ is the axis of the contact ellipse normal to the direction of fluid entrainment.

It was shown by Hertz (1881) that the pressure distribution over the elliptical contact is given by (Johnson, 1992)

$$p = p_H \left[1 - \left(\frac{x}{a} \right)^2 - \left(\frac{y}{b} \right)^2 \right] \tag{8.23a}$$

and

$$p_H = \frac{3w}{2\pi ab}. \tag{8.23b}$$

Here a and b are the lengths of the semi-axes of the contact ellipse, w is the load, and p_H is the maximum value of p.

As mentioned earlier, the values of a and b depend only on geometry; they must be determined before the pressure can be evaluated from Eqs. (8.23). It was Harris (1991) who showed how to calculate a and b. Knowing the curvatures $1/R_x$, $1/R_y$, we can determine the curvature difference, D_c, in Eq. (8.22). D_c, on the other hand, can be expressed purely in terms of the ellipticity parameter, κ as

$$D_c = \frac{(\kappa^2 + 1)\mathcal{E} - 2\mathcal{F}}{(\kappa^2 - 1)\mathcal{E}}, \tag{8.24}$$

where $\mathcal{E}$ and $\mathcal{F}$ are given by are complete elliptic integrals of the second and first kind, respectively:

$$\mathcal{E} = \int_0^{\pi/2} \left[1 - \left(1 - \frac{1}{\kappa^2} \right) \sin^2 \phi \right]^{1/2} d\phi, \tag{8.25a}$$

$$\mathcal{F} = \int_0^{\pi/2} \left[1 - \left(1 - \frac{1}{\kappa^2} \right) \sin^2 \phi \right]^{-1/2} d\phi. \tag{8.25b}$$

Equation (8.24) is, thus, of the form

$$D_c = f(\kappa). \tag{8.26a}$$

Equation (8.26a) enables us to find the value of the curvature difference, D_c, for given value of the ellipticity parameter κ. But this is not what we need. We would like to calculate κ for given value of D_c, i.e., we require the inverse of Eq. (8.26a)

$$\kappa = f^{-1}(D_c). \tag{8.26b}$$

Once in posession of this inverse relationship, Eq. (8.26b), we can, for given D_c, calculate the ellipticity parameter κ, and with κ in hand determine the contact ellipse semi-axes a, b, and the deflection δ according to (Harris, 1991)

$$a = \left[\frac{6\kappa^2 \mathcal{E} w \mathcal{R}}{\pi E'} \right]^{1/3}, \tag{8.27a}$$

$$b = \left[\frac{6\mathcal{E} w \mathcal{R}}{\pi \kappa E'} \right]^{1/3}, \tag{8.27b}$$

$$\delta = \mathcal{F} \left[\frac{9}{2\mathcal{E}\mathcal{R}} \left(\frac{w}{\pi \kappa E'} \right)^2 \right]^{1/3}, \tag{8.27c}$$

and the stress distribution from Eqs. (8.23). Here

$$\frac{1}{E'} = \frac{1}{2}\left[\frac{1-\nu_1^2}{E_1} + \frac{1-\nu_2^2}{E^2}\right]$$

as before, Eq. (1.13).

Unfortunately, $f(\kappa)$ is a transcendental function, thus its inverse in Eq. (8.26b) must be obtained numerically (Hamrock and Anderson, 1973). Here we shall adopt, instead, the approximation

$$\kappa \approx \bar{\kappa} = 1.0339\left(\frac{R_y}{R_x}\right)^{0.636} \tag{8.28a}$$

obtained by Brewe and Hamrock (1977) by curve fitting. In place of Eq. (8.28a), some authorities prefer

$$\bar{\kappa} = \alpha_r^{2/\pi}, \qquad \alpha_r = \left(\frac{R_y}{R_x}\right). \tag{8.28b}$$

Brewe and Hamrock (1977) also supply us with an excellent approximation to the elliptic integrals in Eqs. (8.25):

$$\mathcal{E} \approx \bar{\mathcal{E}} = 1.0003 + \frac{0.5968}{\alpha_r}, \tag{8.29a}$$

$$\mathcal{F} \approx \bar{\mathcal{F}} = 1.5277 + 0.6023\ln\alpha_r. \tag{8.29b}$$

Later Hamrock and Brewe (1983) recommended the simpler forms

$$\bar{\mathcal{E}} = 1 + \left(\frac{\pi}{2} - 1\right)\Big/\alpha_r,$$

$$\bar{\mathcal{F}} = \frac{\pi}{2} + \left(\frac{\pi}{2} - 1\right)\ln\alpha_r.$$

Using Eqs. (8.28) and (8.29) in Eqs. (8.27), we obtain the approximate values $\bar{a}$, $\bar{b}$, and $\bar{\delta}$. This approximation can be used with confidence for $0.01 \le \alpha_r \le 100$.

By taking $\kappa \to \infty$, we get a *nominal line contact* that, on increasing the load, w, from zero, develops into an infinite-strip contact area of width $2b$. Then

$$p = p_H\left[1 - \left(\frac{y}{b}\right)^2\right]^{1/2} \tag{8.30a}$$

and

$$p_H = \frac{2w'}{\pi b} \tag{8.30b}$$

where w' is the load per unit length of the strip. The half-width of the contact area can be calculated from

$$b = \left(\frac{8w'R_x}{\pi E'}\right)^{1/2}. \tag{8.30c}$$

Table 8.2. *Parameters employed in EHL design formulas*

	Type of contact	
Parameter	Point	Line
Film Thickness, H	h/R_x	h/R_x
Load, W	$w/R_x^2 E'$	$w/R_x L E'$
Speed, U	$\mu_0 V/E' R_x$	$\mu_0 \tilde{u}/E' R_x$
Materials, G	$\alpha E'$	$\alpha E'$
Ellipticity, κ	b/a	

8.4   Nondimensional Groups

If we wish to design for continuous lubricant film in a counter-formal contact, we must be able to predict the minimum film thickness. If the predicted film thickness is large in comparison with the rms surface finish, and if sufficient amount of lubricant is made available, the chances are that on application a continuous EHL film will result. If the conditions for continuous film are not satisfied, partial EHL or boundary lubrication will be obtained.

Having identified the minimum film thickness as the primary parameter in EHL design, we will employ the formulas that are currently in use to calculate it.

It was indicated earlier, Eq. (8.16), that dimensional analysis of the EHL problem yields

$$\phi(H, U, W, G, \kappa) = 0. \tag{8.31}$$

Here we extended Eq. (8.16a) by including the ellipticity parameter κ so as to make Eq. (8.31) applicable also to point contacts. For the two types of contacts, viz., nominal line and nominal point, the parameters in Eq. (8.31) have somewhat different definition, as shown in Table 8.2. Here

$$\frac{1}{R_x} = \frac{1}{r_{1x}} + \frac{1}{r_{2x}}, \qquad \frac{1}{E'} = \frac{1}{2}\left[\frac{1-\nu_1^2}{E_1} + \frac{1-\nu_2^2}{E_1}\right],$$
$$\tilde{u} = (u_1 + u_2)/2, \qquad V = (u^2 + v^2)^{1/2}.$$

According to Eq. (8.31), the film thickness variable, H, depends on four independent nondimensional groups $U, W, G,$ and κ. This is a great improvement over a primitive variable representation and, as usual, dimensional analysis facilitates the presentation of results. However, tabulation of H as a function of four nondimensional groups is still a formidable task. Fortunately, it has been noticed by a number of researchers that, with little sacrifice to accuracy of representation, the number of nondimensional groups involved in determining the film thickness can be reduced by one, when employing a new set of nondimensional groups. Elements of this new set are constructed by combining the elements of the set in Eq. (8.31); for nominal point contact they are:

$$\left.\begin{aligned} g_H &= \left(\frac{W}{U}\right)^2 H \\ g_V &= \left(\frac{GW^3}{U^2}\right)^2 \\ g_E &= W^{8/3}U^2 \\ \kappa &= 1.0339\left(\frac{R_y}{R_x}\right)^{0.636} \end{aligned}\right\} \quad \text{nominal point contact} \qquad (8.32)$$

Besides reducing the number of parameters of the problem and thereby facilitating presentation of results, a further advantage in using Eqs. (8.32) is that the new parameters are easily identified with the main characteristics of EHL. Thus, g_H represents film thickness, g_V represents deviation from isoviscosity, g_E represents significance of elastic deformation, and κ represents the geometry of the contact.

We now have the relationship

$$g_H = \Psi(g_V, g_E, \kappa). \qquad (8.33)$$

Our task now is to find the function $\Psi(g_V, g_E, \kappa)$ over practical ranges of its arguments. This task could be accomplished in one of two ways: (1) analytically, by simultaneous solution of all the relevant equations of the system or (2) numerically, by solving the EHL problem for a large number of inputs (g_V, g_E, κ) and then curve fitting to the supersurface $g_H = \Psi(g_V, g_E, \kappa)$ in four-dimensional parameter space.

Lubrication Regimes

Analytical solution of Eq. (8.33) is clearly out of the question, and we proceed by the numerical method (the details of these calculations are presented in the next section). But before doing that, let us first examine asymptotic cases $g_V = g_E = 0$; $g_E = 0$, $g_V \neq 0$; and $g_V = 0$, $g_E \neq 0$, leaving the case $g_V \neq 0$, $g_E \neq 0$, for last.

(1) $g_V = 0$, $g_E = 0$: Rigid-Isoviscous Regime

In this lubrication regime the pressure is low enough that it leaves for the viscosity unaltered and is unable to cause significant elastic deformation of the surfaces. This is the condition encountered in hydrodynamic journal and thrust bearings (see Chapter 3), and, to a lesser extent, in lightly loaded counter-formal contacts. It is the regime of classical hydrodynamic lubrication.

(2) $g_V \neq 0$, $g_E = 0$: Rigid-Piezoviscous Regime

We encounter this lubrication regime in applications that exhibit pressures high enough to effectively change the lubricant's viscosity from its inlet value, yet not so high as to initiate significant elastic deformation in the bearing material.[6]

(3) $g_E \neq 0$, $g_V = 0$: Elastic-Isoviscous Regime

Though this regime is characterized by significant elastic deformation, the pressure is not high enough to affect the viscosity of the particular lubricant employed. An example

---

[6] We should recognize that, by judiciously changing the pressure-viscosity coefficient, α, and the effective Young's modulus, E', we could relocate our process from the $g_V \neq 0$, $g_E = 0$ regime to the $g_V = 0$, $g_E \neq 0$ regime.

for lubrication in the elastic-isoviscous regime is human and animal joints lubricated by sinovial fluid (Dowson and Wright, 1981). Operation in this regime is often termed *soft EHL*. Automobile tires and elastomeric machine elements also operate in this regime.

(4) $g_E \neq 0$, $g_V \neq 0$: *Elastic-Piezoviscous Regime*

This is the regime of full EHL, also called *hard EHL*. The elastic deformation of the surfaces can be orders of magnitude larger than the thickness of the film and the lubricant viscosity can be orders of magnitude higher than its bulk value.

We note here that the parameters g_V and g_E differ in nominal point contacts from that in nominal line contacts. Definition (8.31) is valid for nominal point contacts.

In nominal line contacts, we drop the ellipticity parameter from the list of nondimensional variables and define

$$\left. \begin{aligned} g_H &= \left(\frac{W}{U}\right) H \\[6pt] g_V &= \frac{W^{3/2} G}{U^{1/2}} \\[6pt] g_E &= \frac{W}{U^{1/2}} \end{aligned} \right\} \quad \text{nominal line contact.} \tag{8.34}$$

The various lubrication regimes are depicted in Figures 8.8 to 8.11.

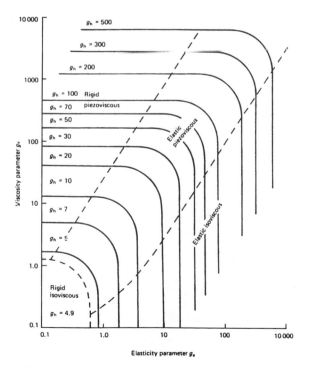

Figure 8.8. Map of lubrication regimes for nominal line. (Reprinted with permission from Arnell, R. D., Davies, P. B., Halling, J. and Whomes, T. L. *Tribology Principles and Design Applications.* Copyright Springer Verlag, © 1991.)

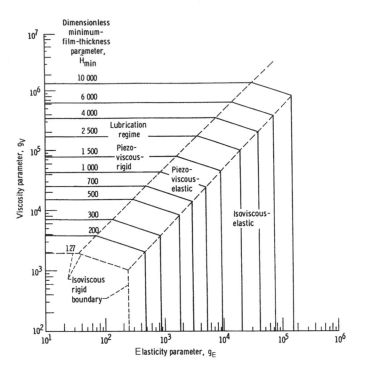

Figure 8.9. Map of lubrication regimes for nominal point contact, $\kappa = 1$. (Hamrock, B. J. and Dowson, D. 1981. *Ball Bearing Lubrication.* Copyright John Wiley & Sons, © 1981. Reprinted by permission of John Wiley & Sons, Inc.)

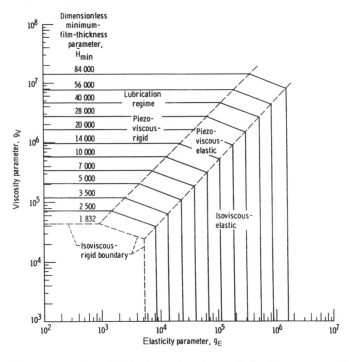

Figure 8.10. Map of lubrication regimes for nominal point contact, $\kappa = 3$. (Hamrock, B. J. and Dowson, D. 1981. *Ball Bearing Lubrication.* Copyright John Wiley & Sons, © 1981. Reprinted by permission of John Wiley & Sons, Inc.)

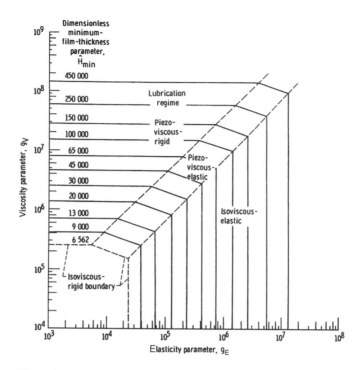

Figure 8.11. Map of lubrication regimes for nominal point contact, $\kappa = 6$. (Hamrock, B. J. and Dowson, D. 1981. *Ball Bearing Lubrication.* Copyright John Wiley & Sons, © 1981. Reprinted by permission of John Wiley & Sons, Inc.)

In the following, we list the formulas applicable in the various regimes:

Film-Thickness Design Formulas

Nominal Line Contact (Arnell, Davis, Halling, and Whomes, 1991)
(1) *Rigid-Isoviscous Regime:*

$$g_{H_{min}} = 2.45.$$ (8.35a)

(2) *Rigid-Piezoviscous Regime:*

$$g_{H_{min}} = 1.05 g_V^{2/3}.$$ (8.35b)

(3) *Elastic-Isoviscous Regime:*

$$g_{H_{min}} = 2.45 g_E^{0.8}.$$ (8.35c)

(4) *Full EHL Regime:*

$$g_{H_{min}} = 1.654 g_V^{0.54} g_E^{0.06}.$$ (8.35d)

Nominal Point Contact (Hamrock, 1990)

(1) *Rigid-Isoviscous regime:*

$$g_{H_{\min}} = 128\alpha_r \lambda_b^2 \left[0.131 \tan^{-1} \left(\frac{\alpha_r}{2} \right) + 1.683 \right]^2,$$ (8.36a)

where

$$\lambda_b = \left(1 + \frac{2}{3\alpha_r} \right)^{-1}.$$

(2) *Rigid-Piezoviscous regime:*

$$g_{H_{\min}} = 141 g_V^{0.375} [1 - e^{-0.0387\alpha_r}].$$ (8.36b)

(3) *Elastic-Isoviscous regime:*

$$g_{H_{\min}} = 8.70 g_E^{0.67} [1 - 0.85 e^{-0.31\kappa}].$$ (8.36c)

(4) *Full EHL regime*

$$g_{H_{\min}} = 3.42 g_V^{0.49} g_E^{0.17} [1 - e^{-0.68\kappa}].$$ (8.36d)

Comparison of Figures 8.9, 8.10, and 8.11 shows the strong effect the value of elliptic eccentricity, κ, has on regime boundaries.

A valuable aid for design of line contacts has been published by ESDU (1985): Film Thickness in Lubricated Hertzian Contacts (EHL): Part I, Item No. 85027, London. See also Harris (1991) and Hamrock (1991).

In the next section, we give details of a numerical scheme that makes calculation of Eq. (8.31) possible, leading to the formulas (8.35) and (8.36).

8.5   Analysis of the EHL Problem

The full EHL problem does not yield to analytical solutions and numerical methods must be employed. Prior to discussing the relevant numerical methods, we will list the set of equations and boundary conditions that must be satisfied simultaneously by any solution of the problem. This set contains equations of fluid dynamics, equations of elasticity, and state equations for the fluid that characterize its viscosity-pressure and density-pressure behavior. We will not list energy conservation equations, however, and deal here only with isothermal processes. As another simplification, we will consider nominal line contact only. The analysis of nominal line contacts is made simpler by the fact that it employs only ordinary differential equations; it does, however, retain most of the features of the EHL problem. For further study, the interested reader should consult the various articles that have appeared in the ASME Journal of Tribology, as well as Ball Bearing Lubrication by Hamrock and Dowson (1981).

Elastic Deformation

Figure 8.12 epicts a semi-infinite elastic body, located in $z > 0$ and loaded along its y-axis. The components of strain resulting from the uniform line loading w' are related

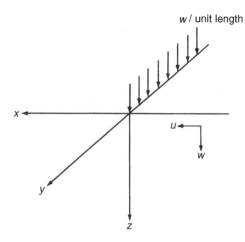

Figure 8.12. Line loading of elastic half-space.

to the components of stress by Hook's law. For the normal components of strain, we have

$$\varepsilon_{xx} = \frac{\partial u}{\partial x} = \frac{1}{E}[T_{xx} - \nu(T_{yy} + T_{zz})], \tag{8.37a}$$

$$\varepsilon_{yy} = \frac{\partial v}{\partial y} = \frac{1}{E}[T_{yy} - \nu(T_{xx} + T_{zz})], \tag{8.37b}$$

$$\varepsilon_{zz} = \frac{\partial w}{\partial z} = \frac{1}{E}[T_{zz} - \nu(T_{xx} + T_{yy})]. \tag{8.37c}$$

Here (u, v, w) are the components of the displacement of the solid, E is Young's modulus of elasticity, and ν is the Poisson ratio. The quantities depicted by $\varepsilon_{xx}, \varepsilon_{yy}, \varepsilon_{zz}$ and T_{xx}, T_{yy}, T_{zz} are the normal components of strain and stress, respectively.

The mixed components of strain are related to the mixed components of stress through the shear modulus, $G = E/2(1 + \nu)$, as

$$\varepsilon_{xy} = \frac{\partial u}{\partial y} + \frac{\partial v}{\partial x} = \frac{1}{G}T_{xy}, \tag{8.38a}$$

$$\varepsilon_{yz} = \frac{\partial v}{\partial z} + \frac{\partial w}{\partial y} = \frac{1}{G}T_{yz}, \tag{8.38b}$$

$$\varepsilon_{zx} = \frac{\partial w}{\partial x} + \frac{\partial u}{\partial z} = \frac{1}{G}T_{zx}. \tag{8.38c}$$

This three-dimensional state of strain can be made simpler by assuming that plane sections defined by $y = $ const. remain plane and will not be displaced in the y direction during loading. But then

$$\varepsilon_{yz} = 0, \qquad \varepsilon_{xy} = 0, \qquad \varepsilon_{yy} = 0. \tag{8.39}$$

Substituting the third of these conditions into Eq. (8.37b) yields

$$T_{yy} = \nu(T_{xx} + T_{zz}), \tag{8.40}$$

so for the plain strain problem Eqs. (8.37) and (8.38) reduce to

$$\frac{\partial u}{\partial x} = \frac{1 - v^2}{E} T_{xx} - \frac{v(1 + v)}{E} T_{zz}, \tag{8.40a}$$

$$\frac{\partial w}{\partial z} = \frac{1 - v^2}{E} T_{zz} - \frac{v(1 + v)}{E} T_{xx}, \tag{8.40b}$$

$$\frac{\partial u}{\partial z} + \frac{\partial w}{\partial x} = \frac{1}{G} T_{xz}. \tag{8.40c}$$

The displacements u, w can be obtained from Eqs. (8.40a) and (8.40b) by integration:

$$u = \frac{1 - v^2}{E} \int T_{xx} \, dx - \frac{v(1 + v)}{E} \int T_{zz} \, dx + f_1(z), \tag{8.41a}$$

$$w = \frac{1 - v^2}{E} \int T_{zz} \, dz - \frac{v(1 + v)}{E} \int T_{xx} \, dz + f_2(x). \tag{8.41b}$$

For the situation depicted in Figure 8.12, the components of stress can be obtained from the stress functions of Boussinesq (Timoshenko and Goodier, 1951; Dowson and Higginson, 1977),

$$\phi = -\frac{w'}{\pi} x \tan^{-1}\left(\frac{x}{z}\right), \tag{8.42}$$

where w' is the load per unit width. The stress function, ϕ, satisfies the compatibility condition $\nabla^4 \phi = 0$ and yields

$$T_{xx} = \frac{\partial^2 \phi}{\partial z^2} = -\frac{2w'}{\pi} \frac{x^2 z}{(x^2 + z^2)^2}, \tag{8.43a}$$

$$T_{zz} = \frac{\partial^2 \phi}{\partial x^2} = -\frac{2w'}{\pi} \frac{z^3}{(x^2 + z^2)^2}, \tag{8.43b}$$

$$T_{xz} = -\frac{\partial^2 \phi}{\partial x \partial z} = -\frac{2w'}{\pi} \frac{xz^2}{(x^2 + z^2)^2}. \tag{8.43c}$$

The displacement due to the line load, w', can now be calculated from Eqs. (8.41), which, on substituting for the stress from Eqs. (8.43a) and (8.43b) gives at any point (x, z)

$$u = -\frac{w'}{\pi} \left\{ \frac{(1 + v)(1 - 2v)}{E} \tan^{-1}\left(\frac{x}{z}\right) - \frac{(1 + v)}{E} \frac{xz}{(x^2 + z^2)} \right\} + f_1(z), \tag{8.44a}$$

$$w = -\frac{w'}{\pi} \left\{ \frac{1 - v^2}{E} \left[\ln(x^2 + z^2) - \frac{z^2}{(x^2 + z^2)} \right] + \frac{v(1 + v)}{E} \frac{x^2}{(x^2 + z^2)} \right\} + f_2(x). \tag{8.44b}$$

We need two conditions to determine the functions $f_1(z)$ and $f_2(x)$. The symmetry conditions of the problem dictate that $-u(-x, z) = u(x, z)$, i.e., that u is an odd function of x. Using this as one of the conditions, we find that $f_1(z) \equiv 0$. The displacements must satisfy Eq. (8.40c); using this as the second condition yields

$$\frac{df_2}{dx} = 0, \qquad f_2 = C = \text{const.}$$

We have no more interest in the horizontal displacement u but are interested in the vertical displacement at the surface $z = 0$:

$$\delta \equiv w|_{z=0} = -\frac{w'}{\pi}\left[\frac{1-v^2}{E}\ln x^2 + \frac{v(1+v)}{E}\right] + C. \tag{8.45}$$

We would now like to develop a displacement formula for normal surface loading over the arbitrary strip $s_1 \le x \le s_2$. To this end, we consider an elemental strip of width ds that is located at distance s, $s_1 < s < s_2$, from the y axis. Let the normal load on the elemental strip be denoted by $p(s)\,ds$. We can get the elemental vertical displacement caused by the loading on the elemental strip from Eq. (8.45) by simply replacing x by $(x - s)$; i.e., by parallel shifting of the coordinate system in the $+x$ direction by the distance s:

$$d\delta = -\frac{p(s)\,ds}{\pi}\left[\frac{1-v^2}{E}\ln(x-s)^2 + \frac{v(1+v)}{E}\right] + C(s). \tag{8.46}$$

Here $d\delta$ is the vertical displacement of the surface at x, due to the elemental strip loading $p(s)\,ds$ at $x = s$.

Integration of Eq. (8.46) over the width $(s_2 - s_1)$ of the strip yields

$$\delta = -\frac{(1-v^2)}{\pi E}\int_{s_1}^{s_2} p(s)\ln(x-s)^2\,ds + \hat{C} \tag{8.47}$$

where

$$\hat{C} = \text{const.} \times (s_2 - s_1) - \frac{v(1+v)}{\pi E}w'$$

The constant $\hat{C}$ can be evaluated in terms of δ_b, the deflection at some $x = b$,

$$\hat{C} = \delta_b + \frac{(1-v^2)}{\pi E}\int_{s_1}^{s_2} p(s)\ln(b-s)^2\,ds \tag{8.48}$$

The total displacement at x caused by the strip load $p(s)$ over $(s_2 - s_1)$

$$\delta = \delta_b - \frac{1-v^2}{\pi E}\int_{s_1}^{s_2} p(s)\ln\left[\frac{x-s}{b-s}\right]^2\,ds. \tag{8.49}$$

Formulation of the Line Contact Problem

In this section, we collect the equations that characterize the tribological interaction of two elastic bodies along a nominal line contact, when a continuous film of lubricant separates the two bodies. Young's modulus and Poisson's ratio of the bodies are E_1, v_1 and E_2, v_2 respectively, so the effective modulus and effective radius are given by

$$\frac{1}{E'} = \frac{1}{2}\left[\frac{1-v_1^2}{E_1} + \frac{1-v_2^2}{E_2}\right], \qquad \frac{1}{R_x} = \frac{1}{r_{1x}} + \frac{1}{r_{2x}}.$$

For this nominal line contact the first integral of the Reynolds equation is

$$\frac{dp}{dx} = 12\mu\tilde{u}\frac{h - h_{\text{cav}}}{h^3}, \tag{8.50}$$

where $\bar{u} = (u_1 + u_2)/2$ is the average of the surface velocities (Table 8.2) and x_0 is the location of the cavitation boundary. This equation is to be solved subject to the Swift-Stieber boundary condition

$$p = 0, \qquad \frac{dp}{dx} = 0 \quad \text{at } x = x_{\text{cav}}. \tag{8.51a}$$

The second of these conditions is redundant since any solution of Eq. (8.50) will necessarily satisfy this boundary condition. The upstream edge of the continuous film is located at $x = x_{\text{min}}$, where we assume the pressure to be atmospheric:

$$p = 0 \quad \text{at } x = x_{\text{min}}. \tag{8.51b}$$

Note that p in Eqs. (8.50) and (8.51) is gauge pressure.

In EHL calculations the viscosity-pressure correlation of Roelands (1966) is usually employed (Houpert and Hamrock, 1986)

$$\mu = \exp\{(\ln \mu_0 + 9.67)[-1 + (1 + 5.1 \times 10^{-9} p)^z]\}, \tag{8.52a}$$

where Z is a material parameter, and the lubricant density is assumed to vary according to the relationship, proposed by Dowson and Higginson (1977),

$$\rho = \rho_0 \left[1 + \frac{0.6 \times 10^{-9} p}{1 + 1.7 \times 10^{-9} p} \right], \tag{8.52b}$$

where p is in Pa.

In this introductory treatment, however, we simplify matters by assuming the viscosity-pressure dependence to be given by the less accurate, but easy to apply, Barus formula

$$\mu = \mu_0 \exp(\alpha p) \tag{8.12}$$

and constrain the lubricant to remain incompressible.

The film thickness distribution is known only within an additive constant h_w and, for two surfaces in contact, is given by

$$h(x) = h_w + \frac{x^2}{2R_x} - \frac{2}{\pi E'} \int_{x_{\text{min}}}^{x_{\text{cav}}} p(s) \ln(x - s)^2 \, ds. \tag{8.53}$$

Here we assumed that the undeformed gap is represented by a parabola, as before. The unknown constant h_w includes the minimum distance of the undeformed surfaces $h_0 = h(0)$ plus the constant term $\hat{C}$

We need one additional constraint to eliminate h_w. This additional relationship is provided by the force balance

$$w' = \int_{x_{\text{min}}}^{x_{\text{cav}}} p(x) \, dx, \tag{8.54}$$

where w' is the external load.

The system of Eqs. (8.12), (8.48), (8.50), and (8.53) and boundary conditions (8.51) form a closed system that characterizes the EHL line contact problem.

The problem is nondimensionalized through the transformation

$$x = b\bar{x}, \qquad s = b\bar{s}, \qquad h = \frac{b^2 \bar{h}}{R_x}, \qquad p = p_H \bar{p}, \tag{8.55a}$$

where the maximum Hertzian pressure, p_H, is given by Eqs. (8.30). The definition $W = w'/R_x E'$ for the load parameter (see Table 8.2) yields

$$b^2 = R_x^2 \frac{8W}{\pi}. \tag{8.55b}$$

In nondimensional form, the film thickness, Eq. (8.53), is given by

$$\bar{h} = \bar{h}_w + \frac{\bar{x}^2}{2} - \frac{1}{2\pi} \int_{\bar{x}_{min}}^{\bar{x}_{cav}} \bar{p}(\bar{s}) \ln(\bar{x} - \bar{s})^2) \, d\bar{s}$$
$$- \frac{1}{2\pi} \ln b^2 \int_{\bar{x}_{min}}^{\bar{x}_{cav}} \bar{p}(\bar{s}) \, d\bar{s}, \tag{8.56}$$

and the nondimensional form of Eq. (8.54) is

$$\frac{\pi}{2} = \int_{\bar{x}_{min}}^{\bar{x}_{cav}} \bar{p}(\bar{s}) \, d\bar{s}. \tag{8.57}$$

Taking into account Eqs. (8.30c) and (8.57), the nondimensional film thickness assumes the form (Houpert and Hamrock, 1986)

$$\bar{h} = \bar{h}_w + \frac{\bar{x}^2}{2} - \frac{1}{2\pi} \int_{\bar{x}_{min}}^{\bar{x}_{cav}} \bar{p}(\bar{s}) \ln(\bar{x} - \bar{s})^2 \, d\bar{s} - \frac{1}{4} \ln \left[R_x^2 \frac{8W}{\pi} \right]. \tag{8.58}$$

The integrated form of the Reynolds equation (8.50) has the nondimensional form

$$\exp\left(G\sqrt{\frac{W}{2\pi}} \bar{p} \right) \frac{d\bar{p}}{d\bar{x}} = \frac{3\pi^2}{4} \frac{U}{W} \frac{\bar{h} - \bar{h}_0}{\bar{h}^3}. \tag{8.59}$$

Equations (8.57), (8.58), and (8.59) and the boundary conditions

$$\bar{p} = 0 \qquad \text{at } \bar{x} = \bar{x}_{min}$$
$$\bar{p} = \frac{d\bar{p}}{d\bar{x}} = 0 \quad \text{at } \bar{x} = \bar{x}_{cav} \tag{8.60}$$

define the line contact problem of EHL in nondimensional form.

Numerical Considerations

There have been several methods of solution of the nominal line contact EHL problem. Here we follow Houpert and Hamrock (1986) in their improvement of Okamura's approach (Okamura, 1982). The main problem is to accurately calculate the film thickness. For high loads, the deformation can be orders of magnitude larger than the minimum film thickness. Accurate calculation of the pressure is made difficult by the increase of viscosity,

which, again, can increase by several orders of magnitude. We now calculate that part of
the film thickness that is due to elastic deformation,

$$\bar{\delta} = \frac{1}{2\pi} \int_{x_{\min}}^{x_{\max}} \bar{p}(\bar{s}) \ln(\bar{x} - \bar{s})^2 \, d\bar{s} - \frac{1}{4} \ln \left[R_x^2 \frac{8W}{\pi} \right], \quad \bar{x}_{\max} \geq \bar{x}_{\mathrm{cav}} \tag{8.61a}$$

The integrand in Eq. (8.61a) is singular at $x = s$, this singularity may be removed using
integration by parts (Kostreva, 1984),

$$\bar{\delta} = -\frac{1}{2\pi} [Ip]_{\bar{x}_{\min}}^{\bar{x}_{\max}} + \frac{1}{2\pi} \int_{\bar{x}_{\min}}^{\bar{x}_{\max}} \frac{d\bar{p}}{ds} I(\bar{x}; \bar{s}) \, d\bar{s} - \frac{1}{4} \ln \left[R_x^2 \frac{8W}{\pi} \right], \tag{8.61b}$$

where

$$I(\bar{x}; \bar{s}) = \int \ln(\bar{x} - \bar{s})^2 \, d\bar{s}$$

$$= -(\bar{x} - \bar{s})[\ln(\bar{x} - \bar{s})^2 - 2]. \tag{8.61c}$$

Since $p(x_i) = p(x_0) = 0$, the boundary term in Eq. (8.61b) vanishes, and we obtain

$$\bar{\delta} = -\frac{1}{2\pi} \int_{\bar{x}_{\min}}^{\bar{x}_{\max}} \frac{d\bar{p}}{d\bar{s}} (\bar{x} - s)[\ln(\bar{x} - \bar{s})^2 - 2] \, d\bar{s} - \frac{1}{4} \ln \left[R_x^2 \frac{8W}{\pi} \right]. \tag{8.62}$$

For the remainder of this section, we drop the overbar, but bear in mind that all variables
have been nondimensionalized at this stage.

We wish to evaluate the integral in Eq. (8.60) numerically. To this end, define a nonuniform partition, with N odd,

$$\Pi: x_{\min} = x_1 < x_2 < x_3 < \cdots < x_{N-2} < x_{N-1} < x_N = x_{\max}, \tag{8.63}$$

and write the integral in Eq. (8.62) as a sum of subintegrals, each evaluated on its own
subinterval,

$$\delta = -\frac{1}{2\pi} \int_{x_1}^{x_3} (.) \, ds + \frac{1}{2\pi} \int_{x_3}^{x_5} (.) \, ds + \cdots + \frac{1}{2\pi} \int_{x_{j-1}}^{x_{j+1}} (.) \, ds + \cdots + \frac{1}{2\pi} \int_{x_{N-2}}^{x_N} (.) \, ds + C. \tag{8.64}$$

where, for brevity, we represent the constant term in Eq. (8.62) by C.

In the first of these subintegrals, the global du mmy variable s ranges from x_1 to x_3, and
we replace it by a local dummy variable $x' = s - x_2$, which has the range $[x_1 - x_2, x_3 - x_2]$.
In the generic interval centered about x_j, j even, the local dummy variable is chosen to
be $x' = s - x_j$, which ranges over the interval $[x_{j-1} - x_j, x_{j+1} - x_j]$, and so forth.
Also, if we substitute x_i for x, Eq. (8.62) calculates the deformation at node i as the
sum of deformations $\delta_{i2}, \delta_{i4}, \ldots, \delta_{i,N-1}$, due to strip loading over the intervals $[x_1, x_3]$,
$[x_3, x_5], \ldots, [x_{N-2}, x_N]$, N odd, so that

$$\delta_i = \delta_{i,2} + \delta_{i,4} + \cdots + \delta_{i,N-1} + C$$

$$= \sum_{j=2,4,\ldots}^{N-1} \delta_{ij} - \frac{1}{4} \ln \left(R_x^2 \frac{8W}{\pi} \right). \tag{8.65}$$

We now evaluate the integrand in Eq. (8.62) for each of the subintervals. The discretized form of Eq. (8.61c), substituting x_i for x and $x_j + x'$ for s, is

$$I_{ij}(x') = -[x_i - (x_j + x')]\{\ln[x_i - (x_j + x')]^2 - 2\}, \qquad (j \text{ even}). \qquad (8.66)$$

Next, we represent $p(s)$ by its nodal values (Dowson and Higginson, 1959). But to make the subintegrals independent of one another, $p(s)$ of a particular subinterval can depend only on the nodal values that are available within that subinterval. For example, in the subinterval centered about node j, j even, we have the nodal values p_{j-1}, p_j, and p_{j+1} available. To interpolate for any $x' \in [x_{j-1} - x_j, x_{j+1} - x_j]$ in terms of the nodal values, we employ Lagrange's quadratic interpolation formula and write[7]

$$p(x') = \frac{x'(x' + x_j - x_{j+1})}{(x_{j-1} - x_j)(x_{j-1} - x_{j+1})} p_{j-1}$$
$$+ \frac{(x' + x_j - x_{j-1})(x' + x_j - x_{j+1})}{(x_j - x_{j-1})(x_j - x_{j+1})} p_j$$
$$+ \frac{x'(x' + x_j - x_{j-1})}{(x_{j+1} - x_{j-1})(x_{j+1} - x_j)} p_{j+1}. \qquad (8.67)$$

Thus, for any point $x' \in [x_{j-1} - x_j, x_{j+1} - x_j]$, the pressure is represented by a parabola that passes through points (x_{j-1}, p_{j-1}), (x_j, p_j), (x_{j+1}, p_{j+1}).

Differentiation of Eq. (8.67) with respect to x' yields the pressure gradient at $x' \in [x_{j-1} - x_j, x_{j+1} - x_j]$,

$$\frac{dp(x')}{dx'} = \frac{2x' + (x_j - x_{j+1})}{(x_{j-1} - x_j)(x_{j-1} - x_{j+1})} p_{j-1}$$
$$+ \frac{2x' + (x_j - x_{j-1}) + (x_j - x_{j+1})}{(x_j - x_{j-1})(x_j - x_{j+1})} p_j$$
$$+ \frac{2x' + (x_j - x_{j-1})}{(x_{j+1} - x_{j-1})(x_{j+1} - x_j)} p_{j+1}. \qquad (8.68)$$

The formula for the pressure derivative, Eq. (8.68), will now be written as

$$\frac{dp(x')}{dx'} = (a_1 x' + a_2) p_{j-1} + (a_3 x' + a_4) p_j + (a_5 x' + a_6) p_{j+1}, \qquad (8.69)$$

where the coefficients $a_1, a_2, \ldots \dot{a}_6$ can be obtained by comparing Eqs. (8.68) and (8.69).

The subintegral centered about node j, j even, in Eq. (8.64) can then be written as

$$\frac{1}{2\pi} \int_{x_{j-1}}^{x_{j+1}} (\cdot) \, ds = p_{j-1} \left\{ \frac{1}{2\pi} \int_{x_{j-1}-x_j}^{x_{j+1}-x_j} (a_1 x' + a_2) I_{ij}(x') \, dx' \right\}$$
$$+ p_j \left\{ \frac{1}{2\pi} \int_{x_{j-1}-x_j}^{x_{j+1}-x_j} (a_3 x' + a_4) I_{ij}(x') \, dx' \right\}$$
$$+ p_{j+1} \left\{ \frac{1}{2\pi} \int_{x_{j-1}-x_j}^{x_{j+1}-x_j} (a_5 x' + a_6) I_{ij}(x') \, dx' \right\}. \qquad (8.70)$$

[7] On substituting $x' = s - x_j$, we revert to the conventional notation of Lagrange's interpolation formula (Gerald, 1973).

The integrals in Eq. (8.70) depend only on the sequence of nodal points, Eq. (8.63). As soon as partition Π is selected in Eq. (8.63), the integrals can be evaluated analytically in a straightforward manner (Houpert and Hamrock, 1986). Let us assume, for simplicity, that this has been done and designate the results by L, M, and U according to

$$L_{i,j-1} = \frac{1}{2\pi} \int_{x_{j-1}-x_j}^{x_{j+1}-x_j} (a_1 x' + a_2) I_{ij}(x') \, dx', \tag{8.71a}$$

$$M_{i,j} = \frac{1}{2\pi} \int_{x_{j-1}-x_j}^{x_{j+1}-x_j} (a_3 x' + a_4) I_{ij}(x') \, dx', \tag{8.71b}$$

$$U_{i,j+1} = \frac{1}{2\pi} \int_{x_{j-1}-x_j}^{x_{j+1}-x_j} (a_5 x' + a_6) I_{ij}(x') \, dx', \quad [j = 2k, k = 1, \ldots, (N-1)/2]. \tag{8.71c}$$

The deformation at node i due to loading on the strip centered about node j is

$$\delta_{i,j} = L_{i,j-1} p_{j-1} + M_{i,j} p_j + U_{i,j+1} p_{j+1}. \tag{8.72}$$

Substituting into Eq. (8.65), the deflection at node i due to loading of all strips is

$$\delta_i = [L_{i,1} p_1 + M_{i,2} p_2 + (U_{i,3} + L_{i,3}) p_3] + \cdots$$

$$+ [(U_{i,2k-1} + L_{i,2k-1}) p_{2k-1} + M_{i,2k} p_{2k} + (U_{i,2k+1} + L_{i,2k+1}) p_{2k+1}] + \cdots$$

$$+ [(U_{i,N-2} + L_{i,N-2}) p_{N-2} + M_{i,N-1} p_{N-1} + U_{i,N} p_N] + C. \tag{8.73}$$

Let us define the *influence coefficients* $D_{i,n}$ as follows

$$D_{i,n} = \begin{cases} M_{i,n} & n \text{ even}, \\ (U_{i,n} + L_{i,n}) & n \text{ odd}, \end{cases}$$

$$U_{i,1} = 0, \qquad L_{i,N} = 0, \qquad 1 \le i \le N. \tag{8.74}$$

The influence coefficient, $D_{i,n}$, calculates deformation at node i due to unit load at node n. We may now write Eq. (8.73) in the concise form

$$\delta_i = \sum_{j=2,4,\ldots}^{N-1} \delta_{ij} - \frac{1}{4} \ln \left(R_x^2 \frac{8W}{\pi} \right)$$

$$= \sum_{n=2}^{N-1} D_{i,n} p_n - \frac{1}{4} \ln \left(R_x^2 \frac{8W}{\pi} \right), \tag{8.75}$$

The unknowns of the boundary value problem, Eqs. (8.57)–(8.60), are the pressure at the nodal points $p_k, 2 \le k \le N-1$, and the constant $\hat{h}_w$, which now also includes the constant term

$$\hat{h}_w = h_w - \frac{1}{4} \ln \left(R_x^2 \frac{8W}{\pi} \right)$$

The position of the cavitation boundary x_{cav} is also unknown at this stage. Instead of evaluating x_{cav}, however, we consider $h_{cav} = h(x_{cav})$ as our unknown and write Eq. (8.56) as

$$0 = h_{cav} - h_w - \frac{1}{2}\xi^2 + \frac{1}{2\pi}\int_{x_i}^{x_M} p(s)\ln[\xi - s]^2\, ds. \tag{8.76}$$

Equation (8.76) is evaluated by interpolation (Okamura, 1983), as $x_{cav} = \max(\xi)$, where ξ satisfies (8.76) and $x_M > x_N$. Determination of the location of the free boundary is, however, difficult. To avoid such difficulties, Bisset and Glander (1988) scale the problem to a fixed interval *a priori*, by using the transformation

$$\hat{x} = \frac{x - x_{min}}{x_{cav} - x_{min}}. \tag{8.77}$$

This will introduce $\hat{h}_{cav} = \hat{h}(1)$ into the equations naturally; $\hat{h}_{cav}$ is now an unknown of the problem.

The system of discretized equations takes the form of a set of nonlinear algebraic equations,

$$F(x) = 0, \qquad x = (u, \lambda), \tag{8.78}$$

where u is the vector of state variables (p_i, h_{cav}, h_w) and λ is a vector of the parameters $U, G, W, \ldots$, etc. In general, if there are n parameters, the solution set of Eq. (8.78) is an n-dimensional manifold. When $n = 1$, the manifold becomes a path.

In the calculations we keep all the parameters constant except one, and use the Gauss-Newton method for local iteration and the method of continuation for tracing the path (Bisset and Gander, 1988; Wang, Al-Sharif, Rajagopal, and Szeri, 1992). Note that Eq. (8.78) has the same form as Eq. (6.28). Application of the Gauss-Newton and path continuation methods have been described in Chapter 6. For application of the highly successful multigrid techniques (Brandt, 1984) to EHL problems, see Venner, ten Napel, and Bosma (1990), and Ai and Cheng (1994).

For simplicity, we have discussed only the nominal line contact problem in detail. By employing this technique, Houpert and Hamrock (1986) were able to obtain solutions up to 4.8 GPa maximum pressure. Extension to nominal point contact is discussed in a series of papers by Hamrock and Dowson (1976a, 1976b, 1997a, 1977b, 1978, 1979). Additional material is to be found in Hamrock (1991).

A great deal of effort has been devoted during the past two decades to the study of rough EHL contacts. Notable recent papers dealing with this subject are Ai and Cheng (1994), Greenwood and Morales-Espejel (1994), Venner and Lubrecht (1994), Ai and Cheng (1996) and Xu and Sadeghi (1996).

8.6   Rolling-Contact Bearings[8]

In contrast to hydrodynamic bearings, which depend for low-friction characteristics on a fluid film between the journal and the bearing surfaces, rolling-element bearings employ a number of balls or rollers that roll in an annular space. To some extent, these rollers help to avoid gross sliding and the high coefficients of friction that are associated with sliding. The term rolling element is used to describe this class of bearing because the

---

[8] Section 8.6 is a reproduction, with permission, of parts of Chapter 8, *Rolling Element Bearings*, by W. J. Anderson, in *Tribology: Friction, Lubrication and Wear*, ed. A. Z. Szeri. Hemisphere Publishing Co., 1980.

contact between the rolling elements and the races or rings consists more of sliding than of actual rolling. A rolling contact implies no interfacial slip; this condition can seldom be maintained, however, because of material deformation and geometric factors. Rolling-element bearings ordinarily consist of two races or rings (the inner race and the outer race), a set of rolling elements (either balls or rollers), and a separator (sometimes called a cage or retainer) for keeping the set of rolling elements approximately equally spaced.

Rolling-element bearings offer the following advantages when compared with hydrodynamic bearings: (1) low starting friction; (2) low operating friction, comparable to that of hydrodynamic bearings at low speeds and somewhat less at high speeds; (3) less sensitivity to interruptions in lubrication than with hydrodynamic bearings; and (4) capability of supporting combined loads. In the latter respect, rolling-element bearings are more versatile than hydrodynamic bearings, which usually can support only radial or thrust loads.

Rolling-element bearings also have disadvantages: (1) they occupy more space in the radial direction than do hydrodynamic bearings and (2) they have a finite fatigue life because of repeated stresses at ball-race contacts – in contrast to hydrodynamic bearings, which usually have an almost infinite fatigue life.

Bearing Types

Ball Bearings
The various types of rolling-element bearings may be placed in two broad categories; the first of these is ball bearings. The most common types of ball bearings are:

(1)   Deep groove or Conrad
(2)   Angular contact
(3)   Self-aligning
(4)   Duplex
(5)   Ball thrust

A typical deep groove ball bearing is shown in Figure 8.13. This bearing has moderately high radial load capacity and moderate thrust load capacity. Figure 8.14 shows an angular contact bearing. This bearing has a higher thrust-load capacity than a deep groove bearing, but it can carry thrust load in only one direction.

A self-aligning ball bearing with the outer-race groove ground to a spherical shape is illustrated in Figure 8.15(a). This bearing has a relatively low load capacity but is insensitive

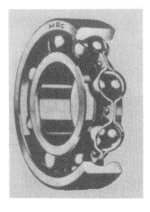

Figure 8.13. Deep grove ball bearing. (Courtesy Marlin Rockwell Corp.)

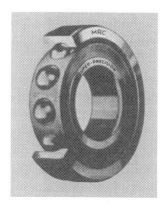

Figure 8.14. Angular contact ball bearing. (Courtesy Marlin Rockwell Corp.)

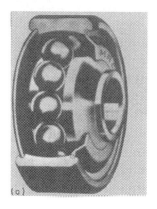

Figure 8.15. Self-aligning ball bearing: (a) grove of outer race ground to a spherical shape, (b) outside diameter of outer race ground to fit spherical housing. (Courtesy Marlin Rockwell Corp.)

to shaft and housing misalignments. Figure 8.15(b) shows a second type of self-aligning ball bearing with the self-aligning feature obtained by grinding the outer-race outside diameter in a spherical shape to fit a spherical housing. This bearing has a higher load capacity than the bearing in Figure 8.15(a), but care must be taken to maintain freedom of movement between the outer race and the housing.

Angular contact bearings are usually used in pairs in duplex mounts. Different types of duplex mounts are shown in Figure 8.16(a) (back to back) and Figure 8.16(b) (face to face). These two arrangements make it possible to carry thrust load in either direction. Bearings are manufactured as matched pairs, so that when they are mated and the races are made flush, each bearing is preloaded slightly. This preloading provides greater stiffness and helps to prevent ball skidding with acceleration at light load. When a high unidirectional thrust load must be carried, a duplex tandem mount [Figure 8.16(c)] is used. With careful manufacture and installation, a tandem bearing pair may have a thrust capacity as much as 1.8 times the capacity of a single bearing.

A thrust ball bearing is shown in Figure 8.17. This bearing has a high thrust capacity but is limited to low speeds because of the high degree of sliding in the ball-race contacts.

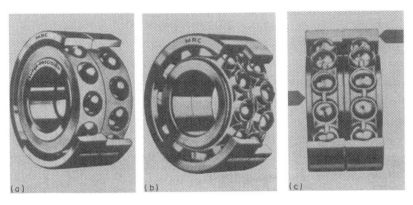

Figure 8.16. Duplex angular contact ball bearing, mounted (a) back to back, (b) face to face, (c) in tandem. (Courtesy Marlin Rockwell Corp.)

Figure 8.17. Thrust ball bearing (Courtesy Marlin Rockwell Corp.).

Roller Bearings

The second broad category is that of roller bearings. Common types are:

(1) Cylindrical
(2) Tapered
(3) Spherical
(4) Needle

Cylindrical roller bearings (Figure 8.18) are best suited of all roller bearing types for high-speed operation. These bearings carry only radial load, and they are frequently used where freedom of movement of the shaft in the axial direction must be provided because of differential expansion.

Tapered roller bearings (Figure 8.19) and spherical roller bearings (Figure 8.20) are high-load-capacity, low-speed roller bearings with combined radial load and thrust load capability.

Needle bearings (Figure 8.21) are capable of carrying high loads and are useful in applications where limited radial space is available.

Although rolling-element bearings are usually equipped with a separator, in some instances they are not. Bearings without separators are usually termed full-complement bearings. A common type of full-complement roller bearing is the needle bearing. In some

Figure 8.18. Cylindrical roller bearing. (Courtesy Marlin Rockwell Corp.)

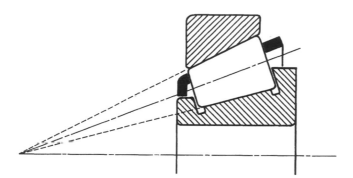

Figure 8.19. Tapered roller bearing.

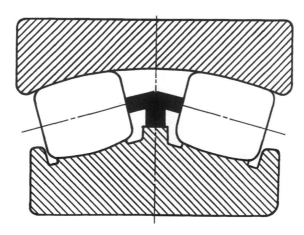

Figure 8.20. Spherical roller bearing.

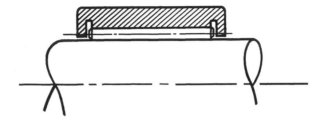

Figure 8.21. Needle bearing with shaft and inner race.

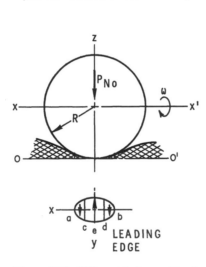

Figure 8.22. Differential slip resulting from curvature of contact ellipse.

low-speed applications where load capacity is of primary importance, full-complement ball bearings are used. In this type of bearing, the annular space between the races is packed with the maximum number of balls.

Rolling Friction

The concepts of rolling friction are important because the characteristics and behavior of rolling-element bearings depend on rolling friction. The theories of Reynolds (1876) and Heathcote (1921) were previously well accepted as correctly explaining the origin of rolling friction for a ball in a groove. The energy lost in rolling was believed to be that required to overcome the interfacial slip that occurs because of the curved shape of the contact area. As shown in Figure 8.22, the ball rolls about the x axis and makes contact with the groove from a to b.

If the groove is fixed, then for zero slip over the contact area no point within the area should have a velocity in the direction of rolling. The surface of the contact area is curved, however, so that points a and b are at different radii from the x axis than are points c and d. For an inelastic ball, points a and b must have different velocities with respect to the x axis than do points c and d because the velocity of any point on the ball relative to the x axis equals the angular velocity, w, times the radius from the x axis. Slip must occur at various points over the contact area unless the body is so elastic that yielding can take place in the contact area to prevent this interfacial slip. Reynolds and Heathcote assumed that this interfacial slip took place and that the forces required to make a ball roll were the forces required to overcome the friction due to this interfacial slip. In the contact area, rolling without slip will occur at a specific radius from the x axis. Where the radius is greater than this radius to the rolling point, slip will occur in the other direction; where it is less than the radius to this rolling point, slip will occur in the other direction. In Figure 8.22, the lines to points c and d represent the approximate location of the rolling bands, and the arrows shown in the three portions of the contact area represent the directions of interfacial slip when the ball is rolling into the paper.

Frictional Losses in Rolling Contact Bearings

Some of the factors that affect the magnitude of friction losses in rolling bearings are:

(1) Bearing size
(2) Bearing type
(3) Bearing design
(4) Load (magnitude and type, either thrust or radial)
(5) Speed
(6) Oil viscosity
(7) Oil flow

In a rolling-contact bearing, friction losses consist of:

(1) Sliding friction losses in the contacts between the rolling elements and the raceways.
(2) Hysteresis losses resulting from the damping capacity of the raceway and the ball material.
(3) Sliding friction losses between the separator and its locating race surface and between the separator pockets and the rolling elements.
(4) Shearing of oil films between the bearing parts and oil churning losses caused by excess lubricant within the bearing.

The relative magnitude of each of these friction losses depends on the bearing type and design, the lubricant, and the type of lubrication. Palmgren (1959) presented friction coefficients for various types of bearings, which may be used for rough calculations of bearing torque. The values were computed at a bearing load that would give a life of 1×10^9 revolutions for the respective bearings. As the bearing load approaches zero, the friction coefficient becomes infinite because the bearing torque remains finite. The following friction coefficients were given by Palmgren:

All these friction coefficients are referenced to the bearing bore.

Table 8.3. *Contact bearing friction coefficients*

Bearing	Friction coefficient
Self-aligning ball	0.0010
Cylindrical roller, with flange-guided short rollers	0.0011
Thrust ball	0.0013
Single-row deep groove ball	0.0015
Tapered and spherical roller, with flange guided rollers	0.0018
Needle	0.0045

Palmgren (1959) and Muzzoli (Wilcock and Booser, 1957) attempted to relate all the factors that influence rolling-bearing torque. Palmgren outlined a method for computing torque for several types of bearings by calculating the zero-load torque and the sliding and hysteresis losses due to the load. Muzzoli's equation for bearing torque is discussed in Wilcock and Booser (1957). A number of factors must be known for these equations to be useful, and their validity depends on the accuracy with which some of the factors are determined.

Astridge and Smith (1972) conducted a thorough experimental study of the power loss in high-speed cylindrical roller bearings. They found that the principal sources of power loss were:

(1) Roller track elastohydrodynamic films (60%)
(2) Roller cage sliding (10%)
(3) Cage-locating surface sliding (10%)
(4) Cage side-chamber wall drag (10%)
(5) Oil flinging from rotating surfaces (7%)
(6) Elastic hysteresis (1%)
(7) Displacement of oil by rollers (1%)

Specific Dynamic Capacity and Life

In ordinary bearing applications where extreme speeds and temperatures are not present, a properly installed and lubricated bearing will eventually fail because of material fatigue. The repeated stresses developed in the contact areas between the rolling elements and the races eventually result in failure of the material, which manifests itself as a fatigue crack. The fatigue crack propagates until a piece of the race or rolling-element material spalls out and produces the failure. Many bearings fail for reasons other than fatigue, but in ordinary applications, these failures are considered avoidable if the bearing is properly handled, installed, and lubricated and is not overloaded.

If a number of similar bearings are tested to fatigue at a specific load, bearing life varies widely among them. For a group of 30 or more bearings, the ratio of the longest to the shortest life may be of the order of 20 or more (Figure 8.23).

A curve of life as a function of the percentage of bearings that failed can be drawn for any group of bearings (Figure 8.24). For a group of 30 or more bearings, the longest life would be of the order of four or five times the average life. The term life, in bearing catalogs, usually means the life that is exceeded by 90% of the bearings. This is the so-called *B*-10

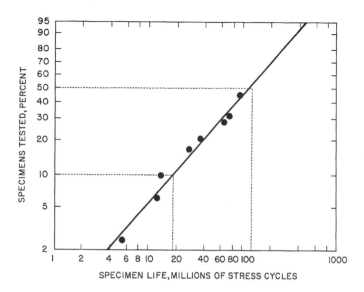

Figure 8.23. Typical Weibull plot of bearing fatigue failures.

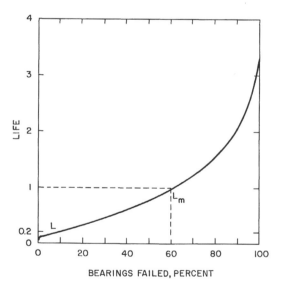

Figure 8.24. Distribution of bearing fatigue failures.

or 10% life. The 10% life is one-fifth the average 50% life for a normal life-dispersion curve.

If two groups of identical bearings are run to fatigue at two different loads, the life varies inversely as the nth power of the load:

$$\frac{L_2}{L_1} = \left(\frac{W_1}{W_2}\right)^n.$$

For nominal point contact $n = 3$, for nominal line contact $n = 4$. For point contact, then,

$$\frac{W_1}{W_2} = \left(\frac{L_2}{L_1}\right)^{1/3}$$

$$W_1 L_1^{1/3} = W_2 L_2^{1/3} = \text{const.}$$

If W is a radial load that acts on a radial bearing and L is 1 million revolutions with rotation of the inner race, then the constant is the *specific dynamic capacity*. The dynamic capacity for thrust bearings is determined if W is a thrust load.

In terms of the specific dynamic capacity, C,

$$L = (C/P)^3,$$

where L is the life in millions of revolutions and P is the equivalent load. In general, the ball bearings support both radial and thrust loads, and formulas for obtaining the equivalent load, P, in terms of radial and thrust loads are given for various bearing types in most bearing catalogs.

The bearing design factors that affect the specific dynamic capacity are the race conformities, the rolling-element dimensions, and the number of rolling elements. Recent research has shown that the bearing material and both the lubricant viscosity and the base stock can have marked effects on fatigue. The original bearing fatigue investigations, which include those of Palmgren, were made before the advent of the extreme temperatures and speeds to which rolling bearings are now subjected. As a result, the great majority of bearings were made of SAE 51100 or SAE 52100 alloys, and wide variations in material fatigue strength were not encountered. In addition, all rolling bearings were lubricated with a mineral oil such as SAE 30 or with a mineral-base soap grease so that the effect of the lubricant on fatigue was not important enough to be included among the parameters affecting fatigue life.

In recent years, a better understanding of rolling-element bearing design, materials, processing, and manufacturing techniques has permitted a general improvement in bearing performance. This is reflected in greater bearing reliability or longer expected life in a particular application. At the same time, operating conditions have become better understood, so that the application factors of the past, which were based primarily on experience, are giving way to quantified environmental factors that permit much better estimates of expected bearing life. These environmental factors, considering application, bearing configuration, and bearing rating, are often exceedingly complex, so that obtaining problem solutions in a reasonable time requires high-speed computers and more information than is readily available to most engineers. It becomes necessary to develop a guide that extends the "engineering approximations," which are illustrated in most bearing manufacturers' catalogs, and provides information that will be of most use to the engineer.

The Anti-Friction Bearing Manufacturers Association (AFBMA) method for determining bearing load rating and fatigue life and/or the basic ratings published in any bearing manufacturer's catalog must be the heart of any design guide. Continuing in the vein of the engineering approximation, it is assumed that various environmental or bearing design factors are, at least for first-order effects, multiplicative. As a result, the expected bearing life, L_A, can be related to the calculated rating life, L_{10}, by

$$L_A = (D)(E)(F)(G)(H)L_{10}$$

$$= (D)(E)(F)(G)(H)\left(\frac{C}{P}\right)^n,$$

where $D \ldots H$ are the life adjustment environmental or bearing design factors, C is the basic load rating, P is the equivalent load, and n is the load-life exponent (3 for ball bearings or 10/3 for roller bearings).

Methods for calculating the life adjustment factors are given in *Life Adjustment Factors for Ball and Roller Bearings* (ASME, 1971). An extensive bibliography is provided in two articles by Tallinn (1992a).

Specific Static Capacity

From considerations of allowable permanent deformation, there is a maximum load that a bearing can support while not rotating. This is called the specific static capacity, C_o. It is arbitrarily defined as the load that will produce a permanent deformation of the race and the rolling element at a contact of 0.0001 times the rolling-element diameter. When permanent deformations exceed this value, bearing vibration and noise increase noticeable when the bearing is subsequently rotated under lesser loads. Specific static capacity is determined by the maximum rolling-element load and the race conformity at the contact. A bearing can be loaded above C_o as long as the load is applied when the bearing is rotating. The permanent deformations that occur during rotation will be distributed evenly around the periphery of the races and will not be harmful until they become more extensive.

Static and dynamic load capacities are normally given for bearings in bearing catalogs.

Fatigue Wear Out

Among the many possible causes of rolling bearing failure, there are two that predominate. These are wear out and fatigue. Study of the factors that cause these failures and means for preventing or delaying them has occupied much of the time of researchers during the past decade. In addition, rolling bearings are being called on to operate at more severe conditions of speed and temperature. The consequent demands for improvements in bearing design and materials and lubricant technology have spawned additional research. The reader should consult Tallinn (1992b).

8.7 Minimum Film Thickness Calculations

Nominal Line Contact

The geometry of the cylindrical roller bearing is depicted in Figure 8.25. Specifications of the problem are adapted from Hamrock and Anderson (1983).

The load on the most heavily loaded roller is estimated from *Stribeck's formula* (Hamrock, 1991),

$$w_{\max} = \frac{4w}{n} = 4.8 \text{ kN}.$$

The radii of curvature at contact on the inner and outer race, respectively, are

$$\frac{1}{R_{x,i}} = \frac{1}{0.08} + \frac{1}{0.032} = \frac{5}{0.032}, \qquad \frac{1}{R_{x,o}} = \frac{1}{0.08} - \frac{1}{0.048} = \frac{5}{0.048},$$

giving $R_{x,i} = 0.0064$ m, and $R_{x,o} = 0.0096$ m.

Table 8.4. *Data for nominal line contact problem*

Inner-race diameter	$d_i = 0.064$ m
Outer-race diameter	$d_o = 0.096$ m
Roller diameter	$d = 0.016$ m
Roller axial length	$L = 0.016$ m
No. Rollers per bearing	$n = 9$
Radial load per roller	$w = 10.8$ kN
Inner-race angular velocity	$\omega_i = 524$ rad/s
Outer-race angular velocity	$\omega_i = 0$
Absolute viscosity	$\mu_0 = 0.01$ Pa.s
Viscosity-pressure coefficient	$\alpha = 2.2 \times 10^{-8} \text{Pa}^{-1}$
Young's modulus (rollers, races)	$E = 207.5$ GPa
Poisson's ratio	$v = 0.3$

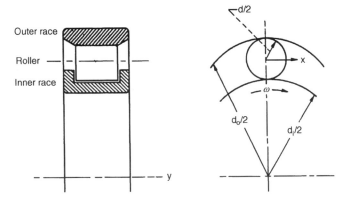

Figure 8.25. Schematic of a roller bearing. (Reproduced with permission from Hamrock, B. J. and Anderson, W. *Rolling Element Bearings.* NASA RP-1105. 1983.)

The effective modulus, E', and the pitch diameter, d_e, are given by

$$E' = \frac{2}{\dfrac{1 - v_1^2}{E_1} + \dfrac{1 - v_2^2}{E_2}} = 228 \text{ GPa}, \qquad d_e = \frac{d_o + d_i}{2} = 0.08 \text{ m}.$$

The surface velocity for cylindrical rollers is calculated from

$$\tilde{u} = \frac{|\omega_i + \omega_o| \left| d_e^2 - d^2 \right|}{4 d_e} = 10.061 \text{ m/s},$$

where we assumed pure rolling.[9]

Calculation will be performed for the *inner-race contact* alone. The speed, load, and

[9] For roller-bearing kinematics, see Anderson (1970), Anderson (1980), Hamrock (1991), and Harris (1991).

Table 8.5. *Date for nominal point contact problem*

Inner-race diameter	$d_i = 0.052291$ m
Outer-race diameter	$d_o = 0.077706$ m
Ball diameter	$d = 0.012700$ m
No. of balls per bearing	$n = 9$
Inner-grove radius	$r_i = 0.006604$ m
Outer-grove radius	$r_o = 0.006604$ m
Contact angle	$\beta = 0$
Radial load	$w_z = 8.9$ kPa
Inner-race angular velocity	$\omega_i = 400$ rad/sec
Outer-race angular velocity	$\omega_o = 0$
Absolute viscosity	$\mu_0 = 0.04$ Pa·s
Viscosity-pressure coefficient	$\alpha = 2.3 \times 10^{-8}$ Pa^{-1}
Young's modulus	$E = 200$ GPa
Poisson's ratio	$\nu = 0.3$

materials parameters are calculated as

$$U = \frac{\mu_0 \tilde{u}}{E' R_{x,i}} = \frac{(0.01)(10.061)}{(2.28 \times 10^{11})(0.0064)} = 6.895 \times 10^{-11},$$

$$W = \frac{w_{\max}}{E' L R_{x,i}} = \frac{4800}{(2.28 \times 10^{11})(0.016)(0.0064)} = 2.0559 \times 10^{-4},$$

$$G = \alpha E' = (2.2 \times 10^{-8})(2.28 \times 10^{11}) = 5.016 \times 10^3.$$

The criteria for the effect of viscous change and elasticity are

$$g_V = \frac{W^{2/3} G}{U^{1/2}} = 1.7804 \times 10^3, \qquad g_E = \frac{W}{U^{1/2}} = 24.759.$$

The point (g_V, g_E) characterizing conditions at the inner contact can be plotted in Figure 8.8, showing that full EHL conditions apply. Thus, the formulas for minimum film thickness calculation are

$$g_{H,\min} = 1.6549 g_V^{0.54} g_E^{0.06} = 114.15,$$

$$H_{\min} = g_{H,\min} \left(\frac{U}{W} \right) = 3.8284 \times 10^{-5},$$

$$h_{\min} = R_{x,i} H_{\min} = 0.245 \mu\text{m}.$$

Employing a slightly different procedure, i.e., approximating from nominal point contact formulas, Hamrock and Anderson find $h_{\min} = 0.32$ μm.

Nominal Point Contact

The geometry of the ball bearing is shown in Figure 8.26. The specifications of the problem are adapted from Hamrock and Dowson (1981):

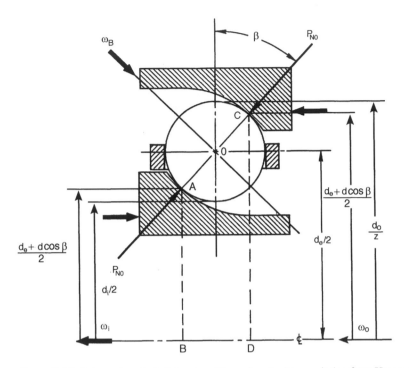

Figure 8.26. Schematics of a ball bearing. (Reproduced with permission from Hamrock, B. J. and Anderson, W. *Rolling Element Bearings*. NASA RP-1105, 1983.)

Calculations will be shown for the *inner-race contact* only. The pitch diameter is

$$d_e = \frac{(d_o + d_e)}{2} = 0.065 \text{ m}.$$

The equivalent radii and curvature sum are

$$R_x = \frac{d(d_e - d\cos\beta)}{2d_e} = 0.00511 \text{ m},$$

$$R_y = \frac{r_i d}{2r_i - d} = 0.165 \text{ m},$$

$$S_C = \frac{1}{\mathcal{R}} = \frac{1}{R_x} + \frac{1}{R_y} = 201.76,$$

yielding $\mathcal{R} = 4.956 \times 10^{-3}$ m, $\alpha_r = R_y/R_x = 32.29$, and an ellipticity parameter

$$\bar{\kappa} = \alpha_r^{2/3} = 9.1348.$$

The approximate formulas for the elliptic integrals give

$$\bar{\mathcal{E}} = 1 + \left(\frac{\pi}{2} - 1\right)/\alpha_r = 1.0177,$$

$$\bar{\mathcal{F}} = \frac{\pi}{2} + \left(\frac{\pi}{2} - 1\right)\ln\alpha_r = 3.5542.$$

Using $E' = 219.8$ MPa for effective modulus and Stribeck's estimate (Anderson, 1980) for the maximum load

$$w_{max} = \frac{5w}{n} = 4.944 \text{ kN}$$

valid for ball bearings, the deformation is calculated from Eq. (8.27c)

$$\delta = \bar{\mathcal{F}}\left[\left(\frac{9}{2\bar{\mathcal{E}}\mathcal{R}}\right)\left(\frac{w_{max}}{\pi\bar{k}E'}\right)^2\right]^{1/3} = 29.087\,\mu\text{m}.$$

Assuming pure rolling, the surface velocity is calculated from (Hamrock, 1991)

$$\tilde{u} = \frac{|\omega_o - \omega_i||d_e^2 - d^2|}{4d_e} = 6.252 \text{ m/s}.$$

The design parameters give

$$U = \frac{\mu_0\tilde{u}}{E'R_x} = \frac{(0.04)(6.252)}{(2.198 \times 10^{11})(5.11 \times 10^{-3})} = 2.227 \times 10^{-10},$$

$$G = \alpha E' = (2.3 \times 10^{-8})(2.198 \times 10^{11}) = 5.055 \times 10^3,$$

$$W = \frac{w_{max}}{E'R_x^2} = \frac{4,944}{(2.198 \times 10^{11})(5.11 \times 10^{-3})^2} = 8.6141 \times 10^{-4}.$$

The minimum film thickness variable is given by

$$g_{H_{min}} = 3.42\, g_V^{0.49}\, g_E^{0.17}[1 - e^{-0.68\kappa}] \tag{8.32}$$

Substituting

$$g_V = \left(\frac{GW^3}{U^2}\right) = 6.5149 \times 10^{13}$$

$$g_E = W^{8/3}U^2 = 1.3545 \times 10^{11}$$

into Eq. (8.32), we find

$$g_{H,min} = 1.5645 \times 10^9,$$

$$H_{min} = g_{H,min}\left(\frac{U^2}{W}\right) = 1.0457 \times 10^{-4},$$

and

$$h_{min} = R_x H_{min} = (0.00511)(1.0457 \times 10^{-4})$$

$$= 0.5345\,\mu\text{m}.$$

Hamrock and Dowson, relying on iteration in place of Stribeck's formula for maximum loading, obtain $h_{min} = 0.557\,\mu\text{m}$. Further details of minimum film thickness calculations can be found in Hamrock and Dowson (1981), Harris (1991), and Hamrock (1991).

8.8   Nomenclature

D_{ij}	influence coefficient
D_c	curvature difference
E'	effective elastic modulus
G	shear modulus
G	material parameter
H	film thickness parameter
$\mathcal{R}$	effective radius
R_x, R_y	relative principal radii of curvature
S_c	curvature sum
U	velocity parameter
$V, \tilde{u}$	effective velocity
W	load parameter
a, b	semi-axes of elliptical contact
b	semi-width of rectangular contact
g_E	elasticity parameter
g_H	film parameter
g_V	viscosity parameter
h	film thickness
h_0	film thickness at $x = 0$
p	pressure
p_H	maximum Hertzian pressure
r_{ix}, r_{iy}	principal radii of curvature
(u, v, w)	elastic displacement
w	normal load
ϕ	stress function
α	pressure-viscosity coefficient
α_r	ratio R_y/R_x
κ	ellipticity parameter
$(\ )_{\min}, (\ )_{\max}$	at inlet, exit
δ	vertical deflection of surface
μ	viscosity
μ_0	viscosity at atmospheric pressure
$(\ )_{cav}$	at cavitation boundary

8.9   References

Ai, X. and Cheng, H. S. 1994. A transient EHL analysis for line contacts with a measured surface roughness using multigrid technique. *ASME Journal of Tribology*, **116**, 549–558.

Ai, X. and Cheng, H. S. 1996. The effects of surface texture on EHL point contacts. *ASME Journal of Tribology*, **118**, 59–66.

Anderson, W. J. 1970. Elastohydrodynamic lubrication theory as a design parameter for rolling element bearings. *ASME Paper 70-DE-19*.

Anderson, W. J. 1980. Rolling-element bearings. In *Tribology: Friction, Lubrication and Wear*, A. Z. Szeri (ed.). Hemisphere.

Archard, G. D., Gair, F. C. and Hirst, W. 1961. The elastohydrodynamic lubrication of rollers. *Proc. Roy. Soc.*, **A 262**, 51.

Arnell, R. D., Davies, P. B., Halling, J. and Whomes, T. L. 1991. *Tribology: Principles and Design Applications*. Springer-Verlag, New York.

ASME. 1971. *Life Adjustment Factors for Ball and Roller Bearings*. ASME Press, New York.

Astridge D. G. and Smith, C. F. 1972. Heat generation in high speed cylindrical roller bearings. *Inst. Mech. Engrs. EHL Symposium*. Leeds, England.

Benjamin, M. K. and Castelli, V. 1971. A theoretical investigation of compliant surface journal bearing. *ASME Trans.*, **93**, 191–201.

Bissett, E. J. and Glander, D. W. 1988. A highly accurate approach that resolves the pressure spike of elastohydrodynamic lubrication. *ASME Journal of Tribology*, **110**, 241–246.

Blok, H. 1950. Fundamental mechanical aspects of thin film lubrication. *Ann. N. Y. Acad. Sci.*, **53**, 779.

Brandt, A. 1984. *Multigrid Techniques: 1984 Guide with Applications to Fluid Mechanics*. GMD-Studien 85.

Brewe, D. E. and Hamrock, B. J. 1977. Simplified solution for elliptical contact deformation between two elastic solids. *ASME Journal of Lubrication Technology*, **99**, 485–487.

Browne, A. L., Whicker, D. and Rohde, S. M. 1975. The significance of thread element flexibility on thin film wet traction. *Tire Sci. Technol.*, **3**, 4.

Carl, T. E. 1964. The experimental investigation of a cylindrical journal bearing under constant and sinusoidal loading. *Proceedings of the 2nd Convention on Lubrication and Wear*, p. 100. Institution of Mechanical Engineers, London.

Cheng, H. S. 1965. A refined solution to the thermal-elastohydrodynamic lubrication of rolling and sliding cylinders. *ASLE Trans.*, **8**, 397–410.

Cheng, H. S. and Sternlicht, B. 1965. A numerical solution for the pressure, temperature and film thickness between two infinitely long lubricated rolling and sliding cylinders, under heavy loads. *ASME Trans., Ser. D*, **87**, 695–707.

Crook, A. W. 1958. The lubrication of rollers. *Philos. Trans. Roy. Soc.*, **A 250**, 387–409.

Dowson, D. 1967. Modes of lubrication in human joints. *Proceedings of the Symposium on Lubrication and Wear in Living and Artificial Human Joints*, paper 12. Institution of Mechanical Engineers, London.

Dowson, D. and Higginson, G. R. 1959. A numerical solution to the elastohydrodynamic problem. *J. Mech. Eng. Sci.*, **1**, 6–15.

Dowson, D. and Higginson, G. R. 1960. Effect of material properties on the lubrication of elastic rollers. *J. Mech. Eng. Sci.*, **2**, 188–194.

Dowson, D. and Higginson, G. R. 1977. *Elastohydrodynamic Lubrication*, 2nd ed. Pergamon, Oxford.

Dowson, D., Higginson, G. R. and Whitaker, A. V. 1962. Elastohydrodynamic lubrication: A survey of isothermal solutions. *J. Mech. Eng.*, **4**, 121–126.

Dowson, D. and Wright, eds. 1981. *The Biomechanics of Joints and Joint Replacements*. Mech. Eng. Pub., Bury St., Edmunds, Suffolk.

Dowson, D., Higginson, G. R. and Whitaker, A. V. 1962. Elastohydrodynamic lubrication: a survey of isothermal solutions. *J. Mech. Eng. Sci.*, **4**, 121–126.

Gatcombe, E. K. 1945. Lubrication characteristics of involute spur-gears – a theoretical investigation. *ASME Trans.*, **67**, 177.

Gerald, C. F. 1973. *Applied Numerical Analysis*. Addison-Wesley, Reading, MA.

Greenwood, J. A. and Morales-Espejel, G. E. 1994. The behavior of transverse roghness in EHL contacts. *Proc. Instn. Mech. Engrs., Part J: J. Engineering Tribology*, **208**, 132.

Hamrock, B. J. 1991. *Fundamentals of Fluid Film Lubrication*. NASA Ref. Pub. 1244.

Hamrock, B. J. and Anderson, W., 1973. Analysis of an arched outer-race ball bearing considering centrifugal forces. *ASME Journal of Lubrication Technology*, **95**, 265–276.

Hamrock, B. J. and Anderson, W., 1983. *Rolling Element Bearings*. NASA RP-1105.

Hamrock, B. J. and Brewe, D. E. 1983. Simplified solution for stresses and deformations. *ASME Journal of Lubrication Technology*, **105**, 171–177.

Hamrock, B. J. and Dowson, D. 1976a. Isothermal elastohydrodynamic lubrication of point contacts, part I – theoretical formulation. *J. Lub. Tech.*, **98**, 223–229.

Hamrock, B. J. and Dowson, D. 1976b. Isothermal elastohydrodynamic lubrication of point contacts, part II – ellipticity parameter results. *J. Lub. Tech.*, **98**, 375–378.

Hamrock, B. J. and Dowson, D. 1977a. Isothermal elastohydrodynamic lubrication of point contacts, part III – fully flooded results. *J. Lub. Tech.*, **99**, 264–276.

Hamrock, B. J. and Dowson, D. 1977b. Isothermal elastohydrodynamic lubrication of point contacts, part IV – starvation results. *J. Lub. Tech.*, **99**, 15–23.

Hamrock, B. J. and Dowson, D. 1978. Elastohydrodynamic lubrication of elliptical contacts for materials of low elastic modulus, part I – fully flooded conjunction. *J. Lub. Tech.*, **100**, 236–245.

Hamrock, B. J. and Dowson, D. 1979a. Elastohydrodynamic lubrication of elliptical contacts for materials of low elastic modulus, part II – starved conjunction. *J. Lub. Tech.*, **101**, 92–98.

Hamrock, B. J. and Dowson, D. 1981. *Ball Bearing Lubrication*. John Wiley & Sons.

Harris, T. A. 1991. *Contact Bearing Analysis*. John Wiley.

Heathcote, H. L. 1921. The ball bearing in the making under test and service. *Proc. Inst. Automot. Eng*, **15**, 569–702.

Houpert, L. G. and Hamrock, B. J. 1986. Fast approach for calculating film thicknesses and pressures in elastohydrodynamically lubricated contacts at high loads. *ASME Journal of Tribology*, **108**, 411–420.

Johnson, K. L. 1984. *Contact Mechanics*. Cambridge University Press.

Jones, W. R. 1975. Pressure-viscosity measurement for several lubricants. *ASLE Trans.*, **18**, 249–262.

Kim, K. H. and Sadeghi, F. 1992. Three-dimensional temperature distribution in EHD lubrication: part I – circular contact. *ASME Journal of Tribology*, **114**, 32–42.

Kostreva, M. M. 1984. Elasto-hydrodynamic lubrication: a non-linear complementary problem. *Int. J. Num. Meth. Fluids*, **4**, 377–397.

Oh, K. P. and Huebner, K. H. 1973. Solution of the elastohydrodynamic finite journal bearing problem. *ASME Trans.*, **95F**, 342–343.

Oh, K. P. and Rohde, S. M. 1977. A theoretical analysis of a compliant shell air bearing. *ASME Trans.*, **99F**, 75–81.

Okamura, H. 1982. A contribution to the numerical analysis of isothermal elastohydrodynamic lubrication. *Tribology of Reciprocating Engines*. Proc. 9th Leeds Lyon Symp. on Tribology, Butterworths, pp. 313–320.

Palmgreen, A. 1959. *Ball and Roller Bearing Engineering*, Third ed. SKF Industries Inc.

Reynolds, O. 1876. On rolling friction. *Philos. Trans. Roy. Soc.*, **166**, 155–174.

Roelands, C. J. A. 1966. *Correlation Aspects of the Viscosity-Pressure Relationship of Lubricating Oils*. Ph.D. Thesis, Delft University of Technology, Netherlands.

Rohde, S. M. and Oh, K. P. 1975. A thermoelastohydrodynamic analysis of finite slider Bearings. *ASME Trans.*, **97F**, 450–460.

Rohde, S. M. 1978. Thick film and transient elastohydrodynamic lubrication problems. *General Motors Res. Publ.* GMR-2742.

Sibley, L. B. and Orcutt, F. K. 1961. Elastohydrodynamic lubrication of rolling contact surfaces. *ASLE Trans.*, **4**, 234–249.

Tallian, T. E. 1992a. Simplified contact fatigue life prediction model, Parts I and II. *ASME Journal of Tribology*, **114**, 207–222.

Tallian, T. E. 1992b. *Failure Atlas for Hertz Contact Machine Elements*. ASME Press, New York.

Tanner, R. I. 1966. An alternative mechanism for the lubrication of sinovial joints. *Phys. Med. Biol.*, **11**, 119.

Timoshenko, S. and Goodier, J. N. 1951. *Theory of Elasticity*. McGraw-Hill, New York.

Venner, C. H., ten Napel, W. E. and Bosma, R. 1990. Advanced multilevel solution of the EHL line contact problem. *ASME Journal of Tribology*, **112**, 426–432.

Venner, C. H. and Lubrecht, A. A. 1994. Transient analysis of surface features in an EHL line contact in the case of sliding. *ASME Journal of Tribology*, **116**, 186–193.

Wang, S. H., Al-Sharif, A., Rajagopal, K. R. and Szeri, A. Z. 1992. Lubrication with binary mixtures: liquid–liquid emulsion in an EHL conjunction. *ASME Journal of Tribology*, **115**, 515–522.

Wilcock, D. F. and Booser, E. R. 1957. *Bearing Design And Application*. McGraw-Hill, New York.

Xu, G. and Sadeghi, F. 1996. Thermal EHL analysis of circular contacts with measured surface roughness. *ASME Journal of Tribology*, **118**, 473–483.

CHAPTER 9

Thermal Effects

Classical lubrication theory predicts bearing performance on the assumption that the viscosity of the lubricant is uniform and constant over the whole film. As the bearing performance is strongly dependent on lubricant viscosity, and as the viscosity of common lubricants is a strong function of temperature (see Figure 9.1), the results of classical theory can be expected to apply only in cases where the lubricant temperature increase across the bearing pad is negligible.

9.1   Effective Viscosity

In many applications (small bearings and/or light running conditions) the temperature rise across the bearing pad, although not negligible, remains small. It is still possible in these cases to calculate bearing performance on the basis of classical theory, but in the calculations one must employ that specific value of the viscosity, called the *effective viscosity*, that is compatible with the average temperature rise in the bearing. This might be realized, for instance, by making an initial guess of the effective viscosity, followed by an iterative procedure, using Figure 9.1, for systematically refining the initial guess. Boswall (1928) calculated the effective viscosity on the basis of the following assumptions:

(1)   All the heat generated in the film by viscous action is carried out by the lubricant.
(2)   The lubricant that leaves the bearing by the sides has the uniform temperature $\Theta = \Theta_i + \Delta\Theta/2$, where $\Delta\Theta = \Theta_o - \Theta_i$ is the temperature rise across the bearing.

Let Q and Q_s represent the volumetric flow rate of the lubricant at the pad leading edge and at the two sides, respectively. Then, a simple energy balance based on the assumptions above yields

$$\frac{\rho c \Delta\Theta}{P} = \frac{4\pi c_\mu}{q\left(1 - \frac{1}{2}Q_s\middle/Q\right)}. \tag{9.1}$$

Here

$$c_\mu = \frac{R}{C}\frac{F_\mu}{W}$$

is the friction variable and $q = Q/RCNL$ is the dimensionless inflow. For short bearings with Gümbel's boundary condition, Eqs. (3.47) and (3.49) give, on substitution,

$$\left(\frac{L}{D}\right)^2 \frac{\rho c \Delta\Theta}{P} = \frac{8\pi(1 - \varepsilon^2)^{3/2}}{\varepsilon^2[\pi^2(1 - \varepsilon^2) + 16\varepsilon^2]^{1/2}}. \tag{9.2}$$

For other bearing geometries, Eq. (9.1) has been tabulated by Raimondi and Boyd (1958).

314

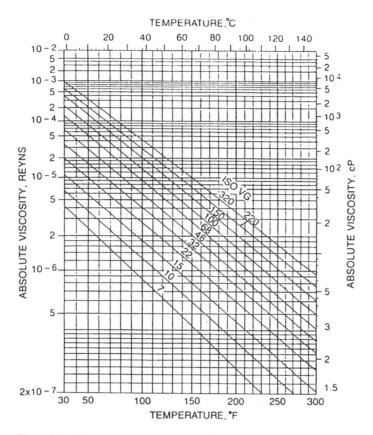

Figure 9.1. Viscosity-temperature curves for typical petroleum oils in ISO viscosity grades. (Reprinted with permission from Booser E. R., *CRC Handbook of Lubrication.* Copyright CRC Press, Boca Raton, Florida. © 1984.)

To indicate the iterative procedure for calculating the effective temperature, T_e, and hence the effective viscosity, $\mu_e(T_e)$, we rewrite Eq. (9.1). Since the right-hand side is a function of the temperature, we put

$$\Delta\Theta = \frac{4\pi c_\mu P}{q\left(1 - \frac{1}{2}Q_s\Big/Q\right)\rho c} \equiv g(\Theta).$$

The iterative procedure is then carried out according to the scheme

$$\Theta_e^{(n)} = \Theta_i + Kg\left[\Theta_e^{(n-1)}\right], \qquad n = 1, 2, 3, \ldots \tag{9.3}$$

Raimondi and Boyd (1958) suggested that $K = 1/2$, and Cameron (1966) recommended that $2/3 < K < 1$. It is questionable, however, whether a universal value of K exists even in small bearings (Seireg and Ezzat, 1972).

The effective viscosity method has been used, for example, on a self-contained bearing in which a disk lubricator supplies lubricant to the bearing (Kaufman, Szeri, and Raimondi, 1978). The bearing is designed to operate with or without external cooling (forced air or water), and carry both radial and axial loads (Figure 9.2).

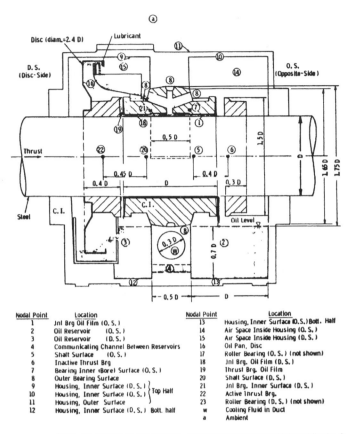

Disc (diam.=2.4 D)   Lubricant

D. S.
(Disc-Side)

O. S.
(Opposite-Side)

Thrust

Steel

Nodal Point	Location	Nodal Point	Location
1	Jnl Brg Oil Film (O. S.)	13	Housing, Inner Surface (O.S.) Bott. Half
2	Oil Reservoir (O. S.)	14	Air Space Inside Housing (O. S.)
3	Oil Reservoir (D. S.)	15	Air Space Inside Housing (D. S.)
4	Communicating Channel Between Reservoirs	16	Oil Pan, Disc
5	Shaft Surface (O. S.)	17	Roller Bearing (O. S.) (not shown)
6	Inactive Thrust Brg	18	Jnl Brg. Oil Film (D. S.)
7	Bearing Inner (Bore) Surface (O. S.)	19	Thrust Brg. Oil Film
8	Outer Bearing Surface	20	Shaft Surface (D. S.)
9	Housing, Inner Surface (D. S.) Top Half	21	Jnl Brg. Inner Surface (D. S.)
10	Housing, Inner Surface (O. S.) Top Half	22	Active Thrust Brg.
11	Housing, Outer Surface	23	Roller Bearing (D. S.) (not shown)
12	Housing, Inner Surface (D. S.) Bott. half	w	Cooling Fluid in Duct
		a	Ambient

location of nodal points for thermal analysis, & principal dimensions as related to journal diameter, D

Figure 9.2. Bearing geometry and location of nodal points. (Reprinted with permission from Kaufman, H. N., Szeri, A. Z. and Raimondi, A. A. Performance of a centrifugal disk-lubricated bearing. *ASLE Trans.*, **21**, 314–322, 1978.)

The bearings, shaft, housing, and lubricant constitute a complex thermal system both in geometry and in boundary conditions. Moreover, the heat transfer characteristics of this system are temperature dependent. In the analysis of Kaufman et al. (1978), the various components of the bearing-lubricant system are represented by nodal points. Once the nodal network is selected, heat conservation equations in finite difference form can be written. The heat transfer coefficients, which become coefficients in the finite-difference equations, are calculated from actual component characteristics. When their dependence on temperature is deemed essential to consider, they are based on temperatures of the previous iteration.

Thus, the bearing-housing-lubricant system, from a heat transfer point of view, is represented by a set of finite-difference conservation equations. These equations are statements of the requirement that the net rate of energy inflow into any nodal point be equal to the rate of energy dissipation at that node. Thus, at the ith nodal point, $i = 1, 2, 3, \ldots, N$,

$$\sum_{j=1}^{N} q_{ji} = -H_i \qquad i \neq j. \tag{9.4}$$

In lumped parameter analyses, such as this, the heat flow rate q_{ji} is given approximately by

$$q_{ji} = A(i, j)(\Theta_j - \Theta_i). \tag{9.5}$$

Here $A(i, j)$ represents the heat transfer coefficient between the ith and the jth node.

Substituting Eq. (9.5) into Eq. (9.4), we obtain

$$\sum_j A(i, j)(\Theta_j - \Theta_i) = -H_i. \tag{9.6}$$

Since in Eq. (9.6) the index i takes values $1, 2, 3, \ldots, N$, Eq. (9.6) is equivalent to a system of simultaneous algebraic equations, which can be written in the matrix form

$$A\Theta = H. \tag{9.7}$$

In Eq. (9.7), the symbol A represents the $N \times N$ matrix,

$$A = \begin{bmatrix} -\sum\limits_{k=1}^{N} A(1, k) & A(1, 2) & \cdots & A(1, N) \\ A(2, 1) & -\sum\limits_{k=1}^{N} A(2, k) & \cdots & A(2, N) \\ \vdots & & & \\ A(N, 1) & A(N, 2) & \cdots & -\sum\limits_{k=1}^{N} A(N, k) \end{bmatrix}, \tag{9.8}$$

while Θ and H represent column vectors,

$$\begin{aligned} \Theta &= (\Theta_1, \Theta_2, \Theta_3, \ldots, \Theta_N)^T, \\ H &= (H_1, H_2, H_3, \ldots, H_N)^T. \end{aligned} \tag{9.9}$$

When all heat transfer coefficients, as well as the viscosity, are kept frozen during a particular iteration, Eq. (9.7) is reduced to a set of linear algebraic equations.

The iterative analysis closely models the transient state of bringing the bearing-housing-lubricant system into thermal equilibrium. It starts with the calculation of heat production in the cold bearing for an assumed mechanical input (bearing losses). The temperature distribution is then obtained from a steady-state heat balance. But, owing to the high viscosity of the oil, the production rates of heat are high, leading necessarily to temperatures that are above the equilibrium temperatures compatible with the given mechanical input. New and improved heat production rates are now calculated, using average oil film temperatures that are based on the just-obtained oil film temperatures and the ones obtained previously. The iteration procedure is continued until sufficient agreement between successively calculated temperatures is reached. This process will converge and give solutions for wide ranges of input parameters.

Figure 9.3 compares theoretical prediction with experimental data on journal bearing temperature, for a range of geometrically similar bearings. Figure 9.4 compares theoretical predictions and experimental data for thrust face temperature under two different unit loads, for a range of bearing size.

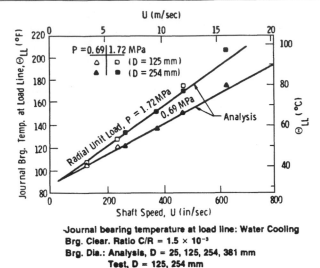

Figure 9.3. Journal bearing temperature at load line. Water cooling, $C/R = 1.5 \times 10^{-3}$, analysis: $D = 25, 125, 254, 381$ mm, test: $D = 125, 254$ mm. (Reprinted with permission from Kaufman, H. N., Szeri, A. Z. and Raimondi, A. A. Performance of a centrifugal disk-lubricated bearing. *ASLE Trans.*, **21**, 314–322, 1978.)

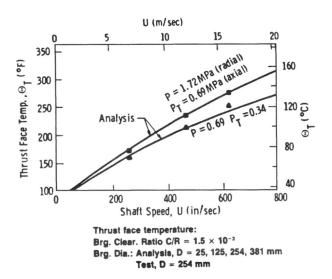

Figure 9.4. Thrust face temperature. $C/R = 1.5 \times 10^{-3}$, analysis: $D = 25, 125, 254,$ 381 mm, test: $D = 254$ mm. (Reprinted with permission from Kaufman, H. N., Szeri, A. Z. and Raimondi, A. A. Performance of a centrifugal disk-lubricated bearing. *ASLE Trans.*, **21**, 314–322, 1978.)

The effective viscosity model can, however, lead to significant errors in predicting bearing performance. It was shown by Seireg and Ezzat (1972) that although the normalized p/p_{max} plots (Figure 9.5) obtained under the different test conditions collapsed into a single curve, the magnitude of the pressures differed significantly depending on inlet temperature, even at constant speed.

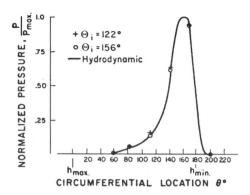

Figure 9.5. Normalized circumferential pressure distribution. (Reprinted with permission from Seireg, A. and Ezzat H. Thermohydrodynamic phenomena in fluid film lubrication. *ASME* Paper No. **72-Lub-25**, 1972.)

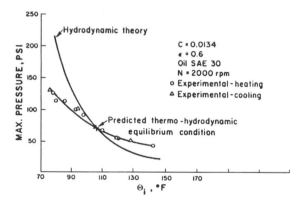

Figure 9.6. Variation of maximum pressure with inlet temperature. (Reprinted with permission from Seireg, A. and Ezzat H. Thermohydrodynamic phenomena in fluid film lubrication. *ASME* Paper No. **72-Lub-25**, 1972.)

Furthermore, the experiments showed that for any particular geometry and oil inlet temperature, there was only one speed at which the effective viscosity model and the experiment gave identical pressure distributions (Figure 9.6). Seireg and Ezzat offer the following conclusion: "The concept of an effective viscosity, although attractive, is infeasible since that value of the viscosity is a function of the bearing geometry, bush and shaft materials, lubricant physical properties, bearing load and speed, and the thermal boundaries."

In large bearings and/or under severe running conditions thermal effects may be significant, to the extent that prediction of bearing performance is no longer possible when based on the assumption of a uniform effective viscosity. Furthermore, in large bearings the usual limit condition in design is the maximum permissible bearing temperature. Early experiments by Gardner and Ulschmid (1974), Gregory (1974), Capitao (1976), and Capitao et al. (1976) are concerned with measuring maximum bearing temperatures.

A simple dimensional analysis will show that the nondimensional maximum bearing temperature, Θ_{max}/Θ_*, where Θ_* is a characteristic temperature and Θ_{max} is the maximum

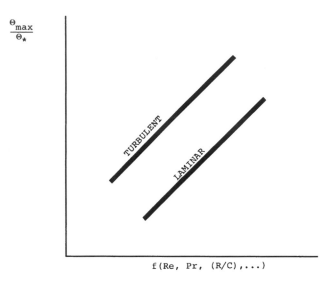

Figure 9.7. Schematics of bearing thermal performance the different flow regimes.

bearing temperature, is dependent on a number of nondimensional groups, such as the Reynolds number, Re, the Prandtl number, Pr, the dissipation number, Λ, the eccentricity ratio (C/R), and so forth. Analyzing a large set of field data on journal bearings of size $D = 1$ inch to $D = 32$ inches, it was found that the data can be arranged to fall on two parallel lines, as indicated schematically in Figure 9.7.

The lower line in Figure 9.7 is valid for laminar flow of the lubricant, while the upper line is characterized by turbulent flow. It is clearly demonstrated here that not only must we take into account the variation of lubricant viscosity with temperature, we must also specify the flow regime, to correctly predict bearing performance. In the next section, we shall discuss the variable viscosity theory, known as thermohydrodynamic theory, of fluid film lubrication. For additional details the reader is urged to consult Pinkus (1990).

9.2   Thermohydrodynamic Theory

Thermohydrodynamic (THD) theory calculates pointwise variations of temperature and viscosity in the lubricant film, then takes these into account when predicting bearing performance. Though there are numerous instances of large temperature rise across bearings operating in the laminar flow regime, THD theory becomes even more important in turbulent lubrication. For this reason, we will derive THD theory for turbulent lubrication; the resulting equations can be made to apply to laminar flow by elementary substitutions, or by specifying $Re^* = 0$ in computation.

The Energy Equation

The energy equation is the mathematical statement of the *Principle of Conservation of Energy: the time rate of energy increase in a body equals the time rate of energy supplied* to it.

$$\frac{d}{dt}(\mathcal{K} + \mathcal{E}) = \mathcal{W} + \mathcal{Q}. \tag{9.10}$$

Here

$$\mathcal{K} = \frac{1}{2}\int_V \rho \mathbf{v} \cdot \mathbf{v}\, dV, \qquad \mathcal{E} = \int_V \rho e\, dV. \tag{9.11}$$

The mechanical energy consists of work done by the surface and the body forces

$$\mathcal{W} = \int_S \boldsymbol{\tau} \cdot \mathbf{v}\, dS + \int_v \rho \mathbf{f} \cdot \mathbf{v}\, dV$$

$$= \int_S T_{lk}v_k n_1\, dS + \int_v \rho f_k v_k\, dV, \tag{9.12}$$

where we made use of Eq. (2.15).

For total energy input, we write

$$\mathcal{Q} = \int_S \mathbf{q} \cdot \mathbf{n}\, dS + \int_v \rho r\, dV, \tag{9.13}$$

where $\mathbf{q}$ is the heat flux across the closed surface $S(t)$ of $V(t)$ directed outward and r is the distributed heat source per unit mass of the body.

Upon substituting Eqs. (9.11), (9.12), and (9.13) into Eq. (9.10), we find

$$\frac{d}{dt}\int_V \rho\left(\frac{1}{2}v_k v_k + e\right)dV = \int_V \rho f_k v_k\, dV + \int_S T_{lk}v_k n_l\, dS + \int_S q_l n_l\, dS + \int_V \rho r\, dV. \tag{9.14}$$

To simplify matters, we assume that there are no internally distributed heat sources, $r = 0$. We also rewrite the left-hand side as follows:[1]

$$\frac{d}{dt}\int_V \left(\frac{1}{2}v_k v_k + e\right)\rho\, dV = \int_V \left(v_k\frac{dv_k}{dt} + \frac{de}{dt}\right)\rho\, dV, \tag{9.15}$$

where dv_k/dt is the kth component of the acceleration vector, Eq. (2.5).

Equation (9.14) contains both surface and volume integrals. We change the surface integrals by the divergence theorem and obtain

$$\int_S (T_{lk}v_k + q_l)n_l\, dS = \int_V [(T_{lk}v_k),_l + q_{l,l}]\, dV$$

$$= \int_V [T_{lk,l}v_k + T_{lk}v_{k,l} + q_{l,l}]\, dV. \tag{9.16}$$

Substituting Eqs. (9.15) and (9.16) into Eq. (9.14), we find

$$\int_V \left[\rho\frac{de}{dt} - T_{lk}v_{k,l} - q_{l,l} - v_k(T_{lk,l} + \rho f_k - \rho a_k)\right]dV = 0. \tag{9.17}$$

---

[1] Equation (9.15) follows from the Reynolds transport theorem $(d/dt)\int_{V(t)} F\, dV = \int_{V(t)}[(dF/dt) + F\,\mathrm{div}\,\mathbf{v}]\, dV$, where d/dt is the material derivative and $V(t)$ is a material volume, and the equation of continuity (2.44a).

The bracketed term multiplying v_k vanishes due to local conservation of linear momentum, Eq.(2.39), and as $V(t)$ is arbitrary the integrand itself must vanish,

$$\rho\frac{de}{dt} = T_{lk}v_{k,l} + q_{l,l}. \tag{9.18}$$

Equation (9.18) is our (local) statement of the *conservation of energy*.

The *stress power*, $T_{lk}v_{k,l}$, can further be simplified, using the Cartesian decomposition (2.20),

$$v_{l,k} = D_{kl} + \Omega_{kl}, \tag{9.19}$$

where $\boldsymbol{D} = (D_{kl})$ is the stretching tensor and $\boldsymbol{\Omega} = (\Omega_{kl})$ is the spin tensor.

It can be shown that the product of a symmetric tensor (stress) and a skew-symmetric tensor (spin) vanishes,[2] so on substituting Eq. (9.19) into Eq. (9.18), only $T_{lk}D_{kl}$ survives. The stress, on the other hand, for a Newtonian fluid is given by

$$T_{lk} = -p\delta_{lk} + 2\mu D_{lk}. \tag{2.35}$$

Substituting Eqs. (9.19) and (2.35) into the energy equation Eq. (9.18), we obtain

$$\rho\frac{de}{dt} = -pv_{k,k} + 2\mu D_{lk}D_{kl} - q_{l,l}, \tag{9.20a}$$

or, in vector notation,

$$\rho\frac{de}{dt} = -p\,\mathrm{div}\,\boldsymbol{v} + 2\mu\boldsymbol{D}:\boldsymbol{D} - \mathrm{div}\,\boldsymbol{q}. \tag{9.20b}$$

Assuming further that Fourier's law of heat conduction

$$\boldsymbol{q} = -k\,\mathrm{grad}\,\Theta$$

holds, where Θ is the temperature and k is the thermal conductivity, we obtain

$$\rho\frac{d(c_v\Theta)}{dt} = -p\,\mathrm{div}\,\boldsymbol{v} + \mu\Phi + \mathrm{div}(k\,\mathrm{grad}\,\Theta). \tag{9.21}$$

Here we put $e = c_v\Theta$ for internal energy density and use the symbol Φ for the *dissipation function*:

$$e = c_v\Theta, \qquad \Phi = 2D_{ij}D_{ji}. \tag{9.22}$$

To further simplify matters, we assume that the lubricant has constant thermal properties, $c_v, k = \mathrm{const.}$ For incompressible fluids, the dilatation work $-p\,\mathrm{div}\,\boldsymbol{v} = 0$ by Eq. (2.44b), $c_v = c_p = c$, and Eq. (9.21) becomes[3]

$$\rho c\left(\frac{\partial\Theta}{\partial t} + u\frac{\partial\Theta}{\partial x} + v\frac{\partial\Theta}{\partial y} + w\frac{\partial\Theta}{\partial z}\right) = k\left(\frac{\partial^2\Theta}{\partial x^2} + \frac{\partial^2\Theta}{\partial y^2} + \frac{\partial^2\Theta}{\partial z^2}\right) + \mu\Phi. \tag{9.23a}$$

[2] $T_{lk}\Omega_{kl} = T_{lk}(-\Omega_{lk}) = -T_{lk}\Omega_{lk} = -T_{kl}\Omega_{lk} = -T_{lk}\Omega_{kl}$, as $\Omega_{kl} = -\Omega_{lk}$, and $T_{kl} = T_{lk}$.
[3] With the aid of the equation of state, $p = \rho\mathfrak{R}\Theta$, it can be shown that $\rho(d/dt)(c_v\Theta) + p\,\mathrm{div}\,\boldsymbol{v} = \rho(d/dt)(c_p\Theta) - dp/dt$, so for perfect gas (9.21) takes the form $\rho(d/dt)(c_p\Theta) = dp/dt + 2\mu\boldsymbol{D}:\boldsymbol{D} - \mathrm{div}\,\boldsymbol{q}$. This equation should not be used for an incompressible fluid.

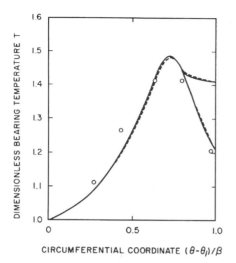

Figure 9.8. Comparison of (○) measured and (——) predicted temperature distributions along the center line of a $D = 10$ cm journal bearing at $N = 900$ revolutions/min. (Reprinted with permission from Suganami, T. and Szeri, A. Z. A thermohydrodynamic analysis of journal bearings. *ASME Journal of Lubrication Technology*, **101**, 21–27, 1979.)

For an incompressible fluid the dissipation function is

$$\phi = 2\left[\left(\frac{\partial u}{\partial x}\right)^2 + \left(\frac{\partial v}{\partial y}\right)^2 + \left(\frac{\partial w}{\partial z}\right)^2\right] + \left(\frac{\partial u}{\partial y} + \frac{\partial v}{\partial x}\right)^2$$
$$+ \left(\frac{\partial v}{\partial z} + \frac{\partial w}{\partial y}\right)^2 + \left(\frac{\partial w}{\partial x} + \frac{\partial u}{\partial z}\right)^2. \tag{9.23b}$$

One of the early applications of Eq. (9.23) to hydrodynamic lubrication was made by Cope (Cope, 1949; Charnes, Osterle, and Saibel, 1952). His work subsequently acquired classical status in lubrication and strongly influenced later research. Cope's model is based on the assumptions of (1) negligible temperature variation across the film and, therefore, (2) negligible heat conduction into the neighboring solids.

All the generated heat is carried out by the lubricant under Cope's assumption (*adiabatic theory*). Furthermore, by neglecting the (second-order) convection terms, the order of the differential equation (9.23a) is lowered. Solutions of the resulting first-order equation cannot satisfy all the boundary conditions of the problem.[4]

The maximum film temperature is always located at the trailing edge in Cope's adiabatic model, whereas measurements locate its position upstream of the trailing edge and just downstream from the position of maximum pressure (Figure 9.8). For a while it was generally accepted that the classical isothermal theory and Cope's adiabatic theory would bracket actual bearing operations. This idea was later discarded (Seireg and Ezzat, 1972; McCallion, Yousif, and Lloyd, 1970).

---

[4] We may draw a parallel between Cope's approximation and the procedure that neglects the viscous terms in the Navier-Stokes equation. The latter yields the equations of motion for an ideal fluid, solutions of which cannot satisfy the no-slip boundary condition.

Dowson and Hudson (1963) were among the first to realize the importance of heat conduction across the film, and their work provided the foundations of *thermohydrodynamic theory*. Notable subsequent papers were those by Dowson and March (1967), McCallion et al., (1970), and Ezzat and Rohde (1972). A common assumption of the work reported in these papers is that heat conduction in the direction of relative motion is unimportant compared to heat convection in the same direction. This permits elimination of the term $\partial^2 \Theta / \partial x^2$ in Eq. (9.23). But in the process of deleting this term, the equation loses some of its generality, as it changes from elliptic to parabolic. The parabolic energy equation is valid only when no *reverse flow* is encountered.

Reverse flow is encountered in journal bearings (see Figure 6.4) near the stationary surface at inlet and usually occurs at high eccentricity ratios – a condition that is also typical of high bearing temperatures. It is advisable, therefore, to retain the conduction term $\partial^2 \Theta / \partial x^2$ in Eq. (9.23), or to make other arrangements to accommodate reverse flow (Suganami and Szeri, 1979; Boncompain, Fillon, and Frêne, 1986).

Before continuing with further discussion of Eq. (9.23), we will cast it in a form that is appropriate for turbulent flow of the lubricant. For this purpose, and in analogue with Eq. (6.29), we write the temperature and the stretching tensor as the sum of the mean and the fluctuation[5]

$$\Theta = \bar{\Theta} + \theta', \quad D_{ij} = \bar{D}_{ij} + d'_{ij}. \tag{9.24}$$

From Eqs. (9.22a) and (7.4) the mean value of the dissipation is

$$\mu \bar{\Phi} = 2\mu D_{ij} D_{ij} = 2\mu \bar{D}_{ij} \bar{D}_{ij} + 2\mu \overline{d'_{ij} d'_{ij}}. \tag{9.25a}$$

For stationary turbulence the last term in Eq. (9.25a) can be obtained from Eq. (7.37)

$$2\mu \overline{d'_{ij} d'_{ij}} = \mu \overline{\left(\frac{\partial v'_i}{\partial x_j} + \frac{\partial v'_j}{\partial x_i} \right) \frac{\partial v'_i}{\partial x_j}}$$

$$= \frac{\partial}{\partial x_i} \overline{v_i \left(p + \rho \frac{q^2}{2} \right)} - \rho \overline{v'_i v'_j} \frac{\partial \bar{V}_j}{\partial x_i} + \mu \frac{\partial}{\partial x_i} \overline{v'_i \left(\frac{\partial v'_i}{\partial x_j} + \frac{\partial v'_j}{\partial x_i} \right)}. \tag{9.25b}$$

Employing the Boussinesq model, Eq. (7.19), for the second term on the right, applying the thin film approximation to simplify the first and third term, and integrating across the film thickness (cf., Vohr in Safar and Szeri, 1974) we find

$$\int_0^h \overline{\mu d'_{ij} d'_{ij}} \, dy \approx \int_0^h \rho \varepsilon_m \bar{D}_{ij} \bar{D}_{ij} \, dy. \tag{9.25c}$$

This is our justification for writing the mean dissipation as

$$\mu \bar{\Phi} \approx 2\bar{\mu} \left(1 + \frac{\varepsilon_m}{\nu} \right) \bar{D}_{ji} \bar{D}_{ij}. \tag{9.26}$$

Substituting Eq. (9.24) into Eq. (9.23), averaging the resulting equation, and taking into account Eq. (9.26), we obtain

$$\rho c \left[\bar{U}_j \frac{\partial \bar{\Theta}}{\partial x_j} + \overline{v'_j \frac{\partial \theta'}{\partial x_j}} \right] = k \frac{\partial^2 \bar{\Theta}}{\partial x_j \partial x_j} + 2\bar{\mu} \left(1 + \frac{\varepsilon_m}{\nu} \right) \bar{D}_{ji} \bar{D}_{ij}. \tag{9.27}$$

---

[5] Note that in Chapter 7 and Chapter 9 the overscore bar signifies statistical average.

Making use of the equation of continuity (7.7b), we can show that

$$\overline{v'_j \frac{\partial \theta'}{\partial x_j}} = \frac{\partial}{\partial x_j} \left(\overline{v'_j \theta'} \right)$$

and write Eq. (9.27) as

$$\rho c \bar{U}_j \frac{\partial \bar{\Theta}}{\partial x_j} = \frac{\partial}{\partial x_j} \left[k \frac{\partial \bar{\Theta}}{\partial x_j} - \rho c \overline{v'_j \theta'} \right] + 2\bar{\mu} \left(1 + \frac{\varepsilon_m}{\nu} \right) \bar{D}_{ji} \bar{D}_{ij}. \tag{9.28}$$

To make use of the particular geometry of the lubricant film, we shall nondimensionalize Eq. (9.28), using Eqs. (7.12) and (9.29), and

$$T = \frac{\bar{\Theta}}{\Theta_*}, \qquad t = \frac{\theta'}{\theta_*}, \qquad \mu = \frac{\bar{\mu}}{\mu_*}, \tag{9.29}$$

where Θ_* and θ_* are the temperature scales for mean and fluctuation, respectively, $\bar{\mu} = \mu(\bar{\Theta})$ and $\mu_* = \mu(\Theta_*)$.

Substituting Eqs. (7.12) and (9.29) into Eq. (9.28), then neglecting terms multiplied by (L_y/L_{xz}), where L_y and L_{xz} are characteristic film dimensions, we find

$$U \frac{\partial T}{\partial \xi} + V \frac{\partial T}{\partial \eta} + W \frac{\partial T}{\partial \zeta}$$

$$= \frac{1}{\text{Pe}} \frac{\partial^2 T}{\partial \eta^2} - \left(\frac{L_{xz}}{L_y} \right) \left(\frac{u_*}{U_*} \right) \left(\frac{\theta_*}{\Theta_*} \right) \left[\frac{\partial \overline{vt}}{\partial \eta} + \left(\frac{L_y}{L_{xz}} \right) \left(\frac{\partial \overline{ut}}{\partial \xi} + \frac{\partial \overline{wt}}{\partial \zeta} \right) \right] \tag{9.30}$$

$$+ \Lambda \mu \left(1 + \frac{\varepsilon_m}{\nu} \right) \left[\left(\frac{\partial U}{\partial \eta} \right)^2 + \left(\frac{\partial W}{\partial \eta} \right)^2 \right].$$

Assuming that $(u_*/U_*) = O(L_y/L_{xz})^{1/2}$ and $\theta_*/\Theta_* = O(L_y/L_{xz})^{1/2}$, conditions that seem to hold in wall turbulence,[6] and letting $L_y/L_{xz} \to 0$, Eq. (9.30) takes the form

$$U \frac{\partial T}{\partial \xi} + V \frac{\partial T}{\partial \eta} + W \frac{\partial T}{\partial \zeta} = \frac{\partial}{\partial \eta} \left[\frac{1}{\text{Pe}} \frac{\partial T}{\partial \eta} - \overline{vt} \right]$$

$$+ \Lambda \mu \left(1 + \frac{\varepsilon_m}{\nu} \right) \left[\left(\frac{\partial U}{\partial \eta} \right)^2 + \left(\frac{\partial W}{\partial \eta} \right)^2 \right]. \tag{9.31}$$

The *Peclet number* and the *dissipation number* have the definition

$$\text{Pe} = \text{Pr} \times \text{Re} \left(\frac{C}{R} \right), \qquad \Lambda = \frac{\mu_* \omega}{\rho c \Theta_*} \left(\frac{L_{xz}}{L_y} \right)^2, \tag{9.32}$$

where $\text{Pr} = c\mu_*/k$ is the *Prandtl number*.

Although it is well recognized that turbulent transport of a scalar quantity is brought about by both gradient-type diffusion caused by small-scale turbulence and by large scale motion

---

[6] See the discussion following Eq. (7.13).

of eddies (Hinze, 1975), it has long been accepted in heat transfer to use the approximation (Kestin and Richardson, 1963)

$$-\rho \overline{v'\theta'} = \rho \varepsilon_H \frac{\partial \bar{\Theta}}{\partial y}, \tag{9.33a}$$

where ε_H is the *eddy viscosity for the transport of heat.* In nondimensional form, Eq. (9.33a) is

$$\overline{vt} = \frac{1}{\mathrm{Re}^*} \frac{1}{\mathrm{Pr}^{(t)}} \frac{\varepsilon_m}{\nu} \frac{\partial T}{\partial \eta}. \tag{9.33b}$$

Here $\mathrm{Re}^* = \mathrm{Re}(C/L)$ is the reduced Reynolds number, and we put $\mathrm{Pr}^{(t)} = \varepsilon_m/\varepsilon_H$, where $\mathrm{Pr}^{(t)}$ is the turbulent Prandtl number.

Substituting Eq. (9.33) into Eq. (9.31), we obtain the final form of the (nondimensional) energy equation applicable to turbulent lubricant films

$$U\frac{\partial T}{\partial \xi} + V\frac{\partial T}{\partial \eta} + W\frac{\partial T}{\partial \zeta} = \frac{1}{\mathrm{Pe}} \frac{\partial}{\partial \eta}\left[\left(1 + \frac{\mathrm{Pr}}{\mathrm{Pr}^{(t)}}\bar{\mu}\frac{\varepsilon_m}{\nu}\right)\frac{\partial T}{\partial \eta}\right]$$
$$+ \Lambda\mu\left(1 + \frac{\varepsilon_m}{\nu}\right)\left[\left(\frac{\partial U}{\partial \eta}\right)^2 + \left(\frac{\partial W}{\partial \eta}\right)^2\right]. \tag{9.34a}$$

Implicit in Eq. (9.34) are the assumptions

$$(u_*/U_*) = O\left(L_y/L_{xz}\right)^{1/2}, \qquad \theta_*/\Theta_* = O(L_y/L_{xz})^{1/2}. \tag{9.34b}$$

In journal bearings, the axial variation of temperature can often be neglected (Dowson and March, 1967), and we employ

$$\left[\left(\frac{\partial U}{\partial \eta}\right)^2 + \left(\frac{\partial W}{\partial \eta}\right)^2\right] \approx \left(\frac{\partial U}{\partial \eta}\right)^2_{\zeta=0}$$

to obtain the approximate equation[7]

$$U\frac{\partial T}{\partial \xi} + V\frac{\partial T}{\partial \eta} = \frac{1}{\mathrm{Pe}} \frac{\partial}{\partial \eta}\left[\left(1 + \frac{\mathrm{Pr}}{\mathrm{Pr}^{(t)}}\mu\frac{\varepsilon_m}{\nu}\right)\frac{\partial T}{\partial \eta}\right] + \Lambda\mu\left(1 + \frac{\varepsilon_m}{\nu}\right)\left(\frac{\partial U}{\partial \eta}\right)^2. \tag{9.35}$$

Although the correct scaling of turbulence appears to be as in Eq. (9.34), leading to Eq. (7.16) for linear momentum and to Eq. (9.35) for energy, the momentum equations are further simplified by most authors by neglecting fluid inertia; this simplification yields Eqs. (7.15) and (9.35) for the characterization of the turbulent flow of incompressible lubricants.

When evaluating ε_m/ν, say from Reichardt's formula, Eq. (7.34), we find the eddy viscosity to be a continuous, monotonic function of the local Reynolds number, $\mathrm{Re}_h = Uh/\nu$; that is, the transition from laminar to fully turbulent flow is smooth and gradual. But, this does not correspond to observation of the physical process of transition. If Re_L and Re_U represent the lower and upper critical Reynolds number of transition, respectively, then we would expect ε_m/ν to vanish whenever $\mathrm{Re}_h < \mathrm{Re}_L$ and to have its full value, as calculated from Reichardt's formula, only when $\mathrm{Re}_h > \mathrm{Re}_U$. [The fact that this obviously incorrect (i.e., in the range $\mathrm{Re} < \mathrm{Re}_U$) eddy viscosity is multiplied in Eq. (9.35) by the laminar Prandtl number only makes things worse, as for typical processes $100 < P_R < 500$.] To compensate for this deficiency of Eq. (9.34), Suganami and Szeri (1979), introduced a

---

[7] Note that $\partial/\partial \xi = \partial/\partial \bar{x} + [\bar{y}\varepsilon \sin \bar{x}/(1 + \varepsilon \cos \bar{x})](\partial/\partial \bar{y})$, where $(\bar{x}, \bar{y}) = (\xi, \eta/H)$.

scaling factor for the eddy viscosity,

$$\vartheta = \begin{cases} 0 & \text{if } \text{Re}_h < 400 \\ 1 - \left(\dfrac{900 - \text{Re}_h}{500}\right)^{1/8} & \text{if } 400 < \text{Re}_h < 900 \\ 1 & \text{if } \text{Re}_h > 900 \end{cases} \qquad (9.36a)$$

and replaced ε_m/ν in Eq. (9.34) by $\vartheta\,\varepsilon_m/\nu$.

It is customary to solve Eq. (9.34) simultaneously with the equation of heat conduction in the bearing and require continuity of both temperature and heat flow rate at the lubricant-bearing interface (Dowson and March, 1967; Ezzat and Rohde, 1972).[8] The remaining boundary conditions are not easily defined and have been a source of intense discussion in the literature.

The idea of a near-isothermal shaft was put forth by Dowson and March (1967) on the basis of experimental observations (Dowson, Hudson, Hunter, and March, 1966). Suganami and Szeri (1979) assumed that the shaft is at a constant temperature at small surface speeds (laminar flow), as a significant portion of the generated heat leaves through the shaft by conduction in that case (Dowson, Hudson, Hunter, and March, 1966). But for large surface velocities (superlaminar flow), they solved Eq. (7.35) on the assumption of a thermally insulated shaft, as now only an insignificant portion of the heat leaves through the shaft. Physically, neither the isothermal shaft nor the insulated shaft assumption is valid. Nevertheless, experimental evidence does not contradict the predictions of Suganami and Szeri (1979), which are based on these assumptions (Figure 9.9). When applying Eq. (9.34) to thrust bearings, constant runner temperature is the accepted boundary condition (Ezzat and Rohde, 1972).

We may specify the temperature on the section of the leading edge that is free from backflow. At the remainder of the leading edge, where backflow is present, as well as at the trailing edge, we do better to prescribe negligible conduction, i.e., zero temperature gradient (Suganami and Szeri, 1979).

In summary, Eq. (9.35) was transformed to the $(\bar{x}, \bar{y})$ coordinate system (see footnote 7) and solved subject to the boundary conditions (Suganami and Szeri, 1979):

$$\left.\begin{aligned} T &= T_i(\bar{y}) & \text{if } u \geq 0 \\ \frac{\partial T}{\partial \bar{x}} &= 0 & \text{if } u \geq 0 \end{aligned}\right\} \quad \text{at } \bar{x} = \bar{x}_1,$$

$$\frac{\partial T}{\partial \bar{x}} = 0 \quad \text{at } \bar{x} = \bar{x}_1 + \beta,$$

$$\left[T + \Gamma(\bar{x})\frac{\partial T}{\partial \bar{y}}\right] = T_a \quad \text{at } \bar{y} = 0, \qquad (9.36b)$$

$$\left.\begin{aligned} T &= T_s & \text{if } (R_h)_{\max} \leq 400 \\ \frac{\partial T}{\partial \bar{y}} &= 0 & \text{if } (R_h)_{\max} > 400 \end{aligned}\right\} \quad \text{at } \bar{y} = H.$$

---

[8] To cut down on the amount of computations, an approximate equation for heat conduction in the bearing may be integrated analytically (Safar and Szeri, 1974; Suganami and Szeri, 1979a; Yu and Szeri, 1975). The matching conditions on temperature and heat flow can then be replaced by an approximate film-bearing interfacial condition.

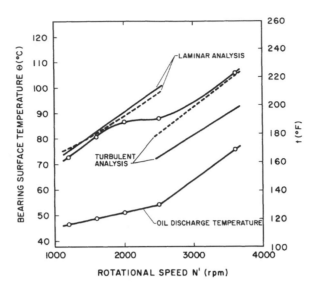

Figure 9.9. Variation of maximum bearing temperature with speed in a pivoted-pad journal bearing with $D = 431.8$ mm and $\beta = 60°$. o, experimental (Gardner and Ulschmid, 1974); — —, constant-temperature shaft; - - - -, insulated shaft, theoretical. (Reprinted with permission from Suganami, T. and Szeri, A. Z. A thermohydrodynamic analysis of journal bearings. *ASME Journal of Lubrication Technology*, **101**, 21–27, 1979.)

The boundary function $\Gamma(\bar{x})$ is obtained from the approximate heat conduction equation for pad temperature Θ_p that neglects conduction in the circumferential direction

$$\frac{1}{r}\frac{\partial}{\partial r}\left(r\frac{\partial \Theta_p}{\partial r}\right) = 0,$$

and matching heat flow and temperature at the film-pad interface.

The Pressure Equation

Derivation of the pressure equation that accounts for pointwise variation of viscosity was first given for laminar flow by Dowson and Hudson (1963). By starting from Eq. (7.15), rather than from the corresponding laminar equations of motion as Dowson and Hudson did, we can combine the ideas of Dowson and Elrod (Elrod and Ng, 1967) and make the derivation apply to the turbulent flow of the lubricant (Safar and Szeri, 1974).

Applying the eddy viscosity hypotheses (7.20), we obtain

$$\frac{\partial \bar{P}}{\partial x} = \frac{\partial}{\partial y}\left[\bar{\mu}\left(1 + \frac{\varepsilon_m}{\nu}\right)\frac{\partial \bar{U}}{\partial y}\right], \tag{9.37a}$$

$$\frac{\partial \bar{P}}{\partial y} = 0, \tag{9.37b}$$

$$\frac{\partial \bar{P}}{\partial z} = \frac{\partial}{\partial y}\left[\bar{\mu}\left(1 + \frac{\varepsilon_m}{\nu}\right)\frac{\partial \bar{W}}{\partial y}\right]. \tag{9.37c}$$

These equations can be integrated formally to yield

$$\bar{U} = \frac{1}{\mu_*}\frac{\partial \bar{P}}{\partial x}\Psi(x,y,z) + U_0\frac{\xi_1(x,y,z)}{\xi_1(x,h,z)}, \tag{9.38a}$$

$$\bar{W} = \frac{1}{\mu_*}\frac{\partial \bar{P}}{\partial z}\Psi(x,y,z). \tag{9.38b}$$

Here

$$\Psi(x,y,z) = \xi_2(x,y,z) - \frac{\xi_2(x,h,z)}{\xi_1(x,h,z)}\xi_1(x,y,z),$$

$$\xi_1(x,\phi,z) = \int_0^\phi \frac{dy}{\mu f(x,y,z)}, \qquad \xi_2(x,\phi,z) = \int_0^\phi \frac{y\,dy}{\mu f(x,y,z)}, \tag{9.38c}$$

and we use the notation

$$f(x,y,z) = \left(1 + \frac{\varepsilon_m}{\nu}\right).$$

Implicit in Eqs. (9.38) are the no-slip velocity boundary conditions

$$\begin{aligned} \bar{U} = \bar{W} = 0 &\qquad \text{at } y = 0, \\ \bar{U} = U_0, \quad \bar{W} = 0 &\qquad \text{at } y = h. \end{aligned}$$

We may now follow the procedure already outlined in the derivation of the classical Reynolds equation. Thus, substituting Eqs. (9.38) into the equation of continuity (7.7a) and integrating with the boundary conditions[9]

$$[\bar{V}]_0^h = \frac{dh}{dt},$$

we obtain

$$\frac{\partial}{\partial x}\left(\frac{\Gamma}{\mu_*}\frac{\partial \bar{P}}{\partial x}\right) + \frac{\partial}{\partial z}\left(\frac{\Gamma}{\mu_*}\frac{\partial \bar{P}}{\partial z}\right) = U_0\frac{\partial}{\partial x}\left(h - \frac{\xi_2(x,h,z)}{\xi_1(x,h,z)}\right) + V_0 \tag{9.39a}$$

for journal bearings and

$$\frac{\partial}{\partial x}\left(\frac{\Gamma}{\mu_*}\frac{\partial \bar{P}}{\partial x}\right) + \frac{\partial}{\partial z}\left(\frac{\Gamma}{\mu_*}\frac{\partial \bar{P}}{\partial z}\right) = U_0\frac{\partial}{\partial x}\frac{\xi_2(x,h,z)}{\xi_1(x,h,z)} + V_0 \tag{9.39b}$$

for slider bearings. Here we employed the notation

$$\Gamma(x,z) = -\int_0^{h(x)} \Psi(x,y,z)\,dy.$$

For isothermal flow, Eq. (9.39) is identical to the pressure equation of Elrod and Ng (1967), and for laminar flow, it reduces to the thermohydrodynamic pressure equation of Dowson and co-workers (Dowson and Hudson, 1963; Dowson and March, 1967).

Although the distribution of the temperature and the viscosity is assumed to be known at the start of the solution of Eq. (9.39), this equation has nonlinear coefficients because

[9] $U_0 = U_2$ for journal bearings and $U_0 = U_1$ for slider bearings, if the bearing surface is stationary. In either case $V_0 = V_2 - V_1$ (see Chapter 2).

the eddy viscosity is dependent on shear stress. When these coefficients are linearized (see Suganami and Szeri 1979), the equation becomes a generalization of the Ng and Pan (1965) linearized turbulent equation. This equation is similar in form to the constant-viscosity turbulent lubrication equation (7.59).

When nondimensionalized for steady state, the turbulent Reynolds equation is given by

$$\frac{\partial}{\partial \bar{x}}\left(H^3 G_{\bar{x}} \frac{\partial P}{\partial \bar{x}}\right) + \left(\frac{D^2}{L}\right) \frac{\partial}{\partial \bar{z}}\left(H^3 G_{\bar{z}} \frac{\partial P}{\partial \bar{z}}\right) = 2\pi \frac{\partial (HF)}{\partial \bar{x}}. \tag{9.40}$$

The turbulence functions $G_{\bar{x}}$ and $G_{\bar{z}}$ are defined by

$$G_{\bar{x}} = \int_0^1 \zeta_1(\eta)\, d\eta - \frac{\zeta_1(1)}{\zeta_2(1)} \int_0^1 \zeta_2(\eta)\, d\eta, \tag{9.41a}$$

$$G_{\bar{z}} = \int_0^1 \zeta_3(\eta)\, d\eta - \frac{\zeta_3(1)}{\zeta_4(1)} \int_0^1 \zeta_4(\eta)\, d\eta, \tag{9.41b}$$

$$F = \frac{1}{\zeta_4(1)} \int_0^1 \zeta_4(\eta)\, d\eta. \tag{9.41c}$$

Here

$$\zeta_1(\beta) = \int_0^\beta \frac{\left(\frac{1}{2} - \eta\right)}{\mu f_c(\eta)} \left[1 - \frac{g_c(\eta)}{f_c(\eta)}\right] d\eta, \tag{9.42a}$$

$$\zeta_2(\beta) = \int_0^\beta \frac{1}{\mu f_c(\eta)} \left[1 - \frac{g_c(\eta)}{f_c(\eta)}\right] d\eta, \tag{9.42b}$$

$$\zeta_3(\beta) = \int_0^\beta \frac{\left(\frac{1}{2} - \eta\right)}{\mu f_c(\eta)}\, d\eta, \tag{9.42c}$$

$$\zeta_4(\beta) = \int_0^\beta \frac{d\eta}{\mu f_c(\eta)}. \tag{9.42d}$$

The functions f_c and g_c are given by Eqs. (6.68) and (6.69), respectively,

$$f_c(\eta) = 1 + \kappa \left[\bar{y} h_c^+ - \delta_\ell^+ \tanh\left(\frac{\bar{y} h_c^+}{\delta_\ell^+}\right)\right], \tag{7.68}$$

$$g_c(\eta) = \frac{1}{2}\kappa \bar{y} h_c^+ \tanh^2\left(\frac{\bar{y} h_c^+}{\delta_\ell^+}\right), \tag{7.69}$$

where $\bar{y} = y/h$ and $h_c^+ = \frac{h}{\nu}\sqrt{\frac{|\tau_c|}{\rho}}$, a local Reynolds number, were defined in Chapter 7.

Equation (9.40) was solved subject to the Swift-Stieber boundary conditions (3.84), and the lubricant viscosity was calculated from Vogel's formula

$$\bar{\mu} = \exp\left[\frac{a}{b + \bar{\Theta}}\right], \tag{9.43}$$

where the material parameters a and b are determined through curve fitting.

Table 9.1. *Test bearing*

Shaft speed	$N = 900$ rpm	$N = 3000$ rpm
Pressure kPa	68.65	46.09
Temperature (°C)	22.40	34.40
Flow rate (cm³/s)	37.60	59.00

$D = 10$ cm, $L = 5$ cm, $\beta = 150°$, $C = 0.01$ cm, $R_b = 15$ cm.

9.3   Journal Bearings

Bearing Temperature

Theoretical predictions by the THD model are compared here with two sets of experimental data.

One of the sets pertain to a pivoted-pad bearing and was taken from Gardner and Ulschmid (1974). Figure 9.9 shows both calculated and experimental bearing temperature data plotted against rotational speed for the $D = 43.2$ cm tilting pad bearing. The graph also displays measured oil discharge temperatures. Transition from laminar to superlaminar flow regime seems to take place in the $2000 < N$ (rpm) < 2500 interval. This corresponds, roughly to $400 < (R_h)_{max} < 600$, where $(R_h)_{max}$ is the maximum value of the local Reynolds number within the 60° bearing arc. To investigate the effect thermal conditions at the shaft-lubricant interface have on lubricant temperature, calculations were performed with both the constant-temperature boundary condition, when the uniform shaft temperature was equated to the oil discharge temperature, and the zero temperature gradient boundary condition. In the laminar regime, there seems to be little difference between the two predictions, as indicated by Figure 9.9, nevertheless the constant-temperature boundary condition shows better agreement with experiment, as was suggested earlier. This is not so in the turbulent regime, here the locally adiabatic shaft (zero normal gradient boundary condition) leads to results that show far better agreement with the experiment, as would be expected on the basis of the previous discussion.

It may also be noticed from Figure 9.9 that laminar theory grossly overestimates the transition region temperatures, whereas turbulent analysis underestimated it. Furthermore, the turbulent analysis does not yield results that coincide with the results of laminar analysis at any finite value of the Reynolds number. The reason for this is that in the turbulent model calculation of the eddy viscosity from Reichardt's model makes for gradual transition, and the eddy viscosity is given a nonzero value even when the flow is laminar, as discussed above. Continuous transition from laminar to turbulent conditions, and agreement with experiments in the transition regime, was achieved on introducing the eddy viscosity scaling factor ϑ from Eq. (9.36).

The second set of data used here for comparison was obtained on a centrally loaded, "viscosity pump" bearing (Suganami and Szeri, 1979a). This bearing has a 210° relief in the top cap and is limited in the axial direction by two deep circumferential oil grooves, one at each end. It is loaded by means of a hydraulic load cell. The temperature measurements were obtained with thermocouples, sunk into the babbit to within 0.5 cm of the bearing surface. Dimensions of the test bearing and relevant feed-oil data are shown in Table 9.1.

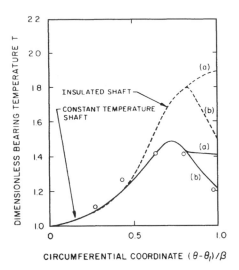

Figure 9.10. Circumferential temperature profile at $N = 900$ rpm, Table 9.1. (Reprinted with permission from Suganami, T. and Szeri, A. Z. A thermohydrodynamic analysis of journal bearings. *ASME Journal of Lubrication Technology*, **101**, 21–27, 1979).

Predicted and measured circumferential temperature profiles are compared for the $D = 10$ cm diameter bearing of Table 9.1 in Figure 9.10. There are two solid curves in the figure. Curve (a) represents solution of the energy equation of the THD model. As this equation is two dimensional, it assumes that the clearance gap is filled with lubricant everywhere. This solution clearly overestimates bearing temperatures. Curve (b) also assumes that a continuous film exists in the diverging gap and cavitation only affects pressure calculations, but now the lubricant is replenished from the oil grooves at the sides. One would expect this solution to underestimate bearing temperatures.

Note that Figure 9.10 displays laminar bearing operations and, in line with previous contention, shows better agreement between theory and experiment when the former is based on constant, temperature boundary condition at the shaft-film interface.

The Role of Nondimensional Parameters

The previous section brings into focus the governing dimensionless parameters of the various lubrication models. If the geometric parameters β and L/D are fixed and if in addition the orientation of the external force, characterized by the value of the parameter α/β where α is the angular position of the load relative to the leading edge, is held constant, the *isothermal model* contains just two parameters. Bearing performance can, in this case, be completely characterized by the parameters

$$\{S, \text{Re}\}.$$

The *adiabatic model* requires only one additional parameter, assuming that the viscosity-temperature dependence of the lubricant is given, as is the reference temperature. Bearing performance can then be characterized by the groups

$$\{S, \text{Re}, \Lambda\}.$$

The THD model requires further addition of parameters. The parameters characterizing THD bearing performance can be conveniently divided into two groups:

$$\left\{ S, \text{Re}, \Lambda, \text{Pe}, \left(\frac{C}{R}\right), \left(\frac{\Theta_s}{\Theta_*}\right) \right\}, \tag{9.44a}$$

$$\left\{ \left(\frac{\Theta_i}{\Theta_*}\right), \left(\frac{\Theta_a}{\Theta_*}\right), \left(\frac{k}{k_\beta}\right), \frac{R_b}{R}, \text{Nu} \right\}. \tag{9.44b}$$

Parameters grouped in expression (9.44b) were found to have only marginal effect on bearing performance and can be left out of consideration in a first approximation to the problem. The dimensionless shaft temperature (Θ_s/Θ_*) is also dropped from further consideration as it is specified only when the flow is laminar. First order effects are then due to

$$\left\{ S, \text{Re}, \Lambda, \text{Pe}, \left(\frac{C}{R}\right) \right\}, \tag{9.45}$$

and we retain these as the essential parameters of the THD problem in rigid bearings. Thermal deformation is also important, as we shall see below. But taking into account thermal deformation of shaft and bearing greatly increases the numerical complexity. The large number of essential parameters precludes the possibility of full parametric study of the THD problem and the representation of its results.

Figure 9.11 shows a plot of the eccentricity ratio versus the Sommerfeld number. There are perhaps two observations to be made here: (1) The adiabatic (ADI) and the isothermal (ISO) models do not bracket actual (THD) bearing performance, and (2) at small to moderate load the turbulent bearing operates at lower eccentricity than does the laminar bearing, but at high load this order is reversed.

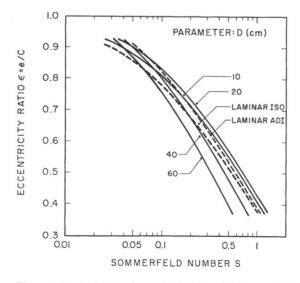

Figure 9.11. Variation of eccentricity ratio with Sommerfeld number at constant bearing diameter, $N = 3600$ rpm, $C/R = 0.002$, $\Theta_i = 50°$. (Reprinted with permission from Suganami, T. and Szeri, A. Z. Parametric study of journal bearing performance: the 80° partial arc bearing. *ASME Journal of Lubrication Technology*, **101**, 486–491, 1979.)

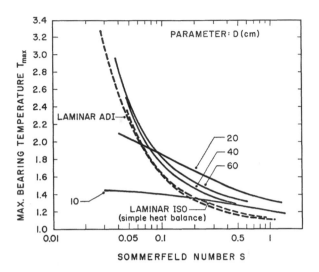

Figure 9.12. Dependence of maximum bearing temperature on Sommerfeld number at constant bearing diameter, $N = 3600$ rpm, $C/R = 0.002$, $\Theta_* = 50°C$. (Reprinted with permission from Suganami, T. and Szeri, A. Z. Parametric study of journal bearing performance: the 80° partial arc bearing. *ASME Journal of Lubrication Technology*, **101**, 486–491, 1979.)

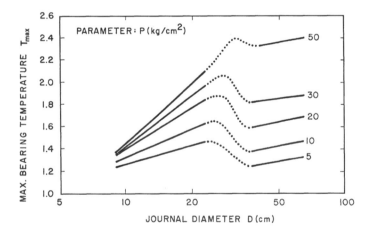

Figure 9.13. Effect of bearing size on maximum bearing temperature at constant specific load, $N = 3600$ rpm, $C/R = 0.002$, $\Theta_* = 50°C$. (Reprinted with permission from Suganami, T. and Szeri, A. Z. Parametric study of journal bearing performance: the 80° partial arc bearing. *ASME Journal of Lubrication Technology*, **101**, 486–491, 1979.)

Figure 9.12 shows maximum bearing temperature plotted versus Sommerfeld number, i.e., inverse of specific bearing load. Maximum bearing temperature in laminar bearings ($D = 10, 20$ cm) seems to be relatively insensitive to specific load; for the smallest of the bearings calculated, the dependence is only marginal. This is not so for large (turbulent) bearings, for which high specific load should be avoided at all cost, as it might lead to overheating.

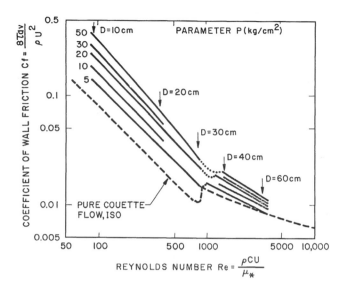

Figure 9.14. Friction factor versus Reynolds number at constant specific load, $N = 3600$ rpm, $C/R = 0.002$, $\Theta_* = 50°C$. (Reprinted with permission from Suganami, T. and Szeri, A. Z. Parametric study of journal bearing performance: the 80° partial arc bearing. *ASME Journal of Lubrication Technology*, **101**, 486–491, 1979.)

Figure 9.13 illustrates the fact that transition to turbulence (taking place under present conditions at $D \approx 25$ cm) is beneficial from the point of limiting bearing temperatures, especially at low loads. This was already demonstrated experimentally by Gardner and Ulschmid (1974). The curves in Figure 9.13 also support a previous contention, viz., that T_{max} is strongly dependent on P when in the turbulent regime.

Friction Factor

Allowing viscosity to decrease with increasing temperature is expected to reduce the wall friction coefficient in pure Couette flow. Increasing the eccentricity and no other change would, on the other hand, tend to increase friction. Figure 9.14 illustrates the result of introducing these two competing tendencies simultaneously; the net effect is an increase in the wall friction coefficient. Only at large Re might this trend be reversed, at least at the smallest of loads tested, as indicated by the curves of Figure 9.14 at $D = 60$ cm. The friction coefficient is strongly dependent on load in the laminar regime. This dependence lessens with increasing Reynolds number.

Journal Locus and Dynamic Coefficients

Figure 9.15 displays various journal center loci. The overall effect of turbulence, and also of viscosity variation, is to displace the journal center in the downstream direction, i.e., to increase the load angle. This shift in ϕ is monotonically increasing with bearing diameter in both the laminar and the turbulent regimes, but the trend seems to be reversed when in the transition regime. Thus, the locus of the $D = 20$ cm bearing is displaced further from its isothermal position, that is, the $D = 40$ cm bearing. This might affect bearing stiffness.

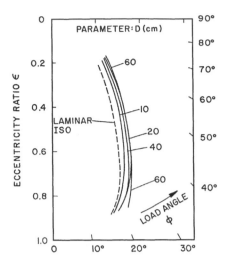

Figure 9.15. Journal locus; $N = 3600$ rpm, $C/R = 0.002$, $\Theta_* = 50°$ C. (Reprinted with permission from Suganami, T. and Szeri, A. Z. Parametric study of journal bearing performance: the 80° partial arc bearing. *ASME Journal of Lubrication Technology*, **101**, 486–491, 1979.)

We saw in Chapter 4 that the incremental oil-film forces that are due to small perturbation of static equilibrium, can be written as[10]

$$\begin{bmatrix} dF_x \\ dF_y \end{bmatrix} = -\frac{W}{C} \begin{bmatrix} \bar{\bar{K}}_{xx} & \bar{\bar{K}}_{xy} \\ \bar{\bar{K}}_{yz} & \bar{\bar{K}}_{yy} \end{bmatrix} \begin{bmatrix} \bar{x} \\ \bar{y} \end{bmatrix} - \frac{W}{C\omega} \begin{bmatrix} \bar{\bar{C}}_{xx} & \bar{\bar{C}}_{xy} \\ \bar{\bar{C}}_{yx} & \bar{\bar{C}}_{yy} \end{bmatrix} \begin{bmatrix} \dot{\bar{x}} \\ \dot{\bar{y}} \end{bmatrix}.$$

The stiffness and damping coefficients of a $\beta = 80°$ fixed-pad partial arc bearing at $\varepsilon = 0.7$ were shown in Table 4.2 under the following conditions: (1) isothermal, laminar flow; (2) adiabatic, laminar flow; (3) THD, laminar flow; (4) isothermal, turbulent flow; and (5) THD, turbulent flow. The calculation involves finding the steady-state equilibrium conditions for the journal (by solving the pressure and energy equations simultaneously) and evaluating the incremental oil film forces while keeping the lubricant at its equilibrium temperature. To perform the perturbation step correctly, one would again have to find simultaneous solutions of the energy and pressure equations while the journal is in its perturbed state. But this would, perhaps unnecessarily, increase the volume of computations.

These coefficients, shown in Table 4.2, were subsequently employed in calculating the stability-threshold speed of a single mass rigid rotor supported at both ends. The threshold speed parameter $\bar{M}$, defined in Eq. (4.41), is plotted against the Sommerfeld number in Figure 4.15.

Table 4.2 shows significant changes in the static and dynamic coefficients. The "cumulative" effect of these changes is a remarkable improvement in threshold speed. Figure 4.15 shows the threshold speed plotted against the Sommerfeld number. The curves for laminar bearings ($D < 20$ cm) seem to bunch together. There is a significant improvement in bearing stability threshold speed in the transition regime ($20 < D < 30$ cm), but the

---

[10] Note that $\bar{\bar{K}}_{xx} = S\bar{K}_{xx}$, etc., where $\bar{K}_{xx}$ is defined in Eq. (4.31).

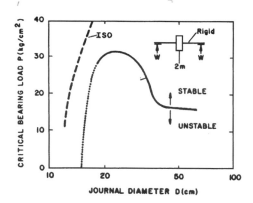

Figure 9.16. Critical bearing load versus shaft diameter for a fixed pad: $N = 3600$ rpm, $C/R = 0.002$, $\Theta_* = 50°C$. (Reprinted with permission from Suganami, T. and Szeri, A. Z. Parametric study of journal bearing performance: the 80° partial arc bearing. *ASME Journal of Lubrication Technology*, **101**, 486–491, 1979.)

dependence on D lessens again when in the turbulent regime ($D > 30$ cm). The threshold speed, $\bar{M}$, becomes meaningful only, however, when it coincides with the shaft rotational speed. When this occurs, a *critical bearing load* is defined for each bearing diameter, under the assumed condition.

Figure 9.16 shows the critical bearing load plotted versus the shaft diameter for a fixed pad. If the specific bearing load exceeds in magnitude the critical bearing load, the bearing is stable. Figure 9.16 indicates that THD effects expand the regime of stable bearing operations, in terms of the bearing specific load P.

Thermal Deformation

Boncompain, Fillon, and Frêne (1986) compared their numerical analysis with experiments from Ferron (1982), and showed significant thermal deformation of the surfaces. Excellent agreement was obtained, though the computations were performed for the mid-plane of the pad only. They estimated the thermoelastic strain in their numerical model from

$$\varepsilon_{ij} = \frac{1 + \nu}{E} T_{ij} - \frac{\nu}{E} T_{kk} + \alpha \Delta \Theta(M) \delta_{ij}. \tag{9.46}$$

Here ε_{ij} is the strain, T_{ij} is the stress, $\Delta \Theta(M) = \Theta(M) - \Theta_*$ is the temperature excess at point M over the reference temperature Θ_*, α is the coefficient of thermal expansion of the solid, and E, ν are elastic moduli. The finite-element method was used to solve the coupled system consisting of the energy equation for the lubricant (9.34), the Reynolds equation with deformation film thickness calculated from Eq. (9.46), Vogel's formula (9.43) for viscosity, and the energy equation for the pad

$$\frac{\partial^2 \Theta_p}{\partial r^2} + \frac{1}{r} \frac{\partial \Theta_p}{\partial r} + \frac{1}{r^2} \frac{\partial^2 \Theta_p}{\partial \varphi^2} = 0. \tag{9.47}$$

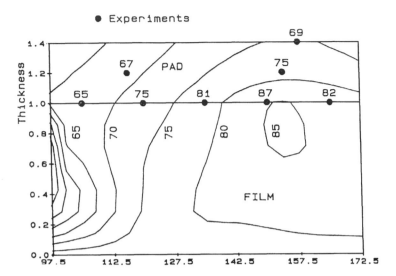

Figure 9.17. $\Theta°C$ = const. surfaces in lubricant film and pad. (Reproduced with permission from Fillon, M. and Frêne, J. Numerical simulation and experimental results on thermo-elasto-hydrodynamic tilting-pad journal bearings. *IUTAM Symposium on Numerical Simulation of Non-isothermal Flow of Viscoelastic Liquids.* 85–99, © 1995 Kluwer Academic Publisher.)

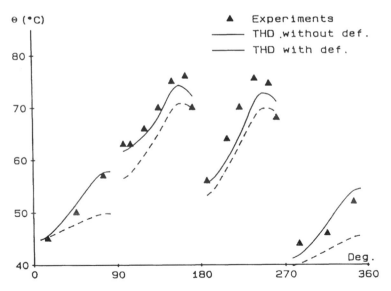

Figure 9.18. Comparison of theoretical and experimental pad temperature for a four-pad bearing. (Reproduced with permission from Fillon, M. and Frêne, J. Numerical simulation and experimental results on thermo-elasto-hydrodynamic tilting-pad journal bearings. *IUTAM Symposium on Numerical Simulation of Non-isothermal Flow of Viscoelastic Liquids.* 85–99, © 1995 Kluwer Academic Publisher.)

Figure 9.17 shows the isotherms in both lubricant film and pad, in a four-pad bearing, indicating that there is considerable heat transfer in the pad in the tangential direction.

As may be judged from Figure 9.18, thermoelastic deformation can have significant effect.

Bouard, Fillon and Frêne (1994) compared the turbulence models of Constantinescu (1972), Elrod and Ng (1967), and Ng and Pan (1965), using the turbulent transition approximation (9.36) of Suganami and Szeri (1979a). As shown in Figure 9.19 for pressure, in Figure 9.20 for temperature, and in Figure 9.21 for coefficient of friction, there is little

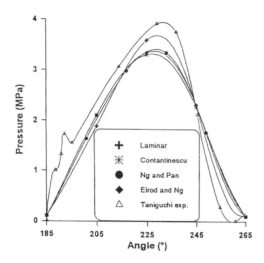

Figure 9.19. Film pressure as predicted by various turbulence models (Bouard, Fillon and Frêne, 1995). (Reproduced from Fillon, M. *Dossier D'Habilitation a Diriger les Recherches.* Universite de Poitiers.)

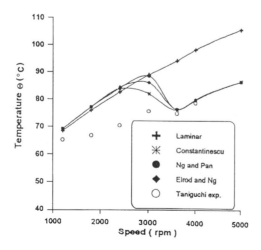

Figure 9.20. Maximum bearing temperature as predicted by the various turbulence models (Bouard, Fillon, and Frêne, 1995). (Reproduced from Fillon, M. *Dossier D'Habilitation a Diriger les Recherches.* Universite de Poitiers.)

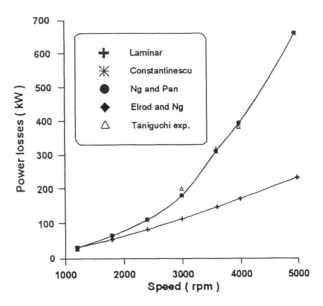

Figure 9.21. Power loss as predicted by the various turbulence models (Bouard, Fillon and Frêne, 1995). (Reproduced from Fillon, M. *Dossier D'Habilitation a Diriger les Recherches.* Universite de Poitiers.)

difference between the predictions of the three theories, under the conditions specified in the paper.

Recent experimental investigations into turbulent lubrication are by Gethin and Medwell (1985); Hopf and Schüler (1989); Taniguchi, Makino, Takeshita, and Ichimura (1990); Simmons and Dixon (1994); Mittwollen and Glienicke (1990); Bouchoule, Fillon, Nicolas, and Barressi (1996); and Monmousseau et al. (1996).

The inlet temperature effect was investigated theoretically by Ettles (1992) and by Ha, Kim, and Kim (1995). An analysis for tilting-pad bearings was published by Bouard, Fillon, and Frêne (1996). Thermal transient analysis was performed by Gadangi, Palazzo, and Kim (1996), and Fillon, Desbordes, Frêne, and Wai (1996).

9.4   Thrust Bearings

One of the sector-shaped pivoted pads of a thrust bearing we want to study is shown in Figure 4.16. (Jeng, Zhou, and Szeri, 1986a). The plane of the runner surface is located at $z = 0$ and the pad surface is located at $z = h(r, \varphi)$ of the cylindrical polar coordinate system $\{r, \varphi, z\}$. The characteristic lengths of the lubricant film are L_z across and $L_{r\theta}$ along the film.

The Pressure Equation

Making use of the fact that $L_z/L_{r\theta} = O(10^{-3})$, in an order of magnitude analysis similar to the one employed Chapter 7, the equations of motion for turbulent flow of the lubricant reduce to[11]

$$\frac{\partial \bar{P}}{\partial r} = \frac{\partial}{\partial z}\left(\bar{\mu}\frac{\partial \bar{U}}{\partial z} - \rho\overline{u'w'}\right), \tag{9.48a}$$

$$\frac{\partial \bar{P}}{\partial z} = 0, \tag{9.48b}$$

$$\frac{1}{r}\frac{\partial \bar{P}}{\partial \varphi} = \frac{\partial}{\partial z}\left(\bar{\mu}\frac{\partial \bar{V}}{\partial z} - \rho\overline{v'w'}\right). \tag{9.48c}$$

Starting from Eqs. (9.48) and repeating the analysis of Ng and Pan (1965), updated for variable viscosity (Jeng, Zhou, and Szeri, 1986a), we obtain

$$\frac{\partial}{\partial \hat{\varphi}}\left[H^3 G_\varphi \frac{\partial P}{\partial \hat{\varphi}}\right] + \beta^2 \hat{\rho} \frac{\partial}{\partial \hat{r}}\left[\hat{\rho} H^3 G_r \frac{\partial P}{\partial \hat{r}}\right] = \beta^2 \hat{\rho}^2 \frac{\partial[H(1-F)]}{\partial \hat{\varphi}}. \tag{9.49}$$

The turbulent functions G_φ, G_r, and F are defined by formulas identical in form to Eqs. (9.41) and (9.42), except that they are functions of $(\hat{r}, \hat{\varphi})$ through f_c, g_c, and H, while G_x, G_z were functions of $(\bar{x}, \bar{z})$.

The boundary conditions that complement equation Eq. (9.49) are

$$P(0, \hat{\varphi}) = P(1, \hat{\varphi}) = 0,$$

$$P(\hat{r}, 0) = P(\hat{r}, 1) = 0, \tag{9.50}$$

$$P = \frac{\partial P}{\partial \hat{\varphi}} = 0, \quad \text{at } \hat{\varphi} = \hat{\varphi}_{\text{cav}}.$$

In arriving at Eq. (9.49), we employed the nondimensionalization

$$r = \Delta\hat{r} + R_1, \quad \varphi = \beta\hat{\varphi}, \quad z = h\hat{z}, \quad \bar{U} = \Delta\omega U, \quad \bar{V} = \Delta\omega V$$

$$\bar{P} = \frac{\mu_*\omega}{\beta}\left(\frac{\Delta}{h_c}\right)^2 P, \quad \bar{\mu} = \mu_*\mu, \quad h = h_C H \tag{9.51}$$

$$\Delta = R_2 - R_1 > 0, \quad \hat{\rho} = \hat{r} + R_1/\Delta, \quad r = \Delta\hat{\rho},$$

where h_C is the film thickness at the geometric center, C, of the pad (Figure 4.16).

The relationship between the normalized coordinate

$$z_c^+ = \frac{z}{\nu}\sqrt{\frac{|\tau_c|}{\rho}} = \hat{z}\frac{h}{\nu}\sqrt{\frac{|\tau_c|}{\rho}} = \hat{z}h_c^+$$

and the local Reynolds number, $R_h = r\omega h/\nu_{av}$ [cf., Eq. (7.76)], now has the form

$$R_h = \int_0^1 \frac{\mu_{av}(z_c^+/\hat{z})^2\,d\bar{z}}{\mu\{1 + \kappa[z_c^+ - \delta_\ell^+ \tanh(z_c^+/\delta_\ell^+)]\}}. \tag{9.52a}$$

For a given R_h, Eq. (9.52a) is a transcendental equation in z_c^+ and must be solved numerically.

---

[11] The order of magnitude analysis based on the limit $L_z/L_{r\theta} \to 0$ yields $\partial P/\partial z = \text{Re}^*(\partial/\partial z)(\overline{w'w'})$ in place of Eq. (9.48b), but past practice has been to also assume $\text{Re}^* \to 0$ [(see Eq. (7.14)]. If the latter assumption is not made, derivation of a Reynolds type pressure equation is not possible.

The local Reynolds number, on the other hand, is calculated from the global Reynolds number, Re, with the aid of the local average viscosity,

$$\mu_{av}(\hat{r}, \hat{\varphi}) = \int_0^1 \mu(\hat{r}, \hat{\varphi}, \hat{z}) \, d\hat{z},$$

from the formula

$$R_h = \frac{H\hat{\rho}}{\nu_{av}} \frac{\Delta}{R_2} \mathrm{Re}, \qquad \mathrm{Re} = \frac{R_2 \omega h_C}{\nu_*}. \tag{9.52b}$$

The eddy viscosity for momentum transfer is

$$\frac{\varepsilon_m}{\nu} = f_c - 1 + \frac{\zeta_4(1)}{\beta^2} \frac{H^2}{\hat{\rho}^2} \frac{\partial P}{\partial \hat{\varphi}} \left[\hat{z} - \frac{1}{2} + \frac{\zeta_1(1)}{\zeta_2(1)} \right] g_c, \tag{9.53}$$

where ζ_1, ζ_2, f_c, and g_c are given by formulas analogous to Eqs. (9.42), (7.68), and (7.69).

When evaluated from Reichardt's formula (Hinze, 1975), the eddy viscosity is a monotonic increasing function of the Reynolds number, having nonzero value in the laminar regime. To remedy this, we introduce a scale factor, ϑ, for eddy viscosity from Eq. (9.36), but now set $R_L = 500$ and $R_U = 900$ (Abramovitz, 1956; Gregory, 1974),

$$\vartheta = \begin{cases} 0.0 & (R_h)_{\max} \le 500 \\ 1 - \left[\dfrac{800 - (R_h)_{\max}}{300} \right]^{1/8} & 500 < (R_h)_{\max} \le 800 \\ 1 & (R_h)_{\max} > 800 \end{cases}$$

and employ $\vartheta (\varepsilon_m/\nu)$ in place of (ε_m/ν) in the calculations.

Film Thickness

Let $\{x, y, z\}$ be an inertial Cartesian coordinate system such that the $z = 0$ plane is located on the runner surface and the z axis is in the vertical direction pointing upward and goes through the geometric center $C\{(R_1 + R_2)/2, \beta/2\}$ of the bearing (Figure 4.16). If ψ_x and ψ_y represent the angular position of the pad in equilibrium, relative to the $\{x, y, z\}$ coordinate system, and if h_C is the film thickness at the pad center C, we have

$$h = h_C - x\psi_y + y\psi_x. \tag{9.54a}$$

Here ψ_x and ψ_y are small angles, positive when measured counterclockwise.

The nondimensional counterpart of Eq. (9.54a) is

$$H = 1 - \frac{\Delta}{R_1} \left\{ \hat{\rho} m_y \sin\left[\beta\left(\frac{1}{2} - \hat{\varphi}\right) \right] + \left[\hat{\rho} \cos\left[\beta\left(\frac{1}{2} - \hat{\varphi}\right) \right] - \frac{R_1 + R_2}{2\Delta} \right] m_x \right\}, \tag{9.54b}$$

where $H = h/h_C$ is the nondimensional film shape. The tilt parameters m_x, m_y have the definition

$$m_x = \frac{R_1 \psi_x}{h_C}, \qquad m_y = \frac{R_1 \psi_y}{h_C}.$$

The Energy Equation

An order of magnitude analysis yields the equation of energy for the lubricant in the form

$$\rho c \left[\bar{U} \frac{\partial \bar{\Theta}}{\partial r} + \bar{V} \frac{\partial \Theta}{r \partial \varphi} + \bar{W} \frac{\partial \bar{\Theta}}{\partial z} + \frac{\partial}{\partial z} (\overline{w'\theta'}) \right]$$

$$= k \left[\frac{1}{r} \frac{\partial}{\partial r} \left(r \frac{\partial \bar{\Theta}}{\partial r} \right) + \frac{1}{r^2} \frac{\partial^2 \bar{\Theta}}{\partial \varphi^2} + \frac{\partial^2 \bar{\Theta}}{\partial z^2} \right] + \bar{\mu} \left(1 + \frac{\varepsilon_m}{\nu} \right) \left[\left(\frac{\partial \bar{U}}{\partial z} \right)^2 + \left(\frac{\partial \bar{V}}{\partial z} \right)^2 \right]. \tag{9.55}$$

The velocity-temperature correlation $\overline{w'\theta'}$ may be approximated via the eddy viscosity hypothesis, where, in analogy to momentum transport, we put

$$-\overline{w'\theta'} = \varepsilon_H \frac{\partial \bar{\Theta}}{\partial z} \tag{9.56}$$

and the energy equation (9.55) is made nondimensional by substituting from Eq. (9.51) and

$$\bar{\Theta} = \Theta_* T, \qquad \mu(\Theta_*) = \mu_*, \qquad \mu = \mu_* \bar{\mu}.$$

When deriving the equation for lubricant pressure, Eq. (9.49), the z dependence was integrated out and the transformation in Eq. (9.55) was from $\{r, \varphi\}$ to $\{\hat{r}, \hat{\varphi}\}$. In the present instance, the mapping is $\{r, \varphi, z\} \rightarrow \{\hat{r}, \hat{\varphi}, \hat{z}\}$, and we must employ more complicated formulas:

$$\frac{\partial}{\partial r} = \frac{1}{\Delta} \left(\frac{\partial}{\partial \hat{r}} - n_r \frac{\partial}{\partial \hat{z}} \right), \qquad \frac{\partial}{\partial \varphi} = \frac{1}{\beta} \left(\frac{\partial}{\partial \hat{\varphi}} - n_\varphi \frac{\partial}{\partial \hat{z}} \right), \qquad \frac{\partial}{\partial z} = \frac{1}{h_C} \frac{\partial}{H \partial \hat{z}}. \tag{9.57}$$

Here we used the notation

$$n_r = \frac{\hat{z}}{H} \frac{\partial H}{\partial \hat{r}}, \qquad n_\varphi = \frac{\hat{z}}{H} \frac{\partial H}{\partial \hat{\varphi}}.$$

The nondimensional energy equation is

$$U \frac{\partial T}{\partial \hat{r}} + \frac{V}{\beta \hat{\rho}} \frac{\partial T}{\partial \hat{\varphi}} + \left(\frac{W}{H} - U n_r - \frac{V}{\beta \hat{\rho}} n_\varphi \right) \frac{\partial T}{\partial \hat{z}}$$

$$= \frac{1}{\mathrm{Pe}} \left\{ \frac{\partial}{\partial \hat{r}} \left(\frac{\partial T}{\partial \hat{r}} - n_r \frac{\partial T}{\partial \hat{z}} \right) - n_r \frac{\partial}{\partial \hat{z}} \left(\frac{\partial T}{\partial \hat{r}} - n_r \frac{\partial T}{\partial \hat{z}} \right) \right.$$

$$+ \frac{1}{(\beta \hat{\rho})^2} \frac{\partial}{\partial \hat{\varphi}} \left(\frac{\partial T}{\partial \hat{\varphi}} - n_\varphi \frac{\partial T}{\partial \hat{z}} \right) + \frac{1}{(\delta H)^2} \frac{\partial^2 T}{\partial \hat{z}^2} - \frac{n_\theta}{(\beta \hat{\rho})^2} \frac{\partial}{\partial \hat{z}} \left(\frac{\partial T}{\partial \hat{\varphi}} - n_\varphi \frac{\partial T}{\partial \hat{z}} \right)$$

$$+ \frac{1}{\hat{\rho}} \left(\frac{\partial T}{\partial \hat{r}} - n_r \frac{\partial T}{\partial \hat{z}} \right) + \frac{\mathrm{Pr}}{\mathrm{Pr}^{(t)}} \frac{1}{\delta^2} \frac{\partial}{\partial \hat{z}} \left(\frac{\mu}{H^2} \frac{\varepsilon_m}{\nu} \frac{\partial T}{\partial \hat{z}} \right) \right\}$$

$$+ \Lambda \left(1 + \frac{\varepsilon_m}{\nu} \right) \frac{\mu}{H^2} \left[\left(\frac{\partial U}{\partial \hat{z}} \right)^2 + \left(\frac{\partial V}{\partial \hat{z}} \right)^2 \right]. \tag{9.58}$$

Pe, Λ, and δ are the Peclet number, the dissipation parameter, and the film thickness parameter, respectively:

$$\text{Pe} = \frac{\rho c \Delta^2 \omega}{k}, \qquad \Lambda = \frac{\mu_* \omega}{\rho c \Theta_* \delta^2}, \qquad \delta = \frac{h_C}{\Delta}. \tag{9.59a}$$

The laminar and turbulent Prandtl numbers are defined through

$$\text{Pr} = \frac{c\mu_*}{k}, \qquad \text{Pr}^{(t)} = \frac{\varepsilon_m}{\varepsilon_H}, \tag{9.59b}$$

and the temperature dependence of viscosity is assumed to be given by Vogel's formula (9.43).

The thermal boundary conditions are:

Runner surface ($z = 0$)

$$T = T_R \ (T_R \text{ the runner surface temperature}) \tag{9.60a}$$

Bearing surface $[z = h(r, \varphi)]$

$$-k\frac{\partial T}{\partial z} = -k_B \frac{\partial T_B}{\partial z_B} \cos \left(\psi_x^2 + \psi_y^2 \right)^{1/2} \tag{9.60b}$$

$$T = T_B \ (T_B \text{ the bearing surface temperature})$$

Leading edge ($\varphi = \beta$)

$$T = T_L \ (T_L \text{ the lubricant inlet temperature}) \tag{9.60c}$$

At the exit planes, i.e., at the trailing edge, $\varphi = 0$, the inside edge, $r = R_1$, and the outside edge, $r = R_2$, of the pad there are no boundary conditions specified in the classical sense by the physics of the problem. In journal bearings, Suganami and Szeri (1979a) assumed zero normal derivative for the temperature at exit planes. This condition is far too restrictive for the present geometry and, eventually, zero second gradient exit boundary conditions were prescribed.

The Heat Conduction Equation

The bearing surface boundary condition, Eq. (9.60b), presupposes simultaneous solution of, or at least iteration between, the equation of energy for the lubricant, Eq. (9.59), and the equation of heat conductivity for the bearing pad:

$$\frac{1}{r_B} \frac{\partial}{\partial r_B} \left(r_B \frac{\partial \Theta_B}{\partial r_B} \right) + \frac{1}{r_B^2} \frac{\partial^2 \Theta_B}{\varphi_B^2} + \frac{\partial^2 \Theta_B}{\partial z_B^2} = 0. \tag{9.61}$$

Here Θ_B is the temperature in the bearing. The z_B axis of the bearing coordinate system $\{r_B, \varphi_B, z_B\}$ is normal to the pad surface and is inclined to the vertical at angle $\alpha = (\psi_x^2 + \psi_y^2)^{1/2}$.

Equation (9.58) requires the z component of the velocity, $W(z)$. This is obtained from the equation of continuity, for we have

$$\bar{W}(z) = -\int_0^z \frac{\bar{U}}{r} dz' - \int_0^z \frac{\partial \bar{U}}{\partial r} dz' - \int_0^z \frac{1}{r} \frac{\partial \bar{V}}{\partial \psi} dz'. \tag{9.62}$$

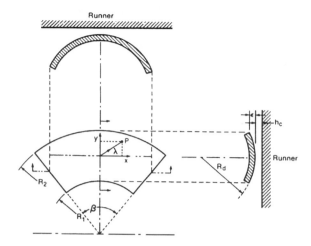

Figure 9.22. Spherical crowning of thrust pad. (Reprinted with permission from Jeng, M. C., Zhou, G. R. and Szeri, A. Z. Thermohydrodynamic solution of pivoted thrust pads, part I: Theory. *ASME Journal of Tribology*, **108**, 195–207, 1986.)

Pad Deformation

For the purposes of this analysis the pad surface is assumed to have spherical crowning; this permits us to characterize pad deformation in a simple manner.

Small bearing pads are supported in a point, and pad deformation is often negligible. In large bearings, however, considerable deformation of the pad is experienced, and simple support in a point will no longer suffice. To minimize deformation, the pad is often supported on a disk via a ring or rings, in the hope that elastic and thermal deformations will counteract one another. In at least in two constructions examined by Jeng, Zhou, and Szeri (1986b), the deformation was found to be near spherical.[12]

With reference to Figure 9.22, let R_d represent the radius of pad-surface curvature and ε be the maximum deviation of the surface from planar along its mid-radius. If λ is the distance between the center of the pad, C, and a generic point, P, then the deformation of point P is given by

$$\delta_h = R_d \left[1 - \sqrt{1 - \left(\frac{\lambda}{R_d} \right)^2} \right] \approx (x^2 + y^2)/2R_d,$$

where $\lambda = \sqrt{x^2 + y^2}$, so that

$$x = r \sin \left(\frac{\beta}{2} - \varphi \right), \qquad y = r \cos \left(\frac{\beta}{2} - \varphi \right) - \frac{R_1 + R_2}{2}.$$

The change, δ_h, in the nondimensional film thickness, $\bar{h}(r, \theta)$, when due to spherical

---

[12] The ratio of the principal radii of curvature was $R_x/R_y = 0.96$ for one of the data and $R_x/R_y = 0.95$ for the other.

Table 9.2. *Comparison of performance data for a thrust pad*

	Experimental	Numerical
$T_{\max}(°C)$	>95	107.61
F (kg)	96,300	96,682
$\hat{\varepsilon}$	0.9	0.7

$\Theta_L = \Theta_* = \Theta_R = 38.5°C$, $\mu(\Theta_*) = 2.62 \times 10^{-5}$ Pa·s, $\bar{R}_2/\Delta = 2.545$, $\beta = 24°$, $\delta = 1.094 \times 10^{-4}$, $m_x = 0.0$, $m_y = 1.0$, Re = 51.398, Pr = 714.69, $\Lambda = 2.1962$, Pe = 1.323×10^8, Nu = 100.0.

crowning, is given by

$$\hat{\delta}_h = \hat{\varepsilon} \left\{ 4(\hat{r} + \hat{R} - 1)^2 + (2\hat{R} - 1)^2 - 4(\hat{r} + \hat{R} - 1)(2\hat{R} - 1) \cos\left[\beta\left(\frac{1}{2} - \hat{\varphi} \right) \right] \right\}.$$

(9.63)

Here

$$\hat{\delta}_h = \frac{\delta_h}{h_C}, \qquad \hat{\varepsilon} = \frac{\varepsilon}{h_C}, \qquad \hat{R} = \frac{R_2}{(R_2 - R_1)}.$$

When the pad is crowned, lubricant pressure might fall below ambient. Liquids, in impure state, can support only negligible tension, and the film will cavitate somewhere in the diverging part of the clearance space.

Table 9.2 compares experimental with numerical data in one particular case; sample isobars and isotherms are displayed in Figure 9.23.

To aid the designer, one may now perform a parameteric study of THD thrust pad performance and tabulate the results. But, the large number of input parameters is discouraging. Making matters worse is the fact that in a thrust bearing the pivot position of the pad remains fixed, while it is the tilt of the pad that changes with a change of the mechanical and/or thermal loading of the bearing. In the course of the numerical work, on the other hand, the analyst begins by specifying the slope of the pad and obtains the performance parameters, including the center of pressure, as a result of calculations. The position of the center of pressure he calculates will not, in general, coincide with the desired pivot location, and the analyst will be required to iterate on the pad tilt angle until agreement is reached. This, of course, is an expensive proposition.

Table 9.3 indicates the effect variable viscosity has on pad performance for two sets of input parameters. The first entries for both test runs were calculated on the assumption that $\mu = h/h_{\max}$; a crude approximation to the equilibrium oil-film viscosity profile. The second entries for both runs correspond to three-dimensional THD solutions.

In Figure 9.24, we display the results of our study of the effects of pad crowning. Again, these solutions have the pivot position fixed, i.e., the results were arrived at via the iteration procedure described previously. The figure suggests a strong effect of crowning on the load capacity and on the lubricant flow. The effect is considerably less on the value of the maximum pad temperature and on the rate of dissipation.

Table 9.3. *Effect of viscosity variation*

	T_{max}	f	j	$\hat{Q}$
Test run 1	4.2154	0.03485	3.50	1.62^a
	(2.7952)	(0.02222)	(2.36)	$(1.88)^b$
Test run 2	3.7472	0.03485	3.49	1.62^a
	(2.6013)	(0.02363)	(2.52)	$(1.85)^b$

$^a \hat{\mu} = h/h_{max}$,
$^b THD$.

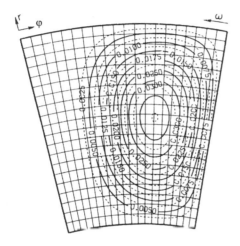

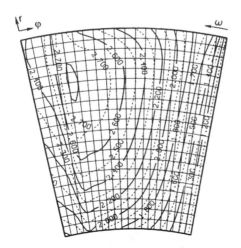

Figure 9.23. Isobars, $P = $ const., (a) and isotherms, $T = $ const. (b) on a deformed pad under the conditions of Table 9.2. (Reprinted with permission from Jeng, M. C., Zhou, G. R. and Szeri, A. Z. Thermohydrodynamic solution of pivoted thrust pads, part I: Theory. *ASME Journal of Tribology*, **108**, 195–207, 1986.)

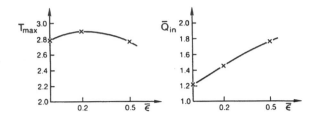

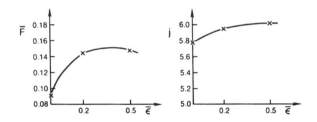

Figure 9.24. Effect of pad crowning on pad performance. (Reprinted with permission from Jeng, M. C., Zhou, G. R. and Szeri, A. Z. Thermohydrodynamic solution of pivoted thrust pads, part I: Theory. *ASME Journal of Tribology*, **108**, 195–207, 1986.)

The dynamic behavior of the tilting thrust pad bearing has been discussed in Chapter 4. The solutions discussed there were based on thermohydrodynamic theory.

9.5   Nomenclature

N	rotational speed
Nu	Nusselt number
P	specific bearing load
Pe	Peclet number
Pr	molecular Prandtl number
$\mathrm{Pr}^{(T)}$	turbulent Prandtl number
Q, Q_s	oil flow, leading edge, side
R	journal radius
R_b	pad outer radius
Re	global Reynolds number
Re*	reduced Reynolds number
R_h	local Reynolds number
S	Sommerfeld number
W	external load
Λ	dissipation number
c	specific heat of lubricant
h	film thickness
k	thermal conductivity, oil
k_B	thermal conductivity, pad
ε	eccentricity ratio
$\varepsilon_m, \varepsilon_h$	eddy diffusivities
λ	heat transfer coefficient

p	pressure
$\Theta, \Theta_p, \Theta_s, \Theta_a$	temperature, lubricant, pad, shaft, ambient
μ	viscosity
ν	kinematic viscosity
ρ	lubricant density
$\{u', v', w'\}$	velocity fluctuation
$\{\bar{U}, \bar{V}, \bar{W}\}$	mean velocity
C	radial clearance
$\bar{\bar{C}}_{xx...}$	dimensionless damping coefficient
c_f	friction factor
$\bar{\bar{K}}_{xx...}$	dimensionless stiffness coefficient
L	bearing axial length
$\bar{M}$	dimensionless stability threshold speed
$\bar{x}_{\text{cavity}}$	position of cavitation boundary
β	pad angle
$\tau_c, \tau_{av}, \tau_w$	Couette, average, wall shear stress
ω	angular velocity
R_1, R_2	pad radius, inner, outer
d	pad thickness
(x, y, z)	coordinates, located in C
Δ	width of pad
$\hat{\rho}$	function of $\bar{r}$
$(\hat{\ })$	nondimensional quantity
$(\bar{\ })$	average quantity
$(\ )'$	fluctuating quantity
$(\ )_*$	reference quantity

9.6   References

Abramovitz, S. 1956. Turbulence in a tilting-pad thrust bearing. *ASME Trans.*, **78**, 7–14.

Bielec, M. K. and Leopard, A. J. 1970. Tilting pad thrust bearings: factors affecting performance and improvements with directed lubrication. *Instn. Mech. Engrs. Tribology Convention*, Paper 13.

Boncompain, R., Fillon, M. and Frêne, J. 1986. Analysis of thermal effects in hydrodynamic bearings. *ASME Journal of Tribology*, **108**, 219–224.

Booser, E. R. 1984. *CRC Handbook of Lubrication*. CRC Press, Boca Raton, FL.

Boswall, R. O. 1928. *The Theory of Fluid Film Lubrication*. Longmans and Green, New York.

Bouard, L., Fillon, M. and Frêne, J. 1995. Comparison between three turbulent models – application to thermohydrodynamic performances of tilting-pad journal bearings. In Fillon, M. 1995. *Habilitation a Diriger les Recherches*. Universite de Poitiers, France.

Bouard, L., Fillon, M. and Frêne, J. 1996. Thermohydrodynamic analysis of tilting-pad journal bearings operating in turbulent flow regime. *ASME Journal of Tribology*, **118**, 225–231.

Bouchoule, C., Fillon, M., Nicolas, D. and Baresi, F. 1996. Experimental study of thermal effects in tilting-pad journal bearings at high operating speeds. *ASME Journal of Tribology*, **118**, 532–538.

Cameron, A. 1966. *The Principles of Lubrication*. Wiley, New York.

Capitao, J. W. 1976. Performance characteristics of tilting pad thrust bearings at high operating speeds. *ASME Trans.*, **96**, 7–14.

Capitao, J. W., Gregory, R. S. and Whitford, R. P. 1976. Effects of high operating speeds on tilting pad thrust bearing performance. *ASME Trans.*, **98**, 73–80.

Charnes, A., Osterle, F. and Saibel, E. 1952. On the energy equation for fluid-film lubrication. *Proc. Roy. Soc.*, **A 214**, 133–136.

Constantinescu, V. N. 1972. Basic relationships in turbulent lubrication and their extension to include thermal effects. *ASME-ASLE Int. Lub. Conf.*, Paper No. **72-Lub-16**, New York.

Cope, W. F. 1949. A hydrodynamical theory of film lubrication. *Proc. Roy. Soc.*, **A 197**, 201–217.

Dowson, D. and Taylor, C. M. 1979. Cavitation in bearings. *Ann. Rev. Fluid Mech.*, **11**, 35–66.

Dowson, D. and Hudson, J. D. 1963. Thermohydrodynamic analysis of the finite slider bearing, I, II. *Instn. Mech. Engrs., Lubrication & Wear Convention*, Paper Nos. 4 & 5.

Dowson, D. and March, C. N. 1967. A thermohydrodynamic analysis of journal bearings. *Proc. Instn. Mech. Engrs.*, **181**, 117–126.

Dowson, D., Hudson, J. D., Hunter, B. and March, C. N. 1966. An experimental investigation of the thermal equilibrium of the steadily loaded journal bearing. *Proc. Instn. Mech. Engrs.*, **181**, 70–80.

Elrod, H. G. and Ng, C. W. 1967. A theory for turbulent fluid films and its application to bearings. *ASME Journal of Lubrication Technology*, **89**, 346–359.

Ettles, C. M. M. 1970. Hot-oil carry-over in thrust bearings. *Proc. Inst. Mech. Engrs*, **184**, 75–81.

Ettles, C. M. M. 1992. The analysis of pivoted pad journal bearing assemblies considering thermoelastic deformation and heat transfer effects. *STLE Tribology Transactions*, **35**, 156–162.

Ezzat, H. A. and Rohde, S. M. 1972. A study of the thermohydrodynamic performance of finite slider bearings. *ASME Journal of Lubrication Technology*, **95**, 298–307.

Ferron, J. 1982. *Contribution a l'etude des phenomenes thermiques dans les paliers hydrodynamiques*. These de 3ᵉ Cycle, Universite de Poitiers, France.

Ferron, J., Frêne, J. and Boncompain, R. 1983. A study of the performance of a plain journal bearing – comparison between theory and experiments. *ASME Journal of Lubrication Technology*, **105**, 422–428.

Fillon, M. 1995. *Habilitation a Diriger les Recherches*. Universite de Poitiers, France.

Fillon, M. and Frêne, J. 1995. Numerical simulation and experimental results on thermo-elasto-hydrodynamic tilting-pad journal bearings. IUTAM Symposium on Numerical Simulation of Non-isothermal Flow of Viscoelastic liquids, pp. 85–99. Kluwer Academic Publisher.

Fillon, M., Desbordes, H., Frêne, J. and Chan He Wai, C. 1996. A global approach to thermal effects including pad deformations in tilting pad journal bearings submitted to unbalance load. *ASME Journal of Tribology*, **118**, 169–174.

Gadangi, R. K., Palazzo, A. B. and Kim, J. 1996. Transient analysis of plain and tilt pad journal bearings including fluid film temperature effects. *ASME Journal of Tribology*, **118**, 423–430.

Gardner, W. W. and Ulschmid, J. G. 1974. Turbulence effects in two journal bearing applications. *ASME Journal of Lubrication Technology*, **96**, 15–21.

Gethin, D. T. and Medwell, J. O. 1985. An experimental investigation into the thermohydrodynamic behavior of a high speed cylindrical bore journal bearing. *ASME Journal of Tribology*, **107**, 538–543.

Gollub, J. P. and Swinney, H. L. 1981. *Transition to Turbulence*. Springer-Verlag, New York.

Gregory, R. S. 1974. Performance of thrust bearings at high operating speeds. *ASME Journal of Lubrication Technology*, **96**, 7–14.

Ha, H. C., Kim, H. J. and Kim C. 1995. Inlet pressure effects on the thermohydrodynamic performance of large tilting pad journal bearing. *ASME Journal of Tribology*, **117**, 160–165.

Hinze, J. O. 1975. *Turbulence*. McGraw-Hill, New York.

Hopf, G. and Schüler, D. 1989. Investigations on large turbine bearings working under transitional conditions between laminar and turbulent flow. *ASME Journal of Tribology*, **111**, 628–634.

Jeng, M. C., Zhou, G. R. and Szeri, A. Z. 1986a. Thermohydrodynamic solution of pivoted thrust pads, part I: Theory. *ASME Journal of Tribology*, **108**, 195–207.

Jeng, M. C., Zhou, G. R. and Szeri, A. Z. 1986b. Thermohydrodynamic solution of pivoted thrust pads, part II: static loading. *ASME Journal of Tribology*, **108**, 208–213.

Kaufman, H. N., Szeri, A. Z. and Raimondi, A. A. 1978. Performance of a centrifugal disk lubricated bearing. *ASLE Trans.*, **21**, 314–322.

Kestin, J. and Richardson, P. D. 1963. Heat transfer across turbulent, incompressible boundary layers. *Int. J. Heat Mass Transfer*, **6**, 147–189.

McCallion, H., Yousif, F. and Lloyd, T. 1970. The analysis of thermal effects in a full journal bearing. *ASME Journal of Lubrication Technology*, **192**, 578–586.

Mittwollen, N. and Glienicke, J. 1990. Operating conditions of multi-lobe journal bearings under high thermal loads. *ASME Journal of Tribology*, **112**, 330–338.

Monmousseau, P., Fillon, M. and Frêne, J. 1996. Transient thermoelastohydrodynamic study of tilting pad journal bearings – comparison between experimental data and theoretical results. *ASME/STLE Tribology conference*, Paper No. **96-trib-29**, San Francisco.

Ng, C. W. and Pan, C. H. T. 1965. A linearized turbulent lubrication theory. *ASME J. Basic Engineering*, **87**, 675–688.

Pinkus, O. 1990. *Thermal Aspects of Fluid Film Tribology*. ASME Press, New York.

Raimondi, A. A. and J. Boyd. 1958. A solution for the finite journal bearing and its application to analysis and design. *ASLE Trans.*, **1**, 159–209.

Safar, Z. and Szeri, A. Z. 1974. Thermohydrodynamic lubrication in laminar and turbulent regimes. *ASME Journal of Lubrication Technology*, **96**, 48–56.

Seireg, A. and Ezzat H. 1972. Thermohydrodynamic phenomena in fluid film lubrication. *ASME Paper No. 72-Lub-25*.

Simmons, J. E. L. and Dixon, S. J. 1994. Effect of load direction, preload, clearance ratio and oil flow on the performance of a 200 mm journal pad bearing. *STLE Tribology Transactions*, **37**, 227–236.

Suganami, T. and Szeri, A. Z. 1979a. A thermohydrodynamic analysis of journal bearings. *ASME Journal of Lubrication Technology*, **101**, 21–27.

Suganami, T. and Szeri, A. Z. 1979b. Parametric study of journal bearing performance: the 80° partial arc bearing. *ASME Journal of Lubrication Technology*, **101**, 486–491.

Taniguchi, S., Makino, T., Takeshita, K. and Ichimura, T. 1990. A thermohydrodynamic analysis of large tilting-pad journal bearing in laminar and turbulent flow regimes with mixing. *ASME Journal of Tribology*, **112**, 542–548.

Yu, T. S. and Szeri, A. Z. 1975. Partial journal bearing performance in the laminar regime. *ASME Journal of Lubrication Technology*, **97**, 94–101.

Lubrication with Non-Newtonian Fluids

Non-Newtonian effects might assume importance in lubrication due mainly to two circumstances: lubrication with process fluids and treatment of the lubricant with polymeric additives. A third circumstance that calls for extension of classical theory to non-Newtonian fluids is relevant to lubrication of elastohydrodynamic (EHD), contacts. When subjected to very rapid rates of shear, viscous fluids exhibit viscoelastic effects. Consider a typical EHD contact, where the pressure might be 2–3 GPa and the rate of shear 10^6 s^{-1}. A lubricant particle will traverse this contact in a millisecond or less. In comparable time, the response of the oil to the applied shear will change from that of a viscous liquid to that of an elastic solid (*glass transition*) and back to a viscous liquid.

To characterize the quality of fluid response to applied shear, viz., its departure from "fluidity," we assign the *Deborah number*, defined by

$$\text{De} = \frac{\text{duration of fluid memory}}{\text{duration of deformation process}}.$$

It is customary to measure duration of fluid memory[1] by the Maxwell *relaxation time* $\lambda = \mu/G$, where μ is the viscosity of the fluid and G is its elastic modulus in shear. Denoting duration of the deformation process, i.e., the *process time*, by τ, we have

$$\text{De} = \frac{\lambda}{\tau}.$$

Purely viscous response, i.e., Newtonian behavior, is characterized by De $= 0$. If De is small, fluids of the differential type (to be defined below) can be used, but if De is large, integral models must be employed to characterize fluid response (Huilgol, 1975).

Evidence for viscoelastic behavior of lubricants was first demonstrated by Barlow et al. (1967, 1972), who subjected the lubricant to oscillating shear. At low frequency, fluid response was characterized by viscous shear, and by predominantly elastic shear as the period of oscillation was decreased to a value less than the relaxation time of the fluid.

Bourgin (1979) classified lubricant flows according to their Deborah number. Denoting the film aspect ratio L_y/L_{xz} (Section 2.2) by ε, he listed the following categories:

(1) De $= O(\varepsilon^2)$: viscous behavior predominates and nonlinear viscous models, such as *power law*, may be employed with success.
(2) De $= O(\varepsilon)$: fluids of the differential type are applicable. All phenomenological functions depend only on tr($\boldsymbol{D}$), where $\boldsymbol{D}$ is the stretching tensor (2.20b).
(3) De $= O(1)$: fluids of the differential type are no longer applicable. Integral representation must be employed (Huilgol, 1975).

Bourgin (1979) found that for a typical journal bearing lubricated with a viscoelastic lubricant De $= O(\varepsilon)$, and for rolling contact bearings, De $= O(1)$.

---

[1] There is ambiguity in associating a time constant with a material, as it will be influenced by the deformation process applied in its evaluation.

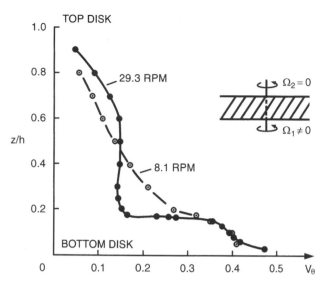

Figure 10.1. Flow of dilute polymer between parallel rotating disks.

For Deborah number De $= 0$, there is no memory, and there is, therefore, no elastic. This purely viscous fluid may, however, posses a viscosity that depends on the local rate of shear (Sirivat, Rajagopal and Szeri, 1988). A striking example of shear thinning, i.e., viscosity decreasing as the fluid is being worked on, is shown in Figure 10.1.

Figure 10.1 plots the tangential velocity of flow induced by rotating parallel disks in dilute separan. Note that at an intermediate layer between disks there is a sudden change of velocity, suggesting that locally the fluid suffers reduction of viscosity there. The flow depicted in Figure 10.1 approximates to the flow occurring in a thrust bearing.

10.1   Hydrodynamic Lubrication

Summary of Previous Work

Various constitutive models have been employed in studying the flow of nonlinear viscous fluids in journal and thrust bearings. One of the early treatments of the problem of lubrication by a non-Newtonian lubricant was by Ng and Saibel (1962). In their regular perturbation treatment of the plane slider, they assumed a stress constitutive equation of the form

$$\tau = \mu \frac{\partial u}{\partial y} + \varepsilon k \left| \frac{\partial u}{\partial y} \right|^2 \frac{\partial u}{\partial y}. \tag{10.1}$$

Equation (10.1) arises when one restricts the flow of a thermodynamically compatible third grade fluid of the differential type (Fosdick and Rajagopal, 1980) to the geometry of Ng and Saibel. In another paper, Hsu and Saibel (1965) use the constitutive equation

$$\tau + k\tau^3 = \mu \frac{\partial u}{\partial y}, \tag{10.2}$$

which is applicable to both pseudoplastic, $k > 0$, and dilatant, $k < 0$, fluids. For the pseudoplastic case, they find that the load capacity is reduced when compared to the performance of a Newtonian lubricant. Fix and Paslay (1967) investigated the lubricant elasticity effects of a Maxwell-type material in a long journal bearing. Their most significant finding is a reduction of the attitude angle, which equals $\pi/2$ under Sommerfeld conditions, due to elasticity of the lubricant. In an experimental paper, Tao and Philippoff (1967) show a pressure distribution for viscoelastic liquids that is flatter than obtained for a Newtonian fluid. They also indicate a shift into the downstream direction of the positive pressure peak. Considering the in-plane oscillation of a flat plate in a linearly viscoelastic fluid, Tanner (1969) argues that the normal stress effect is unimportant in the usual lubrication application. For an elastoviscous liquid, i.e., a nonlinear Reiner-Rivlin fluid generalized to include unequal cross stresses, Hanin and Harnoy (1969) show a load capacity higher than given by a Newtonian fluid.

The Stokesian velocity field is a solution to the equations of motion of a plane, steady, noninertial flow of a second grade fluid (Giesekus, 1963; Tanner, 1966). This fact has been used to extend the results of lubrication theory to flows involving fluids of second grade (Davis and Walters, 1973). Since the velocity field in both problems is identical, the only quantity that is to be calculated is a "modified" pressure field.[2] For a long journal bearing, Davies and Walters find qualitative agreement with the earlier results of Fix and Paslay (1967) concerning the magnitude of the attitude angle. The result of Giesekus and Tanner has been extended to dynamically loaded bearings by Harnoy and Hanin (1974). Tichy (1978) also looks at small-amplitude rapid oscillations and includes temporal inertia for a Lodge rubber-like liquid. Tichy finds perturbation solution to order δ, where δ is the amplitude of plate motion. At low Deborah number, the viscoelastic effect is found to be negligible. However, Tichy warns that bearing operation at high Deborah number could give rise to various resonance effects. He then goes on to state that the often-noted discrepancy between theory and experiment (e.g., Harnoy and Philippoff, 1976) may be in part explained by resonance phenomena.

Beris, Armstrong, and Brown (1983) look at slightly eccentric rotating cylinders; in fact, their perturbation solution is in terms of the eccentricity. The extra boundary conditions, required for a second grade fluid, is the requirement for single valuedness of the solution in the multiply connected domain. Christie, Rajagopal, and Szeri (1987) find solutions at arbitrary eccentricity.

For differential fluids of complexity 1, Bourgin and Gay (1983) investigate the influence of non-Newtonian effects on load capacity in a finite journal bearing and in a Rayleigh bearing (Bourgin and Gay, 1984). Buckholtz looks at the performance of power law lubricants in a plane slider (Buckholtz, 1984) and in a short bearing (Buckholtz and Hwang, 1985). Verma, Sharman, and Ariel (1983) investigated flow of a second-grade, thermodynamically compatible fluid between corotating porous, parallel disks. Flow of a second-grade fluid between parallel disks rotating about noncoincident axes was discussed by Rajagopal (1981).

Lubrication with Power Law Fluid

For De $= 0$, non-Newtonian fluids are often modeled by power law. This section follows the analysis given by Johnson and Mangkoesoebroto (1993) for lubrication with a power law lubricant. The objective is to model shear thinning lubricants that do not exhibit elasticity.

---

[2] The results of Tanner and Giesekus do not address the issue of uniqueness (Fosdick and Rajagopal, 1978; Rajagopal, 1984).

The viscosity for a power law fluid can be written as

$$\mu = \mu (2D_{ij}D_{ij})^{\frac{n-1}{2}}, \tag{10.3a}$$

where $\boldsymbol{D} = \frac{1}{2}(\nabla \boldsymbol{v} + \nabla \boldsymbol{v}^T)$ is the stretching tensor, which is a generalization to three dimensions of (Tanner, 1985)

$$\mu = k \left[\left(\frac{\partial v_1}{\partial x_2} \right)^2 \right]^{\frac{n-1}{2}} = k \left| \frac{\partial v_1}{\partial x_2} \right|^{n-1}. \tag{10.3b}$$

The power law *consistency index*, k, and the power law exponent, n, are material parameters. $n = 1$ yields the Newtonian fluid, for pseudoplastic (shear thinning) fluid, $n < 1$. Shear thinning is characteristic of high polymers, polymer solutions, and many suspensions.

For flow in the (x_1, x_2) plane, $v_3 = 0$ and the equations of motion take the form

$$\rho \left(\frac{\partial v_1}{\partial t} + v_1 \frac{\partial v_1}{\partial x_1} + v_2 \frac{\partial v_1}{\partial x_2} \right) = -\frac{\partial p}{\partial x_1} + \frac{\partial T_{11}}{\partial x_1} + \frac{\partial T_{12}}{\partial x_2},$$

$$\rho \left(\frac{\partial v_2}{\partial t} + v_1 \frac{\partial v_2}{\partial x_1} + v_2 \frac{\partial v_2}{\partial x_2} \right) = -\frac{\partial p}{\partial x_2} + \frac{\partial T_{21}}{\partial x_1} + \frac{\partial T_{22}}{\partial x_2}. \tag{10.4}$$

An application of the lubrication assumption $(L_y/L_{yx}) \ll 1$ to Eq. (10.3a) permits us to write

$$(T_{11}, T_{12}, T_{22}) = \left| \frac{\partial v_1}{\partial x_2} \right|^{n-1} \left(\frac{\partial v_1}{\partial x_1}, \frac{\partial v_1}{\partial x_2}, \frac{\partial v_2}{\partial x_2} \right). \tag{10.5a}$$

Substituting Eq.(10.5a) and the assumption

$$\left(\frac{L_y}{L_{xz}} \right)^{1+n} \frac{L_y^n U_*^{2-n} \rho}{\mu} \ll 1 \tag{10.5b}$$

into Eq. (10.4), we obtain

$$0 = -\frac{\partial p}{\partial x_1} + \frac{\partial T_{12}}{\partial x_2}, \qquad 0 = -\frac{\partial p}{\partial x_2}, \tag{10.6}$$

where T_{12} is given by Eq. (10.5a). These and the following equations are already in nondimensional form.

Note that for Newtonian fluid, $n = 1$, Eq. (10.5b) is a constraint on the reduced Reynolds number, Eq. (2.49):

$$\mathrm{Re}^* = \left(\frac{L_y}{L_{xz}} \right) \frac{L_y U_*}{\nu} \ll 1.$$

Since the pressure is independent of x_2 by Eq. (10.6b), we can integrate the first of Eqs. (10.6a)

$$T_{12} = \left| \frac{\partial v_1}{\partial v_2} \right|^{n-1} \frac{\partial v_1}{\partial x_2} = p'x_2 + c(x_1, t), \tag{10.7}$$

where the prime symbolizes derivative with respect to x_1. Inverting and formally integrating Eq. (10.7), we have

$$v_1 = \int_0^{x_2} |p'x_2 + c|^{\frac{1-n}{n}} (p'x_2 + c)\, dx_2 + U_1, \tag{10.8}$$

where U_1 is the (dimensional) velocity of the surface $x_2 = 0$.

From the equation of mass conservation (2.44b), we have, upon substituting for v_1 from Eq. (10.8) and integrating across the film,

$$v_2 = \frac{1}{n} \int_0^{x_2} (\xi - x_2)|p'\xi - c|^{\frac{1-n}{n}} (P''\xi + c')\, d\xi. \tag{10.9}$$

The boundary conditions are

$$v_1 = U_1(t), \qquad v_2 = 0 \quad \text{at } x_2 = 0,$$
$$v_2 = U_2(t), \qquad v_2 = V(t) \quad \text{at } x_2 = H(x_1, t) = \frac{h(x_1)}{h_0}. \tag{10.10}$$

Satisfaction of conditions (10.10) by Eqs. (10.8) and (10.9) yields two (nondimensional) equations for the unknowns $p'(x_1, t)$ and $c(x_1, t)$:

$$\int_0^1 |Hp'\xi + c|^{\frac{1-n}{n}} (Hp'\xi + c)\, d\xi = \frac{U}{H}, \tag{10.11}$$

$$H^3 p'' \int_0^1 |p'H\xi + c|^{\frac{1-n}{n}} (\xi - 1)\xi\, d\xi + H^2 c' \int_0^1 |p'H\xi + c|^{\frac{1-n}{n}} (\xi - 1)\, d\xi = nV, \tag{10.12}$$

where we put $U = U_2 - U_1$.

Though the integrals indicated in Eqs (8.11) and (8.12) can be evaluated in closed form, Johnson and Mangkoesoebroto (1993) were unable to integrate the coupled nonlinear differential equations numerically, as these equations are extremely stiff for $n < 1$. They did, however, discover an analytical solution based on a nontrivial transformation, reducing Eqs. (10.11) and (10.12) to two coupled algebraic equations for P' and c. The dimensionless pressure is plotted against dimensionless position in Figure 10.2 for a parabolic slider at various values of n, as obtained by Johnson and Mangkoesoebroto (1993).

Fluids of the Differential Type

For a Newtonian fluid, Eq. (2.35), the extra stress, $T + pI$, is a linear function of the current value of the stretching tensor D. The function of proportionality, μ, is a constant, or at most a function of temperature. Newtonian fluids exhibit neither elasticity nor memory effects.

To describe elasticity and memory effects as well as viscous response, we obviously need more general constitutive equations than Eq. (2.35). This leads us to define a class of materials, called *simple materials*, which includes a vast range of nonlinear materials from classical linear elastic to linearly viscous (Truesdell and Noll, 1965).

In Chapter 2, we represented motion of the material point X by the mapping

$$x = \chi(X, t), \tag{2.1}$$

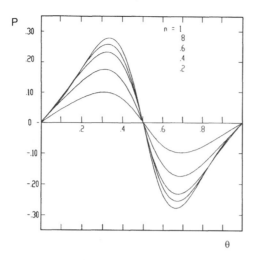

Figure 10.2. Dimensionless pressure versus dimensionless position for parabolic slider $H = 1 + 10.4(x_1 - 5)^2$ lubricated by power law fluid. (Reprinted with permission from Johnson, M. W. and Mangkoesoebroto, S. Analysis of lubrication theory for the power law fluid. *ASME Journal of Tribology*, **115**, 71–77, 1993).

where x is the place occupied by X at time t. We can also define the motion of a neighborhood of X, relative to X, with the help of Eq. (2.1) by the Taylor series expansion

$$x(\hat{X}, t) = x(X, t) + \nabla\chi \cdot (\hat{X} - X) + \cdots \tag{10.13a}$$

where $\hat{X}$ is an arbitrary point in the neighborhood. There is a large class of engineering materials that possess no microstructure, called *simple materials* by Noll, for which the expansion in Eq. (10.13a) can be terminated after the linear term. For simple materials we write

$$dx = F\, dX, \tag{10.13b}$$

where $F = \nabla\chi$ is the deformation gradient, and postulate that stress at a point depends on the deformation of the neighborhood of that point. But to account for memory effects, we consider the whole past history of deformation rather than just its current value. For an incompressible *simple fluid*, therefore, our constitutive equation is

$$T + p\mathbf{1} = \mathfrak{h}^t_{s=-\infty}(F(s)). \tag{10.14}$$

Here $F(s)$, $-\infty < s \leq t$, is the *history of deformation*, up to the present time t, and $\mathfrak{h}$ represents some tensor valued functional of this history. Equation (10.14) expresses the postulate that in a simple fluid, stress at a material point X and time t depends on the past deformation of the neighborhood of that material point.[3]

As it stands, Eq. (10.14) is far too general, however, to be of practical use. The question is, how to proceed? There have been, in the past, two methods of attack:

(1) *Specialize the motion* to a class of motions such that memory is given little opportunity to make itself felt. Such specialization leads to *viscometric flows* (Tanner, 1985), solution of special problems is possible.

---

[3] The Newtonian constitutive equation is obtained from Eq. (10.14) by setting

$$\mathfrak{h}^t_{-\infty}(F(s)) = -\mu\frac{d}{ds}[F(s) + F(s)^T]|_{s=0} = 2\mu D.$$

(2) *Specialize the material* by specializing the constitutive equation (10.14). The question is, how to handle memory dependence?

We shall follow the second of these alternatives but, for starters, restrict fluid memory to an arbitrary short interval. Making this interval infinitesimal, we obtain a class of fluids for which the history of motion prior to an infinitesimal time is irrelevant in determining the stress.

The most important of the fluids with infinitesimal memory are *fluids of the differential type*, those in which the stress at X is determined by the first n material derivatives of the deformation gradient F

$$T + p\mathbf{1} = f\left(\dot{F}, \ddot{F} \ldots, F^{(n-1)}, F^{(n)}\right).\tag{10.15a}$$

By requiring frame indifference, it can be shown that Eq. (10.15a) must be of the form

$$T + p\mathbf{1} = f\left(A_{(1)}, A_{(2)}, \ldots, A_{(n)}\right).\tag{10.15b}$$

Here the $A_{(k)}$ are the coefficients in the Taylor expansion

$$C_t(t') = F_t^T(t')F_t(t') = \mathbf{1} - A_{(1)}(t - t') + \frac{1}{2!}A_{(2)}(t - t')^2 + \cdots$$

and are defined through the recurrence formula

$$A_{(n+1)} = \frac{d}{dt}A_{(n)} + L^T A_{(n)} + A_{(n)}L,\tag{10.16}$$

where

$$A_{(1)} \equiv D = \frac{1}{2}(L + L^T).$$

L is the velocity gradient tensor, Eq. (2.18), d/dt represents material derivative, Eq. (2.5), and the $A_{(k)}$ are the *Rivlin-Ericksen tensors*. The tensor $C_t(t')$ is the relative strain tensor, and $F_t(t')$ is the relative deformation gradient.

The principle of material frame indifference, more specifically the requirement that the constitutive equation be frame indifferent, dictates that f in Eq. (10.16) must be an isotropic function, Eq. (2.32). But isotropic functions of symmetric tensors have polynomial representation (Truesdell and Noll, 1965). As an example, an isotropic function of the first two Rivlin-Ericksen tensors $A_{(1)}$ and $A_{(2)}$ can be written as

$$\begin{aligned}
T + p\mathbf{1} &= f\left(A_{(1)}, A_{(2)}\right) \\
&= \alpha_1 A_{(1)} + \alpha_2 A_{(2)} + \alpha_3 A_{(1)}^2 + \alpha_4\left(A_{(1)}A_{(2)} + A_{(2)}A_{(1)}\right) + \alpha_5 A_{(2)}^2 \\
&\quad + \alpha_6\left(A_{(1)}^2 A_{(2)} + A_{(2)}A_{(1)}^2\right) + \alpha_7\left(A_{(1)}A_{(2)}^2 + A_{(2)}^2 A_{(1)}\right) \\
&\quad + \alpha_8\left(A_{(1)}^2 A_{(2)}^2 + A_{(2)}^2 A_{(1)}^2\right),
\end{aligned}\tag{10.17}$$

where the α_i's depend on the basic invariants of A_1 and A_2. Equation (10.17) defines the *Rivlin-Ericksen fluid* of complexity 2.

The constitutive relation (10.15b) of a fluid of *complexity n* includes the first n Rivlin-Ericksen tensors $A_{(1)}, \ldots, A_{(n)}$. Thus the fluid of complexity 1, from Eq. (10.17), is

$$T + p\mathbf{1} = \alpha_1 A_{(1)} + \alpha_3 A_{(1)}^2.\tag{10.18}$$

We introduce now the concept of slow, or retarded, motion. If $\chi(X, t)$ is a motion, the *retarded motion* is defined as

$$\text{ret}\,\chi(X, t) \equiv \chi(X, rt), \qquad 0 < r < 1. \tag{10.19a}$$

If the motion $\chi(X, t)$ carries particle X into spatial position x at time t, the retarded motion ret $\chi(X, t)$ will carry the same particle into x at the later time $\hat{t} = t/r$. The smaller the value of r, the slower is the motion. It can be shown that

$$\text{ret}\,A_{(n)} = r^n A_{(n)}, \tag{10.19b}$$

so the fluid of complexity 1 in retarded motion is

$$T = -p\mathbf{1} + \alpha_1 r A_{(1)} + \alpha_3 r^2 A_{(1)}^2. \tag{10.19c}$$

The *fluid of grade* 1 is, by definition, of degree 1 in r. Thus the fluid of grade 1 is, from Eq. (10.19c),

$$T = -p\mathbf{1} + \mu A_{(1)}. \tag{10.20a}$$

This is a Newtonian fluid [as $A_{(1)} = D$] if $\mu = $ const.

The grade two fluid has constitutive equation

$$T = -p\mathbf{1} + \mu A_{(1)} + \alpha_1 A_{(2)} + \alpha_2 A_{(1)}^2. \tag{10.20b}$$

We can attach meaning to the concept of complexity and grade as follows: *The lower the complexity, the lower the order of velocity derivatives that are used to determine the stress. The lower the grade, the slower is the motion that can be adequately described by the constitutive equation.*

When a fluid of the differential type is used in journal bearings for which the Deborah number is $\mathcal{O}(10^{-3})$, it is impossible to distinguish between two fluids having both complexity $n > 3$ and degree $d > 1$. Moreover, all phenomenological functions will depend on the trace, $\text{tr}(A_{(1)}^2)$ (Bourgin, 1979).

Lubrication with a Third Grade Fluid

The stress, T, in an incompressible homogeneous fluid of third grade is related to the fluid motion in the following manner (Truesdell and Noll, 1965).

$$\begin{aligned} T = -p\mathbf{1} &+ \mu A_{(1)} + \alpha_1 A_{(2)} + \alpha_2 A_{(1)}^2 + \beta_1 A_{(3)} \\ &+ \beta_2\left[A_{(1)}A_{(2)} + A_{(2)}A_{(1)}\right] + \beta_3\left(\text{tr}\,A_{(1)}^2\right)A_{(1)}. \end{aligned} \tag{10.21}$$

The thermodynamics and stability of a fluid modeled by Eq. (10.21) have been studied in detail by Rajagopal and Fosdick (1980). They find that if all motions of the fluid are to be compatible with thermodynamics, then the following restrictions apply:

$$\mu \geq 0 \; |\alpha_1 + \alpha_2| < \sqrt{24\mu\beta_3}, \qquad \alpha_1 \geq 0,$$

$$\beta_1 = \beta_2 = 0 \quad \text{and} \quad \beta_3 \geq 0.$$

Thus, in the case of a thermodynamically compatible fluid of grade three, the stress constitutive equation takes the simplified form

$$T = -p\mathbf{1} + \mu A_{(1)} + \alpha_1 A_{(2)} + \alpha_2 A_{(1)}^2 + \beta_3\left(\text{tr}\,A_{(1)}^2\right)A_{(1)}. \tag{10.22}$$

Note that if the material constants α_1 and α_2 are zero, then

$$T = -p\mathbf{1} + \left(\mu + \beta_3 \operatorname{tr} A_{(1)}^2\right) A_{(1)}. \tag{10.23}$$

The above model is a special fluid of complexity 1 (Truesdell and Noll, 1965) and in a sense can be thought of as a type of power law model, for in a simple shear flow the shear stress, τ, is related to the velocity gradient, $\partial u / \partial y$, by

$$\tau = \left[\mu + \beta_3 \left(\frac{\partial u}{\partial y}\right)^2\right] \frac{\partial u}{\partial y}. \tag{10.24}$$

Models of the type in Eq. (10.24) have been studied by Bourgin and Gay (1983) within the context of flows in journal bearings. Ng and Saibel (1962) have used the model (10.24) to study the flow occurring in a slider bearing [cf. Eq. (10.8)]. They, however, assumed the coefficient that corresponds to β_3 in our model to be negative. The power law model used by Buckholtz (1985) would reduce to Eq. (10.24) if the parameter N of his model was made equal to 3. The model (10.24) includes the classical linearly viscous fluid, the incompressible homogeneous fluid of second grade, a subclass of the fluids of complexity 1, and several other models that have been used in lubrication as special cases.

On substituting Eq. (10.24) into the balance of linear momentum

$$\operatorname{div} T + \rho b = \rho \frac{d v}{d t}, \tag{2.39}$$

we obtain the following equations of motion (Rajagopal and Fosdick, 1980):

$$\mu \Delta v + \alpha_1 (\Delta \omega \times v) + \alpha_1 (\Delta v)_t + (\alpha_1 + \alpha_2) \left\{ A_{(1)} \Delta v + 2 \operatorname{div}[(\operatorname{grad} v)(\operatorname{grad} v)^T] \right\}$$
$$+ \beta_3 A_{(1)} \left(\operatorname{grad} |A_{(1)}|^2 \right) + \beta_3 |A_{(1)}|^2 \Delta v + \rho b - \rho v_t - \rho(\omega \times v) = \operatorname{grad} P \tag{10.25}$$

where

$$P = p - \alpha_1 v \cdot \Delta v - \frac{1}{4}(2\alpha_1 + \alpha_2)|A_{(1)}|^2 + \frac{1}{2}\rho|v|^2.$$

In the above equation, Δ denotes the Laplacian, $(\cdot)_t$ the partial derivative of the quantity within the parentheses with respect to time, and $|v|$ and $|A_{(1)}|$ the inner product norm and the trace norm of v, and $A_{(1)}$ respectively.

When

$$v = u(x, y)\mathbf{i} + v(x, y)\mathbf{j}$$

for two-dimensional flow is substituted into Eq. (10.25), the equations are made nondimensional in accordance with Eqs. (2.45) and (2.47), and terms of order $(L_y/L_{xz})^2$ and higher are neglected, we arrive at (Kacou, Rajagopal, Szeri, 1987)

$$\frac{\partial^2 \bar{u}}{\partial \bar{y}^2} + \frac{\alpha_1 U}{\mu R} \left\{ \frac{\partial \bar{u}}{\partial \bar{x}} \frac{\partial^2 \bar{u}}{\partial \bar{y}^2} + \bar{v} \frac{\partial^3 \bar{u}}{\partial \bar{y}^3} + \bar{u} \frac{\partial^3 \bar{u}}{\partial \bar{y}^2 \partial \bar{x}} + 3 \frac{\partial \bar{u}}{\partial \bar{y}} \frac{\partial^2 \bar{u}}{\partial \bar{y} \partial \bar{x}} \right\}$$
$$+ 6 \frac{\beta_3 U^2}{\mu} \frac{1}{C^2} \left(\frac{\partial \bar{u}}{\partial \bar{y}} \right)^2 \frac{d^2 \bar{u}}{d \bar{y}^2} = \frac{\partial \bar{P}}{\partial \bar{x}}, \tag{10.26}$$

$$4\frac{\alpha_1 U}{\mu R}\frac{\partial \bar{u}}{\partial \bar{y}}\frac{\partial^2 \bar{u}}{\partial \bar{y}^2} = \frac{\partial \bar{P}}{\partial \bar{y}}.$$ (10.27)

Defining a modified scalar field $\bar{p}^*$ through

$$\bar{p}^* = \bar{P} - 2\frac{\alpha_1 U}{\mu R}\left(\frac{\partial \bar{u}}{\partial y}\right)^2,$$

we can rewrite Eqs. (10.26) and (10.27) as

$$\frac{\partial^2 \bar{u}}{\partial \bar{y}^2} + \gamma\left\{\frac{\partial \bar{u}}{\partial \bar{x}}\frac{\partial^2 \bar{u}}{\partial \bar{y}^2} + \bar{v}\frac{\partial^3 \bar{u}}{\partial \bar{y}^3} + \bar{u}\frac{\partial^3 \bar{u}}{\partial \bar{y}^2 \partial \bar{x}} + \frac{\partial \bar{u}}{\partial \bar{y}}\frac{\partial^2 \bar{v}}{\partial \bar{y}^2}\right\} + \delta\left(\frac{\partial \bar{u}}{\partial \bar{y}}\right)^2 \frac{d^2 \bar{u}}{\partial \bar{y}^2} = \frac{\partial \bar{p}^*}{\partial \bar{x}},$$ (10.28)

$$0 = \frac{\partial \bar{p}^*}{\partial \bar{y}}.$$ (10.29)

In the above equations, the nondimensional parameters

$$\frac{\alpha_1 U}{\mu R} \equiv \gamma \quad \text{and} \quad \frac{\beta_3 U^2}{\mu C^2} \equiv \delta$$

are measures of the non-Newtonian nature of the fluid. Notice that the nondimensional parameter, γ, can be expressed as the ratio of the nondimensional numbers Re, Γ, and (R/C) as

$$\gamma = \frac{\text{Re}}{\Gamma}\left(\frac{R}{C}\right),$$

where

$$\text{Re} \equiv \frac{\rho U C}{\mu} \quad \text{and} \quad \Gamma \equiv \frac{\rho R^2}{\alpha_1}.$$

The nondimensional number, Γ, (Fosdick and Rajagopal, 1978) is referred to as the absorption number. Truesdell (1964) recognized that the term involving α_1 controls the diffusion of vorticity from a boundary. δ is a measure of the ratio of the non-Newtonian shear stress to that of the Newtonian shear stress. It follows from Eqs. (10.28) and (10.29) that

$$\bar{p}^* = \bar{p}^*(x),$$

and thus the modified pressure $\bar{p}^*$ does not vary across the film thickness.

Introducing the stream function $\bar{\psi}(\bar{x}, \bar{y})$ through

$$\bar{u} = \frac{\partial \bar{\psi}}{\partial \bar{y}}, \qquad \bar{v} = \frac{\partial \bar{\psi}}{\partial \bar{x}},$$ (10.30)

we can express Eq. (10.28) as[4]

$$\frac{\partial^4 \bar{\psi}}{\partial \bar{y}^4} + \gamma\left\{-\frac{\partial \bar{\psi}}{\partial \bar{x}}\frac{\partial^5 \bar{\psi}}{\partial \bar{y}^5} + \frac{\partial \bar{\psi}}{\partial \bar{y}}\frac{\partial^5 \bar{\psi}}{\partial \bar{x}\,\partial \bar{y}^4}\right\} + \delta\left\{2\frac{\partial^2 \bar{\psi}}{\partial \bar{y}^2}\left(\frac{\partial^3 \bar{\psi}}{\partial \bar{y}^3}\right)^2 + \left(\frac{\partial^2 \bar{\psi}}{\partial \bar{y}^2}\right)^2 \frac{d^4 \bar{\psi}}{\partial \bar{y}^4}\right\} = 0.$$ (10.31)

---

[4] Equation (10.31) is of higher order than the corresponding equation for the classical linearly viscous fluid, when $\gamma \neq 0$. We thus need additional boundary conditions to solve the problem. We overcome the paucity of boundary conditions by resorting to regular perturbation, although the problem is one of singular perturbation.

We shall assume that the stream function, $\bar{\psi}$, the pressure, $\bar{p}^*$, and the flow rate, $\bar{Q} \equiv Q/CU$, can all be expanded in power series in γ and δ. For the sake of convenience, we shall drop the bars that appear in the nondimensional terms.

Let

$$\psi = \sum_{m=0}^{\infty}\sum_{n=0}^{\infty} \gamma^n \delta^m \psi_{nm}(x, y),$$

$$p^* = \sum_{m=0}^{\infty}\sum_{n=0}^{\infty} \gamma^n \delta^m P_{nm}(x, y),$$ (10.32)

$$Q = \sum_{m=0}^{\infty}\sum_{n=0}^{\infty} \gamma^n \delta^m Q_{nm}(x, y).$$

Substituting Eq. (10.32) into Eq. (10.31) and equating like powers, we obtain the following system of differential equations (first three equations are shown):

$n = 0, m = 0$:

$$\frac{\partial^4 \psi_{00}}{\partial y^4} = 0,$$ (10.33a)

$n = 1, m = 0$:

$$\frac{\partial^4 \psi_{10}}{\partial y^4} = \frac{\partial \psi_{00}}{\partial x}\frac{\partial^5 \psi_{00}}{\partial y^5} - \frac{\partial \psi_{00}}{\partial y}\frac{\partial^5 \psi_{00}}{\partial x \partial y^4},$$ (10.33b)

$n = 0, m = 1$:

$$\frac{\partial^4 \psi_{01}}{\partial y^4} = -2\frac{\partial^2 \psi_{00}}{\partial y^2}\left(\frac{\partial^3 \psi_{00}}{\partial y^3}\right)^2 - \left(\frac{\partial^2 \psi_{00}}{\partial y^2}\right)\left(\frac{\partial^4 \psi_{00}}{\partial y^4}\right).$$ (10.33c)

Although these equations are linear if solved sequentially, they are rapidly becoming cumbersome as n, m increase (Kacou, Rajagopal, and Szeri, 1987).

The boundary conditions that accompany Eqs. (10.33) are

$$\left.\begin{array}{l} \psi_{nm} = 0 \\[2mm] \dfrac{\partial \psi_{nm}}{\partial y} = 0 \end{array}\right\} \qquad n, m = 0, 1, 2, \ldots, \quad \text{at } y = 0,$$ (10.34a)

$$\left.\begin{array}{l} \psi_{nm} = Q_{nm} \\[2mm] \dfrac{\partial \psi_{nm}}{\partial y} = \begin{cases} 1, n = m = 0 \\ 0, n, m = 1, 2, \ldots \end{cases} \end{array}\right\} \qquad n, m = 0, 1, 2, \ldots \quad \text{at } y = H.$$ (10.34b)

The zeroth-order perturbation corresponds to the Newtonian case. It is trivial to show that the solution is

$n = 0, m = 0$:

$$\psi_{00} = \frac{1}{6}\frac{dP_{00}}{dx}\left(y^3 - \frac{3}{2}Hy^2\right) + \frac{1}{2}\frac{y^2}{H}.$$ (10.35a)

The pressure, P_{00}, can be obtained in the standard manner as is done classical lubrication theory:

$$P_{00}(x) = \frac{6\varepsilon \sin x(2 + \varepsilon \cos x)}{(2 + \varepsilon^2)(1 + \varepsilon \cos x)^2}.$$ (10.35b)

Proceeding in a manner identical to that used for determining P_{00} in classical lubrication theory, we can show that

$n = 1, m = 0$:

$$\frac{d^2 P_{10}}{dx^2} + \frac{3}{H}\frac{dH}{dx}\frac{dP_{10}}{dx} + \left(\frac{dP_{00}}{dx}\right)^2\left(\frac{dH}{dx}\right)^2 - \frac{4}{H^3}\frac{d^2 H}{dx^2} - \frac{H}{2}\left(\frac{dP_{00}}{dx}\right)^2\frac{d^2 H}{dx^2}$$

$$- \frac{9}{H^2}\left(\frac{dH}{dx}\right)^2\frac{\partial P_{00}}{dx} + \frac{3}{H}\frac{d^2 H}{dx^2}\frac{dP_{00}}{dx} + \frac{18}{H^4}\left(\frac{dH}{dx}\right)^2 = 0,$$ (10.36)

$n = 0, m = 1$:

$$\frac{d^2 P_{01}}{dx^2} + \frac{3}{H}\frac{dH}{dx}\frac{dP_{01}}{dx} + \frac{H}{5}\frac{dH}{dx}\left(\frac{dP_{00}}{dx}\right)^3$$

$$- \frac{9}{10H}\frac{dH}{dx}\left(\frac{dP_{00}}{dx}\right)^2 + \frac{2}{H^3}\frac{dH}{dx}\frac{dP_{00}}{dx} - \frac{6}{H^5}\frac{dH}{dx} = 0.$$ (10.37)

We are now in a position to determine the pressure up to first-order,

$$p^* = P_{00} + \gamma P_{01} + \delta P_{10} + O(\gamma \delta).$$ (10.38)

The first-order correction, P_{10}, is plotted in Figure 10.3; this profile is very close to the one calculated by Harnoy and Philippoff (1976) for a second-grade fluid, using results due to Giesekus and Tanner. In Figure 10.4, we plot pressure distribution at $\gamma = 0.75$ for various values of δ.

It is fairly clear that viscoelastic effects are not significant at small values of the Deborah number in steady flow. If, however, $\Omega^* > 1$, anomalous resonance effects might occur. There is virtually no information on viscoelastic effects in the flow regime $\Omega^* \sim \mathrm{Re}^* > 1$. Here inertia effects might interact nonlinearly with viscoelastic effects, leading to changed bearing performance.

Among the many models that have been proposed in the literature to describe non-Newtonian fluid behavior at arbitrary De, one that has gained considerable attention is the K-BKZ fluid model (Kay, 1962; Bernstein, Kearsley, and Zapas, 1963). The model has been shown to be consistent with statistical modeling of polymeric fluids. It might, thus, be applicable when investigating process fluid lubrication. K-BKZ fluids have memory and are represented by integral constitutive equations. Because of their memory, K-BKZ fluids might show significant departure from Newtonian fluids in squeeze film applications. Little information is available currently on numerical simulation of fluids with finite memory at high shear rates, but initial studies on special simple integral constitutive models have already provided interesting information regarding the structure of the boundary layer. There is much work to be done here by both the rheologist and the numerical analyst. The properties and predictive capabilities of some non-Newtonian fluids are tabulated by Tanner (1985).

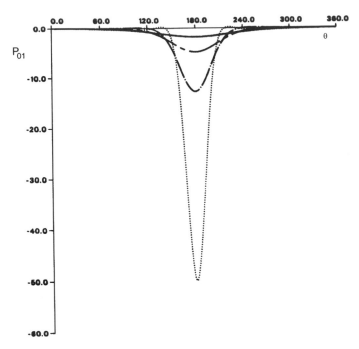

Figure 10.3. Pressure correction P_{01} in a journal bearing lubricated by a fluid of grade three (———, $\epsilon = 0.2$; — — —, $\epsilon = 0.4$; — · —, $\epsilon = 0.6$; ...., $\epsilon = 0.8$). (Reprinted with permission from Kacou, A., Rajagopal, K. R. and Szeri, A. Z. Flow of a fluid of the differential type in a journal bearing. *ASME Journal of Tribology*, **109**, 100–108, 1987.)

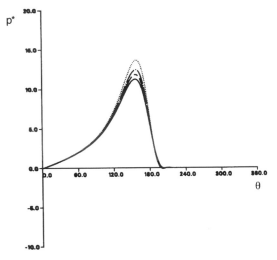

Figure 10.4. Pressure distribution for $\gamma = 0.75$ (— —, Newtonian; — · —, $\gamma = 0, \delta = 0.01$; — • —, $\gamma = 0, \delta = 0.02$; ...., $\gamma = 0, \delta = 0.04$). (Reprinted with permission from Kacou, A., Rajagopal, K. R. and Szeri, A. Z. Flow of a fluid of the differential type in a journal bearing. *ASME Journal of Tribology*, **109**, 100–108, 1987.)

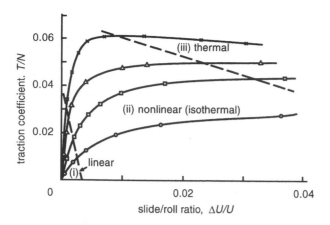

Figure 10.5. Typical traction curves measured on a two disk machine in line contact. (p: x, 1.03 GPa; △, 0.68 GPa; □, 0.5 GPa; ○, 0.4 GPa). (Reprinted by permission of the Council of the Institution of Mechanical Engineers from Johnson, K. L. and Cameron, R. 1967. Shear behavior of EHD oil films. *Proc. Inst. Mech. Eng.*, **182**, 307, 1967.)

10.2 Elastohydrodynamic Lubrication

In discussing the various regimes of lubricant behavior in EHD contacts, we refer to Figure 10.5, where the coefficient of friction is plotted against the ratio of sliding speed to rolling speed for a two-disk machine (Johnson and Cameron, 1967). Under the conditions of the experiment, the ordinate is proportional to shear stress, τ, and the abscissa to shear rate, $\dot{\gamma}$. The various curves are drawn for various constant Hertzian pressure. Each of these curves is seen to have a linear portion, $\tau \propto \dot{\gamma}$, where the fluid is essentially Newtonian. For large p, the region of this type of behavior is limited to very small values of $\dot{\gamma}$. At larger strain rate, the lubricant exhibits nonlinear, shear thinning behavior, though it may still be treated by isothermal theories. At still larger strain rate, the lubricant enters the zone of predominant thermal effects. Here the shear stress shows a slight decrease with increasing strain rate (Johnson and Tevaarwerk, 1977).

The principal features of EHD contacts, such as the sudden contraction of the film or the existence of the second pressure peak, remain largely unchanged by thermal effects if the maximum Hertzian load $p < 0.5$ GPa. If, on the other hand, as it is in practice, $p > 0.5$ GPa, then thermal effects are so severe that lubricant behavior departs significantly from the linear. In these cases traction forces calculated on the assumption of a Newtonian lubricant are an order of magnitude larger than found experimentally.

There have been several theories put forth to explain and model traction in the nonlinear region. Fein (1967) and Harrison and Trachman (1972) suggested that the fluid, during its brief stay in the contact zone, would not have time to assume equilibrium values of density, viscosity, and shear modulus. Dyson (1970) suggested that the apparent decrease of viscosity in steady, continuous shear of a linearly viscoelastic fluid can be used to model the nonlinear part of the traction curve. An alternative explanation, that the shear stress-shear rate is nonlinear for the lubricant, was offered by Trachman and Cheng (1972). Smith (1962) proposed that in the high-pressure contact zone the lubricant was shearing like a plastic solid.

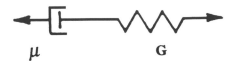

Figure 10.6. Maxwell's constitutive model for a linearly viscoelastic fluid.

Inspired by the constitutive equation for the Maxwell fluid, Johnson and Tevaarwerk (1977) formulated a new rheological model by adding a linear elastic element to a nonlinear viscous element.

Constitutive models

The *Maxwell fluid*, as shown in Figure 10.6, consists of a spring and a dashpot connected in series. The model represents a fluid because the flow in the dashpot will continue as long as force is applied. A sudden change in loading, however, will be accompanied by an instantaneous elastic response due to the spring. To derive the constitutive equation for this fluid, we note that the spring and the dashpot are subjected to the same force τ,

$$\tau = G\gamma_E, \qquad \tau = \mu\dot{\gamma}_v, \tag{10.39a}$$

where G is the elastic modulus in shear and μ is the viscosity, while the total deformation γ is the sum of the deformations γ_E and γ_v in the spring and the dashpot, respectively,

$$\dot{\gamma} = \dot{\gamma}_E + \dot{\gamma}_v$$
$$= \frac{1}{G}\dot{\tau} + \frac{1}{\mu}\tau. \tag{10.39b}$$

For the integral counterpart of Eq. (10.39b), see Tanner (1985).

Experimental data of common lubricants display viscoelastic effects in the linear, small strain rate region of the traction curve (Figure 10.5), such as might be described by Eq. (10.39b), but show nonlinearity in the large strain rate region. To reconcile Eq. (10.39) with both types of fluid response, Johnson and Tevaarwerk (1976) proposed a nonlinear version of Eq. (10.39b),

$$\dot{\gamma} = \frac{1}{G}\dot{\tau} + F(\tau), \tag{10.40}$$

requiring the nonlinear function $F(\tau)$ to reduce to τ/μ for small τ. They also generalized Eq. (10.40) to three dimensions by changing the argument of F to the von Mises equivalent stress

$$\tau_e = \sqrt{\frac{1}{2}\tau_{ij}\tau_{ij}},$$

replacing $\dot{\gamma}$ with the stretching tensor D_{ij} and τ with the shear stress tensor $\tau_{ij} = T_{ij} + p\delta_{ij}$. This generalized form is

$$2D_{ij} = \frac{1}{G}\frac{d\tau_{ij}}{dt} + \frac{\tau_{ij}}{\tau_e}F(\tau_e). \tag{10.41}$$

When applying Eq. (10.41) to situations in which the deformations are large, Tevaarwerk and Johnson (1975) recommend using the "*convected time derivative*," a kind of total derivative that introduces no dependence on a fixed reference frame (Tanner, 1985). But, provided that the recoverable elastic strain τ_e/G remains small, the errors introduced by employing time derivative of the stress in place of its "convected" derivative are small, of order $(\tau_e/G)^2$. Tichy (1996), nevertheless, employs the *convected Maxwell model* in a perturbation analysis in which the Deborah number is the small parameter.

Johnson and Tevaarwerk (1977) used the Eyring hyperbolic sine as the viscous element, as this gave best fit to experimental data. At the same time, it has its roots in molecular theory. The rheological equation proposed by them is

$$2D_{ij} = \frac{1}{G}\frac{d\tau_{ij}}{dt} + \frac{\tau_{ij}}{\tau_e}\frac{\tau_0}{\mu}\sinh\left(\frac{\tau_e}{\tau_0}\right). \tag{10.42}$$

Here the representative stress, τ_0, is a measure of the stress above which the behavior becomes appreciably nonlinear. For $\tau_e \ll \tau_0$, Eq. (10.42) reduces to the linear Maxwell model. This rheological model, Eq. (10.42), was later extended by Houpert (1985) to include thermal effects.

Bair and Winer (1979) modified the Maxwell model by including a limiting shear stress. In earlier rheological models, measurements of contact behavior had to be made in order to predict contact behavior. In the Bair-Winer model, which shows good agreement with experiments, material constants, viz., the low shear stress viscosity μ_0, the limiting elastic shear modulus G_∞, and the limiting yield shear stress τ_L, are all derived from sources other than EHD contact measurements and are functions of pressure and temperature. The Bair-Winer equation relating stress to motion (in one dimension) in the lubricant is[5]

$$\dot{\gamma} = \dot{\gamma}_e + \dot{\gamma}_v = \frac{1}{G_\infty}\frac{d\tau}{dt} - \frac{\tau_L}{\mu_0}\tanh^{-1}\left(\frac{\tau}{\tau_L}\right). \tag{10.43}$$

If the limiting shear stress is very large, then Eq. (10.43) reduces to the classical Maxwell model. If, on the other hand, $(1/G_\infty) \to 0$, the limiting case of viscoelastic behavior, the model behaves fluid like

$$\dot{\gamma} = -\frac{\tau_L}{\mu_0}\tanh^{-1}\left(\frac{\tau}{\tau_L}\right). \tag{10.44}$$

Equation (10.43) can be written in the nondimensional form

$$\frac{\mu_0\dot{\gamma}}{\tau_L} = \mathrm{De}\frac{\partial\bar{\tau}}{\partial\bar{t}} + F(\bar{\tau}), \tag{10.45}$$

where $\bar{\tau} = \tau/\tau_L$ and $De = \mu U/Gl$. If $De \to 0$, signifying absence of viscoelastic effects, Eq. (10.44) results.

There is another factor to consider, solidification of the lubricant. Experiments show that for the synthetic lubricant 5P4E, the glass transition pressure is $\sim$160 MPa at room temperature. Experimentation is very critical near glass transition to guide modeling. In the pressure-shear plate impact experiments of Zang (1995) and Zhang and Ramesh (1996), a projectile carrying a 2-inches diameter plate is accelerated down the barrel of a light gas gun

---

[5] Originally, Bair and Winer specified logarithmic dependence for the nonlinear viscous shear term; this was changed later to the tangent hyperbolic function, as the former is not antisymmetric with respect to zero stress. Najji et al. (1989) offer a slight modification of Eq. (10.43) by writing it in terms of the equivalent stress, τ_e; this, or another stress norm, must be used if extension to three dimensions is required. See also Bair and Winer (1992).

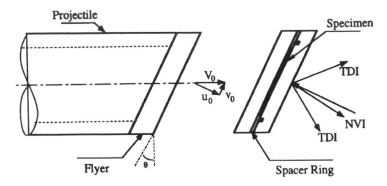

Figure 10.7. Schematic of the pressure-plate impact experiment. (Reprinted with permission from Zang, Y. and Ramesh, K. T. The behavior of an elastohydrodynamic lubricant at moderate pressures and high shear rates. *ASME Journal of Tribology,* **118**, 162–168, 1996.)

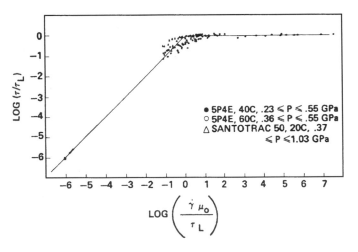

Figure 10.8. Dimensionless shear stress versus dimensionless shear rate. (Reprinted with permission from Bair, S. and Winer, W. O. A rheological model for elastohydrodynamical contacts. *ASME Journal of Lubrication Technology,* **101**, 248, 1979.)

and hits a parallel stationary target. On impact, the normal wave arrives first at the sample and compresses it. The compressed lubricant is then subjected to a shearing deformation. A schematic of the arrangement is shown in Figure 10.7.

In Figure 10.8, the dimensionless shear stress is plotted against the dimensionless shear rate for two lubricants, Santotrac 50 and 5P4E. The limiting shear τ_L is almost independent of shear rate in the relatively high shear rate regime ($\dot{\gamma} > 10^3 \text{ s}^{-1}$), as indicated in Figure 10.8, but is dependent on pressure (Figure 10.9).

Glass transition, or more correctly liquid-solid transition, was first reported by Johnson and Roberts (1974). The critical pressure at which glass transition occurs is strongly temperature dependent, as was first shown by Johnson and Roberts (1974), Alsaad et al. (1978), and Bair and Winer (1979). Several sets of data are shown in Figure 10.10.

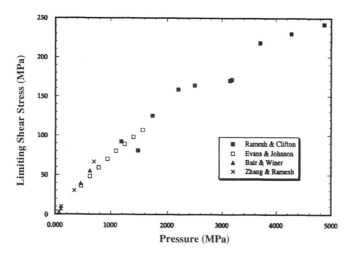

Figure 10.9. Variation of limiting shear with pressure. (Reprinted with permission from Zang, Y. and Ramesh, K. T. The behavior of an elastohydrodynamic lubricant at moderate pressures and high shear rates. *ASME Journal of Tribology*, **118**, 162–168, 1996.)

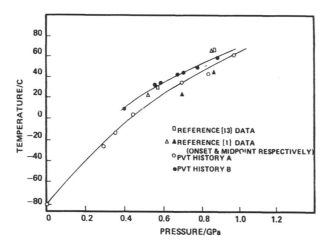

Figure 10.10. Comparison of four sets of liquid-solid transition data. (Reprinted with permission from Bair, S. and Winer, W. O. A rheological model for elastohydrodynamical contacts. *ASME Journal of Lubrication Technology*, **101**, 248, 1979.)

A Generalized non-Newtonian Reynolds Equation for EHL

Following on the work of Najji, Bou-Said, and Berthe (1989), Wolff and Kubo (1996) derived a generalized non-Newtonian Reynolds equation for isothermal EHD line contact. The basic equations that form the starting point of the analysis of Wolff and Kubo are

$$\frac{\partial p}{\partial x} = \frac{\partial \tau_{xy}}{\partial y}, \qquad \frac{\partial p}{\partial y} = 0, \qquad \frac{\partial p}{\partial z} = \frac{\partial \tau_{zy}}{\partial y}, \tag{10.46}$$

$$\frac{\partial \rho}{\partial t} + \mathrm{div}(\rho \boldsymbol{v}) = 0. \tag{2.44a}$$

To take into account several types of rheological laws (Najji, Bou-Said, and Berthe, 1989; Wolff and Kubo, 1996), we put

$$\frac{\partial u}{\partial y} = A\frac{d\tau_{xy}}{dt} + \tau_{xy}F(\tau_e) \tag{10.47}$$

$$\frac{\partial v}{\partial y} = A\frac{d\tau_{xy}}{dt} + \tau_{zy}F(\tau_e), \tag{10.48}$$

where $\tau_e = (\tau_{xy}^2 + \tau_{zy}^2)^{1/2}$, $A = 1/G$ or $A = 0$ if fluid elasticity is ignored, and $F(\tau_e)$ represents the viscosity function.

Substituting Eq. (10.48) into Eq. (10.46), integrating twice with respect to y, and satisfying the no-slip boundary conditions

$$\begin{aligned} u &= u_1, & v &= v_1, & w &= w_1, & \text{for } y = 0, \\ u &= u_2, & v &= v_2, & w &= w_2, & \text{for } y = h, \end{aligned} \tag{10.49}$$

yields the velocity distribution

$$u = \frac{\partial p}{\partial x}\left[\int_0^y F(\tau_e)y\,dy - \frac{f_1}{f_0}\int_0^y F(\tau_e)\,dy\right] + \frac{u_2 - u_1}{f_0}\int_0^y F(\tau_e)\,dy$$

$$+ u_1 + \int_0^y A\frac{d\tau_{xy}}{dt}dy - \frac{k_{x0}}{f_0}\int_0^y F(\tau_e)\,dy, \tag{10.50a}$$

$$v = \frac{\partial p}{\partial z}\left[\int_0^y F(\tau_e)y\,dy - \frac{f_1}{f_0}\int_0^y F(\tau_e)\,dy\right] + \frac{v_2 - v_1}{f_0}\int_0^y F(\tau_e)\,dy$$

$$+ v_1 + \int_0^y A\frac{d\tau_{zy}}{dt}dy - \frac{k_{z0}}{f_0}\int_0^y F(\tau_e)\,dy. \tag{10.50b}$$

The velocity gradients are obtained from Eqs. (10.50) by differentiation

$$\frac{\partial u}{\partial y} = \frac{\partial p}{\partial x}F(\tau_e)\left(y - \frac{f_1}{f_0}\right) + \frac{u_2 - u_1}{f_0}F(\tau_e) + A\frac{d\tau_{xy}}{dt} - \frac{k_{x0}}{f_0}F(\tau_e), \tag{10.51a}$$

$$\frac{\partial v}{\partial y} = \frac{\partial p}{\partial z}F(\tau_e)\left(y - \frac{f_1}{f_0}\right) + \frac{v_2 - v_1}{f_0}F(\tau_e) + A\frac{d\tau_{zy}}{dt} - \frac{k_{z0}}{f_0}F(\tau_e), \tag{10.51b}$$

where

$$k_{x0} = \int_0^h A\frac{d\tau_{xy}}{dt}\,dy,$$

$$k_{z0} = \int_0^h A\frac{d\tau_{zy}}{dt}\,dy,$$

$$f_0 = \int_0^h F(\tau_e)\,dy,$$

$$f_1 = \int_0^h F(\tau_e)y\,dy,$$

From here on, the analysis follows that of Reynolds. We substitute Eqs. (10.50) into the equation of continuity and integrate across the film, as in Eq. (2.63a), to obtain

$$\frac{\partial}{\partial x}\left[m_2\frac{\partial p}{\partial x}\right] + \frac{\partial}{\partial z}\left[m_2\frac{\partial p}{\partial z}\right]$$

$$= h\left[\frac{\partial(\rho_2 u_2)}{\partial x} + \frac{\partial(\rho_2 v_2)}{\partial z}\right] - \frac{\partial}{\partial x}\left[(u_1-u_2)\frac{m_1}{f_0} + u_2\rho_2 h - u_1 m_3\right]$$

$$- \frac{\partial}{\partial z}\left[(v_1-v_2)\frac{m_1}{f_0} + v_2\rho_2 h - v_1 m_3\right] - \frac{\partial}{\partial x}[m_{x4}]$$

$$- \frac{\partial}{\partial z}[m_{z4}] + \int_0^h \frac{\partial \rho}{\partial t}\,dy + \rho_2 w_2 - \rho_1 w_1, \tag{10.52}$$

where

$$m_1 = \int_0^h \left(\rho \int_0^y F(\tau_e)\,d\hat{y}\right)dy,$$

$$m_2 = \frac{f_1 m_1}{f_0} - \int_0^h \left(\rho \int_0^y F(\tau_e)\hat{y}\,d\hat{y}\right)dy,$$

$$m_3 = \int_0^h \rho\,dy,$$

$$m_{x4} = \frac{k_{x0}m_1}{f_0} - \int_0^h \left(\rho \int_0^y A\frac{d\tau_{xy}}{dt}\,d\hat{y}\right)dy,$$

$$m_{z4} = \frac{k_{z0}m_1}{f_0} - \int_0^h \left(\rho \int_0^z A\frac{d\tau_{zy}}{dt}\,d\hat{y}\right)dy.$$

Equation (10.52) is the generalized Reynolds equations of Wolff and Kubo (1996) for elastoviscoplastic lubrication. For viscoplastic fluids A, therefore m_{x4} and m_{z4}, vanish. If the viscosity function $F(\tau_e) = 1/\mu$ and $A = 0$, Eq. (10.52) reduces to the form proposed by Fowles (1970). If in addition to $A = 0$ and $F(\tau_e) = 1/\mu$ we also specify μ, $\rho = $ const, Eq. (10.52) reduce to Eq. (2.64). Equation (10.52) was first derived for $\rho = $ const. by Najji, Bou-Said, and Berthe (1989).

For the two-dimensional flow in a line contact, we set

$$\frac{\partial(\cdot)}{\partial z} = 0, \qquad v_1 = v_2 = 0, \qquad \tau_{xy} = \tau_e = \tau,$$

$$w_1 = 0, \qquad w_2 = u_2\frac{\partial h}{\partial x}, \qquad \frac{d\tau}{dt} = u\frac{\partial \tau}{\partial x},$$

and Eq. (10.51) reduces to

$$\frac{\partial}{\partial x}\left[m_2\frac{dp}{\partial x}\right] = u_1\frac{\partial}{\partial x}[m_3] + (u_2-u_1)\frac{\partial}{\partial x}\left[\frac{m_1}{f_0}\right] - \frac{\partial}{\partial x}[m_4]. \tag{10.53}$$

Equation (10.53) is to be solved subject to the Swift-Stieber boundary conditions

$$p = 0 \qquad \text{at} \quad x = x_{\min},$$
$$p = \frac{dp}{dx} = 0 \quad \text{at} \quad x = x_{\max}. \tag{8.57}$$

We may now use Eq. (10.53) in place of Eq. (8.50) in a scheme to solve the EHD lubrication problem, but for one circumstance, Eq. (10.53) contains a yet undetermined shear stress field. We thus need one additional equation.

To find the required additional equation, we reconsider the first of Eqs. (10.46). As $p = p(x)$ to the approximation employed, dp/dx is independent of y and so is $\partial \tau_{xy}/\partial y$. But then τ_{xy} is linear in y at any x. It is, therefore, enough to evaluate τ_{xy} at $y = 0$ and $y = h$ and the shear-stress distribution is known.

Evaluating Eq. (10.50a) at $y = 0$, $u = 0$ yields an equation in $\tau_{xy}(x, 0)$. An equation in $\tau_{xy}(x, h)$ is obtained when Eq. (10.50a) is evaluated at $y = h$, $u = U$. It is a simple matter then to find the shear-stress distribution from

$$\tau_{xy} = \tau(x, 0) + [\tau(x, h) - \tau(x, 0)] \frac{y}{h}. \tag{10.54}$$

In their calculations for the line contact, Wolff and Kubo (1996) considered the following nonlinear rheological models:

(1) The nonlinear viscous Eyring model (Eyring, 1936),

$$F(\tau) = \frac{\tau_E}{\tau \mu} \sinh(\tau/\tau_E); \tag{10.55a}$$

(2) The nonlinear viscoplastic-like model (Bair and Winer, 1979),

$$F(\tau) = -\frac{\tau_L}{\tau \mu} \ln(1 - \tau/\tau_L); \tag{10.55b}$$

(3) The nonlinear viscoplastic-like model, also called the simplified Bair and Winer model,

$$F(\tau) = -\frac{1}{\mu}(1 - |\tau/\tau_L|)^{-1}; \quad \text{and} \tag{10.55c}$$

(4) The nonlinear viscoplastic-like (circular) model, Lee and Hamrock (1990),

$$F(\tau) = \frac{1}{\mu}[1 - (\tau/\tau_L)^2]^{-1/2}. \tag{10.55d}$$

If the coefficient $\bar{A} \neq 0$, the elastic term can be added to any of these models.

The limiting shear stress was assumed to be linearly dependent on the pressure,

$$\tau_L = \tau_{l0} + \gamma p. \tag{10.56}$$

Wolff and Kubo performed two sets of calculations, one with Santotrac 50 and the other with a P-150 oil.

Figure 10.11 shows the variation of shear stress for several rheological models in Santotrac. The traction coefficient for Santotrac is shown in Figure 10.12. Under pure rolling conditions there is a very small difference in the pressure distribution and the film shape predicted by the Newtonian and the Eyring models, as shown in Figure 10.13.

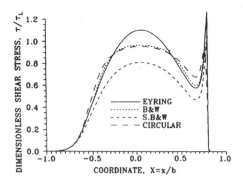

Figure 10.11. Dimensionless shear stress for Santotrac 50 oil; $W = 2.1 \times 10^{-5}$, $U = 10^{-11}$, $S = 0.5$. (Reprinted with permission from Wolff, R. and Kubo, A. A generalized non-Newtonian fluid model incorporated into elastohydrodynamic lubrication. *ASME Journal of Tribology*, **118**, 74–82, 1996.)

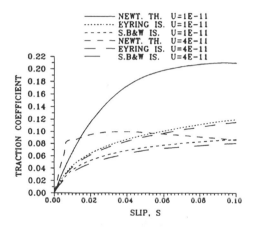

Figure 10.12. Traction coefficient for Santotrac 50 oil, $W = 2.1 \times 10^{-5}$. (Reprinted with permission from Wolff, R. and Kubo, A. A generalized non-Newtonian fluid model incorporated into elastohydrodynamic lubrication. *ASME Journal of Tribology*, **118**, 74–82, 1996.)

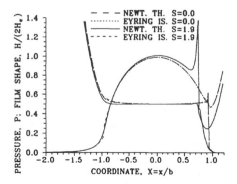

Figure 10.13. Film shape and pressure distribution for Santotrac 50 oil, $W = 8.4 \times 10^{-5}$, $U = 10^{-11}$. (Reprinted with permission from Wolff, R. and Kubo, A. A generalized non-Newtonian fluid model incorporated into elastohydrodynamic lubrication. *ASME Journal of Tribology*, **118**, 74–82, 1996.)

We close this section by quoting the conclusions from Wolff and Kubo's (1966) paper:

(1) Under low load and high rolling velocity conditions the thermal effects caused by high slip have a greater influence on the film shape and pressure distribution than the non-Newtonian effects.

(2) In the low slip range and under moderate and heavy loads the non-Newtonian behavior of an oil has a decisive influence on the traction coefficient, while in the high slip range, the thermal effects may become more important. The influence of the non-Newtonian effects on the traction (in the low slip range) is stronger under heavy load conditions and for high viscosity oils.

(3) For high viscosity oils subjected to heavy load conditions, the nonlinear viscous Eyring model tends to overestimate the traction coefficient.

(4) The viscoplastic models give reasonable traction values over a wide range of loads, slips, and oil viscosity values. The elastic behavior of a lubricant is important only under very low slip conditions. It reduces the traction value, particularly under heavy loads.

(5) In order to predict the traction coefficient over a wide range of parameters, the EHL model has to include not only the non-Newtonian behavior of a lubricant but also the thermal effects. It should also use a proper viscosity formula that fits the measured viscosity well over a wide range of pressure and temperature.

10.3   Lubrication with Emulsions

There are numerous instances of technical importance in which multicomponent lubricants are utilized either by design or by necessity. In many of these cases one of the component is a liquid while the other component is a gas, or both components are liquids; these mixtures do not exhibit Newtonian behavior even when their components themselves are Newtonian, and thus classical lubrication theory is inapplicable to tribological contacts lubricated with mixtures. Our objective in this section is to extend hydrodynamic lubrication theory to lubrication with liquid-liquid and liquid-gas mixtures. The extended theory is able to predict several experimentally observed phenomena, such as oil pooling ahead of an EHD conjunction.

Correct modeling of the flow of multicomponent mixtures is of increasing technical importance far beyond the confines of lubrication. Consequently much effort has been devoted to it; the principal approaches employed in these investigations on the flow of mixtures are (1) the descriptive, (2) the local volume averaging, and (3) the continuum mechanics approaches.

Because of the complexities of multiphase systems, initial attempts to describe the physical processes have relied on descriptive models of a more or less intuitive or empirical nature (Gouse, 1966; Soo, 1967; Butterworth and Hewitt, 1977). These models are generally limited in application to specific multiphase systems and, typically, have a narrow range of validity.

The second approach used to derive multiphase equations employs the technique of volume averaging (Drew, 1971; Gray, 1975; Ishii, 1975). This approach considers the system to be composed of interpenetrating continua, each constituent occupying only part of space and each separated from the others by highly irregular interfaces. Governing equations are obtained by averaging the classical single component equations over some local, representative element of volume.

The basic scientific method utilized in the works discussed in this section is mixture theory, a branch of continuum mechanics that details the behavior of multiphase continua. The theory traces its roots to the pioneering work of Fick and was put on firm mathematical footing by Truesdell (1969). The basic premise of the theory is the assumption that in a mixture of N constituents, each spatial position x is occupied simultaneously by N particles, one from each constituent. As the mass of each constituent of the mixture is continuously distributed in space, it is permissible to write one set of conservation equations for each constituent; these equations must, of course, be coupled, to express the interactions existing between constituent particles.

The notion of overlapping continua is a hypothesis about materials in much the same way as is the notion of a continuum in itself. Models developed using mixture theory are therefore useful, provided that their predictions are interpreted at scales consistent with the mixture hypothesis. Typical droplet size in oil-in-water or water-in-oil emulsions is $2r = 1.5 - 3\,\mu$m (Kimura and Okada, 1987; Schneider, Glossfield, and DeHart, 1986; Nakahara, Makino, and Kyogaka, 1988), representing $0.025 < r/h < 0.05$ in a conventional journal bearing of 0.5-m diameter, at an eccentricity ratio of 0.6. In EHL, we consider results of mixture theory to represent time averaged conditions in the film.

Fundamentals of Mixture Theory

For the sake of completeness, we begin with the following short exposition of the basics of mixture theory (Bowen, 1976; Atkin and Craine, 1976a,b; Truesdell, 1969; Bedford and Drumheller, 1983; Rajagopal and Tao, 1995).

Let X_a, $\alpha = 1$, 2, represent the position of a material point of the αth constituent C_α in its reference configuration, and let x be the spatial point that is occupied at time t by the material point. The motion of a binary mixture C of components C_α, $\alpha = 1$, 2 is defined by

$$x = \chi_\alpha(X_\alpha, t), \qquad t \geq 0, \quad \alpha = 1, 2. \tag{10.57}$$

The range of α for the remainder of this discussion is given in Eq. (10.57) and will not be repeated.

The velocity and the acceleration of the particle X_a are calculated, respectively, from

$$\dot{x}_\alpha = \frac{\partial}{\partial t}\chi_\alpha(X_\alpha, t), \qquad \ddot{x}_\alpha = \frac{\partial^2}{\partial t^2}\chi_\alpha(X_\alpha, t). \tag{10.58}$$

The spatial description of motion follows from Eqs. (10.57) and (10.58), e.g.,

$$v^{(\alpha)}(x, t) = \dot{x}_\alpha\left[\chi_\alpha^{-1}(x, t)\right].$$

Let grad denote differentiation with respect to x, at the configuration of the mixture at time t, then the velocity gradient for C_α at (x, t) is defined by

$$L_{(\alpha)} = \text{grad } v^\alpha(x, t) \tag{10.59}$$

and the stretching tensor by

$$D_{(\alpha)} = \frac{1}{2}\left(L_{(\alpha)} + L_{(\alpha)}^T\right).$$

We denote the *true density* of C_α by γ_α; this is the mass of the αth constituent per unit volume of the αth constituent itself. Distinct from the true density is the *density* (or bulk density) ρ_α, representing the mass of the αth constituent per unit volume of the mixture. The (total) density, ρ, of the mixture is then given by

$$\rho(\mathbf{x}, t) = \sum_\alpha \rho_\alpha(\mathbf{x}, t).$$
(10.60)

The quantity ϕ_α, defined by

$$\phi_\alpha(\mathbf{x}, t) = \frac{\rho_\alpha(\mathbf{x}, t)}{\gamma_\alpha(\mathbf{x}, t)},$$
(10.61)

is the *volume fraction* of the αth constituent. Physically ϕ_α represents the volume of C_α per unit volume of the mixture, therefore

$$\sum_\alpha \phi_\alpha = 1$$
(10.62)

for saturated mixtures (Mills, 1966).

It is convenient to introduce the (mean) velocity of the mixture, $\mathbf{v}$, via the requirement that the total mass flow is the sum of the individual mass flows, so that

$$\mathbf{v} = \frac{1}{\rho} \sum_\alpha \rho_\alpha \mathbf{v}^{(\alpha)}.$$
(10.63)

Next, we introduce material derivatives $d^{(\alpha)}/dt$ and d/dt, following the α constituent and the mixture, respectively, by

$$\frac{d^{(\alpha)}}{dt} = \frac{\partial}{\partial t} + \mathbf{v}^{(\alpha)} \cdot \text{grad}, \qquad \frac{d}{dt} = \frac{\partial}{\partial t} + \mathbf{v} \cdot \text{grad}.$$
(10.64)

Here we do not allow for interconversion of mass. Thus, the local form of the mass conservation equation is

$$\frac{d^{(\alpha)} \rho_\alpha}{dt} + \rho_\alpha \, \text{div} \, \mathbf{v}^{(\alpha)} = 0.$$
(10.65)

The local version of the balance of linear momentum for the αth constituent is

$$\rho_\alpha \frac{d^{(\alpha)} \mathbf{v}^{(\alpha)}}{dt} = \text{div} \, \mathbf{T}_{(\alpha)}^T + \rho_\alpha \mathbf{b}^{(\alpha)} + \boldsymbol{\pi}^{(\alpha)}.$$
(10.66)

The term $\boldsymbol{\pi}^{(\alpha)}$ symbolizes the transfer of momentum per unit volume due to interaction effects, due to relative motion between the constituents. $\boldsymbol{\pi}^{(\alpha)}$ is often referred to as *diffusive body force*. Its exact form is determined by the other components of the mixture and will be specified by a constitutive equation.

If $\mathbf{q}^{(\alpha)}$ is the heat flux, e_α is the internal energy density, and r_α is the external heat supply associated with constituent C_α, the local form of the energy conservation equation for a binary mixture is

$$\rho \frac{de}{dt} = \sum_\alpha \mathbf{T}_{(\alpha)}^T : \mathbf{L}_{(\alpha)} + \boldsymbol{\pi} \cdot \mathbf{V}^{(12)} - \text{div} \left[\frac{\rho_1 \rho_2}{\rho} (e_1 - e_2) \mathbf{V}^{(12)} \right]$$

$$+ \rho r - \text{div} \, \mathbf{q} + \sum_\alpha \rho_\alpha \mathbf{b}^{(\alpha)} \cdot \mathbf{v}^{(\alpha)},$$
(10.67)

where

$$r = \sum_\alpha r_\alpha \rho_\alpha / \rho, \qquad q = \sum_\alpha q^{(\alpha)}, \quad \text{and} \quad \pi = -\pi^{(1)} = \pi^{(2)}.$$

Finally, we postulate the second law of thermodynamics, in the form of the Clausius-Duhem entropy inequality. If η_α represents the *entropy density* for C_α and η is the entropy density for the mixture, then

$$\eta(x, t) = \frac{1}{\rho} \sum_\alpha \rho_\alpha \eta_\alpha(x, t), \tag{10.68}$$

and a local form of the entropy production inequality for the mixture is

$$\rho \frac{d\eta}{dt} + \mathrm{div}\left(\frac{1}{\Theta} q\right) - \frac{\rho r}{\Theta} \geq 0. \tag{10.69}$$

Here we assumed that the flux of entropy due to the heat flux $q^{(\alpha)}$ is $q^{(\alpha)}/\Theta$, where $\Theta(x, t)$ is the common temperature of the constituents, and that the input of entropy due to the heat supply function r_α is r_α/Θ.

Defining the partial Helmholtz free energy density for C_α by $A_\alpha = e_\alpha - \Theta \eta_\alpha$, we can write Eq. (10.69) in the form

$$-\sum_\alpha \left[\frac{d^{(\alpha)} \rho_\alpha A_\alpha}{dt} + \rho_\alpha A_\alpha \, \mathrm{tr}[D_{(\alpha)}] + \rho_\alpha \eta_\alpha \frac{d^{(\alpha)}\Theta}{dt} + \pi \cdot V^{(12)} - T_{(\alpha)} : L_{(\alpha)} \right] - \frac{1}{\Theta} g \cdot q \geq 0.$$

$$\tag{10.70}$$

Constitutive Model

We identify $(12\alpha + 7)$ scalar unknowns $\{\rho_{(\alpha)}, v_{(\alpha)i}, T_{(\alpha)i,j}, \eta_{(\alpha)}, A_{(\alpha)}, \pi_i, q_i, \Theta\}$, $i, j = 1, 2, 3, \alpha = 1, 2$. Of these $(4\alpha + 1)$, viz., $\rho_\alpha, v_{(\alpha)i}$, and Θ are field variables to be calculated from the $(4\alpha + 1)$ scalar conservation equations, Eqs. (10.65)–(10.67). We thus need to specify $(8\alpha + 6)$ constitutive equations for the remaining $(8\alpha + 6)$ scalar unknowns:

$$\{A_\alpha, \eta_\alpha, T_{(\alpha)i,j}, \pi_i, q_i\} \qquad i, j = 1, 2, 3; \quad \alpha = 1, 2. \tag{10.71}$$

For a heat-conducting mixture of incompressible, viscous fluids, it is natural to have the densities, ρ_α or ϕ_α (Bowen, 1980), the temperature, Θ, and the temperature gradient g, as independent variables (Atkin and Craine, 1976a). Density gradients, $h^{(\alpha)}$, are also necessary, since their omission leads to a theory that is too simple (Muller, 1968). In order to include effects connected with the motion of the constituents, the velocities $v^{(\alpha)}$ and velocity gradients $L_{(\alpha)}$ are added to the list of variables; frame indifference then requires that the last two quantities be represented by the relative velocity $V^{(12)}$, the stretching tensors $D_{(\alpha)}$, and the relative spin $\Omega_{(12)} = \Omega_{(1)} - \Omega_{(2)}$. However, it is known from the theory of single materials that velocity gradients characterize viscous effects; viscous effects are nonexistent in the ideal gas component, thus velocity gradient for the liquid component only is considered in gas-liquid emulsions.

Liquid-Liquid Emulsion

Since the mixture is saturated, all processes must be subjected to the volume additivity constraint in Eq. (10.62), which, when differentiated, yields

$$\sum_{\alpha=1}^{2} \left[\frac{d^{(\alpha)} \phi_{\alpha}}{dt} - (v^{(\alpha)} - v) \cdot \operatorname{grad} \phi_{\alpha} \right] \cdot V^{(12)} = 0. \tag{10.72}$$

Using Eq. (10.65), Eq. (10.72) can be rewritten as

$$\sum_{\alpha=1}^{2} \phi_{\alpha} \operatorname{div} v^{(\alpha)} + \left[\frac{\rho_2}{\rho} \operatorname{grad} \phi_1 - \frac{\rho_1}{\rho} \operatorname{grad} \phi_2 \right] \cdot V^{(12)} = 0. \tag{10.73}$$

The thermodynamic theory of constraints (Truesdell and Noll, 1965), which requires introduction of a Lagrange multiplier Π, will be used to account for this constraint in the entropy inequality.

Our choice, and this choice is by no means unique, of independent constitutive variables for a heat-conducting mixture of incompressible fluids is the set (Al-Sharif et al., 1993)

$$\phi_{\alpha}, \Theta, V^{(12)}, g, h^{(\alpha)}, D_{(\alpha)}, \Omega_{(12)}, \tag{10.74}$$

where

$$g = \operatorname{grad} \Theta, \qquad h^{(\alpha)} = \operatorname{grad} \phi_{\alpha}, \qquad \Omega_{(12)} = \Omega_{(1)} - \Omega_{(2)}.$$

Consistent with the axiom of equipresence (Truesdell and Toupin, 1960), each of the quantities in Eq. (10.71) is now assumed to be a function of the variables of Eq. (10.74), but dependence on

$$V^{(12)}, g, h^{(\alpha)}, D_{(\alpha)}, \Omega_{(12)} \tag{10.75}$$

is restricted to be linear, so as to simplify the analysis. This results in a set of constitutive equations for the variables in Eq. (10.71).

Substituting the constitutive equations into the entropy inequality, we now assume that inequality (10.70) must hold for all admissible thermomechanical processes in the mixture (Muller, 1968; Atkin and Craine, 1976a). Standard methods of continuum mechanics yield the reduced constitutive equations (Spencer, 1970; Bowen, 1976). For the diffusive body force and the component stresses, in the case under consideration, we have (Al-Sharif, 1992; Al-Sharif et.al., 1993)

$$\pi = \varpi_1 V^{(12)} + \varpi_4 g + \left[-\rho_2 \frac{\partial A_2}{\partial \phi_1} + \frac{\rho_2}{\rho} \Pi \right] h^{(1)} + \left[\rho_1 \frac{\partial A_1}{\partial \phi_2} - \frac{\rho_1}{\rho} \Pi \right] h^{(2)}, \tag{10.76a}$$

$$T_1 = \left(-\rho_1 + \lambda_1 \operatorname{tr}[D_{(1)}] + \lambda_3 \operatorname{tr}[D_{(2)}] \right) \mathbf{1} + 2\mu_1 D_{(1)} + 2\mu_3 D_{(2)} + \lambda_5 \Omega_{(12)}, \tag{10.76b}$$

$$T_2 = \left(-\rho_2 + \lambda_4 \operatorname{tr}[D_{(1)}] + \lambda_2 \operatorname{tr}[D_{(2)}] \right) \mathbf{1} + 2\mu_4 D_{(1)} + 2\mu_2 D_{(2)} + \lambda_5 \Omega_{(12)}. \tag{10.76c}$$

Equation (10.76b) indicates that, under the constitutive assumption we made, the stress in component C_1 (water) depends on the stretching tensor D_1 of component C_1 but also on the stretching tensor of component C_2 (oil). The diffusive body force π in its simplest form is given by the Stokes resistance law [neglecting dependence on $h^{(\alpha)}$]. We also note that the constitutive equations (10.76) contain a number of material functions, such as the viscosities $\mu_1, \ldots, \mu_4$, that must be determined from experiments, obviously not a simple task.

Liquid-Gas Emulsion

Our choice of constitutive variables for a heat-conducting mixture of an incompressible fluid and an ideal gas is the set

$$\left\{ \rho_\alpha, \Theta, \boldsymbol{V}^{(12)}, \boldsymbol{g}, \boldsymbol{h}^{(\alpha)}, \boldsymbol{D}_{(1)} \right\}. \tag{10.77}$$

If we again assume the constitutive functions to be linear with respect to vector and tensor arguments and substitute into the entropy production inequality, requiring the latter to hold for all admissible thermomechanical processes in the mixture (Muller, 1968), we arrive at the restricted constitutive equations (Chamniprasart et al., 1993). For component stress, e.g., we have

$$\boldsymbol{T}_{(1)} = \left(-p_1 + \lambda_1 \operatorname{tr}\left[\boldsymbol{D}_{(1)}\right]\right)\boldsymbol{1} + 2\mu_1 \boldsymbol{D}_{(1)},$$

$$\boldsymbol{T}_{(2)} = -p_2 \boldsymbol{1}.$$

Lubrication Approximation

Let L_{xz} and L_y represent the length scales, in the "plane" of the film and perpendicular to it, respectively, of the generic lubrication problem. The corresponding velocity scales will be denoted by U_* and $V_* = (L_y/L_{xz})U_*$ and the time scale by $t_* = (L_{xz}/U_*)$. When the equations of motion, Eq. (10.66), are normalized with respect to these characteristic quantities, they will contain various powers of (L_y/L_{xz}). To simplify these equations, we let $(L_y/L_{xz}) \to 0$ while keeping the Reynolds number $L_y U_*/\nu \approx O(1)$ in accordance with the basic premise of lubrication theory.

Liquid-Liquid Emulsion

The result of such manipulations is two sets of equations, one set each for the two constituents of the mixture. When these equations are formally integrated, a generalized Reynolds equation for binary mixtures of Newtonian fluids is obtained (Al-Sharif, 1992). This derivation assumes that the volume fraction, like the pressure, remains constant across the film; the condition $\partial \phi / \partial y = 0$ is shown to hold approximately in the recent experiments of Couet, Brown, and Hunt (1991), who measured local volume fraction for oil dispersed in water.

The extended Reynolds equations for *binary mixtures* of two Newtonian fluids are

$$\frac{\partial}{\partial X}\left[F_1 \frac{\partial P}{\partial X} + F_2 \frac{\partial \phi}{\partial X}\right] + \left(\frac{D}{L}\right)^2 \frac{\partial}{\partial Z}\left[F_1 \frac{\partial P}{\partial Z} + F_2 \frac{\partial \phi}{\partial Z}\right] = 12\pi \frac{\partial (\phi H)}{\partial X}, \tag{10.79a}$$

$$\frac{\partial}{\partial X}\left[F_3 \frac{\partial P}{\partial X} - F_4 \frac{\partial \phi}{\partial X}\right] + \left(\frac{D}{L}\right)^2 \frac{\partial}{\partial Z}\left[F_3 \frac{\partial P}{\partial Z} - F_4 \frac{\partial \phi}{\partial Z}\right] = 12\pi \frac{\partial [(1-\phi)H]}{\partial X}. \tag{10.79b}$$

Equations (10.79) are in a nondimensional form suitable for a journal bearings of diameter D and length L, and we put $\phi = \phi_1 = 1 - \phi_2$. The (nonlinear) coefficients $F_1, \ldots, F_4$ are functions of the volume fraction ϕ and the material parameters such as μ_1, μ_2, μ_3 and μ_4, occurring in the constitutive equations.

Liquid-Gas Emulsion

As a result of lubrication approximation, details of which can be found in Chamniprasart et al. (1993), the component densities ρ_1 and ρ_2 are required to satisfy two nonlinear partial differential equations, the extended Reynolds equations

$$\frac{\partial}{\partial X}\left[\bar{\rho}_1 H^3 \frac{\partial \Xi}{\partial X}\right] + \left(\frac{D}{L}\right)^2 \frac{\partial}{\partial Z}\left[\bar{\rho}_1 H^3 \frac{\partial \Xi}{\partial Z}\right] = 12\Lambda \frac{\partial(\bar{\rho}_1 H)}{\partial X}, \tag{10.80a}$$

$$\frac{\partial}{\partial X}\left[\bar{\rho}_2 H^3 \frac{\partial \Xi}{\partial X}\right] + \left(\frac{D}{L}\right)^2 \frac{\partial}{\partial Z}\left[\bar{\rho}_2 H^3 \frac{\partial \Xi}{\partial Z}\right] = 12\Lambda \frac{\partial(\bar{\rho}_2 H)}{\partial X} + 12H\left[\frac{\partial \Gamma_x}{\partial X} + \left(\frac{D}{L}\right)^2 \frac{\partial \Gamma_z}{\partial Z}\right],$$

$$\tag{10.80b}$$

where

$$\Xi = \Xi(\bar{\rho}_1, \bar{\rho}_2), \qquad \Gamma_1 = \Gamma_1(\bar{\rho}_1, \bar{\rho}_2), \qquad \Gamma_2 = \Gamma_2(\bar{\rho}_1, \bar{\rho}_2).$$

Equations (10.80a) and (10.80b) represent the model of Chamniprasart et al. (1993) for the flow of bubbly oil in a bearing. To fully define the problem, these equations must be supplemented with conditions on ρ_1 and ρ_2 at the film boundaries.

Applications

Journal Bearing

The lubricating action of an emulsion can be understood by analyzing Figures 10.14 and 10.15. Similar to single phase lubricant, a pressure profile is generated such that the resulting flow field satisfies mass conservation of each of the components of the mixture. But in this case the composition of the mixture is changing in response to the stress field, the volume fraction of the higher viscosity constituent increases toward the minimum film thickness zone (Figure 10.14).

According to Hirn (1954), water requires high sliding speed to enter the clearance; at low sliding speeds the shaft will pull an oil-rich film into the clearance. The results in

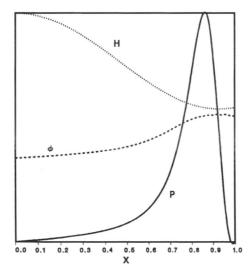

Figure 10.14. Pressure, film thickness and oil volume fraction distribution for oil-water emulsion in a journal bearing. (Reprinted with permission from Al-Sharif, A., Chamniprasart, T., Rajagopal, K. R. and Szeri, A. Z. Lubrication with binary mixtures: liquid-liquid emulsion. *ASME Journal of Tribology*, **115**, 46–55, 1993.)

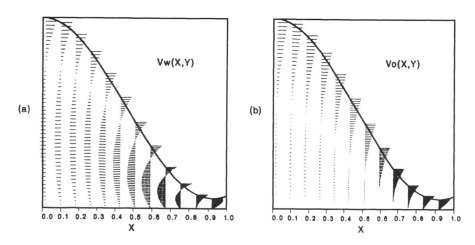

Figure 10.15. Velocity profiles for (a) water and (b) oil in oil-water mixture. (Reprinted with permission from Al-Sharif, A., Chamniprasart, T., Rajagopal, K. R. and Szeri, A. Z. Lubrication with binary mixtures: liquid-liquid emulsion. *ASME Journal of Tribology*, **115**, 46–55, 1993.)

Figure 10.15 reflect this behavior: Poiseuille flow is dominant in the water phase (a), but Couette flow is dominant in the oil phase (b).

Figure 10.16 shows the pressure distribution in water-in-oil emulsion, corresponding to various inlet water volume fraction. On increasing water concentration from zero, the peak lubricant pressure first increases above its value for pure oil, only to exhibit rapid decline on further increasing the volume fraction of the water droplets. On reaching volume fraction of unity, we obtain the pressure distribution of pure water lubricant.

EHL Conjunction

In EHL contacts lubricated with water-in-oil emulsion, agreement with classical EHD film thickness is obtained only if the viscosity of the continuous phase (oil) is used when the diameter of the water droplets is in excess of the film thickness, for then, presumably, only oil passes through the conjunction (Dalmaz and Godet, 1977; Hamaguchi et al., 1977). This is also indicated by the fact that increased bulk viscosity, which results from adding water droplets to the oil lubricant, has no effect on film thickness (Hamaguchi et al., 1977). But if the water particles are small, the EHD film thickness is larger than could be obtained for the base oil; it is stipulated then that the emulsion proper and not just the continuous oil phase enters the conjunction. This yields a thicker film as the emulsion has higher viscosity than the base oil (Dalmaz, 1980; Wan et al., 1984).

The prime concern of the designer of EHL contacts is the film thickness. In EHL with conventional mineral oil lubricant, the film thickness parameter, H, is expressed as a certain function of the velocity parameter, U, the load parameter, W, and the materials parameter, G (Table 8.12). The principal additional parameters which characterize the material behavior of the binary mixture lubricant are (1) the inlet volume fraction of the oil ϕ_i; (2) the surface tension group, $\hat{C} = R_\delta^2/W_e$, where R_δ and W_e are the droplet Reynolds number and the Weber number, respectively; and (3) the relative droplet radius, $\bar{r} = r/C$, of the discretized phase (Wang, Al-Sharif, Rajagopal, and Szeri, 1993).

It was reported by several investigators (Hamaguchi et al., 1977; Wan et al., 1984) that there is no detectable film in the conjunction for an oil-in-water emulsion, unless the emulsion breaks down and an oil pool is formed at the entrance; in this latter case the

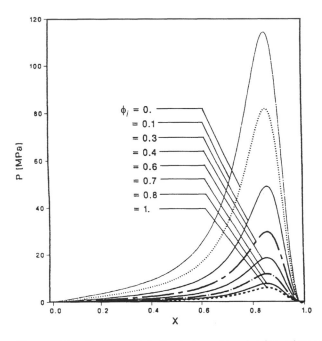

Figure 10.16. Pressure distribution in journal bearing for various values of inlet water volume fraction of water-oil mixture. (Reprinted with permission from Al-Sharif, A., Chamniprasart, T., Rajagopal, K. R. and Szeri, A. Z. Lubrication with binary mixtures: liquid-liquid emulsion. *ASME Journal of Tribology*, **115**, 46–55, 1993.)

film thickness in the EHL conjunction is defined by the viscosity of the oil. Wang et al. were able to demonstrate a phenomena that is not unlike oil pooling. Figure 10.17 displays the film thickness and the pressure distributions in an oil lubricated conjunction giving $H_{min} = 0.3923$ under the conditions noted. Figure 10.17 also shows the film thickness for an oil-in-water emulsion with inlet oil volume fraction $\phi_i = 0.1$. The volume fraction can be seen to increase rapidly with x from its entrance value of 0.1 to 0.834 and as a result of pooling the minimum film thickness achieves $H_{min} = 0.234$, or 62% of its value for pure oil, while the pressures in the two cases are close.

Cold Rolling
Grudev and Razmakhnin (1985a,b) report on extensive experimental investigations using water-based lubricants. Their principal findings are (1) the oil volume fraction, ϕ, increases with increasing reduction ratio, γ, of the workpiece to a limiting value that is largely independent of other conditions; (2) increasing the yield stress of the workpiece increases the tendency for oil-pooling; and (3) for every condition there is a value of γ above which the oil-in-water emulsion acts as pure oil. Wang et al. examined these propositions. Figure 10.18 shows the oil volume fraction increase with the reduction ratio, for the conditions of this paper the limit appears to be $\phi_{max} \sim 0.7$.

At this limiting value of ϕ, the emulsion behaves as the pure oil. Figure 10.19 represents our solution for oil volume fraction distribution at various values of strip yield stress. The inlet value of f is fixed at $\phi_i = 0.3$ in these calculations. At a yield stress of $\sigma_{yp} = 500$ MPa

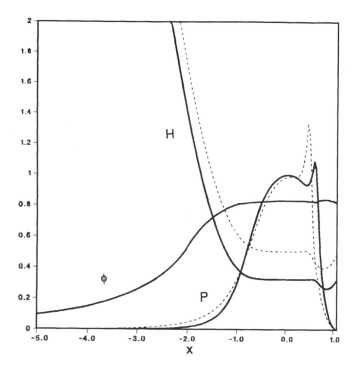

Figure 10.17. Film thickness, pressure, and void fraction profiles. $W = 1.35 \times 10^{-5}$, $U = 6.43 \times 10^{-12}$, $G = 3500$; - - - -, pure oil; — —, oil-water mixture. (Reprinted with permission from Wang, S. H., Al-Sharif, A., Rajagopal, K. R. and Szeri, A. Z. Lubrication with binary mixtures: liquid-liquid emulsion in an EHD conjunction. *ASME Journal of Tribology*, **115**, 515–524, 1993.)

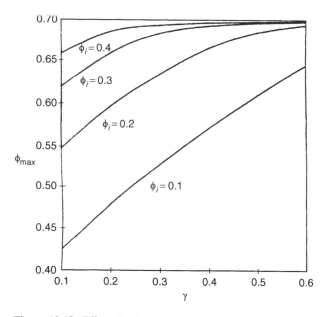

Figure 10.18. Effect of reduction ratio on ϕ_{max}. (Reprinted with permission from Wang, S. H., Szeri, A. Z. and Rajagopal, K. R. Lubrication of emulsion in cold rolling. *ASME Journal of Tribology*, **115**, 523–532, 1993.)

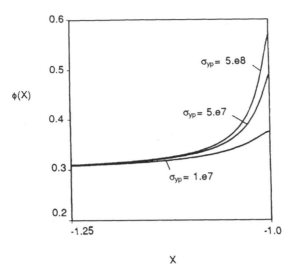

Figure 10.19. Effect of workpiece yield stress on oil volume fraction distribution. (Reprinted with permission from Wang, S. H., Szeri, A. Z. and Rajagopal, K. R. Lubrication of emulsion in cold rolling. *ASME Journal of Tribology*, 115, 523–532, 1993.)

the working zone has $\phi_{max} \approx 0.56$ and considerably thinner film than obtained with $\sigma_{yp} = 10$ MPa, which gives only $\phi_{max} \approx 0.37$. Figures 10.18 and 10.19 thus confirm experimental findings of Grudev and Razmakhnin.

Lubrication with Bubbly Oil

The bearing whose performance we study was operated while fully submerged (Braun and Hendricks, 1981). To simulate fully submerged operation Chamniprasart et al. make two assumptions: (1) the oil bath surrounding the bearing has constant, uniform composition, and (2) the conditions at the edge of the lubricant film are identical to and are given by the conditions in the bath. The first of these assumptions permits specification of a reference air volume fraction, ϕ_{20}, and a mixture reference bulk viscosity, β_0, at bath conditions. The second assumption assigns boundary conditions $\rho_1 = \rho_{10}$ and $\rho_2 = \rho_{20}$ on component densities.

The reference bulk modulus of the mixture, β_0, will vary on varying the gas reference volume fraction, ϕ_{20}. When $\phi_{20} \to 0$, $\beta_0 \to \beta_1$, and we are effectively dealing with an oil, lubricant: the corresponding pressure distribution in the bearing approaches the classical, incompressible lubricant pressure distribution (Figure 10.20).

On increasing ϕ_{20}, on the other hand, we find that the negative pressure loop continually diminishes until, at $\phi_{20} = 1$, we obtain the pressure distribution for a gas lubricant (Raimondi, 1961). The maximum pressure first increases as ϕ_{20} is increased from zero; to get this increase in pressure, we would have to use a viscosity larger than the oil viscosity, μ_1, had we applied the classical Reynolds equation. This puts us in qualitative agreement with Taylor (1932) and Hayward (1961) for small air volume fraction in the mixture.

Figure 10.21 plots experimental data by Braun and Hendricks (1981) along with predictions from mixture theory (Chamniprasart et al., 1993) for centerline pressure, showing pressure variation in the cavitation zone.

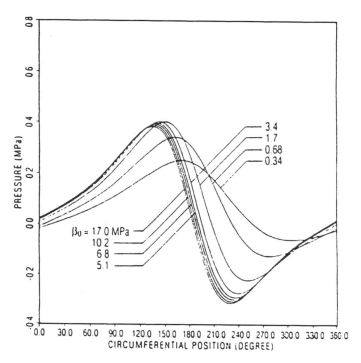

Figure 10.20. Pressure profile for various values of the reference bulk modulus β_0. (Reprinted with permission from Chamniprasart, T., Al-Sharif, A., Rajagopal, K. R. and Szeri, A. Z. Lubrication with bubbly oil. *ASME Journal of Tribology*, **115**, 253–260, 1993.)

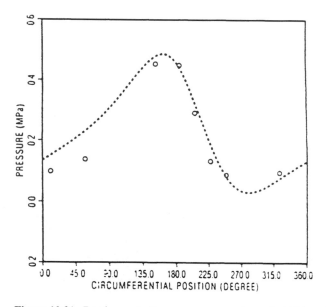

Figure 10.21. Bearing centerline pressure; — —, theoretical (Chamniprasart et al., 1993); o, experimental (Braun and Hendricks, 1981). (Reprinted with permission from Chamniprasart, T., Al-Sharif, A., Rajagopal, K. R. and Szeri, A. Z. Lubrication with bubbly oil. *ASME Journal of Tribology*, **115**, 253–260, 1993.)

10.4 Nomenclature

A	Helmholtz free energy
$\boldsymbol{A}_{(n)}$	Rivlin-Ericksen tensor
C	radial clearance
$\hat{C}$	surface tension group
C, C_α	mixture, αth constituent
D	shaft diameter
$\boldsymbol{D}$	stretching tensor
De	Deborah number
$\boldsymbol{F}$	deformation gradient
G	material parameter
H	film thickness parameter
H	nondimensional film thickness
$\boldsymbol{L}$	velocity gradient
L	journal length
L_{xz}, L_y	characteristic lengths
R	journal radius
Re	Reynolds number
Re*	reduced Reynolds number
R_δ	droplet Reynolds number
$\boldsymbol{T}$	stress tensor
U	velocity parameter
U_*, V_*	characteristic velocities
$\boldsymbol{V}^{(12)}$	relative velocity
W	load parameter
$\boldsymbol{\Omega}$	spin
$\boldsymbol{\Omega}_{(1,2)}$	relative spin
We	particle Weber number
X	reference (material) coordinate
r	external heat supply
$\boldsymbol{b}$	external body force
c	specific heat
$\boldsymbol{g}$	temperature gradient
$\boldsymbol{h}$	density gradient
h	film thickness
k_1, k_2	material functions
k	power law consistency index
n	power law exponent
p, P	lubricant pressure
$\boldsymbol{q}$	heat transfer by conduction
ret(.)	retarded motion
t	time
$\boldsymbol{x}$	Eulerian coordinate
$\boldsymbol{v} = (u, v, w)$	velocity
ϖ_1, ϖ_2	material functions
$\boldsymbol{\omega}$	vorticity

γ	reference density, reduction ratio
e	internal energy density
η	entropy density
Θ	mixture temperature
λ	relaxation time
$\lambda_1, \ldots \lambda_5$	material functions
$\mu_1, \ldots, \mu_4$	viscosities
ν	kinematic viscosity
π	diffusive body force
τ	shear stress
τ_e, τ_L	equivalent, limiting shear stress
τ	residence time
ρ	density
ϕ	volume fraction
χ	deformation function
$()_{(\alpha)}, ()^{(\alpha)}$	pertaining to C_a
$()_*$	reference quantity
$()_o$	reference state
$(\bar{})$	nondimensional
$\text{tr}[.]$	trace
Π	Lagrange multiplier

10.5   References

Al-Sharif, A. 1992. Hydrodynamic lubrication with emulsions. *Ph.D. thesis*, Department of Mechanical Engineering, University of Pittsburgh.

Al-Sharif, A., Chamniprasart, T., Rajagopal, K. R. and Szeri, A. Z. 1993. Lubrication with binary mixtures: liquid-liquid emulsion. *ASME Journal of Tribology*, **115**, 46–55.

Atkin, R. J. and Craine, R. E. 1976a. Continuum theories of mixtures: basic theory and historical development. *Q. J. Mech. Appl. Math.*, **29**, 209.

Atkin, R. J. and Craine, R. E. 1976b. Continuum theories of mixtures: applications. *J. Inst. Math. Appl.*, **17**, 153.

Bair, S. and Winer, W. O. 1979. A rheological model for elastohydrodynamical contacts. *ASME Journal of Lubrication Technology*, **101**, 248.

Bair, S. and Winer, W. O. 1992. The high pressure high shear stress rheology of liquid lubricants. *ASME Journal of Tribology*, **114**, 1–13.

Barlow, A. J., Erginsau, A. and Lamb, J. 1969. Viscoelastic relaxation in liquid mixtures. *Proc. Roy. Soc.*, **A 309**, 473–496.

Barlow, A. J., Harrison, G., Irving, J. B., Kim, M. G. and Lamb, J. 1972. The effect of pressure on the viscoelastic properties of Liquids. *Proc. Roy. Soc.*, **A 327**, 403–412.

Bedford, A. and Drumheller, D. S. 1983. Recent advances: theories of immiscible and structured mixtures. *Int. J. Eng. Sci.*, **21**, 863.

Beris, A. N., Armstrong, R. C. and Brown, R. A. 1983. Perturbation theory for viscoelastic fluids between eccentric rotating cylinders. *J. Non-Newtonian Fluid Mech.*, **13**, 109–148.

Bernstein, B., Kearsley, E. A. and Zapas, L. J. 1963. A study of stress relaxation with finite strain. *Trans. Soc. Rheol.*, **7**, 391–410.

Bourgin, D. 1979. Fluid-film flows of differential fluids of complexity *n* dimensional approach – applications to lubrication theory. *ASME Journal of Lubrication Technology*, **101**, 140–144.

Bourgin, D. and Gay, B. 1983. Determination of the load capacity of a finite width journal bearing by a finite element method in the case of a non-Newtonian Lubricant. *ASME Journal of Lubrication Technology.*

Bowen, R. M. 1976. *Theory of Mixtures, Continuum Physics*, A. C. Eringen (ed.), **3**, Part 1. Academic Press, New York.

Braun, M. J. and Hendricks, R. 1981. An experimental investigation of the vaporous/gaseous cavity characteristics of an eccentric journal bearing. *ASLE Trans.*, **27**, **1**, 1–14.

Buckholtz, R. H. 1985. Effects of power-law, non-Newtonian lubricants on load capacity and friction for plane slider bearing. *ASME/ASLE Joint Conf.* Paper No. 85-Trib-23.

Buckholtz, R. H. and Wang, B. 1985. The accuracy of short bearing theory for Newtonian lubricants. *ASME/ASLE Joint Conf.* Paper No. 85-Trib-49.

Butterworth, O. and Hewitt, G. F. 1977. *Two-Phase Flow Heat Transfer.* Oxford University Press, London.

Chamniprasart, T., Al-Sharif, A., Rajagopal, K. R. and Szeri, A. Z. 1993. Lubrication with bubbly oil. *ASME Journal of Tribology*, **115**, 253–260.

Chamniprasart, K. 1992. A theoretical model of hydrodynamic lubrication with bubbly oil. *Ph.D. dissertation*, Department of Mechanical Engineering, University of Pittsburgh.

Cheng, H. S. 1966. Plastohydrodynamic lubrication. *Friction in Metal Processing*, New Orleans, p. 69.

Christie, I., Rajagopal, K. R. and Szeri, A. Z. 1987. Flow of a non-Newtonian fluid between eccentric rotating cylinders. *Int. J. Eng. Sci.*, **25**, 1029–1047.

Couet, B., Brown, P. and Hunt, A. 1991. Two-phase bubbly droplet flow through a contraction: experiments and a unified theory. *Int. J. Multiphase Flow*, **17**, 291.

Dalmaz, G. 1981. Friction and film thickness measurements of water glycol and a water-in-oil emulsion in rolling-sliding point contacts. *Friction and Wear*, Proc. 7th Leeds-Lyon Symp., 231–242.

Dalmaz, G. and Godet, M. 1978. Film thickness and effective viscosity of some fire resistant fluids in sliding point contact. *ASME Journal of Lubrication Technology*, **100**, 304.

Davis, M. J. and Walters, K. 1973. In *Rheology of Lubricants.* (ed). T. Davenport, 65–80.

Drew, D. A. 1971. Averaged field equations for two-phase flow studies. *Appl. Math.*, **50**, 133.

Dyson, A. 1970. Frictional traction and lubricant rheology in elastohydrodynamic lubrication. *Phil. Trans. Roy. Soc.*, **A 266**, 1170.

Eyring, H. 1936. Viscosity, plasticity and diffusion as examples of absolute reaction rates. *J. Chem. Phys.*, **4**, 283.

Fein, R. S. 1967. *ASME. Journal of Lubrication Technology*, **94**, 306.

Fix, G. J. and Paslay, P. R. 1967. Incompressible elastic viscous lubricants in continuous-sleeve journal bearings. *ASME J. Appl. Mech.*, **34**, 579–582.

Fosdick, R. and Rajagopal, K. R. 1978. Anomalous features in the model of second order fluids. *Arch. Rational Mech. Anal.*, **70**, 145.

Fosdick, R. and Rajagopal, K. R. 1980. Thermodynamics and stability of fluids of third grade. *Proc. Roy. Soc.*, **A 339**, 351.

Giesekus, H. 1963. Die simultane translation und rotations bewegung einer Kugel in einer elastoviskosen Flussigkeit. *Rheol. Acta*, **4**, 59–71.

Gouse, S. W. 1966. *An index to two-phase gas-liquid flow literature.* MIT Press, Cambridge, MA.

Gray, W. G. 1975. A derivation of the equations for multi-purpose transport. *Chem. Eng. Sci.*, **30**, 29.

Grudev, A. P. and Razmakhnin, A. D. 1984a. Principles of formation of lubricating film during cold rolling using emulsions. *Steel in the USSR*, **15**, p. 15.

Grudev, A. P. and Razmakhnin, A. D. 1985b. Cold rolling with emulsion lubricant. *Steel in the USSR*, **15**, 19.

Hamaguchi, H., Spikes, H. A. and Cameron, A. 1977. Elastohydrodynamic properties of water-in-oil emulsions. *Wear*, **43**, 17.

Harnoy, A. and Hanin, M. 1974. Second order, elastico-viscous lubricants in dynamically loaded bearings. *ASLE Trans.*, **17**, **3**, 166–171.

Harrison, G. and Trachman, E. G. 1972. The role of compressional viscoelasticity in the lubrication of rolling contacts. *ASME Journal of Lubrication Technology*, **95**, 306–312.

Hayward, A. T. 1961. The viscosity of bubbly oil. *NEL. Fluids Report No. 99*.

Houpert, L. 1985. Fast numerical calculations of EHD sliding traction forces; application to rolling bearings. *ASME Journal of Lubrication Technology*, **107**, 234–248.

Hsu, Y. C. and Saibel, E. 1965. Slider bearing performance with a non-Newtonian lubricant. *ASLE Trans.*, **8**, 191–194.

Huilgol, R. R. 1975. On the concept of the Deborah number. *Trans. Soc. Rheology*, **19**, 297–306.

Ishii, M. 1975. *Thermo-fluid dynamic theory of two-phase flows*. Eyrolles, Paris.

Johnson, K. L. and Cameron, R. 1967. Shear behavior of EHD oil films. *Proc. Inst. Mech. Eng.*, **182**, 307.

Johnson, K. L. and Roberts, A. D. 1974. Observation of viscoelastic behavior of an elastohydrodynamic lubricant film. *Proc. Roy. Soc.*, **A 337**, 217–242.

Johnson, K. L. and Tevaarwerk, J. L. 1977. Shear behavior of elastohydrodynamic films. *Trans. Roy. Soc.*, **A 356**, 215–236.

Johnson, M. W. and Mangkoesoebroto, S. 1993. Analysis of lubrication theory for the power law fluid. *ASME Journal of Tribology*, **115**, 71–77.

Kacou, A., Rajagopal, K. R. and Szeri, A. Z. 1987. Flow of a fluid of the differential type in a journal bearing. *ASME Journal of Tribology*, **109**, 100–108.

Kaye, A. 1962. *Note No. 134*. College of Aeronautics, Cranfield.

Kimura, Y. and Okada, K. 1987. Film thickness at elastohydrodynamic conjunctions lubricated with oil-in-water emulsions. *Proc IMechE.*, **C176**, 85–90.

Lee, R. T. and Hamrock, B. J. 1990. A circular non-Newtonian fluid model used in transient elastohydrodynamic lubrication. *ASME Journal of Tribology*, **112**, 486–505.

Mills, N. 1966. *Int. J. Eng. Sci.*, **4**, 97.

Muller, I. A. 1968. A thermodynamic theory of mixtures of fluids. *Arch. Rational Mech. Anal.*, **28**, 1–39.

Najji, B., Bou-Said, B. and Berthe, D. 1989. New formulation for lubrication with non-Newtonian fluids. *ASME Journal of Tribology*, **111**, 29–34.

Nakahara, T., Makino, T. and Kyogaka, Y. 1988. Observation of liquid droplet behavior and oil film formation in O/W type emulsion lubrication. *ASME Journal of Tribology*, **110**, 348–353.

Ng, C. W. and Saibel, E. 1962. Non-linear viscosity effects in slider bearing lubrication. *ASME Journal of Lubrication Technology*, **7**, 192–196.

Nunziato, J. W. and Walsh, E. 1980. *Arch. Rat. Mech. Anal.*, **73**, 285.

Raimondi, A. A. 1961. A numerical solution for the gas lubrication full journal bearing of finite length. *ASLE Trans.*, **4**, 131–155.

Rajagopal, K. R. 1984. On the creeping flow of the second order fluid. *J. Non-Newtonian Fluid Mech.*, **15**, 239.

Rajagopal, K. R. and Fosdick, R. L. 1980. Thermodynamics and stability of fluids of third grade. *Proc. Roy. Soc., London*, **A 339**, 351–377.

Rajagopal, K. R. and Tao, L. 1995. Mechanics of Mixtures. World Scientific Publishing, Singapore.

Schneider, W. D., Glossfeld, D. H. and DeHart, A. O. 1986. The effect of coolant contamination in lubricating oil on journal bearings load capacity – a radiometric evaluation. *ASLE Preprint No. 86-TC-2F-1.*

Sirivat, A., Rajagopal, K. R. and Szeri, A. Z. 1988. An experimental investigation of the flow of non-Newtonian fluids between rotating disks. *J. Fluid Mech.*, **186**, 243.

Smith, F. W. 1962. *ASLE Trans.*, **5**, 142.

Smith, L. H., Peeler, R. L. and Bernd, L. H. 1960. *Proceedings of the 16th National Conference on Industrial Hydraulics Annual Meeting*, Oct. 20–21, 179.

Soo, S. L. 1967. *Fluid Dynamic of Multiphase Systems*. Olasidell, Waltham, MA.

Spencer, A. J. 1970. *Theory of Invariants, Continuum Physics*, A. C. Eringen (ed.), Vol. 1, Part III. Academic Press, New York.

Takahashi, T. 1966. Friction and Lubrication in Metal-Processing. *ASME*, New Orleans, p. 137.

Tanner, R. I. 1966. Plane creep flows of incompressible second order fluid. *Phys. Fluids*, **9**, 1246–1247.

Tanner, R. I. 1969. Increase of bearing loads due to large normal stress differences in viscoelastic lubricants. *ASME J. Appl. Mech.*, **36**, 634–635.

Tanner, R. I. 1985. *Engineering Rheology*. Clarendon Press, Oxford.

Tao, F. F. and Philippoff, W. 1967. Hydrodynamic behavior of viscoelastic liquids in a simulated journal bearing. *ASLE Trans.*, **100**, 302–315.

Taylor, G. I. 1932. The viscosity of fluid containing small drops of another fluid. *Proc. Roy. Soc.*, **A138**, 41.

Tevaarwerk, J. L. and Johnson, K. L. 1979. The influence of fluid rheology on the performance of traction drives. *ASME Journal of Lubrication Technology*, **101**, 266–274.

Tichy, J. A. 1978. The behavior of viscoelastic squeeze films subject to normal oscillations, including the effect of fluid inertia. *Appl. Sci. Res.*, **33**, 501–517.

Tichy, J. A. 1996. Non-Newtonian lubrication with the corrected Maxwell model. *ASME Journal of Tribology*, **118**, 344–348.

Trachman, E. G. and Cheng, H. S. 1972. Thermal and non-Newtonian effects on traction in elastohydrodynamic lubrication. Paper **C 37/72**. Symp. EHD Lub. *Inst. Mech. Engrs.*, **142**.

Truesdell, C. 1964. The natural time of a viscoelastic fluid: its significance and measurement. *Phys. Fluids*, 7, 1134–1142.

Truesdell, C. 1969. *Rational Thermodynamics*. McGraw-Hill, New York.

Truesdell, C. and Noll, W. 1965. The Non-Linear Field Theories of Mechanics. *Handbuch der Physik*, **III/3**. Springer-Verlag, Berlin.

Truesdell, C. and Toupin, R. 1960. The Classical Field Theories. *Handbuch der Physik*, S. Flugge (ed.), **III/1**. Springer-Verlag, Berlin.

Verma, P. D., Sharman, P. R. and Ariel, P. D. 1984. Applying quasi-linearization to the problem of steady laminar flow of a second grade fluid between two rotating porous disks. *ASME Journal of Lubrication Technology*, **106**, 448–455.

Wan, G. T. Y., Kenny, P. and Spikes, H. A. 1984. Elastohydrodynamic properties of water-based fire-resistant hydraulic fluids. *Tribology Int.*, **17**, 309–315.

Wang, S. H., Al-Sharif, A., Rajagopal, K. R. and Szeri, A. Z. 1993. Lubrication with binary mixtures: liquid-liquid emulsion in an EHD conjunction. *ASME Journal of Tribology*, **115**, 515–524.

Wang, S. H., Szeri, A. Z. and Rajagopal, K. R. 1993. Lubrication of emulsion in cold rolling *ASME Journal of Tribology*, **115**, 523–532.

Wolff, R. and Kubo, A. 1996. A generalized non-Newtonian fluid model incorporated into elastohydrodynamic lubrication. *ASME Journal of Tribology*, **118**, 74–82.

Zang, Y. 1995. EHD lubricant behavior at moderate pressures and high shear rates. Masters Essay, The Johns Hopkins University, Baltimore.

Zang, Y. and Ramesh, K. T. 1996. The behavior of an elastohydrodynamic lubricant at moderate pressures and high shear rates. *ASME Journal of Tribology*, **118**, 162–168.

CHAPTER 11

Gas Lubrication

The qualitative difference in performance between liquids and gases, in general, vanishes as $M \to 0$, where the *Mach number*, M, is the ratio of the fluid velocity to the local velocity of sound. This general conclusion also holds for bearings, and at low speeds the behavior of gas film lubricated bearings is similar to liquid-lubricated bearings – in fact, many of the liquid film bearings could also be operated with a gas lubricant. This similarity between liquid and gas films no longer holds at high speeds, however, the main additional phenomenon for gas bearings being the compressibility of the lubricant.

Perhaps the earliest mention of air as a lubricant was made by Hirn in 1854. Kingsbury (1897) was the first to construct an air-lubricated journal bearing. But the scientific theory of gas lubrication can be considered as an extension of the Reynolds lubrication theory. This extension was made soon after Reynolds' pioneering work: Harrison in 1913 published solutions for "long" slider and journal bearings lubricated with a gas. Nevertheless, the study of gas lubrication remained dormant until the late 1950s, when impetus for the development of gas bearings came mainly from the precision instruments and the aerospace industries.

In self-acting bearings, whether lubricated by liquid or gas, lubrication action is produced in a converging narrow clearance space by virtue of the viscosity of the lubricant. As the viscosity of gases is orders of magnitude smaller than that of commonly used liquid lubricants, gas bearings generally must have smaller clearances and will produce smaller load capacities than their liquid-lubricated counterparts.

Despite the smaller clearances, however, under normal conditions viscous heating in the gas can be neglected. The equation for the conservation of energy for a constant property gas is given by Eq. (9.20b). Here we retain the adiabatic compression term $-p \operatorname{div} \boldsymbol{v}$ on account of compressibility. Elrod and Burgdorfer (1959) simplified Eq. (9.21) for a two-dimensional film of negligible internal energy, flowing between walls held at constant temperature, and found that

$$\frac{\Theta_{\max}}{\Theta_0} - 1 = O(M^2),$$

where Θ_0 is the temperature of the walls and the Mach number, $M = u/a$, is the ratio of local gas velocity to the velocity of sound and $\Theta_{\max} = \max(\Theta)$. Consequently, if the bearing Mach number is small, the usual case, the temperature variation across the lubricant film thickness can be neglected.

Another way to estimate the appropriateness of the isothermal approximation is by comparing the transit time of the gas through the clearance space to the time scale of conduction (Ausman, 1966). Let the temperature of the gas, contained between parallel surfaces, exceed the temperature of the walls by $\Delta\Theta$. This temperature difference will, at the centerline, decrease to $\Delta\Theta/3$ in a time $\tau_c = O(c_p \rho h^2 / k)$, where c_p, ρ, and k, are the specific heat, density, and heat conductivity, respectively, for the gas. The transit time for the gas through the bearing is $\tau_t = O(2R/U)$, where R is the characteristic dimension of the bearing in the direction of motion. If $\tau_t \gg \tau_c$, heat conduction from gas into bearing will occur at a high enough rate to keep the gas film virtually at the temperature of the bearing, and the latter will be at near uniform temperature on account of the large heat conductivity of metals. The criteria for isothermal operation is, therefore, $\tau_t \gg \tau_c$, or $\text{Pr} \, \text{Re}^* \ll 1$. For gas-lubricants, the Prandtl number $\text{Pr} = \mu c_p / k = O(1)$ and in gas-lubricated bearings conditions are such that $\text{Re}^* \ll 1$; typically, $U_* = O(10^3), C = O(10^{-4}), \text{Re} = O(1)$, giving $\text{Pr} \, \text{Re}^* = O(10^{-5})$, thus the condition $\tau_t \gg \tau_c$ is, in general, satisfied.

Another simplifying feature of gas lubricant films is the absence of cavitation. However, we trade the linearity of the boundary conditions[1] for nonlinearity of the equations, and record no gain on this account. In gas films, however, rarely do we have to consider turbulence. A typical $R = 1$ cm air-lubricated journal bearing operating with a clearance ratio of $(C/R) = 10^{-3}$ at a speed of $N = 5 \times 10^4$ rpm would have a Reynolds number less than 100, far below the critical value of $\text{Re}_{CR} \approx 500-800$. We may therefore neglect inertia terms, i.e., apply the Reynolds lubrication theory, and assume laminar flow.

Some of the advantages of employing gas bearings are:

(1)   Chemical stability of the lubricant
(2)   No fire hazard
(3)   Small thermal gradients
(4)   No ecological contamination

However, gas films have shortcomings as well. Under this heading we may mention often unavoidable metal-to-metal contact that increases friction instantaneously severalfold. To minimize wear under dry contact, both bearing and runner surfaces must be hard. Hard surfaces and tight clearances do not easily accommodate debris. It should be considered that though gas bearing specific loads are small, these bearings often operate at speeds 1–2 orders of magnitude higher than liquid lubricated bearings – in consequences gas bearings are more vulnerable to thermal/mechanical distress than are liquid bearings (Gross, 1962; Pan, 1980).

Gas bearings run on thin films, $h = 2.5 \, \mu$m is not uncommon. Such tight clearances require at least $0.025 \, \mu$m rms roughness and must be at least $0.125 \, \mu$m flat, as the self-correcting action of running-in is not available to gas bearings. Near-perfect alignment is also necessary, and elastic/thermal deflection must be limited to less than $0.5 \, \mu$m.

Particularly in magnetic recording applications, where the read-write head has a minimum separation from the disk of order 100 nm or less, we must take into account the clearance height relative to the mean free path of the gas molecules. This ratio is defined as the *Knudsen number*

$$\text{Kn} = \frac{\lambda}{h},$$

where h is the film thickness and λ is the *mean free path* ($\lambda \approx 60$ nm for air under standard conditions). An approximate rule of thumb concerning flow regimes is

$\text{Kn} < 0.01$	continuum flow,
$0.01 \leq \text{Kn} < 15$	slip flow,
$\text{Kn} \geq 15$	molecular flow.

---

[1] Recall that for liquid-lubricated bearings the Reynolds equation is linear in p, but the boundary conditions $p = dp/d\theta = 0$ are not.

As will be shown below, there have been several attempts made to extend the Reynolds equation above Kn > 0.01.

As in usual circumstances $\text{Re}^* = O(C/R)$, when we take formally the limit $(C/R) \to 0$ in the equations of motion while treating the viscosity as a constant, we again arrive at Eqs. (2.56), (2.57), and (2.61). In fact, our starting point in deriving the Reynolds equation for gas lubricant will be Eqs. (2.61) and (2.44a), the equation of continuity for compressible fluids.

11.1   Reynolds Equation for Gas Lubricant

The Reynolds equation is derived here for no-slip boundary conditions. Under the lubrication assumptions the in-plane velocity components of the film are given by Eq. (2.61):

$$u = \frac{1}{2\mu}\frac{\partial p}{\partial x}(y^2 - hy) + \left(1 - \frac{y}{h}\right)U_1 + \frac{y}{h}U_2,$$

$$w = \frac{1}{2\mu}\frac{\partial p}{\partial z}(y^2 - hy). \tag{2.61}$$

The velocity satisfies no-slip conditions at the boundaries

$$\begin{aligned}
u &= U_1, & v &= V_1, & w &= 0 \quad \text{at } y = 0, \\
u &= U_2, & v &= V_2, & w &= 0 \quad \text{at } y = h.
\end{aligned} \tag{2.60}$$

In the manner of Section 2.2, the velocity components are next substituted into the equation of continuity that has been integrated across the film. The equation of continuity for a compressible fluid is given by

$$\frac{\partial p}{\partial t} + \text{div}(\rho \boldsymbol{v}) = 0. \tag{2.44a}$$

Rearrangement of Eq. (2.44a) and integration across the film yields

$$[\rho v]_0^{h(x,t)} = -\int_0^{h(x,t)} \frac{\partial(\rho u)}{\partial x}\,dy - \int_0^{h(x,t)} \frac{\partial(\rho w)}{\partial z}\,dy - \int_0^{h(x,t)} \frac{\partial \rho}{\partial t}\,dy. \tag{11.1}$$

In analogy with Eq. (2.63b), we have

$$[\rho v]_0^{h(x,t)} = \rho \frac{dh}{dt}. \tag{11.2}$$

Substituting from Eqs. (2.61) and (11.12) into Eq. (11.1), we obtain

$$\begin{aligned}
[\rho v]_0^h = {}&-\frac{1}{2}\frac{\partial}{\partial x}\left[\frac{\partial p}{\partial x}\int_0^h \frac{\rho}{\mu}y(y-h)\,dy\right] - \frac{1}{2}\frac{\partial}{\partial z}\left[\frac{\partial p}{\partial z}\int_0^h \frac{\rho}{\mu}y(y-h)\,dy\right] \\
&-\frac{\partial}{\partial x}\int_0^h \rho\left[\left(1 - \frac{y}{h}\right)U_1 + \frac{y}{h}U_2\right]dy + \rho U_2 \frac{\partial h}{\partial x} - \int_0^h \frac{\partial p}{\partial t}\,dy.
\end{aligned} \tag{11.3}$$

Since p is not a function of y, neither is ρ for an ideal gas. Assuming further that $\mu = \mu(x, z)$ at most and applying Leibnitz's rule for differentiation under the integral sign, Eq. (11.3) is written as

$$\frac{\partial}{\partial x}\left(\frac{\rho h^3}{12\mu}\frac{\partial p}{\partial x}\right) + \frac{\partial}{\partial z}\left(\frac{\rho h^3}{12\mu}\frac{\partial p}{\partial z}\right) = \frac{\partial}{\partial x}\left(\rho h\frac{U_1 + U_2}{2}\right) - \rho U_2\frac{\partial h}{\partial x} + h\frac{\partial \rho}{\partial t} + [\rho v]_0^h.$$

(11.4)

For thrust bearings and, in general, for bearing surfaces that undergo rigid body translation but no rotation, we have [Eq. (11.2)]

$$[\rho v]_0^h = \rho(V_2 - V_1) = \rho\frac{\partial h}{\partial t} \quad \text{as } U_{2,r} = 0$$

(11.5)

and

$$\frac{\partial}{\partial x}\left(\frac{\rho h^3}{\mu}\frac{\partial p}{\partial x}\right) + \frac{\partial}{\partial z}\left(\frac{\rho h^3}{\mu}\frac{\partial p}{\partial z}\right) = 6\frac{\partial}{\partial x}(\rho h U_0) + 12\frac{\partial(\rho h)}{\partial t}.$$

(11.6)

Here we followed the notation of Chapter 3 and put $U_0 = U_1 - U_2$.

For journal bearings, one must consider both rigid body rotation and rigid body translation, and we put (following Section 2.2)

$$[\rho v]_0^h = \rho\frac{dh}{dt} = \rho\frac{\partial h}{\partial t} + \rho U_{2,r}\frac{\partial h}{\partial x}$$

$$\approx \rho\frac{\partial h}{\partial t} + \rho U_2\frac{\partial h}{\partial x}$$

(11.7)

as $U_2 = U_{2,r}[1 + O(C/R)] \approx U_{2,r}$ by Eq. (2.70c). Substituting Eq. (11.7) into Eq. (11.6) and collecting terms, we obtain an equation formally identical to Eq. (11.7), but now we have $U_0 = U_1 + U_2$.

For isothermal processes, common to the majority of gas bearing applications,

$$\frac{p}{\rho} = \text{const.}$$

(11.8)

If, further, we have steady conditions, $\partial(\rho h)/\partial t = 0$, Eq. (11.6) takes the form

$$\frac{\partial}{\partial x}\left(\frac{h^3}{\mu}\frac{\partial p^2}{\partial x}\right) + \frac{\partial}{\partial z}\left(\frac{h^3}{\mu}\frac{\partial p^2}{\partial z}\right) = 12U_0\frac{\partial(ph)}{\partial x},$$

(11.9)

where $U_0 = U_2$ for journal bearings and $U_0 = U_1$ for sliders. Equation (11.9) is specified for various bearing geometries by Pan (1980).

Gas lubrication is frequently applied to the head-disk interface in computer hard disk drives. In these applications the Knudsen number is in the 0.01–15 range. When the clearance becomes much smaller than the mean free path, the Reynolds equation with no-slip boundary conditions predicts a shear stress that is too high. It seems, however, that it is not the continuum assumption that is at fault but the no-slip boundary conditions (Anaya, 1996). Chan and Horn (1985) found that the drainage rate of a thin film of fluid between two crossed molecularly smooth mica cylinders was adequately predicted by the continuum Reynolds equation to about $h = 30$ nm. At thinner gaps, good correlation with experiment was obtained by simply adding a fictitious rigid layer to the mica surfaces in the Reynolds equation model (Tichy, 1996).

Table 11.1. *Variants of the Reynolds*
equation for gas-lubricated sliders

Model	Q	c_1	c_2
No-Slip [Eq. (11.20)]	1	0	0
First-Order Slip	1	1	0
Second-Order Slip	1	1	1
Boltzman-Reynolds	Q(Kn)	0	0
1.5-Order Slip	1	1	4/9

The models currently used to predict pressures in head-disk interface in computer hard disk drives are (Anaya, 1996) the first-order slip theory of Burgdorfer (1959), the second-order slip model of Hsia and Domoto (1983), the Boltzman-Reynolds approach of Fukui and Kaneko (1988), and the 1.5-order slip equation of Mitsuya (1993). The governing equation for each of these models and the no-slip model, Eq. (11.9), can be written in the concise form (Anaya, 1996)

$$\frac{\partial}{\partial \bar{x}}\left[\Psi \frac{\partial \bar{p}}{\partial \bar{x}}\right] + \frac{\partial}{\partial \bar{z}}\left[\Psi \frac{\partial \bar{p}}{\partial \bar{z}}\right] = \Lambda \left(\frac{\partial}{\partial \bar{x}} + 2\frac{\partial}{\partial \bar{t}}\right)(\bar{p}\bar{h}), \qquad (11.10a)$$

where

$$\Psi = Q\bar{p}\bar{h}^3 + 6c_1 \text{Kn}_0 \bar{h}^2 + 6c_2 \text{Kn}_0^2 \frac{\bar{h}}{\bar{p}}. \qquad (11.10b)$$

Equations (11.10) are nondimensional, written for a slider of dimension, B, in the direction of motion and length, L, flying over a plane at constant speed, U_0. The nondimensional variables have the definition

$$\bar{h} = \frac{h}{h_0}, \qquad \bar{p} = \frac{p}{p_0}, \qquad \bar{x} = \frac{x}{B}, \qquad \bar{z} = \frac{z}{B}, \qquad \bar{t} = \frac{U_0 t}{B},$$

where p_0, h_0 represent reference pressure and reference film thickness, respectively.

The bearing compressibility number, Λ, and the reference Knudsen number are given by

$$\Lambda = \frac{6\mu U_0 B}{p_0 h_0^2}, \qquad \text{Kn}_0 = \frac{\lambda_0}{h_0}. \qquad (11.11)$$

Here λ_0 is the molecular mean free path at the reference pressure p_0.

Depending on which form of the Eqs. (11.10) is being used, the values of Q, c_1, c_2 are as given in Table 11.1.

In order to study the wear problem arising at intermittent contact of rigid disk and slider, it is essential to calculate the contact pressure. Anaya (1996) presents detailed analysis of the various models in this context.

The validity of slip-flow theory was experimentally confirmed by Hsia and Domoto (1983), with helium as a working fluid, in an effort to separate the high Knudsen number and high bearing number effects. Theoretical load was calculated using first-order slip theory, which is based on momentum transfer between gas and plate, down to 75 nm. However, most of the experimental data falls below the theoretical load curve when the spacing is below 250 nm. They recommended a second-order slip theory as an extension. The Reynolds equation obtained from the linearized Boltzman equation, derived by Fakui and Kaneko (1988), yields load results between the first- and the second-order slip theories. Mitsuya (1993) derived, based on kinetic theory, a higher-order slip-flow model, the 1.5-order slip flow model. Predictions from Mitsuya's theory for load seem to fall between data from the Boltzman-Reynolds model and the second-order slip theory.

11.2 Self Acting Gas Bearings

It is convenient to normalize Eq. (11.9) by the substitution

$$X = \frac{x}{a}, \qquad Z = \frac{z}{a}, \qquad P = \frac{p}{p_a}, \qquad H = \frac{h}{\Delta}, \qquad \Lambda = \frac{6\mu a U_0}{p_a \Delta^2}. \qquad (11.12)$$

For journal bearings $a = R$, $\Delta = C$, and $U_0 = R\omega$. For plane thrust bearing, $a = B$ and Δ is a representative film thickness (e.g., depth of recess). The transformed (nondimensional) equation is

$$\frac{\partial}{\partial X}\left(H^3 P \frac{\partial P}{\partial X}\right) + \frac{\partial}{\partial Z}\left(H^3 P \frac{\partial P}{\partial Z}\right) = \Lambda \frac{\partial(PH)}{\partial X}. \qquad (11.13)$$

For long bearings, we set $\partial(\cdot)/\partial Z \to 0$ and, by integrating Eq. (11.13), obtain

$$\frac{dP}{dX} = \frac{\Lambda}{H^3 P}(PH - K), \qquad (11.14)$$

where K is a constant of integration.

The pressure gradient, $\partial P/\partial X$, in Eq. (11.14) must remain bounded under all conditions, for otherwise the pressure would increase to physically unacceptable levels within a short distance of the inlet. But for very large speeds, the left-hand side of Eq. (11.14) will remain bounded only if $PH \to K$, leading to

$$PH = K = P_i H_i, \qquad \Lambda \to \infty. \qquad (11.15)$$

This high-speed asymptote of gas bearing operation has been determined solely from the forcing term, i.e., the right-hand side of the Reynolds equation, which is independent of the length of the bearing. This, of course, suggests that there is negligible leakage at high speed[2] and finite-length and long bearings behave similarly at the limit $U_0 \to \infty$.

Next, consider gas bearing lubrication at very small velocities, $U_0 \to 0$. From Eq. (11.14), it follows that

$$\partial P/\partial X = O(\Lambda) \quad \text{as} \quad \Lambda \to 0. \qquad (11.16a)$$

Thus, for the right-hand side of Eq. (11.13), we may write

$$\Lambda \frac{\partial(PH)}{\partial X} = \Lambda P \frac{\partial(H)}{\partial X} + O(\Lambda^2), \qquad (11.16b)$$

---

[2] The same conclusion can be reached by comparing characteristic times of fluid transport in the direction of relative motion (shear flow) and in the axial direction (pressure flow).

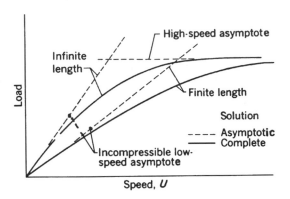

Figure 11.1. Variation of load as function of speed for self-acting gas bearings. (Reprinted with permission from Ausman, J. S. Gas-lubricated bearings. In *Advanced Bearing Technology* by E. E. Bisson and W. J. Anderson. NASA SP-38, 1964.)

and, on substituting from Eq. (11.16) into Eq. (11.13), we obtain

$$\frac{\partial}{\partial X}\left(H^3\frac{\partial P}{\partial X}\right) + \frac{\partial}{\partial Z}\left(H^3\frac{\partial P}{\partial Z}\right) = \Lambda\frac{\partial H}{\partial X} + O(\Lambda^2), \qquad \Lambda \to 0. \qquad (11.16c)$$

Equation (11.16c) has the form of the incompressible Reynolds equation.

From Eqs. (11.15) and (11.16), we draw the important conclusions, following Ausman (Ausman, 1966):

(1) At low speeds, the behavior of gas lubricant is approximated by the behavior of liquid lubricant.[3]

(2) At high speed, the product ph = const. and the load capacity becomes independent of speed and depends only on the inlet (ambient) pressure.

(3) At high speeds, side leakage becomes negligible as both long bearing and finite bearing solutions approach the asymptote ph = const. (Scheinberg, 1953).

The above three fundamental characteristics of self acting gas bearing are shown schematically in Figure 11.1.

Journal Bearings

For journal bearings, we put $a = R$, $X = \theta$, $\Delta = C$ in Eq. (11.13). The applicable Reynolds equation is

$$\frac{\partial}{\partial\theta}\left(H^3 P\frac{\partial P}{\partial\theta}\right) + \frac{\partial}{\partial Z}\left(H^3 P\frac{\partial P}{\partial Z}\right) = \Lambda\frac{\partial(PH)}{\partial\theta}, \qquad (11.17a)$$

$$\Lambda = \frac{6\mu\omega}{p_a}\left(\frac{R}{C}\right)^2,$$

---

[3] Note, however, that liquid lubrication and gas lubrication are not completely identical even at low speed. While there is cavitation in liquid bearings in the diverging portion of the clearance, gas films remain continuous from inlet to outlet (Constantinescu, 1969).

and the boundary conditions are

$$P = 1 \quad \text{at} \quad Z = \pm \frac{L}{D}. \tag{11.17b}$$

We also require periodicity in θ.

Equation (11.17a) is nonlinear, and no general closed-form solutions of it exist. Numerical solutions are well documented, however, and are available in the literature (Elrod and Malanoski, 1960; Raimondi, 1961). Although closed form solutions are not available for arbitrary values of the parameters, we are able to obtain analytical solutions in asymptotic cases.

The pressure equation may be linearized at small values of the eccentricity ratio, ε, by assuming

$$P = 1 + \varepsilon P_1. \tag{11.18}$$

Substituting Eq. (11.18) into Eq. (11.17a), we obtains the first order perturbation equation as follows:

$$\frac{\partial}{\partial \theta} \left[\left(\frac{\partial}{\partial \theta} - \Lambda \right) P_1 \right] + \frac{\partial^2 P_1}{\partial Z^2} = -\Lambda \sin \theta, \tag{11.19}$$

This formulation is known as the *linearized p* solution (Ausman, 1959). Using the notation

$$\Xi \equiv \frac{L}{D} \sqrt{1 + \Lambda^2} \left(\cosh 2\sigma \frac{L}{D} + \cos 2\xi \frac{L}{D} \right),$$

$$\Gamma_1 \left(\Lambda, \frac{L}{D} \right) \equiv \frac{(\sigma - \xi \Lambda) \sin 2\xi \frac{L}{D} - (\sigma \Lambda + \xi) \sinh 2\sigma \frac{L}{D}}{\Xi},$$

$$\Gamma_2 \left(\Lambda, \frac{L}{D} \right) \equiv \frac{(\sigma - \xi \Lambda) \sinh 2\sigma \frac{L}{D} + (\sigma \Lambda + \xi) \sin 2\xi \frac{L}{D}}{\Xi},$$

the load capacity is found to be

$$\frac{F_R}{p_0 L D} = \frac{\pi \varepsilon \Lambda}{2(1 + \Lambda^2)} \left[\Lambda + \Gamma_1 \left(\Lambda, \frac{L}{D} \right) \right], \tag{11.20a}$$

$$\frac{F_T}{p_0 L D} = \frac{\pi \varepsilon \Lambda}{2(1 + \Lambda^2)} \left[1 - \Gamma_2 \left(\Lambda, \frac{L}{D} \right) \right], \tag{11.20b}$$

where

$$\left. \begin{array}{c} \sigma \\ \xi \end{array} \right\} = \sqrt{\frac{(1 + \Lambda^2)^{1/2} \pm 1}{2}} \qquad \text{(positive roots)}.$$

Note from Eqs. (11.20) that the load is linearly dependent on the eccentricity ratio in this approximation. In reality the increase in load is far more rapid than linear, once $\varepsilon \approx 0.3$ has been passed; the linear approximation is valid for only ε less than this value. Figure 11.2 shows the (dimensionless) total load capacity and the attitude angle for a gas journal bearing at small eccentricities. The prediction from Eq. (11.20) is compared with numerical solutions in Figure 11.3.

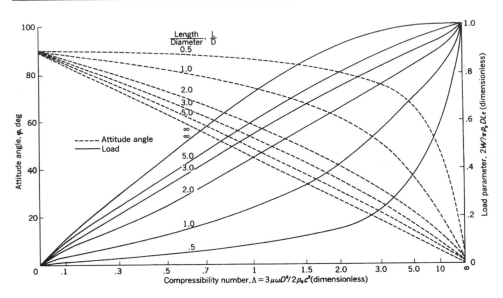

Figure 11.2. Isothermal first-order perturbation solution for journal bearings. (Reprinted with permission from Ausman, J. S. Gas-lubricated bearings. In *Advanced Bearing Technology* by E. E. Bisson and W. J. Anderson. NASA SP-38, 1964.)

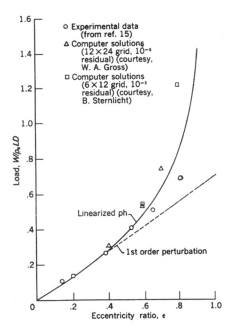

Figure 11.3. Comparison of perturbation solutions and computer solution with experimental data. (Reprinted with permission from Ausman, J. S. An improved analytical solution for self-acting, gas lubricated journal bearings of finite length. *ASME J. Basic Eng.*, **83**, 188–194, 1961.)

In an effort to assure stronger than a linear dependence of load on ε, Ausman considered the product $ph \equiv \Psi$, rather than the pressure itself, as the dependent variable, and linearized the Reynolds equation with respect to Ψ.

To derive this approximation (Constantinescu, 1969), known as the *linearized ph solution*, we first write the pressure equation in a form that gives prominence to the function $\bar{\Psi} = \Psi/p_0\Delta$:

$$H\bar{\Psi}\left(\frac{\partial^2\Psi}{\partial X^2} + \frac{\partial^2\Psi}{\partial Z^2}\right) - \bar{\Psi}^2\left(\frac{\partial^2 H}{\partial X^2} + \frac{\partial^2 H}{\partial Z^2}\right) - \Lambda\frac{\partial\bar{\Psi}}{\partial X}$$

$$= \bar{\Psi}\left(\frac{\partial H}{\partial X}\frac{\partial\bar{\Psi}}{\partial X} + \frac{\partial H}{\partial Z}\frac{\partial\bar{\Psi}}{\partial Z}\right) - H\left[\left(\frac{\partial\bar{\Psi}}{\partial X}\right)^2 + \left(\frac{\partial\bar{\Psi}}{\partial Z}\right)^2\right]. \tag{11.21}$$

For $\Lambda \to \infty$, we have, by Eq. (11.15), $PH = \bar{\Psi} \to$ const. and the right-hand side of Eq. (11.21) vanishes. This right-hand side also vanishes for $\Lambda \to 0$, as now $p \to p_0$, $P \to 1$, and $\Psi \to H$. Ausman made the assumption that the right-hand side of Eq. (11.21) vanishes not only at the limits but over the whole range of Λ. But then Eq. (11.21) can be written in the approximate form

$$H\bar{\Psi}\nabla^2\bar{\Psi} - \Lambda\frac{\partial\bar{\Psi}}{\partial X} = \bar{\Psi}^2\nabla^2 H. \tag{11.22}$$

Equation (11.22) is still not linear, however. To remedy this, Ausman made the further assumption that when $\bar{\Psi}$ is a coefficient, then $\bar{\Psi} \approx H$, so that $H\bar{\Psi} \approx \bar{\Psi}^2 \approx H^2$. The resulting equation will now be linear and free of coefficient, but for

$$\frac{\Lambda}{H^2} = \frac{6\mu\omega}{p_0}\left(\frac{R}{C}\right)^2\left(\frac{C}{h}\right)^2$$

$$\approx \frac{6\mu\omega}{p_0}\left(\frac{R}{h_{\text{ave}}}\right)^2$$

$$\approx \Lambda.$$

For journal bearings $H = 1 + \varepsilon\cos\theta$, and Eq. (11.22) reduces to

$$\frac{\partial}{\partial\theta}\left[\left(\frac{\partial}{\partial\theta} - \Lambda\right)\bar{\Psi}\right] + \frac{\partial^2\bar{\Psi}}{\partial Z^2} = -\cos\theta. \tag{11.23}$$

To solve this equation, we first make it homogeneous via the substitution

$$\bar{\Psi}(X, Z) = \bar{\Psi}_\infty(X) - \bar{\Psi}^*(X, Z)$$

and use separation of variables to solve the resulting equation in $\bar{\Psi}^*(X, Z)$. The high degree of similarity between Eqs. (11.19) and Eq. (11.23) means that the load components calculated from Eq. (11.23) will be but a function of ε times the load components calculated from the linearized p solution

$$F_R|_\Psi = \frac{2}{\varepsilon^2}\left[\frac{1 - \sqrt{1 - \varepsilon^2}}{\sqrt{1 - \varepsilon^2}}\right]F_R|_p, \tag{11.24a}$$

$$F_T|_\Psi = \frac{2}{\varepsilon^2}[1 - \sqrt{1 - \varepsilon^2}]F_T|_p. \tag{11.24b}$$

Here $(\cdot)|_{\psi}$ and $(\cdot)|_p$ refer to the linearized Ψ and linearized p solutions, respectively. Figure 11.3 compares linearized p, linearized Ψ, and computer solutions.

For large Λ, Eq. (11.17a) may be written as

$$\frac{\partial H}{\partial \theta} = \frac{1}{\Lambda}\left[\frac{\partial}{\partial \theta}\left(H^3 P \frac{\partial P}{\partial \theta}\right) + \frac{\partial}{\partial Z}\left(H^3 P \frac{\partial P}{\partial Z}\right)\right]. \tag{11.25a}$$

Neglecting the right-hand side upon taking the limit $\Lambda \to \infty$ would lead to

$$\lim_{\Lambda \to \infty} \frac{\partial \Psi}{\partial \theta} = 0 \tag{11.25b}$$

and satisfaction of the boundary conditions

$$\frac{\partial \Psi}{\partial \theta} = \varepsilon \sin \theta, \quad \text{at } Z = \pm \frac{L}{D},$$

would not be possible. Clearly, the small parameter $1/\Lambda$ multiplies the highest derivatives in Eq. (11.25a), and one is faced with a *singular perturbation* problem (Nayfeh, 1973). To overcome the difficulty, Pan (1980) uses *matched asymptotic expansion*.[4] The key for matching inner and outer solutions is the *mass content rule* (Elrod and Burgdorfer, 1959; Pan, 1980)

$$\frac{1}{2\pi}\int_0^{2\pi} H^3 P^2 \, d\theta = 1 + \frac{3}{2}\varepsilon^2,$$

which is obtained by integrating Eq. (11.25a) over the circumference, taking into account the periodicity of both P and H.

Infinitely Long Step Slider

The pressure differential equation (11.9) reduces to

$$\frac{d}{dx}\left(\frac{Uph}{2} - \frac{1}{12\mu}h^3 p \frac{\partial p}{\partial x}\right) = 0, \tag{11.26a}$$

and the boundary condition is

$$p(0) = p(L) = p_a. \tag{11.26b}$$

The slider is shown schematically in Figure 11.4.

The equation is nondimensionalized as in Eq. (11.12) to give

$$\frac{d}{dX}\left(\Lambda PH - H^3 P \frac{\partial P}{\partial X}\right) = 0, \tag{11.27}$$

$$P(0) = P(1) = 1.$$

Integration of Eq. (11.27) yields

$$PH\left(1 - \frac{H^2}{\Lambda}\frac{dP}{dX}\right) = \hat{K}, \tag{11.28a}$$

---

[4] According to Pan (1980), Eq. (11.25b) is to be employed in the open domain $|Z| < L/D$ as the *outer solution*. The *inner solution*, which satisfies the edge conditions, is obtained from a rescaling of Eq. (11.25a). To achieve this rescaling, the axial coordinate is stretched in the ratio $\sqrt{\Lambda}$.

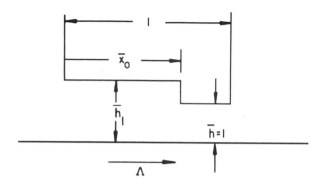

Figure 11.4. Step slider geometry.

where $\hat{K}$, a constant of integration, equals the value of PH at X^*, where $dP/dX = 0$:

$$\hat{K} = PH|_{X=X^*}. \tag{11.28b}$$

Employing Eq. (11.28a) in Eq. (11.27), we get

$$\frac{dP}{dX} = \lambda\left(1 - \frac{\gamma}{P}\right), \tag{11.29}$$

where $\gamma = \hat{K}/H$ and $\lambda = \Lambda/H^2$ are constants in each interval $0 \leq X \leq X_0$ and $X_0 < X \leq 1$. The integral of Eq. (11.29) is

$$P + A + \gamma \ln(P - \gamma) - \lambda X \tag{11.30}$$

Applying the boundary condition at $P = 1$ at $X = 0$, gives $A = -1 - \ln(1 - \gamma)$, and if $\gamma_1 = \hat{K}/H_1$ and $\lambda_1 = \Lambda/H_1^2$, where $H_1 = H(1)$, we find in the range $0 \leq X \leq X_0$:

$$P - 1 + \gamma_1 \ln\left(\frac{P - \gamma_1}{1 - \gamma_1}\right) = \lambda_1 X. \tag{11.31a}$$

Equation (11.30) can also be evaluated in the range $X_0 \leq X \leq 1$:

$$P - 1 + \hat{K} \ln\left(\frac{P - \hat{K}}{1 - \hat{K}}\right) = \Lambda(X - 1), \tag{11.31b}$$

where Eq. (11.31b) satisfies $P = 1$ at $X = 1$.

At $X = X_0$, $P = P_0 \equiv P(X_0)$, and we use this condition to match the pressures at X_0:

$$H_1(P_0 - 1) + \hat{K} \ln\left(\frac{P_0 H_1 - \hat{K}}{H_1 - \hat{K}}\right) = \frac{\Lambda}{H_1} X_0, \tag{11.32a}$$

$$P_0 - 1 + \hat{K} \ln\left(\frac{\hat{K} - P_0}{\hat{K} - 1}\right) = \Lambda(X_0 - 1). \tag{11.32b}$$

The system of Eqs. (11.32) contains two unknowns, P_0 and $\hat{K}$. These equations cannot be solved analytically, and one has to resort to numerical methods. For the limiting cases $\Lambda \to 0$ and $\Lambda \to \infty$, however, we are able to obtain closed form solutions.

Asymptotic case $\Lambda \to 0$:

    We look for small perturbation of the incompressible case and put

$$P_0 = 1 + \Lambda \pi_0 + O(\Lambda^2). \tag{11.33}$$

Substituting for P_0 into Eq. (11.31) and collecting terms multiplied by Λ, we obtain $\hat{K}$ and π_0 as (Pan, 1980)

$$\hat{K} = \left[\frac{(1 - X_0)H_1^2 + X_0}{(1 - X_0)H_1^3 + X_0} \right] H_1,$$

$$\pi_0 = \frac{X_0(1 - X_0)(H_1 - 1)}{(1 - X_0)H_1^3 + X_0}.$$

The pressure distribution can now be written for small Λ:

$$P = 1 + \Lambda \pi_0 \frac{X}{X_0}, \qquad 0 \le X \le X_0, \tag{11.34a}$$

$$P = 1 + \Lambda \pi_0 \frac{1 - X}{1 - X_0}, \qquad X_0 \le X \le 1. \tag{11.34b}$$

Asymptotic case $\Lambda \to \infty$:

    From Eq. (11.31), it can be shown that $\Lambda \to \infty$ requires $\hat{K} \to H_1$ and $P_0 \to H_1$. To signify this, we put for large Λ

$$\hat{K} = H_1 - \delta H, \qquad \delta H / H_1 \ll 1,$$
$$P_0 = H - \delta P, \qquad \delta P / P_0 \ll 1, \tag{11.35a}$$

and substitute into Eq. (11.31)

$$H_1(H_1 - 1 - \delta P) + (H_1 - \delta P) \ln \left(\frac{H_1(H_1 - 1) + \delta H - H_1 \delta P}{\delta H} \right) = \frac{\Lambda X_0}{H_1},$$

$$H_1 - 1 - \delta P + (H_1 - \delta H) \ln \left(\frac{\delta P - \delta H}{H_1 - 1 - \delta H} \right) = -\Lambda(1 - X_0). \tag{11.35b}$$

    We recognize that on the left-hand sides the logarithmic terms dominate as $\Lambda \to \infty$. Furthermore, at the limit, $H_1(1 - \delta H / H_1) \to H_1$. Then, Eq. (11.35b) becomes

$$H_1 \ln \left(\frac{H_1(H_1 - 1) + \delta H - H_1 \delta P}{\delta H} \right) = \frac{\Lambda X_0}{H_1}, \tag{11.36a}$$

$$H_1 \ln \left(\frac{\delta P - \delta H}{H_1 - 1 - \delta H} \right) = -\Lambda(1 - X_0). \tag{11.36b}$$

The above equations can be solved for $\delta H, \delta P$ (Pan, 1980)

$$\begin{bmatrix} \delta H \\ \delta P \end{bmatrix} = \frac{H_1 - 1}{1 + E_0} \left\{ \begin{array}{c} H_1 E_1(1 - E_2) \\ H_1 E_1(1 - E_2) + E_2(1 - E_1) \end{array} \right., \tag{11.37}$$

where

$$E_1 = \exp\left[-\frac{\Lambda X_0}{H_1^2}\right], \qquad E_2 = \exp\left[-\frac{\Lambda(1-X_0)}{H_1}\right],$$

$$E_0 = E_1(H_1 - 1 - H_1 E_2).$$

On substituting for δH, δP in Eq. (11.32), the pressure at large Λ is obtained:
$0 \leq X \leq X_0$:

$$P = 1 + \frac{(H_1 - 1)(1 - E_2)}{1 + E_0}\left\{\exp\left[-\frac{\Lambda(X_0 - X)}{H_1^2}\right] - E_1\right\} \qquad (11.38a)$$

$X_0 \leq X \leq 1$:

$$p = 1 + \frac{(H_1 - 1)(1 - E_1)}{1 + E_0}\left\{1 - \exp\left[-\frac{\Lambda(1 - X)}{H_1^2}\right]\right\}. \qquad (11.38b)$$

Sample results for $h = 1.5$ and $X_0 = 0.5$ are given in Figure 11.5. The primary effect of compressibility is related to the pressure peak. For small Λ the above ambient value of the peak pressure is independent of the value of the ambient pressure and is proportional to the product of speed and viscosity, as with incompressible lubricant. At large Λ, on the other hand, the peak pressure is a constant multiple of the ambient pressure, where the multiplier is numerically equal to the gap ratio H_1.

Figure 11.5 displays the variation of the peak pressure P_0 with Λ. The (normalized) pressure distribution for various values of Λ are also shown in the insert. Note that for the incompressible case P_0 varies linearly with Λ, but for higher values the curve deviates more and more from the incompressible case, finally reaching a limiting value as $\Lambda \to \infty$, in accordance with earlier assertions.

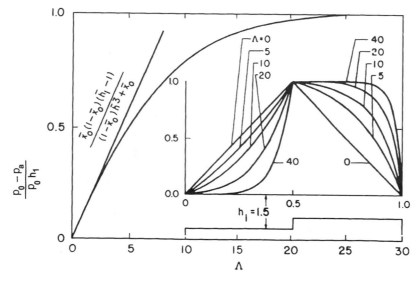

Figure 11.5. Pressure distribution in step slider (Pan, 1980).

11.3   Nomenclature

B	integration constant
C	radial clearance
F_r, F_T	radial, tangential force
H	film thickness(dimensionless)
$\hat{K}$	integration constant
Kn	Knudsen number
L	bearing length
M	Mach number
P	pressure (dimensionless)
R	bearing radius
Re	Reynolds number
Re*	reduced Reynolds number
U, U_1, U_2	surface velocity
X, Z	coordinates (dimensionless)
Θ	temperature
a	characteristic length
c_p	specific heat
h	film thickness
k	heat conductivity
p	pressure
p_a	ambient pressure
t	time
u, v, w	velocity components
Λ	bearing (compressibility) number
ε	eccentricity ratio
ω	shaft angular velocity
λ	mean free path
μ	viscosity
τ_c, τ_t	characteristic times

11.4   References

Anaya Dufresne, M. 1996. *On the Development of a Reynolds Equation for Air Bearings with Contact.* Ph.D. Dissertation, Carnegie Mellon University.

Ausman, J. S. 1959. Theory and design of self-acting gas-lubricated journal bearings including misalignment effects. *Proc. 1st Int. Symp. Gas-Lub. Bearings*, **ACR-49**, 161–192, ONR Washington, D.C.

Ausman, J. S. 1961. An improved analytical solution for self-acting, gas lubricated journal bearings of finite length. *ASME J. Basic Eng.*, **83**, 188–194.

Ausman, J. S. 1966. Gas-lubricated bearings. *Advanced Bearing Technology*, E. E. Bisson and W. J. Anderson (ed.). NASA, Washington, D.C.

Burgdorfer, A. 1959. The influence of molecular mean free path on the performance of hydrodynamic gas lubricated bearings. *ASME J. Basic Eng.*, **81**, 94–100.

Chan, D. Y. C. and Horn, G. 1985. *J. Chemical Physics*, **83**, 5311.

Constantinescu, V. N. 1969. Gas lubrication. *The American Society of Mechanical Engineers*, New York.

Elrod, H. G. Jr. and Burgdorfer, A. 1959. Refinement of the theory of gas lubricated journal bearing of infinite length. *Proc. 1st Int. Symp. Gas-Lub. Bearings*, **ACR-49**, 93–118, ONR, Washington, D.C.

Elrod, H. G. Jr. and Malanoski, S. B. 1960. *Theory and Design Data for Continuous Film, Self Acting Journal Bearings of Finite Length*. Rep. I-A 2049–13 and Rep. I-A 2049–17.

Fukui, S. and Kaneko R. 1988. Analysis of ultra-thin gas film lubrication based on linearized Boltzman equation including thermal creep. *ASME Journal of Tribology*, **110**, 253–262.

Gross, W. A. 1962. *Gas Film Lubrication*. Wiley, New York.

Harrison, W. J. 1913. The hydrodynamical theory of lubrication with special reference to air as a lubricant. *Trans. Cambridge Phil. Soc.*, **22**, 39–54.

Hirn, G. A. 1984. Sur les principaux phenomenes .... *Bull. Soc. Ind. Mulhouse*, **26**, 188.

Hsia, Y.-T. and Domoto, G. A. 1983. An experimental investigation of molecular rarefaction effects in gas lubricated bearings at ultra-low clearances. *ASME Journal of Lubrication Technology*, **105**, 120–130.

Kingsbury, A. 1897. Experiments with an air lubricated journal. *J. Am. Soc. Naval Engineers*, **9**, 267–292.

Mitsuya, Y. 1993. Modified Reynolds equation for ultra-thin film gas lubrication using 1.5-order slip-flow model and considering surface accommodation coefficient. *ASME Journal of Tribology*, **115**, 289–294.

Nayfeh, A. H. 1973. *Perturbation Methods*. Wiley, New York.

Pan, C. H. T. 1981. Gas bearings. In *Tribology: Friction, Lubrication and Wear*, A. Z. Szeri (ed.). McGraw-Hill, New York.

Pinkus O. and Sternlicht, B. 1961. *Theory of Hydrodynamic Lubrication*. McGraw-Hill, New York.

Raimondi, A. A. 1961. A numerical solution of the gas lubricated full journal bearing. *ASME Trans.*, **4**, 131–155.

Scheinberg, S. A. 1953. Gas lubrication of slider bearings. *Friction and Wear of Machines*, **8**, 107–204. *Inst. Machine Sci., Academy Sci.*, USSR.

Tichy, J. A. 1996. Modeling thin film lubrication. *ASME/STLE Tribology Congress*, San Francisco.

Index

409

Printed in the United States
By Bookmasters